# Biology of Plant-Microbe Interactions

Edited by
**Gary Stacey, Beth Mullin,
and Peter M. Gresshoff**

---

**Proceedings of the 8th International Symposium
on Molecular Plant-Microbe Interactions
Knoxville, Tennessee, July 14-19, 1996**

---

Published by the
International Society for Molecular Plant-Microbe Interactions
St. Paul, Minnesota, USA

Library of Congress Catalog Card Number: 96-78699
International Standard Book Number: 0-9654625-0-1

©1996 by the International Society for Molecular Plant-Microbe Interactions
Second printing, 1997

Printed in the United States of America on acid-free paper

International Society for Molecular Plant-Microbe Interactions
3340 Pilot Knob Road
St. Paul, Minnesota 55121-2097, USA

# BIOLOGY OF PLANT-MICROBE INTERACTIONS

EDITORS: Gary Stacey, Ph.D.
Center for Legume Research
Department of Microbiology
Department of Ecology and Evolutionary Biology
M409 Walters Life Science Building
The University of Tennessee
Knoxville, TN        37996-0845
Phone: 423-974-4041
Fax: 423-974-4007
Email: GSTACEY@utk.edu

Beth Mullin, Ph.D.
Center for Legume Research
Department of Botany
437 Hesler Biology Building
The University of Tennessee
Knoxville, TN        37996
Phone: 423-974-6203
Fax: 423-974-0978

Peter M. Gresshoff, Ph.D., D.Sc.
Center for Legume Research
Plant Molecular Genetics
Institute of Agriculture
267 Ellington Plant Science Bldg.
The University of Tennessee
Knoxville, TN        37901-1071
Phone: 423-974-8841
Fax: 423-974-2765

# PREFACE

Plant-microbe interactions are important in many agricultural settings. These interactions can be both beneficial and detrimental. Because of their importance, many plant-microbe interactions are being studied throughout the world using modern molecular methods. This book represents a collection of papers that were given at the Eighth International Symposium on Molecular Plant-Microbe Interactions which was held in Knoxville, Tennessee, USA in July, 1996. This meeting brought together approximately 930 scientists from over thirty countries. The breadth of chapters in this book attests to the wide variety of plant-microbe interactions being studied and the scope of these investigations. Work varies from molecular biology, chemistry, biochemistry, cell biology, etc. Therefore, the 'Biology of Plant-Microbe Interactions' is a fitting title for this volume. Topics covered include molecular studies of plant symbioses, *Agrobacterium*, bacterial and fungal pathogens, microbial signals and signal transduction pathways, biocontrol, and applications to biotechnology. This book also contains a section focusing on new, emerging areas of research on plant-microbe interactions.

The chapters are written by world authorities and represent the latest information on the subjects covered. The authors have given special attention to information gathered in the last two years. This volume should be useful to researchers, scholars, and students interested in the most recent advances in the study of plant-microbe interactions. The editors hope that this volume will serve as a stimulus to encourage young scientists to enter this rapidly expanding research field.

The Eighth International Symposium on Molecular Plant-Microbe Interactions would not have been possible without the generous support of the University of Tennessee, several federal agencies, private corporations and foundations. A special thanks is also given to the International Society for Molecular Plant-Microbe Interactions under whose auspices the meeting was organized and to the Society Staff, Officers and Members of the Board for their support. The editors would also like to thank the staff of the University of Tennessee Conference Center for their excellent assistance in preparing the meeting. Our appreciation also goes to the members of the International Advisory Board whose assistance was invaluable in preparing the meeting program. Finally, we would also like to thank the faculty of the Center for Legume Research at the University of Tennessee who served as the local organizing committee.

Knoxville, Tennessee       Gary Stacey
August 1996         Beth Mullin
              Peter M. Gresshoff

# TABLE OF CONTENTS

**Page**

**Signal Transduction**

**Plant Resistance**

## Genetics of Fungal Pathogenicity

## Virology

**Plant-Microbe Symbioses**

### Diversity and Ecology of Plant-Associated Microbes

**Emerging Areas of Research**

**International Symposium on Molecular Plant-Microbe
Interactions: Meeting Summary**

# Olfaction in Plants: Specific Perception of Common Microbial Molecules

Thomas Boller and Georg Felix

Friedrich Miescher-Institute, P.O. Box 2543, CH-4002 Basel, Switzerland

It is often stated that "recognition" is the key initial event in the response of plants to microbes, and that recognition is based on the perception of chemical signals produced by the microbes. For example, in a given plant-pathogen interaction, the plant may possess specific chemoreceptors for pathogen-derived elicitors. If such an elicitor signal is "recognized", i.e. detected by the corresponding receptor and appropriately transduced, the plant may activate its defense responses and ward off the attack. Without recognition, the pathogen may invade the plant and cause disease. Similarly, in the interaction of a plant with a mutualistic symbiont such as *Rhizobium*, the plant may require specific chemoreceptors for symbiotic signals. If such a symbiotic signal is percieved and transduced, the plant may initiate an appropriate response and enter symbiosis. Without recognition, the symbiosis may fail to be established.

In recent years, many chemoperception systems for microbial substances in plant cells have been identified. They are typically highly specific and selective for the microbial compounds in question, and they resemble more the olfactory sense with its multitude of specific odorant receptors than the taste sense with its small number of broad taste qualities (Boller, 1995).

A biochemical approach to these receptors involves essentially four steps. First, the microbial signal that is recognized by the plant has to be identified and obtained in a chemically pure form. Second, binding assays with a labelled derivative of the microbial signal are needed to identify a binding site. Third, the binding site has to be solubilized if necessary, purified to homogeneity, and cloned. Fourth, the receptor function of the binding protein has to be demonstrated, for example by expression of the corresponding cDNA in a system lacking the receptor. Despite of much effort in several laboratories, no plant receptor for microbial substances has thus far been identified on the basis of all these four steps.

1

In contrast, a number of disease resistance genes have been cloned (review: Dangl, 1995). Among the proteins encoded by these resistance genes, several ones contain domains tantalizingly similar to structures expected in receptors, such as leucine-rich repeats and/or protein kinase domains, although biochemical evidence for a receptor function is lacking. It will be interesting to find out whether the receptors for known microbial signals bear any resemblance with resistance gene products.

## Bioassays

### BIOASSAYS WITH INTACT PLANT TISSUES

Bioassays are required to detect and characterize the plants' perception systems for microbial substances. When the focus is on the _biological responses_ to microbial signals, bioassays involving the intact plant or its parts are useful. Elicitors were initially defined as microbial substances inducing phytoalexins, and therefore, phytoalexin accumulation in plant parts has frequently been used as a bioassay for pathogen-derived signals (Darvill and Albersheim, 1984). Root hair coiling has served well as a bioassay for the symbiotic signals produced by rhizobia, the Nod factors (Dénarié et al., 1992). A certain disadvantage of these systems is that the biological responses can often be recorded only hours or even days after addition of the signals, making conclusions about signal perception difficult.

### BIOASSAYS WITH CELL CULTURES

More recently, cell suspension cultures have often been used in bioassays. Cell cultures have a disadvantage too: They do not represent the intact biological system, and one cannot be sure that their response corresponds to the reaction of the intact plant. It has been difficult, for example, to examine gene-for-gene interactions in cell cultures (De Wit, 1992). However, when the focus is on the _perception systems_ themselves, cell suspension cultures have a distinct advantage since in this case, the earliest detectable reactions, such as changes in ion fluxes, are most interesting for bioassays. These changes occur in seconds or minutes after stimulation and are comparable to the action potentials monitored to study animal olfaction. In suspension cultures, a large number of cells can be stimulated at exactly the same time, so that changes in ion fluxes occur in a synchronous manner and can be assayed very simply, for example by monitoring the pH in the growth medium (Felix et al., 1993). Typically, in this type of bioassay, a transient or more permanent increase in pH occurs in response to microbial substances ("alkalinization response").

It is important to distiguish the alkalinization response elicited by signal substances from unspecific effects. Alkalinization of the growth medium can also occur in an unspecific manner upon addition of bases or membrane-active compounds such as detergents. However, there are two distinct characteristics of these unspecific changes. First, they occur without an apparent lag, while the alkalinization induced by microbial signals starts only after a lag of at least 20 seconds. Second, they are insensitive to K-252a, an inhibitor of protein kinases, while alkalinization induced by microbial signals is blocked by K-252a.

*Involvement of protein phosphorylation in the "alkalinization response".* Inhibition of the alkalinization response by K-252a, even in mid-course of the response, indicates an involvement of rapid changes in protein phosphorylation in signal transduction (Felix et al. 1991). We have used pulse-labelling of suspension-cultured cells with radioactive phosphate to study these changes (Felix et al., 1991, 1993, 1994). These studies show that a number of proteins are hyperphosphorylated within minutes after addition of the microbial stimuli. The first detectable changes appear to occur at the same time as the pH in the medium starts to increase, and they are blocked by K-252a (Felix et al., 1991). Interestingly, calyculin A, an inhibitor of protein phosphatases, rapidly induces the alkalinization response and the same type of changes in protein phosphorylation in the absence of microbial signals. This indicates that the phosphoproteins critical for the response are continually phosphorylated and dephosphorylated in the non-elicited state, and that inhibition of dephosphorylation is sufficient to initiate the response (Felix et al., 1994).

Over the last years, we have studied perception of microbial signals in a tomato cell suspension culture, initially using stimulation of ethylene production and later the alkalinization response as bioassays. Here, an overview is provided on the microbial signals active in this system.

## Signals Derived from Fungi

CELL WALL POLYSACCHARIDES

*Glucans.* A classic elicitor is the branched $\beta$-1,3-$\beta$-1,6-heptaglucoside, isolated as the smallest elicitor-active compound from cell walls of *Phytophthora megasperma* f.sp. *glycinea*, a fungal pathogen of soybean, (Darvill and Albersheim, 1984). This heptaglucoside, which is half-maximally active at a concentration of ~ 3 nM in the soybean cotyledon bioassay, is inactive in bioassays with tomato cells, even when tested in micromolar concentrations (Felix, unpublished results).

*Chitin fragments.* Chitin is a major constituent of the cell walls of most higher fungi. It also occurs in arthropods and many other invertebrates but is not present in plants. Tomato cells have a highly sensitive perception system for chitin oligomers, as revealed in bioassays involving the alkalinization response (Felix et al., 1993): Chitin oligomers with a degree of polymerization (DP) $\geq$ 4 are half-maximally active at concentrations of 0.1 nM; the trisaccharide is about 1000-fold less active, and the dimer and monomer, which potentially could be present in plants, are essentially inactive (Felix et al., 1993). A high-affinity binding site for chitin fragments is present on the surface of the tomato cells and in microsomal membranes (Baureithel et al., 1994). The relative affinities of different chitin derivatives for the binding site correspond to their activities in the bioassay, as expected for a binding site representing the biological receptor.

Chitin and chitin oligomers act as elicitors also in various other plant systems (review: Boller, 1995).

EXTRACELLULAR PROTEINS AND GLYCOPEPTIDES

*Glycopeptides.* Using stimulation of ethylene biosynthesis as a bioassay, fungal glycopeptides were found to act as signals in the tomato cells (Basse and Boller, 1992). Chemically defined glycopeptide elicitors, obtained from purified yeast invertase cleaved with chymotrypsin, are also active (Basse et al., 1992). The structure of the *N*-linked glycan side chains is decisive for activity: Glycopeptides with 8 mannosyl groups have little elicitor activity but the ones with 9 - 11 groups are potent elicitors (half-maximal activity at ~ 3 nM). Interestingly, the side chains with 8 mannosyl groups represent a "core" present in all eukaryotes including plants but the larger ones have their ninth mannosyl residue attached to this core at a position unique for fungi; thus, the plant's chemoperception system is highly specific for "non-self" fungal glycopeptides (Basse et al., 1992).

A high-affinity binding site for glycopeptides is present on tomato cells and in tomato microsomes (Basse et al., 1993). Its affinity for various glycopeptides is similar to their biological activities, indicating that the binding site might be the biological receptor. The binding site has been solubilized and partially purified (Fath and Boller, 1996).

*Extracellular proteins from Phytophthora.* A crude cell wall preparation from *Phytophthora megasperma* acts as an elicitor in tomato cells (Basse and Boller, 1992). The activity appears to be associated with a glycoprotein but has not been further characterized.

Extracellular proteins from *Phytophthora* species have been more fully characterized as elicitors in other systems. For example, the so-called elicitins, a group of highly related, non-glycosylated 10 kDa proteins produced by most *Phytophthora* species, have been identified as elicitors of necrosis in tobacco leaves (Ricci et al., 1989). In addition, an elicitin-related glycoprotein has been purified from *Phytophthora* that elicits necrosis in tobacco leaves (Bailleul et al. 1995). Both proteins are active at nanomolar levels in appropriate bioassays. None of these proteins induces the alkalinization response in tomato cells (G. Felix, unpublished results).

Parsley cells recognize a 42 kDa glycoprotein as the elicitor-active component in cell wall preparations of *Phytophthora megasperma* (Nürnberger et al., 1994). A short non-glycosylated peptide generated from this protein by proteolytic cleavage is equally active and induces short-term ion fluxes as well as defense gene transcription and phytoalexin accumulation half-maximally at ~ 2 nM (Nürnberger et al., 1994). This peptide is inactive as an elicitor of the alkalinization response in tomato cells (G. Felix, unpublished results).

Taken together, these results indicate that different plant species recognize different extracellular proteins of *Phytophthora*, using for recognition different but similarly highly sensitive, selective and specific chemoperception systems.

 *Xylanase:* Xylanase from the saprophyte *Trichoderma viride* stimulates ethylene biosynthesis in tobacco leaves; it seems to be be perceived directly and not *via* its enzymatic products (Sharon et al., 1993). This xylanase is also a strong elicitor in tomato cells (Felix et al., 1994).

MEMRANE COMPONENTS

*Ergosterol.* Tomato cells perceive ergosterol, the main sterol in most higher fungi, with exquisite sensitivity, as demonstrated using the alkalinization response as bioassay (Granado et al., 1995). Ergosterol is half-maximally active at ~ 0.01 nM. In contrast, plant and animal sterols are completely inactive even at micromolar concentrations. The extreme selectivity and sensitivity of this perception system is reminiscent of steroid perception in animals (Granado et al., 1995).

## Signals Derived from Bacteria

OLIGOSACCHARIDES

*Lipo-chitooligosaccharides.* Rhizobia produce nodulation factors (Nod factors) which act as chemical signals in roots of their leguminous host plants and are essential for the formation of nodules in which symbiotic nitrogen fixation takes place (Dénarié et al., 1992). Nod factors are lipo-chitooligosaccharides, i.e. derivatives of chitin oligomers (usually DP 4 or DP 5) which carry a *N*-fatty acyl group instead of an *N*-acetyl group on the glucosamine at the non-reducing end. They are generally carry further "decorations" important for host specificity, such as a sulfatyl, methylfucosyl or arabinosyl group at the reducing end or an acetyl, methyl or carba-moyl residue at the non-reducing end. In the most sensitive bioassays such as root hair deformation, Nod factors are half-maximally active in host plants at concentrations of 0.1 - 1 nM, while unmodified chitin fragments are inactive (Dénarié et al., 1992).

Surprisingly, the perception system for chitin fragments in tomato cells also recognizes Nod factors (Staehelin et al., 1994). Perception of Nod factors in tomato has probably no functon; however, one might speculate that the receptors for Nod factors in legumes, with their expected high selectivity, have evolved from a less specific receptor for chitin fragments such as the one observed in tomato (Staehelin et al., 1994). Alternatively, lipochitooligosaccharides might play a role in plant development in general, as indicated by the finding that lipochitooligosaccharides, at femtomolar concentrations, induce cell division in the absence of hormones in tobacco cells (Rohrig et al., 1995).

SURFACE PROTEINS

*Harpins.* Phytopathogenic bacteria often induce a hypersensitive response in non-host cells. A cluster of conserved genes, the *hrp* cluster, is essential for induction of this *h*ypersensitive *r*esponse in non-host plants as well as for *p*athogenesis in susceptible host plants (Goodman and Novacki, 1994). The majority of the *hrp* genes appear to encode elements of a type III protein secretion system, based on their similarity to genes responsible for the secretion of pathogenicity factors in *Yersinia* (Van Gijsegem et al., 1995). The products of the *hrpN* gene from *Erwinia amylovora* (Wei et al., 1992) and of the *hrpZ* gene from *Pseudomonas syringae* pv. *syringae* (He et al., 1993), the so-called harpins, have been found to induce a hypersensitive response in tobacco cells. Both harpins have been purified from an extract of boiled bacteria; they resemble each other physically since both are resis-tant to boiling but highly sensitive to proteases; however, they share no sequence homology (He et al. 1993).

*Flagellins.* We tried to determine whether harpin-type molecules were also active in tomato cells. A preparation of the *Erwinia* harpin was only very weakly active in a bioassay involving the alkalinization response although boiled extracts from *Erwinia*, *Pseudomonas* and also *Escherichia* were highly active. We prepared extracts from boiled *Pseudomonas syrin-gae* pv. *tabaci* and found that this preparation induced a strong alkaliniza-tion response (S. Volko and G. Felix, unpublished results). When the prepa-ration was submitted to SDS-PAGE electrophoresis, a protein band of ~ 31 kDa carried most of the inducing activity. This protein band was submitted to N-terminal amino acid sequencing; surprisingly the sequence had no homology with any harpin but was highly similar to the N-terminus of the flagellin from *Pseudomonas aeruginosa* (J. Duran and G. Felix, unpublished results). Flagellins from bacteria are highly variable in size and sequence but contain a highly conserved domain of about 20 amino acids in the N-terminus which is practically identical in flagellins from bacteria as different as *Pseudomonas*, *Bacillus* and *Escherichia*. A synthetic peptide corresponding to this domain was even more active than flagellin in the alkalinization bioassay; subnanomolar concentrations were saturating for the response (G. Felix, unpublished results). Thus, tomato cells possess a recognition system that specifically detects the most conserved stretch of bacterial flagellins, a key surface protein of motile bacteria. Interestingly, the flagellins of *Agrobacterium* and *Rhizobium* are unusually divergent in the N-terminal domain. Peptides corresponding to the conserved domains of these flagellins showed no activity in the bioassay, even in micromolar concentrations (G. Felix, unpublished results). Thus, these plant-associated bacteria may avoid the plant's perception system for flagellins.

## Conclusion

In addition to the extremely specialized recognition systems underlying gene-for-gene relationships with pathogens (De Wit, 1992, Dangl, 1995), plants have evolved highly sensitive, specific and selective perception systems for common surface molecules of fungi and bacteria. Many of these molecules, such as chitin and ergosterol, occur in pathogenic microbes as well as in saprophytes and mutualistic symbionts. Thus, it is unlikely that recognition of these molecules induces an all-out defense. Rather, these chemoperception systems might have an orienting role, allowing the plant to "smell" the presence of microbes and of "non-self" organisms in its environment (Boller, 1995). It will be interesting to find out whether and how these perception systems for characteristic microbial surface molecules may interact with the more specialized recognition systems involved in gene-for-gene relationships.

## References

Baillieul, F., Genetet, I., Kopp, M., Saindrenan, P., Fritig, B. and Kauff-mann, S. (1995) *Plant J*, **8**, 551-560.

Basse, C.W., Bock, K. and Boller, T. (1992) *J Biol Chem* **267**, 10258-10265.

Basse, C.W. and Boller, T. (1992) *Plant Physiol* **98**, 1239-1247.

Basse, C.W., Fath, A. and Boller, T. (1993) *J Biol Chem* **268**, 14724-14731.

Baureithel, K., Felix, G. and Boller, T. (1994) *J Biol Chem* **269**, 17931-17938.

Boller, T. (1995) *Annu Rev Plant Physiol Plant Mol Biol* **46**, 189-214.

Dangl, J.L. (1995) *Cell* **80**, 363-366.

Darvill, A.G. and Albersheim, P. (1984) *Annu Rev Plant Physiol* **35**, 243-275.

De Wit, P.J.G.M. (1992) *Annu Rev Phytopathol* **30**, 391-418.

Dénarié, J., Debellé, F. and Rosenberg, C. (1992) *Annu Rev Microbiol* **46**, 494-531.

Fath, A. and Boller, T. (1996) *Plant Physiol*, in press.

Felix, G., Grosskopf, D.G., Regenass, M. and Boller, T. (1991) *Proc Natl Acad Sci USA* **88**, 8831-8834.

Felix, G., Regenass, M. and Boller, T. (1993) *Plant J* **4**, 307-316.

Felix, G., Regenass, M., Spanu, P. and Boller, T. (1994) *Proc Natl Acad Sci USA* **91**, 952-956.

Goodman, R.N. and Novacky, A.J. (1994) *The Hypersensitive Reaction in Plants to Pathogens. A Resistance Phenomenon.* American Phytopathological Society, St. Paul.

Granado, J., Felix, G. and Boller, T. (1995) *Plant Physiol* **107**, 486-490.

He, S.Y., Huang, H.-C. and Collmer, A. (1993) *Cell* **73**, 1255-1266.

Nürnberger, T., Jabs, D., Nennstiel, D., Sacks, W.R., Hahlbrock, K. and Scheel, D. (1994) *Cell* **78**, 449-460.

Ricci, P., Bonnet, P., Huet, J.-C., Sallantin, M., Beauvais-Cante, F., Bruneteau, M., Billard, V., Michel, G. and Pernollet, J.-C. (1989) *Eur J Biochem* **183**, 555-563.

Rohrig, H., Schmidt, J., Walden, R., Czaja, I., Miklasevics, E., Wieneke, U., Schell, J. and John, M. (1995) *Science* **269**, 841-843.

Sharon, A., Fuchs, Y. and Anderson, J.D. (1993) *Plant Physiol* **102**, 1325-1329.

Staehelin, C., Granado, J., Müller, J., Wiemken, A., Mellor, R.B., Felix, G., Regenass, M., Broughton, W.J. and Boller, T. (1994) *Proc Natl Acad Sci USA* **91**, 2196-2200.

Van Gijsegem, F., Gough, C., Zischek, C., Niqueux, Arlat, M., Genin, S., Barberis, P., German, S., Castello, P. and Boucher, C. (1995) *Mol Microbiol* **15**, 1095-1114.

Wei, Z.-M., Laby, R.J., Zumoff, C.H., Bauer, D.W., He, S.Y., Collmer, A. and Beer, S.V. (1992) *Science* **257**, 85-88.

# SIGNAL RECOGNITION AND TRANSDUCTION IN BACTERIAL SPECK DISEASE RESISTANCE OF TOMATO

Gregory Martin, Xiaoyan Tang, Jianmin Zhou,
Reid Frederick, Yulin Jia, and Ying-Tsu Loh

Department of Agronomy
Purdue University
West Lafayette, IN 47907-1150 USA

*Bacterial speck disease resistance and fenthion sensitivity*

Resistance to bacterial speck disease in tomato is governed by a "gene-for-gene" interaction in which a single resistance locus (*Pto*) in the plant responds to the expression of a specific avirulence gene (*avrPto*) in the pathogen *Pseudomonas syringae* pv. *tomato*. Disease results if either *Pto* or *avrPto* are lacking from the corresponding organisms. Interestingly, tomato cultivars containing the *Pto* region develop a hypersensitive-like response upon exposure to an organophosphorous insecticide, fenthion, that is similar to the resistance response caused by the avirulent bacterium.

The *Pto* and *Fen* genes were isolated by a map-based cloning approach and found to be members of a clustered gene family with similarity to serine/threonine protein kinases (Martin et al., 1993, 1994). Recently, a mutagenesis/map-based cloning strategy has led to the isolation of another gene near *Pto*, named *Prf*, that is required for both *Pto*-mediated disease resistance and fenthion sensitivity (Salmeron et al., 1994, 1996). The Pto protein shares 80% identity (87% similarity) with Fen and both are functional protein kinases which specifically phosphorylate serine and threonine residues (Loh & Martin, 1995). We have therefore hypothesized that a protein phosphorylation cascade is involved in both *Pto*-mediated disease resistance and *Fen*-mediated insecticide sensitivity.

*Pto-interacting proteins*

Further elucidation of the Pto and Fen signaling pathway(s) would benefit from the identification of proteins that physically interact with these kinases. As an alternative approach to mutagenesis, which to date has uncovered only two loci involved in tomato bacterial speck resistance (*Pto* and *Prf*), we have utilized the yeast two-hybrid system developed by Roger Brent's group to directly clone genes encoding proteins which interact physically with Pto (http://xanadu.mgh.harvard.edu). We screened approximately $2 \times 10^6$ cDNAs derived from tomato leaves inoculated with *P. s. tomato* using the two-hybrid system and isolated 10 distinct classes of cDNA clones encoding Pto-interacting (Pti) proteins that specifically interact with Pto (Zhou et al., 1995; and unpublished). Sequence analysis of representatives of the 10 *Pti* classes revealed homology, in some cases, with genes of known function. To date, we have characterized Pti1 which encodes a protein kinase because of its obvious potential role in signal transduction (Zhou et al., 1995), and we have begun characterization of the Pti4, Pti5, and Pti6 clones which encode putative DNA-binding proteins.

## Pto interacts with putative transcription factors

Pti4/5/6 share extensive homology with the recently identified ethylene-responsive-element binding proteins (EREBPs; Ohme-Takagi and Shinshi, 1995; Zhou et al., 1996). Four EREBPs were isolated from tobacco and shown to specifically bind an 11-base pair sequence (TAAGAGCCGCC) that is conserved in the promoter regions of many ethylene-responsive pathogenesis-related (PR) genes (Ohme-Takagi and Shinshi, 1995). At the amino acid level, Pti4 is 86% and 77% similar, whereas Pti5 is 62% and 67% similar to EREBP1 and EREBP2, respectively. Gene products encoded by Pti4/5/6 all contain EREBP-like DNA binding domains and two clones (Pti4 and Pti6) also contain putative nuclear localization sequences and acidic regions which may act as transcriptional activation domains (Zhou et al., 1996). These characteristics, along with the physical interaction with Pto, suggest that Pti4/5/6 represent a link between the specific recognition of the pathogen expressing *avrPto* and the regulation of certain plant defense genes (Zhou et al., 1996).

## Characterization of the L. esculentum pto and fen alleles.

An understanding of how *Pto* and *Fen* function in recognition and signal transduction would benefit from the study of alleles that are unable to confer disease resistance or fenthion sensitivity. Towards this objective, we isolated and characterized the alleles of *Pto* and *Fen* from cultivated tomato, *L. esculentum*, and designated them *pto* and *fen*. High conservation of genome organization between the two tomato species allowed us to identify the *pto* and *fen* alleles from among the cluster of closely related *Pto* gene family members (Jia et al., 1996).

The *pto* and *fen* alleles were found to be transcribed and to have uninterrupted open reading frames which code for predicted proteins that are 87% and 98% identical to the Pto and Fen protein kinases, respectively (Jia et al., 1996). Interestingly, *in vitro* auto-phosphorylation assays revealed that both pto and fen proteins encoded active kinases. In addition, the pto and fen kinases phosphorylate the Pti1 serine/threonine kinase. However, the pto kinase shows impaired interaction with Pti1 and with several previously isolated Pto-interacting proteins in the yeast two-hybrid system (Jia et al., 1996). The observation that pto and fen are active kinases and yet do not confer disease resistance or fenthion sensitivity suggests that the amino acid substitutions which distinguish them from Pto and Fen may interfere with recognition of the corresponding signal molecule or with protein-protein interactions involved in *Pto* and *Fen*-mediated signaling (Jia et al., 1996).

## Recognition specificity of the Pto and Fen kinases

The Pto and Fen proteins share a high degree of sequence similarity but respond to two very different stimuli. In order to delineate the region(s) within the Pto and Fen proteins that are involved in specific recognition of either the bacterial signal molecule or fenthion we have developed a series of chimeric proteins (**Figure 1**).

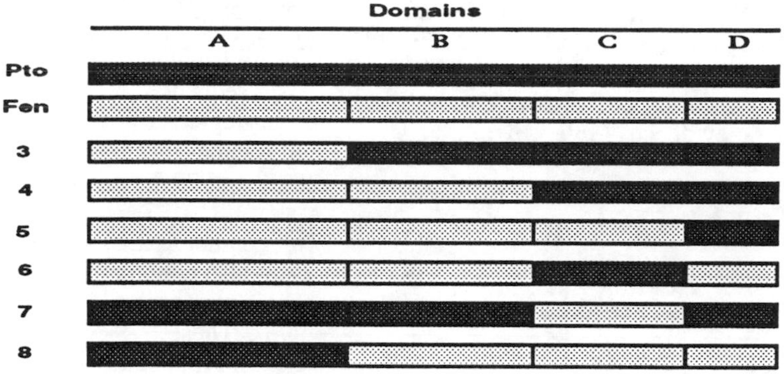

**Domains**

*Figure 1.* *Pto and Fen chimeric proteins. Darkened regions are derived from Pto, while open boxes are from Fen.*

Each construct was used to generate transgenic tomato plants which are being tested directly for resistance to *P. s. tomato* (*avrPto*) and sensitivity to fenthion in order to localize the region(s) involved in specific elicitor recognition. The constructs are also being tested for interaction with Pti1 and other Pto- and Fen-interacting proteins in the yeast two-hybrid system.

Chimeric-MBP fusion proteins were expressed in *E. coli* and are being tested for their ability to autophosphorylate and to cross-phosphorylate purified Pti1 protein. These experiments should identify the region(s) within the Pto protein necessary for the specific cross-phosphorylation of Pti1. Once the domain(s) required for specific elicitor recognition and the physical interaction and cross-phosphorylation of Pti1 are localized, oligonucleotide site-directed mutagenesis will be used to determine the amino acid residues involved in each of these reactions.

## Interaction of AvrPto and Pto in the yeast two-hybrid system

In the past few years, there have been several reports of similarities between virulence mechanisms of plant pathogens and certain mammalian pathogens. Disease-causing strains of *Yersinia*, for example, employ a secretion system that allows them to directly introduce virulence proteins into the mammalian host cell. This so-called Type III secretion system also appears to be present in bacterial pathogens of plants (Barinaga, 1996). Based on these recent reports, we speculated that the AvrPto protein itself may be introduced into the plant cell and serve as a signal molecule during the host-pathogen interaction. If this were the case, one possibility is that AvrPto might interact directly with the Pto kinase. To test this possibility, we examined the two proteins in the yeast two-hybrid system.

We first determined that neither the *avrPto* gene nor the *Pto* gene, when introduced separately into yeast as the prey or the bait plasmid respectively, was able to activate reporter gene transcription. However, when *avrPto* and *Pto* were introduced into the same yeast cell we observed very strong activation of both the LEU and *lacZ* reporter genes indicating tight binding of these two proteins (**Figure 2**). Tests of AvrPto in combination with several other kinases including Pti1, Fen, Pelle or Raf showed no evidence of interaction.

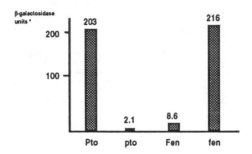

**Figure 2.** *The interaction of AvrPto with the Pto, pto, Fen and fen kinases in the yeast two-hybrid system. Numbers are the average of three independent transformants each with three replications.*

The interaction of AvrPto with the products of the *pto* and *fen* recessive alleles was also tested. Interestingly AvrPto did not interact with the pto kinase but did interact with the fen protein (**Figure 2**). This suggested that the region of Pto that interacts with AvrPto may share certain critical residues with fen and not with the Fen or pto proteins.

Tests of AvrPto interactions with each of the chimeric proteins demonstrated that the avirulence gene product interacted specifically with chimeric #7 (chimeric #3, for unknown reasons, cannot be expressed in *E. coli* or yeast; we were thus unable to test its interaction with AvrPto). Comparison of chimeric #7 with the others suggests that the "B" region of Pto is involved in AvrPto binding. These results prompted us to examine the alignment of the B region of Pto, pto, Fen and fen (**Figure 3**). This analysis showed that the only amino acids that distinguish Pto and fen from pto and Fen are a threonine and a methionine at positions 133 and 134 in the Pto protein (**Figure 3**). We are currently testing the role of these residues in the AvrPto interaction by site-directed mutagenesis followed by functional analysis in the two-hybrid system and in plants.

If AvrPto and Pto do, in fact, physically interact in the plant cell it would provide a molecular explanation for pathogen recognition in "gene-for-gene" interactions. The question remains, of course, as to how the binding of AvrPto to Pto would activate the disease resistance response. It is possible that AvrPto acts similarly to many ligands which interact with receptor kinases and causes the formation of a dimer between two Pto molecules. Subsequent cross-phosphorylation of the two Pto kinase proteins could then initiate a phosphorylation cascade. A second possibility is that the binding of AvrPto to a single Pto molecule may effect a conformational change in the Pto kinase thus enhancing its activity. Finally, it is plausible that AvrPto in some way mediates the interaction between Pto and Prf and this interaction activates Pto kinase activity. We are currently examining these models using *in vitro* kinase assays, the yeast "trihybrid" system, and immunoprecipitation from leaf extracts with and without pathogen infection. Finally, we are using saturation mutagenesis of both the *avrPto* and *Pto* genes in order to determine those amino acid residues that are involved in the physical interaction.

```
Pto ...DLPTMSMSWE QRLEICIGAA RGLHYLHTRA IIHRDVKSIN ILLDENFVPK
fen ...DLPTMSMSWE QRLEICIGAA RGLHYLHTNG VIHRDVKCTN ILLDENFVPK
Fen ...DLP..SMSWE QRLEICIGAA RGLHYLHKNA VIHRDVKCTN ILLDENFVPK
pto ...DLP..SMSWE QRLEICIGAA RGLHYLHTNG VMHRDVKSSN ILLDENFVPK

Pto ITDFGISKKG TELDQTHLST VVKGTLGYID PEYFIKGRLT EKSDVY...
fen ITDFGISKTM PELDLTHLST VVRGNIGYIA PEYALWGQLT EKSDVY...
Fen ITDFGISKTM PELDQTHLST VVRGNIGYIA PEYALWGQLT EKSDVY...
pto ITDFGLSKTR PQLYQT...T DVKGTLGYID PEYFIKGRLT EKSDVY...
```

*Figure 3.*        *Alignment of the "B" region of Pto, Fen and the pto, fen alleles.*

## A Model for the Pto/Fen-mediated signaling pathways

Our current model for the Pto and Fen signaling pathways is shown below **(Figure 4)**. In this model, we posit that the avirulent bacterium uses a Type III secretion system to directly introduce the AvrPto protein into the plant cell. Once inside the cell, AvrPto binds to the Pto kinase and this interaction stimulates a phosphorylation cascade leading to the hypersensitive response. A separate pathway involving the Pti4/5/6 transcription factors leads to the early induction of various pathogenesis-related genes.

The Fen kinase probably shares some components of the Pto pathway (including Prf) but may also involve other distinct partner proteins. The mechanism whereby fenthion is sensed by the plant cell remains unclear. Once activated, the Fen kinase may also interact with, and specifically phosphorylate, a Pti1 homolog (which we term Fen-interacting protein 1, Fni1). These multiple pathways regulate defense gene expression and other defense responses, including the HR, all of which contribute to limiting the growth of the invading pathogen and producing the localized necrosis displayed in fenthion sensitivity. The role of *Prf* in both the *Pto*- and *Fen*-mediated signal transduction pathways remains unclear **(Figure 4)**. It is possible that Prf acts in concert with Pto and Fen, perhaps mediating an association with the plasma membrane. Alternatively, it is conceivable

13

that Prf functions downstream of Pto and Fen, likely after Pti1 and the putative Pti1 homolog, Fni1.

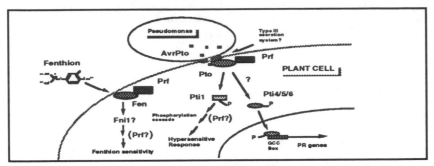

**Figure 4.** *Model of Pto and Fen signal transduction pathways.*

### Acknowledgments
This research was supported by National Science Foundation grant MCB-93-03359. GBM is a David and Lucile Packard Foundation Fellow.

### References
Barinaga, M. (1996). A shared strategy for virulence. Science 272, 1261-1263.

Jia, Y., Loh, Y.-T., Zhou, J. and G. B. Martin (1996). Alleles of *Pto* and *Fen* occur in bacterial speck-susceptible and fenthion-insensitive tomato lines and encode functional protein kinases. (submitted).

Loh, Y.-T., and Martin, G.B. (1995). The *Pto* bacterial resistance gene and the *Fen* insecticide sensitivity gene encode functional protein kinases with serine/threonine specicificity. Plant Physiol. 108, 1735-1739.

Martin, G.B., Brommonschenkel, S.H., Chunwongse, J., Frary, A., Ganal, M.W., Spivey, R., Wu, T., Earle, E.D., and Tanksley, S.D. (1993). Map-based cloning of a protein kinase gene conferring disease resistance in tomato. Science 262, 1432-1436.

Martin, G.B., Frary, A., Wu, T., Brommonschenkel, S., Chunwongse, J., Earle, E.D., and Tanksley, S.D. (1994). A member of the tomato *Pto* gene family confers sensitivity to fenthion resulting in rapid cell death. Plant Cell 6, 1543-.

Ohme-Takagi, M. and Shinshi, H. (1995). Ethylene-inducible DNA binding proteins that interact with an ethylene-responsive element. Plant Cell 7, 173-.

Salmeron, J. M., Barker, S. J., Carland, F. M., Mehta, A. Y., and Staskawicz, B. J. (1994). Tomato mutants altered in bacterial disease resistance provide evidence for a new locus controlling pathogen recognition. Plant Cell 6, 511-.

Salmeron, J.M., Oldroyd, G.E.D., Rommens, C.M.T., Scofield, S., Kim, H.S., Lavelle, D.T., Dahlbeck, D., and Staskawicz, B.J. (1996). Tomato *Prf* is a member of the leucine-rich repeat class of plant disease resistance genes and lies embedded within the *Pto* kinase gene cluster. Cell, in press

Zhou, J., Loh, Y.-T., and Martin, G.B. (1995). The tomato gene *Pti1* encodes a serine/threonine kinase that is phosphorylated by Pto and is involved in the hypersensitive response. Cell 83, 925-935.

Zhou, J., Tang, X., and Martin, G.B. (1996). The Pto kinase interacts with proteins that bind a promoter element in several pathogenesis-related genes. (submitted).

# A comparison of methods for the determination of the oxidative burst in whole plants

Ann T. Schroeder*, Gregory Martin[#] and Philip S. Low*

Departments of Chemistry* and Agronomy[#]
Purdue University
West Lafayette IN USA

Rapid production of hydrogen peroxide and other active oxygen species in response to stimulation with pathogens or elicitors is commonly termed the oxidative burst (for review, see Low and Schroeder, 1996.) The oxidative burst was first identified in plant cell culture when soybean cells were treated with a fungal extract or with cell wall fragments of oligogalact-uronic acid (Apostol et al., 1989). While the burst has been characterized in cell suspension cultures from a variety of plants, its properties have not yet been examined in whole plants. Such studies are crucial, since defense-related processes such as cell wall stabilization, phytoalexin biosynthesis, defense gene expression, and hypersensitive cell death are stimulated or augmented in cell culture upon production of hydrogen peroxide, and questions naturally arise as to whether similar processes occur *in planta*. Nevertheless, burst assays in whole plants have been slow to develop due to the unstable nature of reactive oxygen species and difficulties in delivering and viewing probes of oxidant generation in intact tissues.

Several lines of evidence support a role for hydrogen peroxide production in mounting resistance reactions in whole plants. First, potato tuber slices treated with an incompatible pathogen, *Phytophthora infestans*, show a marked increase in superoxide production (Doke 1983), suggesting oxidant generation can be triggered by pathogen interaction. Further, potato plants which express glucose oxidase in their intercellular spaces maintain significantly increased levels of hydrogen peroxide in their tissues, and simultaneously show increased resistance to fungal infections (Wu et al., 1995). Finally, a time dependent increase in superoxide levels has been observed upon infection of tomato cotyledons with race-specific elicitors of *Cladosporium fulvum*, suggesting gene-for-gene resistance reactions include oxidant biosynthesis (May et al., 1996).

A variety of indicators and protocols have been developed to detect reactive oxygen species in plant cell suspension cultures. The aim of this paper is to examine the advantages and disadvantages of available methods for detecting and localizing the oxidative burst *in planta*. For this purpose, the four gene-for-gene combinations of *Psuedomonas syringae* pv. tomato (+/- avrPto) and Rio Grande +/-Pto tomato plants are evaluated for production of active oxygen species, since only the incompatible interaction results in prolonged oxidant biosynthesis in cell culture (Chandra et al., 1996).

EVALUATION OF METHODS FOR DETECTING THE OXIDATIVE BURST

*Starch/KI*. One of the most facile methods for detecting hydrogen peroxide production is the starch-potassium iodide assay. Hydrogen peroxide oxidizes iodide ions to iodine, which in turn complexes with starch to form a blue-purple color. This simple histochemical test has been adapted to a variety of applications, including examination of hydrogen peroxide generation in fresh-cut sections of stem and root tissue (Olson and Varner, 1993; Wu et al. 1995), and identification of $H_2O_2$-generating cells of whole tissue sections on nitrocellulose which has been impregnated with starch and KI (Schopfer, 1994). While such methods allow for crude localization of oxidant production in sections which can exude $H_2O_2$, application of this technique in leaves is limited due to active degradation processes and physical barriers to secretion. Other disadvantages of the starch/KI assay for *in situ* evaluation of $H_2O_2$ production were found to be the high back-ground that arises from air oxidation of the iodide, and the minimum concentration of 500 μM $H_2O_2$ found to be necessary for visualization. Based upon extra-polations from cell culture data, accumulation of up to 1 mM $H_2O_2$ could conceivably occur *in planta* if detoxification were prevented (Legendre et al., 1992); however, we were unable to detect reliable hydrogen peroxide production in the resistant combination using inoculated leaves either embedded in starch/KI-containing agarose or blotted onto starch/KI coated nitrocellulose.

*Nitro Blue Tetrazolium (NBT)*. While a variety of indicators specifically detect $H_2O_2$, NBT is selective for superoxide. Upon reduction, the yellow NBT forms a blue formazan precipitate. The NBT itself, however, is highly soluble in water, so it may be introduced into plant tissue either by soaking more succulent tissue (Doke, 1983) or vacuum infiltrating leaves (May et al., 1996). We initially attempted to transpire NBT into leaves; however, the high level of oxidant production in the xylem and phloem consumed a majority of the dye, and the resulting blue precipitate blocked transpiration, greatly weakening the plants. However, when the NBT (0.5 mg/mL) was vacuum infiltrated into the leaves of the tomato plants at various times fol-lowing inoculation with *P. syringae* (avrPto) and the oxidation reaction was allowed to proceed, a blue color developed at sites of oxidant biosynthesis. It seems that the intensity and dimensions of this blue coloration reflect the magnitude of the *in situ* oxidative burst, since large quantities of the formazan precipitate were only observed in the resistant combination; i.e. the three susceptible interactions had only background levels of the blue precipitate.

It is curious that other attempts to exploit the NBT assay for visualization of oxidant production *in situ* have required co-administration of NADPH (Doke, 1983; May et al., 1996). If the oxidant generating system is located in the cell wall, such as the touch-activated NADPH oxidase described by Engelberth et al. (1996), such exogenous NADPH may indeed be necessary. However, if the oxidant generating system is positioned in the plasma membrane, as is the case with human neutrophils and probably the defense-related oxidative burst in plants (Dwyer et al., 1996; Tenhaken et al., 1995), then exogenous NADPH should have no effect, since the source of reductant for these reactions is thought to be intracellular NADPH.

*CeCl$_3$.* While the techniques outlined above allow for localization of reactive oxygen species to specific cell types in heterogeneous tissue sections, they lack the resolving power to localize H$_2$O$_2$ production to subcellular organelles. For this application, a CeCl$_3$-based assay has been developed (Thomas and Trelease, 1981). CeCl$_3$ is an electron-dense heavy metal that remains soluble in aqueous solution unless oxidized to form a cerium peroxide precipitate at sites of H$_2$O$_2$ biosynthesis. Deposition of the peroxide at such sites then allows for identification of loci involved in oxidant biosynthesis by electron microscopy. Unfortunately, the cumbersome requirement to thin section the desired tissue and prepare the sample for electron microscopy renders this method undesirable for high throughput analyses of the oxidative burst in whole plants.

*Ti(IV)Cl$_4$.* Similar to the cerium chloride detection system, the titanium chloride indicator is also based on the formation of a perhydroxide complex. In contrast, however, the titanium complex has a bright yellow color, allowing its use as a colorimetric assay reagent. Consequently, the Ti(IV)Cl$_4$ technique has successfully been employed to evaluate changes in hydrogen peroxide content upon cold shock (Okuda et al., 1991) and senescence in pear fruit (Brennan and Frenkel, 1977). However, due to numerous plant pigments in leaf tissue which also absorb at the wavelength (450 nm) of the titanium complex, it became necessary to homogenize the stimulated plant tissue and extract the endogenous pigments with activated charcoal, or alternatively, to remove them using a strong anion exchange resin. The choice of extraction method is important, as extraction with an organic solvent such as acetone does not allow for determination of background absorbance at 450 nm by removal of all endogenous H$_2$O$_2$ with catalase. We have found that grinding the leaves in liquid nitrogen, a minimal volume of 5% trichloroacetic acid, and activated charcoal followed by neutralization with ammonium hydroxide yields the most satisfactory decolorization. This procedure also rapidly inactivates virtually all enzymes, preserving the endogenous H$_2$O$_2$ for subsequent analysis with Ti(IV)Cl$_4$.

Table I.

| Indicator | Application Method | Species Detected | Sensitivity Threshold | Advantages (+) and disadvantages (-) |
|---|---|---|---|---|
| Starch/KI | Agarose plates; tissue printing | oxidants | 500 μM | + localization<br>-oxidant must be exuded from the plant<br>-high background in tissue printing from air oxidation<br>-difficult to implement in "dry" tissue such as leaves |
| CeCl$_3$ | Electron microscopy | oxidants | | +Allows cellular and sub-cellular localization<br>-Expensive technique and equipment requirements |
| Ti(IV)Cl$_4$ | colorimetric assay with plant extracts | oxidants | 50 μM | +quantitation<br>-plant pigments interfere with assay; must decolorize with activated charcoal or strong anion exchange chromatography prior to assay |
| Nitroblue tetrazolium (NBT) | infiltration | $O_2^{\cdot-}$ | | +localization<br>+specific for $O_2^{\cdot-}$ |
| Diamino-benzidine (DAB) | infiltration | $H_2O_2$ | | +localization |
| luminol | luminescence in plant extracts | $H_2O_2$ | <1 μM | +quantitation |
| Fluorescent techniques | fluorescence assay with plant extracts | varies | varies | +quantition<br>-endogenous fluorescent compounds |

18

Disadvantages of this technique include not only the aforementioned interference from endogenous pigments, but also the inability to exploit the methodology for *in situ* localization of oxidant generation. Furthermore, Ti(IV)Cl$_4$ is air sensitive and reacts violently with water unless stabilized in an acid solution. Consequently, a more sensitive and less hazardous technique for detecting active oxygen species in extracts is preferable.

*Luminescent techniques.* The most sensitive and specific method for detection of hydrogen peroxide in plant tissue extracts is the chemiluminescent reaction of luminol catalyzed by an endogenous oxidant plus ferricyanide (Warm and Laties, 1982). While this method unfortunately requires homogenization of the infected plant tissue in some type of metabolic stopping solution, its ability to detect sub-micromolar changes in hydrogen peroxide allows for quantitative measurement of the oxidative burst in plant extracts. The resultant light can be measured with a luminometer. While we were not able to detect statistically significant changes in hydrogen peroxide content with the titanium assay, significant changes were readily seen with the luminol assay. Furthermore, both the first and second phases of the oxidative burst could be observed, while with the NBT assay, only the second burst was easily measured.

CONCLUSIONS

A summary of the various techniques can be found in Table I.

LITERATURE CITED

Apostol, I., Heinstein, P.F., and Low P.S. 1989. Rapid stimulation of an oxidative burst during elicitation of cultured plant cells. Plant Physiol. 90:109 116.

Brennan, T., and Frenkel, C. 1977. Involvement of hydrogen peroxide in the regulation of senescence in pear. Plant Physiol. 59:411-416.

Chandra, S., Martin, G. and Low, P.S. 1996. The Pto kinase mediates the oxidative burst in tomato. Proc. Natl. Acad. Sci. In press.

Doke, N. Involvement of superoxide anion generation in the hypersensitive response of potato tuber tissues to infection with an incompatible race of Phytophthora infestans and to the hyphal wall components. Physiol. Plant Path. 23:345-357.

Dwyer, S.C., Legendre, L.L., Low, P.S. and Leto, T. 1996. Plant and human neutrophil oxidative burst complexes contain immunologically related proteins. Biochim. Biophys. Acta 1289:231-237.

Engelberth, J., Bockelmann, C., Liβ, H, and Weiler, E.W. 1996. Enzymes in the cell wall and plamalemma of mechanically induced tendrils: evidence for a functional role of NADH-oxidases in the process of touch perception. Plant Physiol. in press.

Legendre L., Rueter, S., Heinstein, P.F., and Low, P.S. 1993. Characteriza tion of the oligogalacturonide-induced oxidative burst in cultured soybean cells. Plant Physiol. 102: 233-240.

May, M.J., Hammond-Kosack, K.E., and Jones, J.D.G. 1996. Involvement of reactive oxygen species, glutathione metabolism, and lipid peroxidation in the *Cf*-gene-dependent defense response of tomato cotyledons induced by race-specific elicitors of *Cladosporium fluvum*. Plant Physiol. 110:1367-1379

Okuda, T., Matsuda, Y., Yamanaka, A, and Sagisaka, S. 1991. Abrupt increase in the level of hydrogen peroxide in leaves of winter wheat is caused by cold treatment. Plant Physiol. 97:1265-1267.

Olson, P.D., and Varner, J.E. 1993. Hydrogen peroxide and lignification. Plant J. 4:887-892.

Schopfer, P. 1994. Histochemical demonstration and localization of $H_2O_2$ in organs of higher plants by tissue printing on nitrocellulose paper. Plant Physiol. 104:1269-1275.

Tenhaken, R., Levine, A., Brisson, L.F., Dixon, R.A., and Lamb, C. 1995. Function of the oxidative burst in hypersensitive disease resistance. Proc. Natl. Acad. Sci. USA 92:4158-4163.

Thomas, J., and Trelease, R.N. 1981. Cytochemical localization of glyco-late oxidase in microbodies (glyoxysomes and peroxisomes) of higher plant tissues with the $CeCl_3$ technique. Protoplasma. 108:39-53.

Warm, E. and Laties, G.G. 1982. Quantification of hydrogen peroxide in plant extracts by the chemiluminescence reaction with luminol. Phytochem. 21: 827-831.

Wu, G., Shortt, B.J., Lawrence, E.B., Levine, E.B., Fitzsimmons, K.C., and Shah, D.M. 1995. Disease resistance conferred by expression of a gene encoding $H_2O_2$-generating glucose oxidase in transgenic potato plants. Plant Cell 7:1357-1368.

# Signal Perception and Intracellular Signal Transduction in Plant Pathogen Defense

Wolfgang Wirtz[1], Dirk Nennstiel[1], Thorsten Jabs[2], Sabine Zimmermann[3], Dierk Scheel[1] and Thorsten Nürnberger[1].

[1]Institut für Pflanzenbiochemie, Halle/Saale, Germany, [2]Rheinisch-Westfälische Technische Hochschule, Aachen, Germany, [3]Max-Planck-Institut für Molekulare Pflanzenphysiologie, Golm, Germany.

Initiation of defense responses requires recognition of either plant- or pathogen-derived signals, termed elicitors, by the plant. This is believed to be mediated by receptors which specifically bind these molecules and thereby initiate intracellular transmission of the signal. Subsequent conversion of this signal through a transduction cascade results in induction of a particular cell response(s), such as defense gene activation. High affinity receptors for both race-specific and general elicitors derived from phytopathogenic fungi have been shown to reside in plant plasma membranes (Cosio et al. 1990; Nürnberger et al. 1994a; Wendehenne et al. 1995; Kooman-Gersmann et al. 1996). Intracellular signal conversion and transduction include changes in the ion permeability of the plasma membrane, generation of reactive oxygen species, and alterations in the phosphorylation status of various proteins (Dixon et al. 1994; Ebel and Cosio 1994). These elements of signaling chains have been found in plants establishing either cultivar-specific or species-specific resistance, which reinforces the idea that the molecular mechanisms underlying signal perception and transduction may be similar in plants undergoing either type of resistance.

The non-host resistance response of parsley leaves infected with zoospores of the soybean pathogen, *Phytophthora sojae*, comprises lesion formation, callose apposition in the cell wall, phytoalexin production, and transcriptional activation of defense-related genes (Schmelzer et al. 1989). A part of this complex defense response, namely phytoalexin production and defense gene activation, can be also observed in suspension-cultured parsley cells or protoplasts upon treatment with fungal cell wall material. Zoospores from of another phytopathogenic fungus, *Phytophthora infestans*, have been found to induce hypersensitive cell death in parsley plants (Gross et al.

1993) and cell wall preparations of this fungus stimulate phytoalexin accumulation and defense gene activation in parsley cell cultures. We have employed elicitors from both fungi to study signal perception mechanisms and early elicitor-induced signaling events in parsley.

## ELICITORS

Two elicitor glycoproteins have been isolated from the culture filtrate of the phytopathogenic fungi, *P. sojae* and *P. infestans*, due to their ability to induce accumulation of furanocoumarin phytoalexins in parsley cell cultures. Elicitor activity of a 42-kDa glycoprotein from *P. sojae* resides in the protein moiety and is not affected by heat denaturation, indicating that this activity is determined by primary sequence information rather than secondary structure motifs (Parker et al. 1991). An internal peptide of 13 amino acids (Pep-13) was found to be sufficient to stimulate the same responses as the crude fungal cell wall elicitor and the purified glycoprotein (Nürnberger et al. 1994a). These responses include transcriptional activation of defense-related genes, reduction of transcript levels of non-defense-related genes encoding parsley histones, cdc2 and cyclin, stimulation of ethylene and phytoalexin biosynthesis, and an increase in extracellular chitinase activity (Nürnberger et al. 1994a; Logemann et al. 1995). We have not yet detected any response of parsley cells that is stimulated by the fungal cell wall elicitor, but not by Pep-13.

The amino acid sequences of the oligopeptide and the intact protein elicitor did not show any significant homology to sequences contained in several data bases (Sacks et al. 1995). Pep-13 was demonstrated to be sufficient and necessary for elicitor activity of the intact glycoprotein (Nürnberger et al. 1994). Deletion of one N- and one C-terminal amino acid yielded the minimum peptide with full elicitor activity. Substitution analysis, in which individual amino acids of Pep-13 were progressively replaced by alanine, identified two residues (W-2, P-5) critical for activity. All other exchanges exerted little or no effect on the elicitor activity of this peptide. [1]H-NMR studies revealed a random-coil-like structure of the oligopeptide in aqueous solution.

Recently, an 85-kDa glycoprotein elicitor was purified to homogeneity from the culture filtrate of *P. infestans*. This protein induced synthesis of the same pattern of phytoalexins in parsley cells as is elicited by the *P. sojae*-derived Pep-13. Similarly, the elicitor activity of this protein resided in (glyco)peptides smaller than 10 kDa. Polyclonal antisera raised against the 42-kD glycoprotein elicitor from *P. sojae* or Pep-13 did not recognize the *P. infestans* elicitor. Immunological analysis revealed mannose residues (high-mannose-type N-glycosylation) to be the major constituent of the carbohydrate moiety of the protein, representing approximately 20% of the formula weight of the entire molecule.

The cDNA encoding the protein part of the glycoprotein was isolated. The deduced amino acid sequence was found to contain 5 putative N-glycosylation sites as well as an N-terminal sequence motif reminiscent of a leader peptide. Sequence comparison demonstrated significant homologies at the amino acid level between the *P. infestans* elicitor and mammalian glucocerebrosidases and a bacterial xylanase. Strikingly, two amino acid residues known to constitute the active site of these enzymes were found to be conserved in the sequence of the elicitor protein as well, suggesting hydrolase activity of this fungal elicitor. Comparison of peptide sequences derived from fragments of this protein or from the corresponding cDNA with the respective sequences of the *P. sojae* glycoprotein revealed no homologies between the two elicitor proteins.

## ELICITOR RECEPTORS

The cultivar or species specificity of purified fungal elicitors, their ability to induce plant defense responses at low nanomolar concentrations, and the precisely defined signal structures required for elicitor-mediated activation of these reactions suggest the involvement of receptors in elicitor recognition and subsequent intracellular signal generation (Ebel and Cosio 1994). A receptor/ligand complex in the parsley system could be visualized on the protoplast surface by incubation with a fluoresceinated Pep-13 derivative and silver enhanced immunogold labeling with anti-fluorescein antibodies as viewed by epipolarization microscopy (Diekmann et al. 1994). However, no indications for receptor-mediated endocytosis of the receptor/ligand complex upon ligand binding could be obtained. Kinetic characterization of elicitor-binding was performed using parsley microsomal membrane preparations or protoplasts and radioiodinated Pep-13. A saturable single-class high-affinity binding site for this ligand was found in both preparations (Nürnberger et al. 1994a). Binding was reversible and specific with respect to both structural features of the ligand and the plant species that recognized the ligand. To demonstrate a functional link between ligand binding and activation of phytoalexin production a series of structural derivatives of Pep-13 were tested in primary structure/activity relationship studies. The same structural requirements were generally found to be responsible for both efficient competition of binding of Pep-13 to the binding site and for stimulation of the plant defense response. Furthermore, the magnitudes of the $EC_{50}$- and $IC_{50}$-values obtained with the various peptides investigated were consistently shown to be similar, indicating that activation of plant defense is receptor-mediated. Interestingly, the 85-kDa glycoprotein from *P. infestans* was incapable of competing receptor binding of Pep-13, a finding that suggests the existence of a distinct recognition system for this elicitor in the plasma membrane of parsley cells.

Radioiodinated Pep-13 could be covalently coupled to its binding site in parsley microsomal or plasma membranes by each of three homobifunctional chemical crosslinkers (Nürnberger et al. 1995). SDS-PAGE and autoradiography of membrane proteins demonstrated labeling of a 91-kDa protein, which could not be detected after either addition of unlabeled Pep-13 or omission of the crosslinker from the reaction mixture. Detergent solubilization of membrane proteins did not affect the ability of the receptor to bind Pep-13. Furthermore, crosslinking of Pep-13 to solubilized receptor revealed that the 91kD protein appears to be sufficient for ligand binding.

In an attempt to isolate the receptor/ligand complex we employed a biologically active, biotinylated derivative of radioiodinated Pep-13 in chemical crosslinking experiments and subsequent affinity chromatography on avidin agarose. SDS-PAGE/autoradiography analysis of eluted proteins demonstrated the potential applicability of this experimental strategy for the isolation of the Pep-13 binding site. Alternative approaches to isolate the elicitor receptor as well as the corresponding receptor cDNA (affinity chromatography on Pep-13-agarose; ligand affinity screening of a mammalian cell line transfected with a size-enriched parsley cDNA library) are under way.

## SIGNAL TRANSDUCTION

The effects of elicitors on ion fluxes across the plasma membrane have been considered to be part of elicitor-specific signal transduction leading to the induction of plant defense responses in several plant/pathogen interactions (Ebel and Cosio 1994). Among the earliest events detectable after treatment of parsley cells with Pep-13 are changes in ion permeability of the plasma membrane (influxes of $H^+$ and $Ca^{2+}$, effluxes of $K^+$ and $Cl^-$). Four lines of evidence strongly suggest a functional link between binding of Pep-13 to its receptor and the rapidly induced ion fluxes as elements of the signal transduction cascade triggering pathogen defense in parsley. (i) Omission of calcium from the culture medium blocked defense-related gene activation and phytoalexin formation (Nürnberger et al. 1994a). Furthermore, elicitor treatment of cultured parsley cells in the presence of $Ca^{2+}$-channel blockers such as flunarizine significantly inhibited both the targeted ion flux and phytoalexin production (Nürnberger et al. 1994b). (ii) Initial changes in ion fluxes were detectable within 2-5 min after addition of the elicitor and peaked after 20-30 min, thereby clearly preceding transcriptional activation of defense-related genes and phytoalexin formation. (iii) Similar elicitor concentrations were needed for half-maximal stimulation of the various plant responses. (iv) Identical structural properties of the oligopeptide were found to be required for efficient stimulation of all four ion fluxes and furanocoumarin synthesis, as well as for binding of the elicitor to its binding site (Nürnberger et al. 1994a).

In patch-clamp experiments with parsley protoplasts an inward-rectifying calcium-permeable ion channel was detected. Electrophysiological investigation of this channel revealed a unitary conductance of 140 pS, unusually large for plant calcium-permeable ion channels. Exhibiting low, voltage-independent activity in the resting state, channel activity could be strongly increased by addition of either cell wall or peptide elicitor. This activation was due solely to an increase in the number of channels that opened simultaneously, since no changes in the single-channel conductance were observed upon elicitor application. This channel may therefore contribute to the elicitor-induced increase in cytosolic $Ca^{2+}$-concentration observed in parsley cells (Sacks et al. 1993).

Synthesis and extracellular accumulation of hydrogen peroxide (oxidative burst) upon elicitor treatment of cultured parsley cells is stimulated within 5 min upon addition of elicitor (Nürnberger et al. 1994a). This response is dependent on extracellular $Ca^{2+}$ and could be inhibited by the same set of ion channel blockers that were found to inhibit both elicitor-induced ion fluxes and phytoalexin formation. Additionally, the non-selective ionophore, Amphotericin B, stimulated not only the transcriptional activation of the same pattern of defense genes as did the elicitor, but also the oxidative burst. These findings indicate that elicitor-induced ion fluxes are required for the induction of the oxidative burst in cultured parsley cell as well. Diphenylene iodonium, an inhibitor of the superoxide anion-forming enzyme, NADPH oxidase, did not only block the elicitor-induced oxidative burst, but also accumulation of defense-related transcripts and phytoalexin production, whereas proton influx, constitutive gene expression and cell viability remained unaffected. Thus, reactive oxygen species appear to be involved in signaling leading to defense gene activation.

In summary, binding of Pep-13 to its receptor in the parsley plasma membrane appears to initiate processes that involve plasma membrane-located ion channels and production of reactive oxygen species, which subsequently result in transient activation of defense-related genes and phytoalexin production.

This work was supported by the European Community (Biotech and HC&M Programs), the Deutsche Forschungsgemeinschaft (grants Sche 235/3-1 and 3-2), Bayer Chemical Company, Leverkusen, and Fonds der Chemischen Industrie.

Cosio, E. G., Frey, T., Verduyn, R., van Broom, J., and Ebel, J. 1990. High-affinity binding of a synthetic heptaglucoside and fungal glucan phytoalexin elicitors to soybean membranes. FEBS Lett. 271:223-226.

Diekmann, W., Herkt, B., Low, P. S., Nürnberger, T., Scheel, D., Terschüren, C., and Robinson, D. G. 1994. Visualization of elicitor-binding loci at the plant cell surface. Planta 195:126-137.

Dixon, R. A., Harrison, M. J., and Lamb, C. J. 1994. Early events in the activation of plant defense responses. Annu. Rev. Phytopathol. 32:479-501.

Ebel, J. and Cosio, E. G. 1994. Elicitors of plant defense responses. Int. Rev. Cytol. 148:1-36.

Gross, P., Julius, C., Schmelzer, E., and Hahlbrock, K. 1993. Translocation of cytoplasm and nucleus to fungal penetration sites is associated with depolymerization of microtubules and defense gene activation in infected, cultured parsley cells. EMBO J. 12:1735-1744.

Kooman-Gersmann, M., Honee, G., Bonnema, G., and de Wit, P. J. G. M. 1996. A high affinity binding site for the AVR9 peptide elicitor of *Cladosporium fulvum* is present on plasma membranes of tomato and other solanaceous plants. Plant Cell 8:929-940.

Logemann, E., Wu, S. C., Schröder, J., Schmelzer, E., Somssich, I. E., and Hahlbrock, K. 1995. Gene activation by UV light, fungal elicitor or fungal infection in *Petroselinum crispum* is correlated with repression of cell cycle-related genes. Plant J. 8:865-876.

Nürnberger, T., Nennstiel, D., Jabs, T., Sacks, W., Hahlbrock, K., and Scheel, D. 1994a. High affinity binding of a fungal oligopeptide elicitor to parsley plasma membranes triggers multiple defense responses. Cell 78:449-460.

Nürnberger, T., Colling, C., Hahlbrock, K., Jabs, T., Renelt, A., Sacks, W. R., and Scheel, D. 1994b. Perception and transduction of an elicitor signal in cultured parsley cells. Biochem. Soc. Symp. 60:173-182.

Nürnberger, T., Nennstiel, D., Hahlbrock, K., and Scheel, D. 1995. Covalent cross-linking of the *Phytophthora megasperma* oligopeptide elicitor to its receptor in parsley membranes. Proc. Natl. Acad. Sci. USA 92:2338-2342.

Parker, J. E., Hahlbrock, K., and Scheel, D. 1988. Different cell wall components from *Phytophthora megasperma* f. sp. *glycinea* elicit phytoalexin production in soybean and parsley. Planta 176:75-82.

Sacks, W. R., Ferreira, P., Hahlbrock, K., Jabs, T., Nürnberger, T., Renelt, A., and Scheel, D. Pages 485-495 in: Advances in Molecular Genetics of Plant-Microbe Interactions. E. W. Nester and D. P. S. Verma, eds., Kluwer Acad. Publ., Dordrecht, Netherlands.

Sacks, W. R., Nürnberger, T., Hahlbrock, K., and Scheel, D. 1995. Molecular characterization of nucleotide sequences encoding the extracellular glycoprotein elicitor from *Phytophthora megasperma*. Mol. Gen. Genet. 246:45-55.

Schmelzer, E., Krüger-Lebus, S., and Hahlbrock, K. 1989. Temporal and spatial patterns of gene expression around sites of attempted fungal infection on parsley leaves. Plant Cell 1:993-1001.

Wendehenne, D., Binet, M.-N., Blein, J.-P., Ricci, P., and Pugin, A. 1995. Evidence for specific, high-affinity binding sites for a proteinaceous elicitor in tobacco plasma membranes. FEBS Lett. 374:203-207.

# SYSTEMIC ACQUIRED RESISTANCE IN ARABIDOPSIS THALIANA

K. Summermatter, Th. Birchler, L. Sticher, B. Mauch-Mani, M. Schneider and J.P. Métraux

Institut de Biologie végétale, Université de Fribourg, 1700 Fribourg, Switzerland

## INTRODUCTION

In many plants, necrotizing infections can result in a broad spectrum protection against subsequent infections at the same or at distal parts of the plant (reviewed in Schneider et al. 1996). Salicylic acid (SA) was proposed as a putative signaling molecule for the induction of systemic acquired resistance (SAR) (reviewed in Schneider et al. 1996). We have further evaluated the biological role of SA by searching for SA-insensitive mutants. Plants having lost the ability to respond to SA might be of particular interest to understand the role of this molecule in the development of SAR. *Arabidopsis thaliana* is a suitable genetic system for such studies since it displays SA-mediated SAR (Delaney et al. 1994, Mauch-Mani and Slusarenko 1994, Summermatter et al. 1995), and responds to SA by an increase in resistance against various pathogens (Schneider et al. 1996). We now report on a mutant, called SI (salicylic acid insensitive), selected for its survival on toxic levels of SA.

## MATERIALS AND METHODS

Culture conditions for *A. thaliana* as well as the procedures for the cultivation, inoculation and disease quantification with *Pseudomonas syringae* pv *syringae* were previously described (Summermatter et al. 1995). Free and bound salicylic acid was determined by HPLC using *o*-anisic acid as previously described (Meuwly and Métraux 1993). T-DNA mutagenized *A. thaliana* seeds (ecotype Wassilewskija, (WS), Feldman collection) were grown in petri dishes containing MS-1 medium amended with toxic levels of SA (100 µM). When scored 4 weeks after sowing, wild type plants displayed abnormal white color and inhibited growth while mutants retained normal growth and green color. Leaf tissue (1-2 cm$^2$) from *Pseudomonas*-inoculated and adjacent non-inoculated regions were fixed in glutaraldehyde/osmium tetroxyde, embedded in Araldit, and stained with lead citrate/uranyl acetate. Observations were carried out with a Philipps CM 100 TEM.

RESULTS

*SAR in Arabidopsis.* Fig.1 shows the decrease in lesion size observed 11 d after preinoculation of the lower leaves with *P. syringae*. The lesions had a necrotic hypersensitive reaction-like appearance and were restricted to the site of bacterial inoculation.

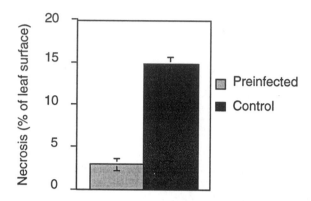

Fig. 1. SAR in *A. thaliana* induced by a preinoculation of *P. syringae* pv *syringae* on the lower leaves of the rosette and challenged in the upper leaves with the same bacteria. Lesions were determined 11 d after primary inoculation (n = 6; ± SD).

One interesting feature is that in this experiment a relatively high titer of bacteria was used to challenge the upper leaves ($10^6$ cells/ml). As can be seen in Fig. 2, the bacterial titers remained similar in challenged upper leaves from control compared to plants preinfected on the lower leaves despite necrotization which starts within the 24-48 h after inoculation. Ultrastructural observations of leaf tissue in the inoculated region of pretreated or control plants showed no marked difference during the early phase of the challenge infection. In particular, bacteria did not appear to be encased at the cell wall as has been reported in the case of *P. syringae* mutants which do not cause a hypersensitive necrosis (Bestwick et al. 1995). It seems likely that the effect of a pretreatment on the lower leaf results in the suppression of symptoms which may result from containment or histological localisation of the bacteria.

*SA insensitive mutant screening.* One mutant was selected on the basis of its apparent SA insensitivity. It was backcrossed with wild type WS plants and selfed. Of 117 plants, 83 scored for a SA insensitive and 32 for a SA sensitive phenotype, indicating that the mutation is dominant.

We then determined if survival on SA-amended medium was simply the result of lack of SA uptake from the medium. Fig. 3 shows the levels of both free and bound SA in leaves of SI and wild type (WS) plants treated with 1mM SA (drench application in the soil) for 48 h. Both SI and WS plants accumulated high levels of free and bound SA, indicating that survival on SA is unlikely to be the result of lack of SA uptake from the medium.

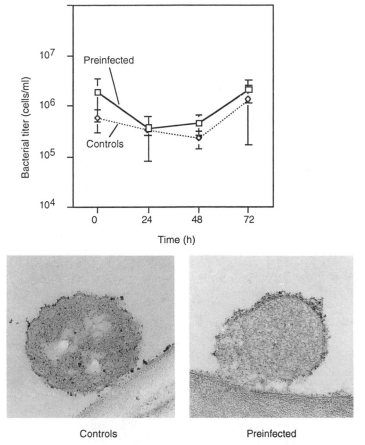

Fig.2. Bacterial titers and appearance of bacteria in challenged leaves of control and preinfected plants (x 41'000) (n = 4;± SD).

In wild type plants, treatments with SA or with synthetic resistance inducers such as dichloroisonicotinic acid (INA) resulted in enhanced resistance against pathogens.

We compared resistance induced in wild type and SI plants after treatment with SA or INA against *P. syringae* pv *syringae*. Interestingly, SI plants loose responsiveness to SA while INA still is effective in inducing resistance (data not shown), indicating that the SI mutant is not impaired in its general ability to defend itself against pathogens.

We determined if SAR was still deployed against *P. syringae* pv *syringae* after preinfection with *P. syringae* pv *syringae.* (Fig. 4).

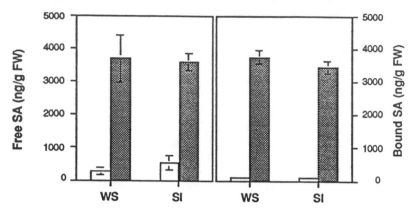

Fig. 3. Free and bound SA levels in SI and wild type (WS) plants drench-treated with SA (n=4; ± SD). White histograms: controls; dark histograms: 1 mM SA.

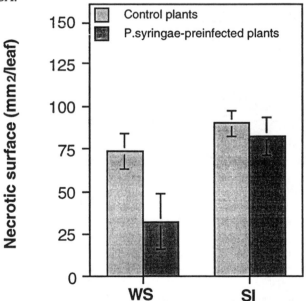

Fig. 4. SAR induced by a preinfection with *P. syringae* pv *syringae* against a challenge with the same pathogen 48 h after preinfection of the lower leaves. Symptoms were determined 7 days after challenge of the upper leaves of the rosette (n = 5; + SD).

In contrast to their wild type counterparts, SI mutants were unable to induce SAR (Fig. 4). This behaviour seems to reflect the insensitivity of SI towards SA. However, SI mutants were still able to fully induce the local and systemic accumulation of SA in response to a local pretreatment with *P. syringae* pv *syringae* on the lower leaves (Fig. 5).

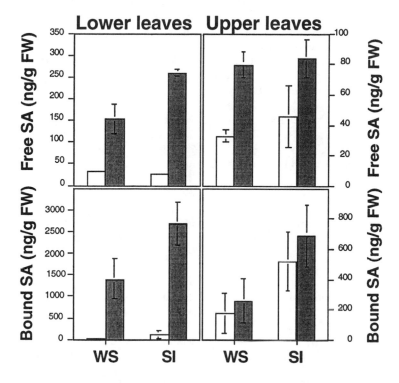

Fig. 5. Local and systemic accumulation of free and bound SA 48 h after infection of the lower leaves with *P. syringae* pv *syringae*. White histograms: controls; dark histograms: leaves from infected plants (n=4; ± SD).

After preinfection with *P. syringae* pv *syringae*, both free and bound SA increased between 1.2 and 3 times in SI compared to WS plants (Fig. 5).

DISCUSSION

Preinfection of *A. thaliana* with *P. syringae* pv *syringae* results in a decrease in lesion size upon challenge infection with the same organism. EM observations showed no distinctive differences between bacteria in SAR and in control plants. Suppression of symptoms might perhaps involve detoxification rather than histological containment.

We were able to select a mutant which is unable to display SAR against *P. syringae* pv *syringae*. SI is not impaired in the ability to take up SA from the medium nor to produce SA locally or systemically after local infection. Thus the mutation is likely to be at a step after SA perception. Pretreatments with INA did induce resistance indicating that the potential for defence has not been impaired by the mutation. The experiments described above support the notion that SA is an important component in the induction of SAR.

A model for the mode of action of SA and INA was proposed recently whereby SA inhibits the activity of catalase thus leading to an increase in active oxgen species which in turn could act directly or as signals for defense reactions (Chen et al., 1993). Results obtained here suggest that SA and INA have different mode of actions.

## ACKNOWLEDGEMENTS

This work was supported by grant 34098.92 from the Swiss National Science Foundation and by financial support from CIBA.

## REFERENCES

Bestwick, C.S., Bennett, M.H., and Mansfield, J.W. 1995. Hrp mutant of *Pseudomonas syringae* pv *phaseolica* induces cell wall alterations but not membrane damage leading to the hypersensitive reaction in lettuce. Plant Physiol. 108, 503-516.

Chen, Z., Silva, H., and Klessig, D. F. 1993. Active oxygen species in the induction of plant systemic acquired resistance by salicylic acid. Science 262, 1883-1885.

Delaney, T. P., Uknes, S., Vernooij, B., Friedrich, L., Weymann, K., Negretto, D., Gaffney, T., Gut-Rella, M., Kessmann, H., Ward, E., and Ryals, J. 1994. A central role of salicylic acid in plant disease resistance. Science 266, 1247-1249.

Malamy, J., Carr, J. P., Klessig, D. F., Raskin, I. 1990. Salicylic acid a likely endogenous signal in the resistance response of tobacco to viral infection. Science 250, 1002-1004.

Mauch-Mani, B., and Slusarenko, A. J. 1994. Systemic acquired resistance in *Arabidopsis thaliana* induced by a predisposing infection with a pathogenic isolate of *Fusarium oxysporum*. Molec. Plant - Microbe Interact. 7, 378-383.

Meuwly, P., and Métraux, J. P. 1993. Ortho-anisic acid as internal standard for the simultaneous quantitation of salicylic acid and its putative biosynthetic precursors in cucumber leaves. Anal. Biochem. 214, 500-505.

Schneider M., Schweizer P., Meuwly P. and Métraux J. P. 1996. Systemic induced resistance in plants. Int. Rev. Cytology, 168, 303-340.

Summermatter K., Sticher L., and Métraux J. P. 1995. Systemic responses in Arabidopsis thaliana infected and challenged with *Pseudomonas syringae* pv *syringae*. Plant Physiol, 108, 1379-1385.

# Studies of the Salicylic Acid Signal Transduction Pathway

Daniel F. Klessig, Jörg Durner, Zhixiang Chen, Marc Anderson, Uwe Conrath, He Du, Ailan Guo, Yidong Liu, Jyoti Shah, Herman Silva, Hideki Takahashi, and Yinong Yang, Waksman Institute and Department of Molecular Biology and Biochemistry, Rutgers, The State University of New Jersey, Piscataway, N.J. 08855, U.S.A.

An ever increasing body of evidence suggests that salicylic acid (SA) plays an important role in the activation of defense responses against microbial attack in dicotyledonous plants. In addition to helping establish SA's role in disease resistance, we have attempted to identify components of the SA signal transduction pathway(s). Below, the results of our recent studies are briefly summarized.

### Role of Catalase and Ascorbate Peroxidase
During the past several years, we have attempted to elucidate the mechanisms of SA action in plant disease resistance, and in so doing, have identified a SA-binding protein in tobacco leaves. The SA-binding protein is a family of catalases, whose $H_2O_2$-degrading activity is blocked by SA and those analogues of SA which are biologically active for the induction of $PR$ genes and enhanced resistance. As SA treatment of tobacco also led to a moderate increase in $H_2O_2$ levels, we proposed that the elevated levels of $H_2O_2$, or other reactive oxygen species (ROS) derived from $H_2O_2$, are involved in the activation of plant defenses (Chen et al., 1993). In support of this model, it was discovered that prooxidants, which lead to accumulation of ROS, partially mimic the action of SA. Moreover, 2,6-dichloroisonicotinic acid (INA; a synthetic inducer of $PR$ genes and enhanced resistance) and its biologically active analogues also inhibit tobacco catalase in vivo (Conrath et al., 1995). Finally, SA and INA inhibit ascorbate peroxidase, the other key $H_2O_2$-scavenging enzyme (Durner and Klessig, 1995).

Recently, the involvement of catalase inhibition by SA and elevated $H_2O_2$ levels in plant defense responses has been called into question. We, as well as the groups of John Ryals (Neuenschwander et al., 1995) and John Draper (Bi et al., 1995), observed that $PR-1$ gene activation by prooxidants such as $H_2O_2$ or 3-aminotriazole is suppressed in NahG transgenic tobacco, in which the SA signal is destroyed. This result, together with the finding by León et al. (1995) and Neuenschwander et al. (1995), that high levels of $H_2O_2$ induce production of SA, argues that $H_2O_2$ acts upstream of SA in the signal transduction pathway rather than, or in addition to, acting downstream of SA. However, we have discovered that NahG plants are less responsive to INA for induction of $PR-1$ genes than wild type tobacco. Since Malamy et *al.* (1996) and Vernooij *et al.* (1995) demonstrated that INA does not act through SA, this result suggests that NahG and wild type plants differ in more than just the loss of the SA signal. The product of *nahG*-encoded salicylate hydroxylase is catechol, which is a potent antioxidant. We have found that catechol and other antioxidants can suppress SA or INA activation of *PR-1* genes. These findings, therefore, argue that *PR-1* gene activation may be modulated through the plant's redox state.

To provide a more direct test of the involvement of catalase and ROS in the activation of defense responses, we have constructed transgenic tobacco plants in which the level of catalase is depressed due to expression of the tobacco catalase 1 (*cat1*) or catalase 2 (*cat2*) gene in an antisense orientation. Very dramatic reductions in catalase levels (≥ 90%) were accompanied by induction of *PR-1* genes and enhanced resistance to tobacco mosaic virus (TMV). However, the concurrent development of chlorosis or necrosis complicates the interpretation of these results (Takahashi *et al.*, 1996). Three lines, with severely depressed catalase levels, have been crossed with NahG plants. Some of these ASCAT1/NahG F1 plants show similar levels of reduction in total catalase activity. Preliminary analyses suggest that in contrast to the parental ASCAT1 plants, the F1 plants do not (i) develop necrosis, (ii) express the *PR-1* gene at high levels, or (iii) show enhanced resistance to TMV. If confirmed, these results suggest that in the absence of SA accumulation, reduction in catalase activity and the resulting elevation of $H_2O_2$ is not sufficient for the induction of *PR-1* gene expression and enhanced resistance.

To analyze the inhibition of catalase by SA in more detail, catalase from tobacco leaves was purified to homogeneity. There were ten or more isoforms of tobacco leaf catalase, which is a tetrameric enzyme made up of four identical or similar subunits. These isoforms were separated by chromatofocusing and found to exhibit similar sensitivity to SA, regardless of their subunit composition. Significant inhibition was only observed at 100 μM SA or more (Durner and Klessig, 1996).

### Mechanism of SA Inhibition of Catalase

Detailed analyses of SA's interaction with tobacco and mammalian catalases indicate that SA acts as an electron-donating substrate for the peroxidative reaction, which is a secondary activity of catalase (Durner and Klessig, 1996). The catalitic cycle, in which $H_2O_2$ is converted to $H_2O$ and $O_2$ involves steps 1 and 2 shown in Figure 1 and the enzyme intermediate called compound (cpd) I. The peroxidative cycle involves steps 1, 3 and 4, the enzyme intermediates cpds I and II and an electron-donating substrate like SA (denoted as AH in Figure 1). SA inhibits catalase by siphoning cpd I from the extremely rapid catalitic cycle into the

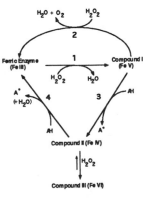

Figure 1

slow peroxidative cycle (> 1000X slower) by providing an electron to cpd I, converting it to cpd II. By promoting the peroxidative reaction, however, SA can also activate or protect catalase. Normally, at high $H_2O_2$ concentrations, much of the enzyme will be inactivated by its conversion to cpds II and III. Since SA can provide an electron to convert cpd II to active ferric enzyme (FE), the presence of SA prevents the formation of cpd III, thus protecting and/or restoring catalase activity.

How might SA function within an infected plant? In infected tissues, SA levels can approach 100 µM, a concentration sufficient to cause considerable inhibition of catalase and ascorbate peroxidase or protection of catalase, depending on the conditions. Whether SA positively or negatively modulates catalase activity will depend on the redox status of the cell. In the healthy tissue surrounding but not immediately adjacent to the infection site, $H_2O_2$ concentrations will be low to moderate and elevated SA levels likely inhibit catalase by promoting the slow peroxidative cycle (Figure 2). The resulting rise in $H_2O_2$ could induce *PR* genes and stimulate synthesis of more SA.

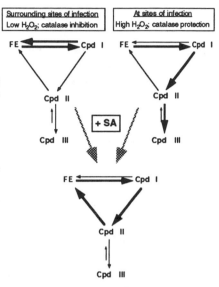

Figure 2

In the infected cells and immediately adjacent tissue, high levels of ROS resulting from the oxidative burst associated with the hypersensitive response (HR) could lead to substantial inactivation of catalase by conversion of cpd I to cpds II and III (Figure 2). Under conditions of such oxidative stress, SA likely helps to maintain a basal level of catalase activity again by acting as an electron-donating substrate to convert inactive enzyme intermediates such as cpd II back to the active ferric enzyme. This protective, antioxidative property of SA might serve to limit the impact of oxidative processes associated with development and spread of the necrotic lesion.

### SA Free Radical and Lipid Peroxidation

What, if any, might be the role of catalase (and ascorbate peroxidase) in systemic acquired resistance (SAR)? In uninfected leaves and other tissues of the infected plant, the levels of SA appear to be far below the concentration required to effectively inhibit catalase and ascorbate peroxidase, unless SA is highly concentrated in a subcellular compartment such as the peroxisome. If one assumes equal distribution of SA within cells, SA's role in SAR development is unlikely to involve elevated levels of $H_2O_2$ resulting from SA inhibition of catalase (and ascorbate peroxidase), as we originally proposed. Nonetheless, SA induction of SAR may be mechanistically coupled to its interaction with catalase and peroxidases. When SA donates an electron to catalase and peroxidases, it becomes a free radical since it now contains an unpaired electron (Figure 1). Free radicals can initiate lipid peroxidation and the resultant formation of lipid peroxides. Indeed, our preliminary studies indicate that SA induces lipid peroxidation and that several naturally occurring lipid peroxides activate *PR-1* genes in tobacco cells. Since lipid peroxidation is a chain reaction, a small amount of SA free radical could result in the formation of an effective lipid peroxide signal, without readily discernible inhibition of catalase or ascorbate peroxidase. Lipid peroxides may be involved in transducing the SA signal in infected, as well as uninfected, tissues. The lipid peroxide and $H_2O_2$ signals might act independently or in concert to transduce this signal.

### SA-Inducible *myb* Oncogene Homologue

To better understand the SA-mediated signal transduction pathway, we are identifying other signaling components. A TMV-inducible *myb* oncogene homologue (*myb1*) has been isolated and characterized from tobacco (Yang and Klessig, 1996). It is activated in TMV-infected resistant tobacco during development of the HR and SAR following the rise of endogenous SA. It is also rapidly induced by exogenously applied SA, INA and their biologically active analogues. This induction precedes activation of *PR* genes. Furthermore, the recombinant Myb1 protein specifically binds to a consensus Myb-binding site found in the promoter of the *PR-1a* gene. Taken together, these results suggest that the tobacco *myb1* gene encodes a signaling component downstream of SA which is likely to participate in transcriptional activation of *PR* genes and plant disease resistance.

### Protein Dephosphorylation Mediates
### SA Induction of *PR-1* Genes

Our studies with protein kinase and phosphatase inhibitors suggest that there are two or more phosphoproteins which function in the SA signaling pathway leading to *PR-1* gene activation (Conrath *et al.*, 1996). The protein phosphatase inhibitors, okadaic acid and calyculin A, were found to block SA-mediated activation of *PR-1* genes, implying the involvement of a phosphoprotein downstream of SA. The protein kinase inhibitors, K-252a and staurosporine, induced *PR-1* gene expression. Surprisingly, this induction was also partially suppressed in NahG transgenic tobacco plants. Moreover, K-252a stimulated production of SA and its glucoside, arguing that another phosphoprotein acts upstream of SA. Together, these results suggest that *PR-1* gene activation can be mediated by dephosphorylation of serine/ threonine residue(s) of two or more unidentified phosphoproteins.

### Putative *Arabidopsis* Mutants in the SA Signaling Pathway

We, as well as several other groups, are attempting to identify other components of the SA signal transduction pathway using genetic approaches. Two classes of mutants have been obtained in *Arabidopsis*, which appears to contain a very similar SA signaling pathway as characterized in TMV-infected tobacco. The first class is SA insensitive (*sai*; Shah *et al.*, 1996). These mutants were selected for lack of activation by SA of the chimeric *tms2* (*iaah*) gene driven by the tobacco *PR-1a* promoter. The *tms2* gene encodes an amidohydrolase which converts biologically inactive α-naphthalene acetamide to the biologically active auxin, α-naphthalene acetic acid. Elevated levels of auxin are toxic to root growth in germinating *Arabidopsis* seedlings. Transgenic *Arabidopsis* lines carrying the chimeric *PR-1a:tms2* gene were constructed, mutagenized, and then subjected to selection for normal root growth even in the presence of SA and α-napthalene acetamide. *sai1* is the prototypic member of this class of mutants. In the presence of SA, it fails to express the chimeric transgene and three endogenous *PR* genes (*PR-1*, *PR-2*, and *PR-5*). *sai1* also does not respond to the synthetic inducer INA. The nonresponsive phenotype is recessive and is due to a mutation at a single genetic locus. These results suggest that the mutation is in a component of the pathway which may act as a positive regulator downstream of SA.

The second class of mutants constitutively express *PR-1* genes even in the absence of infection or inducers like SA. A library of T-DNA insertional mutants was screened for mutants with high levels of *PR-1* mRNA. One of these mutants, called *cep1* for constitutive expression of *PR* genes, expresses *PR-1*, *PR-2* and *PR-5* genes at high levels in the absence of an inducer. The elevated levels of SA and its glucoside found in *cep1* suggest that the mutation may be in a component of the pathway upstream of SA. Experiments are in progress which will lead to the eventual identification and characterization of this component through cloning of the mutated gene.

## A Diverse Group of Genes Whose Induction is Distinct From That of *PR 1* Genes and May Be Independent of SA

To gain further insights into defense mechanisms and the signal transduction pathway(s) leading to resistance, a variety of genes have been cloned by differential screening of a cDNA library constructed from TMV-infected tobacco. Northern analysis showed that 7 groups of genes are induced in TMV-infected tobacco during development of both the HR and SAR. Five of the groups encode well characterized enzymes involved in hormone metabolism (1-aminocyclopropane-1-carboxylate [ACC] oxidase and ACC deaminase), isoprenoid synthesis (3-hydroxy-3-methylglutaryl-coenzyme A [HMG-CoA] reductase), oxidative stress (ascorbate peroxidase), and pathogen attack (acidic chitinase). The sixth group, consisting of at least 5 members, corresponds to the SAR8.2 family which is associated with SAR (Alexander *et al.*, 1992). The seventh group is composed of a novel gene termed *pin-1* for pathogen induced gene. The *pin-1* insert encodes a 185 amino acid protein.

Exogenously applied SA induced all of these genes, but only transiently. However, temperature shift experiments showed that activation of these genes precedes both the dramatic rise in SA levels and the induction of *PR-1* genes. Moreover, expression of these genes, but not the *PR-1* genes, was induced by TMV infection of NahG plants. These data indicate that activation of these genes during defense responses is distinct from induction of *PR-1* genes and may be independent of SA accumulation.

## Literature Cited

Alexander, D., Stinson, J., Pear, J., Glascock, C., Ward, E., Goodman, R. M., and Ryals, J. 1992. A new multigene family inducible by tobacco mosaic virus or salicylic acid in tobacco. Mol. Plant-Microbe Interact. 5:513-515.

Bi, Y. M., Kenton, P., Mur, L., Darby, R., and Draper, J. 1995. Hydrogen peroxide does not function downstream of salicylic acid in the induction of PR protein expression. Plant J. 8:235-245.

Chen, Z., Silva, H., and Klessig, D. F. 1993. Active oxygen species in the induction of plant systemic acquired resistance by salicylic acid. Science 262:1883-1886.

Conrath, U., Chen, Z., Ricigliano, J. R., and Klessig, D. F. 1995. Two inducers of plant defense responses, 2,6-dichloroisonicotinic acid and salicylic acid, inhibit catalase activity in tobacco. Proc. Natl. Acad. Sci. USA. 92:7143-7147.

Conrath, U., Silva, H., and Klessig, D. F. 1996. Protein dephosphorylation mediates salicylic acid-induced expression of *PR-1* genes in tobacco. Plant J. In press.

Durner, J. and Klessig, D. F. 1995. Inhibition of ascorbate peroxidase by salicylic acid and 2,6-dichloroisonicotinic acid, two inducers of plant defense responses. Proc. Natl. Acad. Sci. USA. 92:11312-11316.

Durner, J. and Klessig, D. F. 1996. Salicylic acid is a modulator of plant and animal catalases. J. Biol. Chem. In review.

León, J., Lawton, M. A., and Raskin, I. 1995. Hydrogen peroxide stimulates salicylic acid biosynthesis in tobacco. Plant Phys. 108:1673-1678.

Malamy, J., Sánchez-Casas, P., Hennig, J., Guo, A., and Klessig, D. F. 1996. Dissection of the salicylic acid signaling pathway for defense responses in tobacco. Mol. Plant-Microbe Interact. In press.

Neuenschwander, U., Vernooij, B., Friedrich, L., Uknes, S., Kessmann, H., and Ryals, J. 1995. Is hydrogen peroxide a second messenger of salicylic acid in systemic acquired resistance? Plant J. 8:227-233.

Shah, J., Tsui, F., and Klessig, D. F. 1996. Identification and characterization of a salicylic acid insensitive mutant, *sai1*, of *Arabidopsis thaliana*. Mol. Plant-Microbe Interact. In review.

Takahashi, H., Chen, Z., Liu, Y., and Klessig, D. F. 1996. Development of necrosis and activation of disease resistance in transgenic tobacco plants with severely reduced catalase levels. Plant J. In review.

Yang, Y. and Klessig, D. F. 1996. Isolation and characterization of a TMV-inducible *myb* oncogene homologue from tobacco. Proc. Natl. Acad. Sci. USA. In review.

Vernooij, B., Friedrich, L., Ahl-Goy, P., Staub, T., Kessmann, H., and Ryals, J. 1995. 2,6-Dichloroisonicotinic acid-induced resistance to pathogens without the accumulation of salicylic acid. Mol. Plant-Microbe Interact. 8:228-234.

# Genetic Interactions Between Genes Controlling Cell Death and Pathogen Recognition in Arabidopsis

Jeffery L. Dangl, Robert A. Dietrich, J.-B. Morel, Douglas C. Boyes, Thorsten Jabs *, John M. McDowell, Murray R. Grant #, Susanne Kjemtrup and Scott Kaufman

Dept. of Biology, University of North Carolina, Chapel Hill, N.C. 27599 USA, * RWTH, Aachen, Germany, and # Botany Dept., Univ. of Leicester, U.K.

Programmed cell death in plants occurs during two apparently distinct aspects of plant life cycles. First, cell death in the context of disease resistance is the topic of this chapter. Second, cell death accompanying terminal differentiation, for example formation of secondary xylem and tracheary cells, is critical at specific points during plant development. As well, there are several other peculiarities of plant development which also probably utilize some form of cell death (recently reviewed in Jones and Dangl, 1996). Yet, a lack of genetic analysis has hindered assignment of causal roles to any of the observed molecuar events correlated with this type of cell death. How do these two modes of cell death in plants compare and contrast? And are there analogies to models of cell death control in other eukaryotes? Our naiveté regarding cell death and its control in plants is a glaring omission from developing paradigms for this critical aspect of eukaryotic cell and developmental biology.

## Cell Death and Disease Resistance

Because virtually every specialized cell type in a plant is a potential target for infection, and since there is no circulating surveillance system, every plant cell must be principally capable of recognizing any potential pathogen. Plants have developed very sophisticated genetic strategies for recognition of "non-self", most of which result in localized, very rapid, cell death at the site of attempted pathogen ingress. This cell death, termed the hypersensitive reaction (HR), is found in nearly all responses mediated by single, dominant or co-dominantly acting plant "Disease Resistance Genes" (R-genes;Flor, 1971), and is also a hallmark of "non-host" resistance. Cell death in response to pathogen is triggered by recognition, mediated directly or indirectly by plant R-gene products, of either the direct or indirect product of pathogen-encoded avirulence (avr) genes (Crute, 1985; Keen, 1990; Long and Staskawicz, 1993; Dangl, 1994).

Subsequent signal transduction, rapid generation of active oxygen species, and biosynthesis or release of potential antimicrobial effector molecules, are thought to contribute to both host and pathogen cell death

(Dixon et al., 1994; Godiard et al., 1994; Baker and Orlandi, 1995; Low and Merida, 1996). It should be stressed, however, that not all disease resistance reactions are accompanied by cell death, and that the HR may be a consequence, and not a primary cause of resistance in some cases (Király et al., 1972; Görg et al., 1993).

Genetic and molecular analyses are currently being employed to dissect the resistance reaction, isolate and characterize $R$-genes and the recent cloning of several $R$ genes promises rapid advancement in this field (Bent et al., 1994; Jones et al., 1994; Mindrinos et al., 1994; Whitham et al., 1994; Grant et al., 1995; Lawrence et al., 1995; Song et al., 1995; Dixon et al., 1996; Salmeron et al., 1996). As well, other loci necessary for $R$-gene function (Cao et al., 1994; Freialdenhoven et al., 1994; Hammond-Kosack et al., 1994; Salmeron et al., 1994; Century et al., 1995; Delaney et al., 1995; Freialdenhoven et al., 1996) and those with roles in control of cell death per se have recently been defined (e.g Dietrich et al., 1994; Greenberg et al., 1994).

Structural features of proteins encoded by $R$ genes cloned to date certainly suggest function: Other than the *Pto/Fen* family of kinases described in the chapters by G. Martin and B. Staskawicz, the rest of these molecules encode varying numbers of leucine rich repeats (LRR). These have both ligand binding and protein docking functions in other systems. Of the LRR-containing class, one subset is probably extracellular and anchored through the plasma membrane (e.g *Cf-2, Cf-5, Cf-9,* and *Xa-21*-- which in addition displays a cytoplasmic kinase domain), and the other is potentially intracellular. The intracellular sort can be further subdivided by the nature of predicted domains at their amino termini. One class (*RPM1, RPS2, PRF*) contains potential leucine zippers, while the other *(N, L6, RPP5*) contain homology to a signaling domain encoded by proteins in other systems. Both of the subclasses contain nucleotide binding sites, which are presumably involved in signaling. These points are discussed in additional chapters in this volume (F. Ausubel, B. Baker, D. Jones) and have been detailed in reviews by Dangl, 1995; Staskawicz et al., 1995; and Boyes et al., 1996.

Key outstanding questions center around: 1) demonstration of function for these predicted domains, 2) determination of which domain determines specificity (and whether that domain directly interacts with the *avr*-encoded signal), 3) the number and nature of additional steps in the signaling pathway leading to resistance, and 4) mechanisms driving evolution of new $R$ gene specificities.

We originally described Arabidopsis accessions that lacked *RPM1* activity (Grant et a., 1995). In fact, these accessions lack the entire *RPM1* gene. The absence of the gene is associated with a putative deletion event which includes addition of 98bp of novel repetitive DNA into the postion inhabited by the *RPM1* gene in resistant accessions. We term these alleles *rpm1-null*. The deletion endpoints, and the 98bp insertion, are identical in each of five *rpm1-null* accessions analyzed, suggesting that the event occurred only once during *Arabidopsis thaliana* evolution. The sequences flanking the deletion are also exhibit multiple nucleotide substitutions, suggesting rapid evolution even in non-coding DNA flanking *RPM1*. This phenomenon is perhaps generalizable, as it appears to have happened inBrassica as well (M. R. Grant et al, unpublished).

Rapid cell death at an infection site also can result in systemic signaling giving rise to "systemic acquired resistance" (SAR) in distal tissue (Ryals et al., 1994; see also chapters by J. Ryals and J.-P. Métraux this volume). Salicylic acid (SA) is required locally at both the primary infection site, and in the distal secondary tissue, for establishment and maintenance of SAR (Gaffney et al., 1993; Delaney et al., 1994; Vernooij et al., 1994). The nature of the systemic signal molecule remains elusive.

## Cell Death Mutants

That the cell death generated during a disease resistance reaction is a "programmed cell death" is clear. In several plant species, mutants have been observed with a visible phenotype that resembles the lesions caused by pathogen attack. These have been identified in corn (disease lesion mimics, Walbot et al., 1983; Pryor, 1987; Johal et al., 1994a; Johal et al., 1994b) tomato (autogenous necrosis, Langford, 1948) and Arabidopsis (*lsd1-lsd5*, Dietrich et al., 1994; *acd1*, Greenberg and Ausubel, 1993; *acd2*, Greenberg et al., 1994, all non-allelic). The phenotypes associated with these mutants hint that they represent steps along normal disease resistance response pathways.

Alternatively, we suggested that these mutant phenotypes could represent perturbations of normal metabolism, which the cell senses, and which leads to rapid cell death. In either case, this mutant class suggests that host factors alone are sufficient to generate cell death potentially analogous to that associated with resistance to pathogen attack. The high frequency of non-allelic mutations in maize suggests that a large number of loci can exhibit these mutant phenotypes (Walbot et al., 1983).

We showed that the *lsd* mutants of Arabidopsis faithfully mimic the resistance response observed during pathogen triggered cell death (Dietrich et al., 1994), and similar analyses were reported for *acd2* (Greenberg et al., 1994) and the newly discovered *lsd6* mutant (Weyman et al., 1995). The *lsd* mutants develop disease resistance-like cell death lesions and express markers of authentic disease resistance responses, including cytological and biochemical indicators of plant defense, high expression levels of SAR-related genes, and significant resistance to normally virulent plant pathogens. They also exhibit very clear developmental and cell-type specific preference for onset of cell death. Our results, and the known developmental control of several maize lesion mimic mutant phenotypes, beg the question of the inter-relationship between developmental and environmental cues for plant cell death.

Establishment of transgenic tobacco and Arabidopsis which constitutively degrade SA (Gaffney et al., 1993; Delaney et al., 1994) define local SA accumulation at infection sites (Enyedi et al., 1992 ) as a "linchpin" in the disease resistance pathway. Double mutant analysis between these plants (so-called *nahG* plants) and both cell death and resistance response mutants in Arabidopsis are useful in building conjectural models of how *R* gene function relates to SA accumulation and disease resistance, and where both pathogen-dependent and -independent cell death fit onto this scheme. For example, both *lsd2* and *lsd4* in combination with *nahG* still form lesions of dead cells, but no longer accumulate mRNA for marker genes whose expression is a marker for SAR onset. They are also now fully susceptible to pathogens which can cause disease on the *lsd* parent (M. Hunt, R. A. Dietrich, et al., unpublished). Thus, these two genes lie upstream of the accumulation of salicylic acid. In contrast, *lsd6* (Weyman, et al., 1996) and *lsd1* (unpublished) define genetically an amplification loop feeding from post-SA accumulation back into cell death. Our recent demonstration that superoxide is a key regulator of the *lsd1* phenotype, and that its action precedes cell death, supports the idea that reactive oxygen intermediates are also involved in setting the cell death program into motion (see also chapter by C. Lamb, this volume).

An outstanding question is what is the inter-relationship (if any) between the disease resistance pathway and the loci defined by the *lsd* mutants and others of this ilk. We have isolated extragenic suppressors of *lsd1* and *lsd5* to begin to answer this question. Of 6 independent *lsd5* suppressors, we know that at least two of them are impaired in *RPM1* and *RPS2* function (J.B. Morel and JLD, unpublished). One other can cross-suppress *lsd4* and *lsd6* (J. B. Morel et al., unpublished). These analyses are not yet complete. Because we know that *lsd4* is upstream of SA accumulation, and that *lsd6* functions in an SA-dependent feedback loop, the *lsd5* suppressor which suppresses three *lsd* mutations must define a key control point common to each, and important in the SA-dependent pathway. As well, the *lsd1* suppressor for which analysis is most advanced (it is recessive and reverts all aspects of the *lsd1* phenotype to wild type), can be tentatively positioned downstream of the negative regulatory function of wild type *LSD1*.

These analyses have validated the original idea that study of *lsd*-type mutants would teach us something about both cell death control and the disease resistance pathway. Further studies, and isolation of the genes defined by both the *lsd* mutants and their extragenic suppressors will continue to complement other genetic approaches for analysis of disease resistance.

REFERENCES

**Baker, C. J. and Orlandi, E. W.** (1995). Active oxygen in plant pathogenesis. Annu. Rev. Phytopathol. **33**, 299-322.

**Bent, A. F., Kunkel, B. N., Dahlbeck, D., Brown, K. L., Schmidt, R., Giraudat, J., Leung, J. and Staskawicz, B. J.** (1994). *RPS2* of *Arabidopsis thaliana*: A leucine-rich repeat class of plant disease resistance genes. Science **265**, 1856-1860.

**Boyes, D. C., McDowell, J. M. and Dangl, J. L.** (1996). Many roads lead to resistance. Curr. Biol. **6**, 634-637.

Cao, H., Bowling, S. A., Gordon, S. and Dong, X. (1994). Characterization of an Arabidopsis mutant that is non-responsive to inducers of systemic acquired resistance. Plant Cell **6**, 1583-1592.

Century, K. S., Holub, E. B. and Staskawicz, B. J. (1995). *NDR1*, a locus of *Arabidopsis thaliana* that is required for disease resistance to both a bacterial and a fungal pathogen. Proc. Natl. Acad. Sci., USA **92**, 6597-6601.

Crute, I. R. (1985). The genetic bases of relationships between microbial parasites and their hosts. In Mechanisms of Resistance to Plant Diseases., R. S. S. Fraser, eds. (Dordrecht: Kluwer Academic Press), pp. 80-143.

Dangl, J. L. (1994). The enigmatic avirulence genes of phytopathogenic bacteria. In Bacterial Pathogenesis of Plants and Animals: Molecular and Cellular Mechanisms., J. L. Dangl, eds. (Heidelberg: Springer Verlag), pp. 99-118.

Dangl, J. L. (1995). Pièce de Résistance:  Novel classes of plant disease resistance genes. Cell **80**, 363-366.

Delaney, T., Uknes, S., Vernooij, B., Friedrich, L., Weymann, K., Negrotto, D., Gaffney, T., Gut-Rella, M., Kessman, H., Ward, E. and Ryals, J. (1994). A central role of salicylic acid  in plant disease resistance. Science **266**, 1247-1250.

Delaney, T. P., Friedrich, L. and Ryals, J. A. (1995). *Arabidopsis* signal transduction mutant defective in chemically and biologically induced disease resistance. Proc. Natl. Acad. Sci., USA **92**, 6602-6606.

Dietrich, R. A., Delaney, T. P., Uknes, S. J., Ward, E. J., Ryals, J. A. and Dangl, J. L. (1994). Arabidopsis mutants simulating disease resistance response. Cell **77**, 565-578.

Dixon, M. S., Jones, D. A., Keddie, J. S., Thomas, C. M., Harrison, K. and Jones, J. D. G. (1996). The tomato *Cf-2* disease resistance locus comprises two functional genes encoding leucine-rich repeat proteins. Cell **84**, 451-460.

Dixon, R. A., Harrison, M. J. and Lamb, C. J. (1994). Early events in the activation of plant defense responses. Annu. Rev. Phytopathol. **32**, 479-501.

Enyedi, A. J., Yalpani, N., Silverman, P. and Raskin, I. (1992). Localization, conjugation and function of salicylic acid in tobacco during the hypersensitive reaction to tobacco mosaic virus. Proc. Natl. Acad. Sci., USA **89**, 2480-2484.

**Flor, H. H.** (1971). Current status of the gene-for-gene concept. Annu. Rev. Phytopathol. **9,** 275-296.

**Freialdenhoven, A., Peterhansel, C., Kurth, J., Kreuzaler, F. and Schulze-Lefert, P.** (1996). Identification of genes required for the function of non-race-specific *mlo* resistance to powdery mildew in barley. Plant Cell **8,** 5-14.

**Freialdenhoven, A., Scherag, B., Hollricher, K., Collinge, D., Christensen, H.-T. and Schulze-Lefert, P.** (1994). *Nar-1* and *Nar-2*, two loci required for *Mla-12*-specified race resistance to powdery mildew in barley. Plant Cell **6,** 983-994.

**Gaffney, T., Friedrich, L., Vernooij, B., Negrotto, D., Nye, G., Uknes, S., Ward, E. and Ryals, J.** (1993). Requirement for salicylic acid for the induction of systemic acquired resistance. Science **261,** 754-756.

**Godiard, L., Grant, M. R., Dietrich, R. A., Kiedrowski, S. and Dangl, J. L.** (1994). Perception and response in plant disease resistance. Curr. Opin. Genet. and Dev. **4,** 662-671.

**Görg, R., Hollricher, K. and Schulze-Lefert, P.** (1993). Functional analysis and RFLP-mediated mapping of the *Mlg* resistance locus in barley. Plant J. **3,** 857-866.

**Grant, M. R., Godiard, L., Straube, E., Ashfield, T., Lewald, J., Sattler, A., Innes, R. W. and Dangl, J. L.** (1995). Structure of the *Arabidopsis RPM1* gene enabling dual specificity disease resistance. Science **269,** 843-846.

**Greenberg, J. T. and Ausubel, F. M.** (1993). Arabidopsis mutants compromised for the control of cellular damage during pathogenesis and aging. Plant J. **4,** 327-342.

**Greenberg, J. T., Guo, A., Klessig, D. F. and Ausubel, F. M.** (1994). Programmed cell death in plants: A pathogen-triggered response activated coordinately with multiple defense functions. Cell **77,** 551-564.

**Hammond-Kosack, K. E., Jones, D. A. and Jones, J. D. G.** (1994). Identification of two genes required in tomato for full-*Cf-9*-dependent resistance to *Cladisporium fulvum*. Plant Cell **6,** 361-374.

**Johal, G. S., Hulbert, S. H. and Briggs, S. P.** (1994a). Disease lesion mimics of maize: A model for cell death in plants. Bioessays **17,** 685-692.

**Johal, G. S., Lee, E. A., Close, P. S., Coe, E. H., Neuffer, M. G. and Briggs, S. P.** (1994b). A tale of two mimics: transposon mutagenesis and characterization of two disease lesion mimic mutations in maize. Maydica **39**, 69-76.

**Jones, D. A., Thomas, C. M., Hammond-Kosack, K. E., Balint-Kurti, P. J. and Jones, J. D. G.** (1994). Isolation of the tomato *Cf-9* gene for resistance to *Cladosporium fulvum* by transposon tagging. Science **266**, 789-793.

**Keen, N. T.** (1990). Gene-for-gene complementarity in plant-pathogen interactions. Annu. Rev. Genet. **24**, 447-463.

**Király, Z., Barna, B. and Érsek, T.** (1972). Hypersensitivity as a consequence, not the cause, of plant resistance to infection. Nature **239**, 456-458.

**Langford, A. N.** (1948). Autogenous necrosis in tomatoes immune from *Cladosporium fulvum* Cooke. Can. J. Res. **26**, 35-64.

**Lawrence, G. J., Finnegan, E. J., Ayliffe, M. A. and Ellis, J. G.** (1995). The *L6* gene for flax rust resistance is related to the Arabidopsis bacterial resistance gene *RPS2* and the tobacco viral resistance gene *N*. Plant Cell **7**, 1195-1206.

**Long, S. R. and Staskawicz, B. J.** (1993). Procaryotic plant parasites. Cell **73**, 921-935.

**Low, P. S. and Merida, J. R.** (1996). The oxidative burst in plant defense: Function and signal transduction. Physiol. Plant. **96**, 533-542.

**Mindrinos, M., Katagiri, F., Yu, G.-L. and Ausubel, F. M.** (1994). The *A. thaliana* disease resistance gene *RPS2* encodes a protein containing a nucleotide-binding site and leucine-rich repeats. Cell **78**, 1089-1099.

**Pryor, T. P.** (1987). Stability of alleles at Rp (resistance to *Puccinia sorghi*). Maize. Genet. Newslett. **61**, 37-38.

**Ryals, J., Uknes, S. and Ward, E.** (1994). Systemic acquired resistance. Plant Physiol. **104**, 1109-1112.

**Salmeron, J. M., Barker, S. J., Carland, F. M., Mehta, A. Y. and Staskawicz, B. J.** (1994). Tomato mutants altered in bacterial disease resistance provide evidence for a new locus controlling pathogen recognition. Plant Cell **6**, 511-520.

**Salmeron, J. M., Oldroyd, G. E. D., Rommens, C. M. T., Scofield, S. R., Kim, H. S., Lavelle, D. T., Dahlbeck, D. and Staskawicz, B. J.** (1996). Tomato *Prf* is a member of the leucine-rich repeat class of plant disease resistance genes and lies embedded within the *Pto* kinase gene cluster. Cell **in press.**

**Song, W.-Y., G.-L., W., Chen, L.-L., Kim, H.-S., Pi, L. Y., Holsten, T., Gardner, J., Wang, B., Zhai, W.-X., Zhu, L.-H., Fauquet, C. and Ronald, P. C.** (1995). A receptor kinase-like protein encoded by the rice disease resistance gene, *Xa21*. Science **270,** 1804-1806.

**Staskawicz, B. J., Ausubel, F. M., Baker, B. J., Ellis, J. and Jones, J. D. G.** (1995). Molecular genetics of plant disease resistance. Science **268,** 661-667.

**Vernooij, B., Friedrich, L., Morse, A., Reist, R., Kolditz-Jawhar, R., Ward, E., Uknes, S., Kessmann, H. and Ryals, J.** (1994). Salicylic acid is not the translocated signal responsible for inducing systemic acquired resistance, but is required in signal transduction. Plant Cell **6,** 959-965.

**Walbot, V., Hoisington, D. A. and Neuffer, M. G.** (1983). Disease lesion mimics in maize. In Genetic Engineering of Plants, T. Kosuge and C. Meredith, eds. (New York: Plenum Publishing Co.), pp. 431-442.

**Weyman, K., Hunt, M., Uknes, S., Neuenschwander, U., Lawton, K., Steiner, H.-Y. and Ryals, J.** (1995). Suppression and restoration of lesion formation in Arabidopsis *lsd* mutants. Plant Cell **12,** 2013-2022.

**Whitham, S., Dinesh-Kumar, S. P., Choi, D., Hehl, R., Corr, C. and Baker, B.** (1994). The product of the Tobacco Mosaic Virus resistance gene *N* : similarity to to Toll and the Interleukin-1 receptor. Cell **78,** 1101-1115.

# Arabidopsis thaliana Enhanced Disease Susceptibility (eds) Mutants

Elizabeth E. Rogers, Jane Glazebrook*, Sigrid Volko, Frederick M. Ausubel

Department of Genetics, Harvard Medical School and Department of Molecular Biology, Massachusetts General Hospital, Boston, MA 02114 USA

*Current Address: Center for Agricultural Biotechnology, 1105 Ag/Life Sciences Surge Building, University of Maryland, College Park, MD 20742-3351 USA

## Introduction

The responses by which eukaryotic hosts defend themselves from pathogen attack have not been subjected to rigorous genetic analysis in any organism. This sort of analysis requires a tractable genetic system for the host, and well-characterized host-pathogen interactions. Whereas vertebrates are generally unsuitable for such studies, plants offer many opportunities for the genetic analysis of host-pathogen interactions.

Plants respond in a variety of ways to pathogenic microorganisms (Lamb, 1994; Lamb et al., 1989). Infected cells undergo rapid programmed cell death called the hypersensitive response (HR). Cell walls are reinforced by lignification, suberization, callose deposition, and cross-linking of hydroxyproline-rich proteins. A variety of hydrolytic enzymes, including so-called pathogenesis-related (PR) proteins, and low molecular weight antibiotics (called phytoalexins) are synthesized. A membrane-associated oxidative burst occurs that results in the NADPH-dependent production of $O_2^-$ and $H_2O_2$. A large body of physiological, biochemical and molecular evidence suggests that particular plant defense responses play a direct role in conferring resistance to pathogens. For example, phytoalexins and many PR proteins like chitinases and β-1,3-glucanases directly inhibit pathogen growth *in vitro* (Mauch et al., 1988; Paxton, 1981; Ponstein et al., 1994; Schlumbaum et al., 1986; Sela-Buurlage et al., 1993; Terras et al., 1992; Woloshuk et al., 1991). In addition, constitutive expression in transgenic plants of PR genes or constitutive expression of certain phytoalexin biosynthetic genes has been

47

shown to decrease disease susceptibility in a limited number of cases (Alexander et al., 1993; Broglie et al., 1991; Hain et al., 1993; Liu et al., 1994; Terras et al., 1995; Zhu et al., 1994).

Isolation of plant defense-response mutants would not only help elucidate the roles of known pathogen-induced responses in combating particular pathogens, but would also facilitate the identification of plant defense mechanisms not already correlated with a known biochemical or molecular genetic response. Unfortunately, however, most of the plant hosts that have been used in the past for host-pathogen studies are not suited for genetic analysis due to large or polyploid genomes and long generation times. However, with the development of well-characterized host-pathogen systems involving the model plant *Arabidopsis thaliana* as the host, comprehensive genetic analysis of host defense responses has recently become feasible (Crute et al., 1994; Kunkel, 1996). *A. thaliana* (referred to as "*Arabidopsis*" for simplicity) offers several advantages compared to other plants that have been used previously to study plant-pathogen interactions. The small stature, fast generation time, copious production of tiny (20 μg) seeds, and small (~100 Mb) genome relatively free of repeated sequences of *Arabidopsis* facilitate the use of genetic strategies to identify defense-related mutants and the use of gene tagging or map-based positional cloning strategies to isolate the corresponding genes (Meyerowitz, 1989). Until the development of pathosystems involving *Arabidopsis* as a host, it was difficult to provide definitive evidence that a particular plant defense response plays a major role in combating pathogen attack. In *Arabidopsis*, however, it is possible to study specific plant defense responses by removing them by mutation. Moreover as described in this chapter, our laboratory has demonstrated the feasibility of conducting a general screen for *Arabidopsis* mutants that display enhanced susceptibility to pathogen attack, allowing the identification of defense response components that are not already correlated with a known biochemical or molecular genetic response (Glazebrook et al., 1996).

Although in general there is a paucity of genetic analysis in the study of plant-pathogen interactions, genetic analysis has played a crucial role in the breeding of pathogen resistance in economically important plants. In the 1950s, H.H. Flor developed a powerful explanatory genetic model that explains the observation that some races (strains) of a particular pathogen elicit a strong defense response (usually involving a hypersensitive programmed cell death response) on a given cultivar of a host species whereas other races (strains) of the same pathogen proliferate and cause disease (Flor, 1971). (A pathogen that elicits a strong defense response is said to be avirulent on that host, the host is said to be resistant, and the plant-pathogen interaction is said to be incompatible. In contrast, strains which cause disease on a particular host are said to be virulent, the host is said to be susceptible, and the plant-pathogen interaction is said to be compatible.) In many cases, the molecular basis of incompatibility appears to be due to a gene-for-gene correspondence between pathogen "avirulence" *(avr)* genes and host "resistance" *(R)* genes (Flor, 1971). A plant carrying a particular resistance gene will be resistant to pathogens carrying the corresponding *avr* gene. A simple molecular explanation for this gene-for-gene correspondence between *avr* and *R* genes is that *avr* genes generate signals for which resistance genes encode the cognate receptors. A signal transduction pathway then carries the *avr*-generated signal to a set of target genes which initiates the HR and other host defenses (Gabriel and Rolfe, 1990; Keen, 1992; Lamb et al., 1989). During the past few years, a variety of R genes have been identified in Arabidopsis that correspond to viral, bacterial, and fungal pathogens and several have recently been cloned (reviewed in Kunkel, 1996).

One well-studied consequence of gene-for-gene interactions is a defense response known as systemic acquired resistance or SAR. SAR is a phenomenon wherein infection of a plant with an avirulent pathogen leads to accumulation of PR proteins in the uninfected leaves, which concomitantly become resistant to a variety of normally virulent pathogens (Enyedi et al., 1992; Malamy and Klessig, 1992). Salicylic acid plays an important role as a signaling compound in SAR (Gaffney et al., 1993), and treatment of plants with salicylic acid leads to PR protein accumulation and pathogen resistance (Enyedi et al., 1992; Malamy and Klessig, 1992). SAR-associated PR proteins include chitinases, β-glucanases, and chitin-binding proteins (Essenberg et al., 1992; Ponstein et al., 1994), as well as others with unknown activities (Dixon and Lamb, 1990; Lamb et al., 1989). All of the important features of SAR have been observed in *Arabidopsis* (Uknes et al., 1992; Uknes et al., 1993)

Recently, genetic analysis in *Arabidopsis* has been used help identify components of the SAR signalling pathway. Specifically, Cao et al. 1994 constructed transgenic *Arabidopsis* plants carrying a fusion of the SAR-induced gene *BGL2* (which encodes a β-1,3 glucanase) to the *E. coli uidA* reporter gene (which encodes β–glucuronidase and which is commonly referred to as the *"GUS"* gene). Cao et al. carried out screens for *Arabidopsis* mutants that either expressed the *BGL2 GUS* reporter constitutively in the absence of SA treatment or failed to express the fusion following treatment with SA. These screens lead to the identification of a series of mutants called *npr* and *cpr* (for non-expression of PR genes and constitutive expression of *PR* genes, respectively) which define genes that appear to be involved both in the regulation of *BGL2* specifically and SAR in general (Bowling et al., 1994; Cao et al., 1994). Mutants which exhibit an *npr1* mutant phenotype were also isolated in an independent study (Delaney et al., 1995).

### A general Screen for *Arabidopsis* mutants that exhibit enhanced disease susceptibility to *Pseudomonas syringae*

Our laboratory utilizes the interaction between *Arabidopsis* and *Pseudomonas syringae* pv. *maculicola (Psm)* strain ES4326 as a model system. *Psm* ES4326 proliferates extensively in *Arabidopsis* leaves and causes the development of disease symptoms (Davis et al., 1991; Dong et al., 1991). In contrast, *Psm* ES4326 expressing the *avr* gene *avrRpt2*, cloned from a different *P. syringae* strain (Dong et al., 1991; Whalen et al., 1991), elicits an HR and only grows about 1% as much in *Arabidopsis* leaves as *Psm* ES4326 not expressing *avrRpt2*. Until recently, the primary focus in our laboratory has been to carry out a genetic analysis of the gene-for gene interaction between the *Arabidopsis RPS2* disease resistance gene and the *P. syringae* avirulence gene *avrRpt2*. This culminated in the cloning of *RPS2* (Mindrinos et al., 1994). We are now in the process of shifting our focus to the genetic analysis of *Arabidopsis* genes that are not directly involved in gene-for-gene interactions but rather are involved in the defense response to <u>virulent</u> strains of *P. syringae* such as *Psm* ES4326 which cause disease lesions instead of eliciting a strong defense response.

It does not appear correct to conclude that because *Psm* ES4326 is virulent and causes severe disease that *Arabidopsis* does not mount a vigorous defense response. As shown in Figure 1, the final density to which *Psm* ES4326 will grow in an *Arabidopsis* leaf is directly related to the dose at which *Psm* ES4326 is infiltrated, suggesting that there is a specific mechanism for limiting *in planta* growth. The phenotypes of two recently isolated categories of Arabidopsis mutants supports this conclusion. Our laboratory previously reported the isolation of *Arabidopsis* mutants that accumulate reduced levels of the phytoalexin called camalexin, the only phytoalexin that has been found in significant quantities in *Arabidopsis* (Glazebrook and Ausubel, 1994; Tsuji et al., 1992). Importantly, *Psm* ES4326 causes more severe disease symptoms and grows to higher titers in some of these *pad* (**p**hyto**a**lexin **d**eficient) mutants. Similarly, *npr1* mutants (described above ) also exhibit a similar enhanced susceptibility phenotype as *pad* mutants (Cao et al., 1994). The enhanced disease susceptibility phenotypes of *pad* and *npr1* mutants  suggested that a comprehensive screen for *Arabidopsis* mutants that allowed enhanced growth of *Psm* ES4326 would probably uncover previously unknown defense mechanisms.

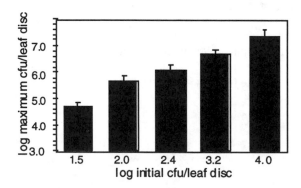

**Figure 1.** Maximal growth of *Psm* ES4326 in wild-type *Arabidopsis* leaves as a function of inoculation dose.

We have therefore carried out different screens to isolate additional *eds* (<u>e</u>nhanced <u>d</u>isease <u>s</u>usceptibility) mutants. These screens used plants grown from EMS mutagenized M2 seed of the Columbia ecotype. Two leaves of each M2 plant were infected at a dose of strain *Psm* ES4326 at which wild-type plants show very weak symptoms manifested as small chlorotic spots three days after infection, whereas *pad* and *npr1* mutants show large areas of chlorosis. A total of 15 *eds* mutants that reproducibly allow at least one half log more growth of *Psm* ES4326 as compared to wild type were identified among 12,500 plants screened.

Figure 2 illustrates the enhanced growth of *Psm* ES4326 in a typical *eds* mutant and Table 1 summarizes what we know about a subset of these *eds* mutants. Most phenotypic testing has been carried out on plants with *eds* mutant phenotypes that were backcrossed at least once to wild-type to remove unlinked mutations. Because some *pad* mutants as well as *npr1* mutants have the same enhanced susceptibility phenotype with respect to *Psm* ES4326 as the *eds* mutants (Glazebrook et al., 1996), we tested whether the *eds* mutants synthesize wild-type levels of camalexin in response to infection by *Psm* ES4326 (*pad* phenotype) and whether *PR1* gene expression can be induced by salicylic acid (*npr1* phenotype). Several *eds* mutants displayed *pad* or *npr1* phenotypes and some of these have been crossed with *pad1-1*, *pad2-1*, *pad3-1*, and *npr1-1* plants to determine whether these *eds* mutants are allelic to previously isolated *PAD* genes or to *NPR1*. The results of these analyses, summarized in Table 1, showed that at least two of the *eds* mutants (*eds-5* and *eds-53*) contain mutations that are allelic to *npr1-1*, that *eds-47* contains a mutation that is allelic to *pad2-1*, and that *eds-9* defines a previously unknown *PAD* gene *(PAD4)* (Glazebrook et al., 1996). The fact that some of the *eds* mutants contained mutations that were alleles of genes previously known to be involved in defense responses shows that the screen does yield bona fide defense response mutants.

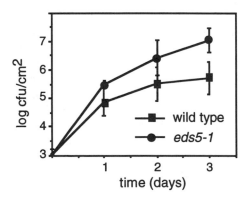

**Figure 2.** Growth of *Psm* ES4326 in *eds5-1* and wild-type plants. Other *eds* mutants as well as *pad 1, pad2,* or *npr1* mutants give similar results.

To date, a total of nine *eds* mutants have been crossed to each other and to *Arabidopsis* wild-type (Col-0) in pairwise combinations. The resulting F1 plants and F2 plants were compared to each parent, and to wild-type plants, using the *Psm* ES4326 growth assay and visual inspection of infected leaves. All *eds* mutants examined thus far are recessive and nine of ten segregate as single Mendelian loci. As summarized in Table 1, the *eds* mutants define at least eight complementation groups. Additionally, all *eds* mutants are being repeatedly backcrossed to wild-type to eliminate unlinked mutations.

In preliminary results several *eds* mutants tested so far appear to be more sensitive to other *Pseudomonas* species which are *Arabidopsis* pathogens such as *P. syringae* pv. *tomato* DC3000 and *Pseudomonas aeruginosa* strain UCBPP-PA14. *P. syringae* strain DC3000 behaves similarly to *Psm* ES4326 on *Arabidopsis* in essentially all pathogenicity-related assays (Kunkel et al., 1993; Whalen et al., 1991). *P. aeruginosa* strain PA14 is pathogenic in both *Arabidopsis* and mice (Rahme et al., 1995). In contrast to the observation that several *eds* mutants are more susceptible to other *Arabidopsis Pseudomonas* pathogens, these same *eds* mutants do not appear to be more sensitive to the *Arabidopsis* "non-host" pathogen *P. syringae* pv. *phaseolicola* strain NPS3121, a well-studied bean pathogen that is very weakly pathogenic in *Arabidopsis* (Yu et al., 1993).

**Table 1**: Phenotypes of and other information about *eds* mutants

| Mutant number | Gene designation and allele number | Log enhanced growth of *Psm* ES4326 in *eds* mutant in comparison to wild-type | Camalexin Induction by *Psm* ES4326[a] | *PR1* Induction by salicylic acid[b] |
|---|---|---|---|---|
| *eds-2* | *eds2-1* | 1.1 | + | + |
| *eds-4* | *eds3-1* | 1.7 | + | + |
| *eds-5* | *npr1-2* | 2.2 | + | - |
| *eds-6* | *eds4-1* | 1.3 | + | + |
| *eds-8* | *eds4-2* | 1.6 | + | + |
| *eds-9* | *pad4-1* | 1.1 | - | + |
| *eds-11* | *eds5-1* | 1.3 | + | + |
| *eds-12* | *eds9-1* | 1.4 | + | + |
| *eds-13* | *eds6-1* | 1.2 | + | + |
| *eds-18* | *eds7-1* | 1.0 | + | + |
| *eds-40* | | 1.1 | + | + |
| *eds-42* | | 0.6 | + | + |
| *eds-47* | *pad2-2* | 1.9 | - | + |
| *eds-48* | *eds8-1* | 1.6 | + | + |
| *eds-53* | *npr1-3* | 1.9 | + | - |

[a]"+" indicates same level of camalexin as wild-type plants; - indicates less camalexin accumulation than in wild-type.

[b]"+" indicates same level of *PR1* induction as in wild-type; - indicates no *PR1* induction.

## Conclusions

Previous screens for *Arabidopsis* defense-related mutants were either based on identifying specific resistance gene mutations corresponding to a characterized *avr* gene (Bisgrove et al., 1994; Century et al., 1995; Kunkel et al., 1993; Yu et al., 1993) or they were based on the identification of mutants that failed to activate a biochemically defined defense response (Cao et al., 1994; Delaney et al., 1995; Glazebrook and Ausubel, 1994). In contrast, the screen that we carried out for *eds* mutants is the first reported direct screen for mutants with quantitative pathogen-susceptibility phenotypes in any multicellular organism.

It seems likely that additional categories of *eds* mutants could be isolated by direct screening for mutants that exhibit enhanced susceptibility to pathogens other than *Psm* ES4326. Indeed, a mutant called *eds1* has been isolated which is susceptible to several different avirulent strains of the fungal pathogen *Peronospora parasitica* that are recognized by different resistance genes (Jane Parker, personal communication).

We are continuing our screen for *eds* mutants. To date we have isolated a total of 26 different mutants. The low frequency of allelic pairs among the *eds* mutations studied to date indicates that the screen for eds mutants is not fully saturated and that there are most likely many additional *EDS* genes yet to be discovered..

## Acknowledgments

The work on *eds* mutants was supported by National Research Initiative Competetive Grants Program grant 940-1199 and National Institutes of Health grant 48707, awarded to F.M.A.

## Literature Cited

Alexander, D., Goodman, R.M., Gut-Rella, M., Glascock, C., Weymann, K., Friedrich, L., Maddox, D., Ahl-Goy, P., Luntz, T., Ward, E. and Ryals, J. 1993. Increased tolerance to two oomycete pathogens in transgenic tobacco expressing pathogenesis-related protein 1a. Proc. Natl. Acad. Sci. USA 90:7327-7331.

Bisgrove, S.R., Simonich, M.T., Smith, N.M., Sattler, A. and Innes, R.W. 1994. A disease resistance gene in *Arabidopsis* with specificity for two different pathogen avirulence genes. Plant Cell 6:927-933.

Bowling, S.A., Guo, A., Cao, H., Gordon, A.S., Klessig, D.F. and Dong, X. 1994. A mutation in *Arabidopsis* that leads to constitutive expression of systemic acquired resistance. Plant Cell 6:1845-1857.

Broglie, K., Chet, I., Holliday, M., Cressman, R., Biddle, P., Knowlton, S., Mauvais, C.J. and Broglie, R. 1991. Transgenic plants with enhanced resistance to the fungal pathogen *Rhizoctonia solani*. Science 254:1194-

1197.

Cao, H., Bowling, S.A., Gordon, S. and Dong, X. 1994. Characterization of an *Arabidopsis* mutant that is nonresponsive to inducers of systemic acquired resistance. Plant Cell 6:1583-1592.

Century, K.S., Holub, E.B. and Staskawicz, B.J. 1995. *NDR1*, a locus of *Arabidopsis thaliana* that is required for disease resistance to both a bacterial and a fungal pathogen. Proc. Natl. Acad. Sci. USA 92:6597-6601.

Crute, I., Beynon, J., Dangl, J., Holub, E., Mauch-Mani, B., Slusarenko, A., Staskawicz, B. and Ausubel, F. 1994. Microbial pathogenesis of *Arabidopsis*. 705-747 in: *Arabidopsis*, E.M. Meyerowitz and C.R. Somerville, ed. Cold Spring Harbor Laboratory Press, Cold Spring Harbor, NY.

Davis, K.R., Schott, E. and Ausubel, F.M. 1991. Virulence of selected phytopathogenic pseudomonads in *Arabidopsis thaliana*. Mol. Plant-Microbe Interact. 4:477-488.

Delaney, T.P., Friedrich, L. and Ryals, J.A. 1995. *Arabidopsis* signal transduction mutant defective in chemically and biologically induced disease resistance. Proc. Natl. Acad. Sci. USA 92:6602-6606.

Dixon, R.A. and Lamb, C.J. 1990. Molecular communication in interactions between plants and microbial pathogens. Ann. Rev. Plant Physiol. Plant Mol. Biol. 41:339-367.

Dong, X., Mindrinos, M., Davis, K.R. and Ausubel, F.M. 1991. Induction of *Arabidopsis* defense genes by virulent and avirulent *Pseudomonas syringae* strains and by a cloned avirulence gene. Plant Cell 3:61-72.

Enyedi, A.J., Yalpani, N., Silverman, P. and Raskin, I. 1992. Signal molecules in systemic plant resistance to pathogens ands pests. Cell 70:879-886.

Essenberg, M., Pierce, M.L., Hamilton, B., Cover, E.C., Scholes, V.E. and Richardson, P.E. 1992. Development of fluorescent, hypersensitively necrotic cells containing phytoalexins adjacent to colonies of *Xanthomonas campestris* pv *malvacearum* in cotton leaves. Phys. Mol. Plant Path. 41:85-99.

Flor, H.H. 1971. Current status of gene-for-gene concept. Ann. Rev. Phytopathol. 9:275-296.

Gabriel, D.W. and Rolfe, B.G. 1990. Working models of specific recognition in plant-microbe interactions. Annu. Rev. Phytopathol. 28:365-391.

Gaffney, T., Friedrich, L., Vernooij, B., Negrotto, D., Nye, G., Uknes, S., Ward, E., Kessmann, H. and Ryals, J. 1993. Requirement of salicylic acid for the induction of sytemic aquired resistance. Science 261:754-756.

Glazebrook, J. and Ausubel, F.M. 1994. Isolation of phytoalexin-deficient mutants of *Arabidopsis thaliana* and characterization of their interactions with bacterial pathogens. Proc. Natl. Acad. Sci. USA 91:8955-8959.

Glazebrook, J., Rogers, E.E. and Ausubel, F.M. 1996. Isolation of *Arabidopsis* mutants with enhanced disease susceptibility by direct screening. Genetics 143:973-982.

Hain, R., Reif, H.-J., Krause, E., Langebartels, R., Kindl, H., Vornam, B., Wiese, W., Schmelzer, E., Schreier, P.H., Stocker, R.H. and Stenzel, K. 1993. Disease resistance results from foreign phytoalexin expression in a novel plant. Nature 361:153-156.

Keen, N.T. 1992. The molecular biology of disease resistance. Plant Mol. Biol. 19:109-122.

Kunkel, B.N. 1996. A useful weed put to work: Genetic analysis of disease resistance in *Arabidopsis thaliana*. Trends in Genetics 12:62-69.

Kunkel, B.N., Bent, A.F., Dahlbeck, D., Innes, R.W. and Staskawicz, B.J. 1993. *RPS2*, an *Arabidopsis* disease resistance locus specifying recognition of *Pseudomonas syringae* strains expressing the avirulence gene *avrRpt2*. Plant Cell 5:865-875.

Lamb, C.J. 1994. Plant disease resistance genes in signal perception and transduction. Cell 76:419-422.

Lamb, C.J., Lawton, M.A., Dron, M. and Dixon, R.A. 1989. Signals and transduction mechanisms for activation of plant defenses against microbial attack. Cell 56:215-224.

Liu, D., Raghothama, K.G., Hasegawa, P.M. and Bressan, R.A. 1994. Osmotin overexpression in potato delays development of disease symptoms. Proc. Natl. Acad. Sci. USA 91:1888-1892.

Malamy, J. and Klessig, D.F. 1992. Salicylic acid and plant disease resistance. Plant J. 2:643-654.

Mauch, F., Mauch-Mani, B. and Boller, T. 1988. Antifungal hydrolases in pea tissue. II Inhibition of fungal growth by combinations of chitinase and β-1,3-glucanase. Plant Physiol. 88:936-942.

Meyerowitz, E.M. 1989. *Arabidopsis*, a useful weed. Cell 56:263-269.

Mindrinos, M., Katagiri, F., Yu, G.-L. and Ausubel, F.M. 1994. The *A. thaliana* disease resistance gene *RPS2* encodes a protein containing a nucleotide-binding site and leucine-rich repeats. Cell 78:1089-1099.

Paxton, J.D. 1981. Phytoalexins- a working redefinition. Phytopathol. Z 101:106 109.

Ponstein, A.S., Bres-Vloemans, S.A., Sela-Buurlage, M.B., Elzen, P.J.M.v.d., Melchers, L.S. and Cornelissen, B.J.C. 1994. A novel pathogen- and wound-inducible tobacco (*Nicotiana tabacum*) protein with antifungal activity. Plant Physiol 104:109-118.

Rahme, L.G., Stevens, E.J., Wolfort, S.F., Shao, J., Tompkins, R.G. and Ausubel, F.M. 1995. Common virulence factors for bacterial pathogenicity in plants and animals. Science 268:1899-1902.

Schlumbaum, A., Mauch, F., Voegli, U. and Boller, T. 1986. Plant chitinases are potent inhibitors of fungal growth. Nature 324:365-367.

Sela-Buurlage, M.B., Ponstein, A.S., Bres-Vloemans, S.A., Melchers, L.S., van den Elzen, P.J.M. and Cornelissen, B.J.C. 1993. Only specific tobacco (*Nicotiana tabacum*) chitinases and β-1,3-glucanases exhibit antifungal activity. Plant Physiol. 101:857-863.

Terras, F.R.G., Eggermont, K., Kovaleva, V., Raikhel, N.V., Osborn, R.W., Kester, A., Rees, S.B., Torrekens, S., Leuven, F.V., Vanderleyden, J., Cammune, B.P.A. and Broekaert, W.F. 1995. Small cysteine-rich antifungal proteins from radish: Their role in host defense. Plant Cell 7:573-588.

Terras, F.R.G., Schoofs, H.M.E., DeBolle, M.F.C., Van Leuven, F., Rees, S.B., Vanderleyden, J., Cammue, B.P.A. and Broekaert, W.F. 1992. Analysis of two novel classes of plant antifungal proteins from radish (*Raphanus sativus* L.) seeds. J. Biol. Chem. 267:15301-15309.

Tsuji, J., Jackson, E.P., Gage, D.A., Hammerschmidt, R. and Somerville, S.C. 1992. Phytoalexin accumulation in *Arabidopsis thaliana* during the hypersensitive reaction to *Pseudomonas syringae* pv.*syringae*. Plant Physiol. 98:1304-1309.

Uknes, S., Mauch-Mani, B., Moyer, M., Potter, S., Williams, S., Dincher, S., Chandler, D., Slusarenko, A., Ward, E. and Ryals, J. 1992. Acquired resistance in *Arabidopsis*. Plant Cell 4:645-656.

Uknes, S., Winter, A.M., Delaney, T., Vernooij, B., Morse, A., Friedrich, L., Nye, G., Potter, S., Ward, E. and Ryals, J. 1993. Biological induction of systemic acquired resistance in *Arabidopsis*. Mol. Plant-Microbe Interact. 6:692-698.

Whalen, M.C., Innes, R.W., Bent, A.F. and Staskawicz, B.J. 1991. Identification of *Pseudomonas syringae* pathogens of *Arabidopsis* and a bacterial locus determining avirulence on both *Arabidopsis* and soybean. Plant Cell 3:49-59.

Woloshuk, C.P., Meulenhoff, J.S., Sela-Buurlage, M., Elzen, P.J.M.v.d. and Cornelissen, B.J.C. 1991. Pathogen-induced proteins with inhibitory activity toward *Phytophthora infestans*. Plant Cell 3:619-628.

Yu, G.-L., Katagiri, F. and Ausubel, F.M. 1993. *Arabidopsis* mutations at the *RPS2* locus result in loss of resistance to *Pseudomonas syringae* strains expressing the avirulence gene *avrRpt2*. Mol. Plant-Microbe Interact. 6:434-443.

Zhu, Q., Maher, E.A., Masoud, S., Dixon, R. and Lamb, C.J. 1994. Enhanced protection against fungal attack by constitutive co-expression of chitinase and glucanase genes in transgenic tobacco. Bio/technology 12:807-812.

# Molecular, Genetic and Physiological Analysis of *Cladosporium* Resistance Gene Function in Tomato

David A. Jones, Penny Brading, Mark Dixon, Kim Hammond-Kosack, Kate Harrison, Kostas Hatzixanthis, Martin Parniske, Pedro Piedras, Miguel Torres, Saijun Tang, Colwyn Thomas and Jonathan D. G. Jones.

The Sainsbury Laboratory, John Innes Centre, Norwich, UK.

The object of our research, like that of many others in the field of molecular plant pathology, is to bridge the gap in our understanding of the recognition and signalling processes that occur between challenge by an incompatible pathogen carrying an avirulence gene and the resistance response by a plant carrying the corresponding resistance gene. The system in which we have chosen to investigate these processes is the interaction between the leaf mould fungus, *Cladosporium fulvum*, and the tomato, *Lycopersicon esculentum*. A major reason for choosing this system is the fact that two avirulence proteins and their corresponding genes, *Avr9* and *Avr4*, have already been isolated from *C. fulvum* (van den Ackerveken et al., 1992; Joosten et al., 1994). We have now isolated four genes for resistance to *C. fulvum* from tomato, including the *Cf-9* and *Cf-4* genes corresponding to *Avr9* and *Avr4*, respectively, as well as *Cf-2* and *Cf-5*.

## The predicted nature of the Cf proteins

*Cf-9* encodes an 863 amino acid, membrane-anchored, glycoprotein with a large extracytoplasmic domain containing 27 leucine-rich repeats (LRRs) and a small cytoplasmic domain (Fig. 1) (Jones et al., 1994). *Cf-4*, tightly linked to *Cf-9* on the short arm of chromosome *1*, encodes a protein very similar to that of *Cf-9* (Thomas et al., unpublished) (Fig. 1). It differs by two fewer LRRs, two other small deletions and a number of amino acid substitutions in the amino-terminal half of the protein. The carboxyl-terminal half of Cf-4 is identical to that of Cf-9, suggesting that resistance specificity is determined by the amino-terminal portions of these proteins and that the carboxyl-terminal portions probably interact with a common signalling partner. The *Cf-2* locus contains two, independently-functional genes each encoding a 1112 amino acid protein similar in structure to Cf-9, but carrying 38 LRRs (Dixon et al., 1996) (Fig. 1). The two Cf-2 proteins differ from one another by only three amino acids. The LRRs of Cf-2 differ from those of Cf-9, being much more regular in length and showing alternating repeats of two distinct LRR motifs (Fig. 1). The carboxyl-terminal third of Cf-2 shows sequence conservation with the carboxyl-terminal half of Cf-9, suggesting that the carboxyl-terminal portions of Cf-9 and Cf-2 might interact with similar, but perhaps distinct, signalling partners. *Cf-5*, tightly linked to *Cf-2* on the short arm of chromosome *6*, encodes a protein very similar to that of *Cf-2* (Dixon et al., unpublished) (Fig. 1). It differs from Cf-2 by six fewer LRRs from the alternating repeat region and a number of amino acid substitutions predominantly in the amino terminal two thirds of the protein. The carboxyl-terminal half of Cf-5 is highly conserved compared to Cf-2. Like Cf-9 and Cf-4, this suggests that resistance specificity is determined by the amino-terminal portions of these proteins and that the carboxyl-terminal portions probably interact with a common signalling partner.

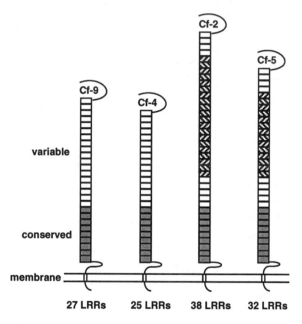

Fig. 1. A diagrammatic summary of the predicted structural domains of the Cf-9, Cf-4, Cf-2 and Cf-5 proteins. The boxes represent individual LRR motifs. Cross hatching is used to indicate the alternating LRR motifs of Cf-2 and Cf-5 and grey shading to indicate conserved LRR motifs.

## Genetic evidence for signalling partners

A number of mutations have been obtained indicating that other genes are required for *Cf* genes to function and might encode signalling partners.

EMS (ethyl methane sulphonate) mutagenesis of plants carrying *Cf-9* yielded two recessive mutations at loci unlinked to *Cf-9* or to one another (Hammond-Kosack et al., 1994). These two mutations, designated *rcr-1* and *rcr-2* (required for *Cladosporium* resistance), allowed significantly more growth of the pathogen in the incompatible *Cf-9/Avr9* interaction than would otherwise occur, but did not allow sporulation of the fungus. Subsequent EMS and di-epoxy butane mutagenesis experiments, with plants carrying *Cf-2*, yielded four additional mutations (Dixon et al., unpublished). Three of these are recessive mutations that confer complete susceptibility in the normally incompatible *Cf-2/Avr2* interaction. Two of these mutations are allelic and have been designated *rcr-3$^A$* and *rcr-3$^B$*. It is not yet known if the third mutation is also an *rcr-3* mutation, but in the interim it has been designated *rcr-6*. The fourth mutation, designated *Rcr-5*, is a dominant mutation that confers partial susceptibility on both the incompatible *Cf-2/Avr2* and *Cf-9/Avr9* interactions. The effects of *rcr-3* and *rcr-6* on the *Cf-9/Avr9* interaction, or of *rcr-1* and *rcr-2* on the *Cf-2/Avr2* interaction are currently being determined. The fact that the *rcr-5* gene is required for both *Cf-2* and *Cf-9* to function, suggests that even if the *Cf-9* and *Cf-2* signalling pathways are distinct they may converge.

## The earliest resistance response

Identification of the resistance proteins involved in recognition of the avirulence proteins provides a starting point from which to dissect the signal transduction pathway leading to resistance. An alternative and complementary approach is to work from a defined endpoint in the opposite direction. The problem with this approach is identifying the right endpoint. There are many different resistance responses following challenge by incompatible pathogens. These include an oxidative burst, lipid peroxidation, altered redox status, increased lipoxygenase activity, salicylic acid (SA) biosynthesis, activation of pathogenesis-related (PR) genes, ethylene biosynthesis, phytoalexin biosynthesis and the hypersensitive response (HR). Some of these are likely to be primary events activated directly by the R/Avr signalling pathway, but others are likely to be secondary events activated indirectly. To identify an appropriate signal transduction endpoint, we have been assessing the various resistance responses that occur in the *Cf-9 /Avr9* and *Cf-2/Avr2* interactions to distinguish those which are the earliest and most likely causal events from those which are later and more likely consequential events (Hammond-Kosack et al., 1996; May et al., 1996). The responses are largely the same in both interactions, but their timing is different. The earliest detectable responses of cotyledons of resistant tomato plants to injections with crude preparations of the corresponding avirulence peptides are an oxidative burst, lipid peroxidation, raised glutathione levels and altered redox status, which occur two to six hours after injection in the *Cf-9 /Avr9* interaction and four to eight hours after injection in the *Cf-2/Avr2* interaction. Lipoxygenase activity rises, cell viability is lost and SA and ethylene are synthesized eight to twelve hours after injection in the *Cf-9/Avr9* interaction, but these responses are more delayed in the *Cf-2/Avr2* interaction. Although lipoxygenase activity starts to rise eight hours after injection, SA synthesis does not occur until after twelve hours, ethylene synthesis after sixteen hours and cell viability is not lost until after 24 hours. Grey necrosis in the injected area is observed after sixteen hours in the *Cf-9/Avr9* interaction, as opposed to chlorosis after 72 hours in the *Cf-2/Avr2* interaction. An important difference in the response, which may contribute to the difference in timing and the nature of the final macroscopic symptoms, is the supraoptimal opening of stomata leading to increased water loss. This occurs two to six hours after injection in the *Cf-9/Avr9* interaction, but does not occur at all in the *Cf-2/Avr2* interaction. Consistent with this interpretation, both lipid peroxidation and the macroscopic symptoms could be abolished by maintaining the injected cotyledons at 98% relative humidity to ameliorate the water loss. Furthermore, no HR is observed in a challenge by incompatible *C. fulvum* if the infected cotyledons or leaves are maintained at 98% relative humidity, although fungal growth is still halted. Several other resistance responses have also been shown to be unnecessary to restrict fungal growth. For example in resistant plants carrying the salicylic acid hydroxylase (*nahG*) transgene, SA accumulation is abolished, but the plants remain capable of restricting the growth of incompatible *C. fulvum* (Brading et al., unpublished). Similarly, resistant plants homozygous for the never ripe (*Nr*) gene of tomato cannot respond to ethylene yet the growth of incompatible *C. fulvum* is still restricted (Brading et al., unpublished). However, these are all late events and are unlikely to be primary responses. The oxidative burst, lipid peroxidation, raised glutathione levels and altered redox status are the earliest resistance responses, but of these, the latter three are themselves symptoms of oxidative stress, and probably a consequence of the oxidative burst. Therefore, under these assay conditions, the oxidative burst is likely to be a primary response to the signal transduction pathway activated by interaction between tomato Cf proteins and *C. fulvum* Avr peptides, and an appropriate endpoint from which to work back up the signal transduction pathway. Further evidence to support this conclusion is provided by more recent experiments using cell-suspension cultures of transgenic tobacco plants carrying *Cf-9* (Piedras et al., unpublished). Cells from these cultures produce an oxidative burst within minutes after exposure to Avr9 peptide. Furthermore, this oxidative burst is inhibited by diphenylene iodonium, which in mammals is a specific inhibitor of the NADPH oxidase, which reduces molecular oxygen to superoxide anions.

## Bridging the gap

Now that we have identified potential anchor points at each end of the signal transduction chain leading to resistance in the tomato - *C. fulvum* interaction, how do we proceed? At the level of pathogen recognition, we still need to determine whether Avr9 peptide binds directly to Cf-9 protein. We then need to identify the next step in the signal transduction chain. We predict this to be a transmembrane protein with an extracytoplasmic domain able to recognise the putative Cf-9 protein/Avr9 peptide complex, and a cytoplasmic signalling domain (Fig. 2). How do we isolate such a component? We are pursuing two approaches; positional cloning of *Rcr* genes and interactive cloning. Positional cloning is feasible, but time consuming, because of the genetic complexity involved in manipulating the *Cf* and *Rcr* genes as well as polymorphic DNA markers. Nevertheless, we have already established *Rcr-1* and *Rcr-2* mapping populations for AFLP analysis (Thomas et al., unpublished). Interactive cloning might be quicker, but may be less feasible, because conventional approaches such as the yeast two hybrid system, often used to identify interacting cytoplasmic proteins, may not enable identification of interacting extracytoplasmic proteins. Nevertheless, we are attempting to produce recombinant Cf-9 protein to use as a molecular probe for interacting proteins in a tomato cDNA expression library (Tang et al., unpublished). At the level of the primary response, we, like others (e.g. Low and Dwyer, 1994; Tenhaken et al., 1995), predict the oxidative burst to resemble that observed for mammalian phagocytes (reviewed by Segal and Abo, 1993) (Fig. 2). Assuming this to be true, we then need to identify the components of the plant NADPH oxidase complex. How do we isolate these components? One approach is to identify likely candidates from the available plant EST databases based on homology to the genes encoding components of the mammalian NADPH oxidase.

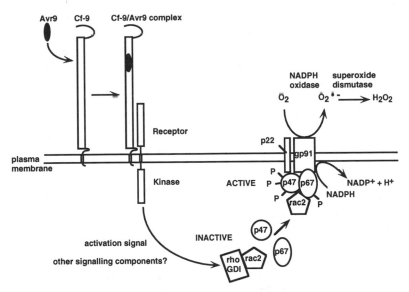

Fig. 2. A working model for the activation of an oxidative burst in response to interaction between Cf-9 and Avr9, based on the mammalian oxidative burst (adapted from Low and Dwyer, 1994).

Some success has already been achieved with the identification and subsequent isolation of a rice homologue of gp91 (Groom et al., 1996). The sequence information gained from the identification of components of the *Arabidopsis* and rice NADPH oxidase complexes may facilitate the isolation of the components of the tomato NADPH oxidase e.g. through PCR techniques. However, some components of the NADPH complex, such as gp91, are structural/enzymatic components, whereas others such as p47, p67, rho GDI and rac2 are regulatory components responsive to signalling. Plant homologues of these components, if they exist, would seem likely candidates to receive the signal generated by R/Avr interaction and their identification and isolation is a high priority.

## Acknowledgements

We thank the Gatsby Charitable Foundation, the UK BBSRC and the EEC for grants that supported the work described here.

## Literature cited

Dixon, M. S., Jones, D. A., Keddie, J. S., Thomas, C. M., Harrison, K. and Jones, J. D. G. 1996. The tomato *Cf-2* disease resistance locus comprises two functional genes encoding leucine-rich repeat proteins. Cell 84:451-459.

Groom, Q. J., Torres, M. A., Fordham-Skelton, A. P., Hammond-Kosack, K. E., Robinson, N. J. and Jones, J. D. G. 1996. *RbohA*, a rice homologue of the mammalian gp91$^{phox}$ respiratory burst oxidase gene. Plant J. In press.

Hammond-Kosack, K. E. Jones, D. A. and Jones, J. D. G. 1994. Identification of two genes required in tomato for full *Cf-9*-dependent resistance to *Cladosporium fulvum*. Plant Cell 6:361-374.

Hammond-Kosack, K. E., Silverman, P., Raskin, I. and Jones, J. D. G. 1996. Race-specific elicitors of *Cladosporium fulvum* induce changes in cell morphology and the synthesis of ethylene and salicylic acid in tomato plants carrying the corresponding *Cf* disease resistance gene. Plant Physiol. 110:1381-1394.

Jones, D. A., Thomas, C. M., Hammond-Kosack, K. E., Balint-Kurti, P. J. and Jones, J. D. G. 1994. Isolation of the tomato *Cf-9* gene for resistance to *Cladosporium fulvum* by transposon tagging. Science 266:789-793.

Joosten, M. H. A. J., Cozijnsen, T. J. and de Wit, P. J. G. M. 1994. Host resistance to a fungal tomato pathogen lost by a single base-pair change in an avirulence gene. Nature 367:384-386.

Low, P. S. and Dwyer, S. C. (1994). Comparison of the oxidative burst signaling pathways of plants and human neutrophils. Adv. Mol. Genet. Plant-Microbe Interact. 3:361-369.

May, M. J., Hammond-Kosack, K. E. and Jones, J. D. G. 1996. Involvement of reactive oxygen species, glutathione metabolism and lipid peroxidation in theCf-gene-dependent defense response of tomato cotyledons induced by race-specific elicitors of *Cladosporium fulvum*. Plant Physiol. 110:1367-1379.

Segal, A. W. and Abo, A. 1993. The biochemical basis of the NADPH oxidase of phagocytes. Trends Biol. Sci. 18:43-47.

Tenhaken, R., Levine, A., Brisson, L. F., Dixon, R. A. and Lamb, C. (1995). Function of the oxidative burst in hypersensitive disease resistance. Proc. Natl Acad. Sci. USA 92:4158-4163.

van den Ackerveken, G. F. J. M., van Kan, J. A. L. and de Wit, P. G. J. M. 1992. Molecular analysis of the avirulence gene *avr9* of the fungal tomato pathogen *Cladosporium fulvum* fully supports the gene-for-gene hypothesis. Plant J. 2:359-366.

# The *N* Gene of Tobacco Confers Resistance to Tobacco Mosaic Virus in Transgenic Tomato

Barbara Baker*†‡, Steve Whitham*, and Sheila McCormick*†

*Department of Plant Biology, University of California, Berkeley, CA 94720
†Plant Gene Expression Center, University of California and United States Department of Agriculture-Agriculture Research Service, Albany, CA 94710
‡To whom reprint requests should be addressed.

## Introduction

The *N* gene of tobacco confers a 'gene-for-gene' resistance to the viral pathogen Tobacco Mosaic Virus (TMV) and most other members of the tobamovirus family. *N* is a member of a class of disease resistance genes whose predicted protein products possess a putative nucleotide binding site (NBS) and leucine-rich repeat (LRR) region (Whitham, et al. 1994). The members of the NBS/LRR class of resistance genes confer resistance to taxonomically diverse pathogens including; viruses, bacteria, and fungi (Staskawicz, et al. 1995). Additionally, the amino-terminus of N possesses similarity to the cytoplasmic domain of the Toll and Interleukin-1 receptors. *N*-mediated resistance to TMV is characterized by the formation of localized necrotic lesions at the sites of viral infection, which is known as the hypersensitive response (HR). TMV becomes localized to the cells in and immediately surrounding the HR lesions. The molecular identity of the viral product that triggers the *N*-mediated HR has not been well established. However, the 126 kDa replicase associated protein may be necessary for the induction of the N-mediated HR (Padgett and Beachy 1993).

Some TMV strains can infect over 200 plant species including most members of the Solanaceae (Watterson 1993). Solanaceous plants, including tomato and pepper, are closely related to tobacco and are agriculturally important crop species. TMV can cause significant yield losses in tomato and pepper and efforts to identify TMV resistance in these plants have been successful (reviewed in (Watterson 1993)). For example, the *Tm-1*, *Tm-2*, and *Tm-2a* loci confer resistance to most strains of TMV in tomato. In pepper, four genes (*L1* to *L4*) confer resistance to most strains of TMV. Tomato and pepper cultivars that lack these resistance genes are highly susceptible to systemic infection by TMV. Like *N*, the *Tm-2/2a* and *L* genes mediate an HR in response to infection by most viral strains. However, the TMV movement and coat proteins trigger host defense responses in tomato plants that carry *Tm-2/2a* genes and pepper plants bearing the L3 respectively. The TMV encoded elicitors of the *L1*, *2*, and *4*-mediated resistances in pepper have not been determined. The *Tm-1* gene of tomato does not mediate an HR in response to TMV infection, but reduces TMV replication in infected cells. Significantly, none of the tomato or pepper TMV resistance genes alone are effective against as many strains of tobamoviruses as the *N* gene. To date, only one tobamovirus strain, TMV-ob, can overcome *N*-mediated resistance. The ability of TMV-ob to overcome *N* may be conditional because at temperatures less than 19°C, TMV-ob can induce a hypersensitive response on plants bearing the *N* gene. If the *N* gene retained its resistance properties in tomato or pepper, then it could provide an additional means to combat tobamovirus diseases in these crops. In addition to crop protection, tomato plants expressing a functional *N* gene would provide an excellent genetic system to dissect the signal transduction pathway leading to HR and inhibition of viral replication and movement.

To test whether *N* could confer TMV resistance in another crop plant species, a TMV-susceptible tomato cultivar was transformed with the *N* gene. Here, we demonstrate that *N* confers resistance to TMV in transgenic tomato. As in tobacco, *N* mediates a hypersensitive response, a common resistance response, characterized by the formation of localized necrosis at infection sites that is correlated with inhibition of viral replication and movement. This is significant for two reasons: first, the transfer of a member of the NBS/LRR class of resistance genes to another species demonstrates the utility of this class of genes in engineering crop resistance. Second, TMV and other related tobamoviruses can be devastating pathogens of Solanaceous crops, including tomato and pepper, and we provide an additional, effective means to combat this disease genetically.

## Results

### THE *N* GENE MEDIATES A LOCALIZED HYPERSENSITIVE RESPONSE TO TMV IN TRANSGENIC TOMATO PLANTS

Systemic TMV infection of susceptible tomato cultivars is characterized by mottling of the leaves, stunted growth, yield reduction and reduced fruit quality . To test if the *N* gene could confer resistance to TMV in tomato, we transformed the TMV-susceptible cultivar VF36 (McCormick 1991) NA construct (pTG34) bearing the *N* gene and recovered four VF36, pTG34 transformants named VTG34-1, 2, 3, and 4. We confirmed the presence of the *N* gene in these four transformants by Southern blot hybridization of genomic tomato DNA with a probe from the *N* gene. The VTG34-1,2,3, and 4 tomato lines carried 3-4, 1, 1, and 3 copies, respectively, of the pTG34 T-DNA construct.

In tobacco cultivars bearing the *N* gene, resistance to TMV is characterized by the formation of localized necrotic lesions known as the hypersensitive response (HR). TMV becomes localized to cells in and immediately surrounding the HR lesions and does not move systemically. We tested whether the transgenic VTG34 tomato plants could elaborate an HR to TMV by inoculating leaves with the U1 strain of TMV (TMV-U1). At approximately 5-7 days postinoculation, localized necrosis was visible on the VTG34-1 leaves that were inoculated with TMV-U1 whereas leaves from untransformed VF36 displayed no signs of necrosis. The response of the other three transformants was identical to VTG34-1 with respect to lesion timing and appearance. The HR of tomato developed as circular black lesions with irregular margins surrounded by chlorosis. By comparison, HR on a leaf from the TMV resistant tobacco cultivar Samsun NN appears as tan, circular lesions with discrete dark brown margins and no chlorosis. In contrast to the transgenic tomato plants, TMV-induced HR lesions were observed at approximately 48 hours postinoculation on Samsun NN, which is typical for tobacco. Interestingly, transgenic tomato leaves that have been excised from plants, inoculated with TMV-U1, and sealed in a petri dish on wetted Whatman paper can develop HR lesions at 2 to 3 days postinoculation, as do tobacco leaves subjected to the same treatment. This suggests that tomato has the capacity to mediate HR with approximately the same timing as tobacco. However, onset of the HR of tomato may be influenced by different environmental and physiological factors than tobacco, such as light intensity, water potential, or humidity. Leaves of mock inoculated Samsun NN and VTG34-1 plants do not develop localized necrosis demonstrating that HR is dependent on the presence of TMV and is not due to the inoculation procedure.

The *N* gene confers resistance to many tobamoviruses and has not been observed to be overcome by any tobamovirus in the field. Thus, the tobamovirus resistance conferred by *N* is extremely durable in tobacco. To test if *N* could confer resistance to other tobamoviruses in tomato, we inoculated transgenic tomato leaves with two additional tobamoviruses, Tomato Mosaic Virus (ToMV) and the crucifer strain of TMV (TMV-Cg). ToMV is the most common tobamovirus infecting tomato in the field. The nucleic acid sequence of ToMV is approximately 80% identical to TMV-U1 (Ohno, et al. 1984). TMV-Cg is not usually found in tomato, but was used here because it is more distantly related to the TMV-U1 strain than ToMV. An alignment of the nucleic acid sequence of TMV-Cg (GenBank accession D38444) to TMV (GenBank accession V01408) and ToMV (GenBank accession X02144) using the Bestfit program (Genetics Computer Group) shows that TMV-Cg is related to both TMV-U1 and ToMV by approximately 60% nucleic acid sequence identity. Both ToMV and TMV-Cg elicit HR on tobacco plants bearing *N*. We tested ToMV and TMV-Cg for their ability to elicit HR on the leaves of transgenic tomato plants. HR lesions developed at 5-7 days postinoculation and were similar in appearance to those induced by TMV-U1. As observed with TMV-U1 infection, there were no differences in the HR lesion response of the other three tomato transformants. However, the onset of HR lesion appearance was more rapid in response to ToMV infection than TMV-U1 or TMV-Cg. ToMV-induced HR could be observed at 6-12 hours prior to the appearance of HR lesions in response to the other tobamoviruses. This observation may be due to ToMV being more virulent in tomato than other tobamoviruses and thus inducing $N$-mediated defense responses more rapidly.

To confirm that the TMV-dependent HR observed on the VTG34-1 plant was due to the presence of the pTG34 T-DNA construct bearing the $N$ gene, we testcrossed TMV-susceptible VF36 to VTG34-1. Testcross progeny were expected to segregate for HR(+) and HR(-) phenotypes, and if the HR(+) phenotype was due to the presence of $N$, then only HR(+) plants should possess the pTG34 T-DNA. We inoculated 48 testcross progeny with TMV-U1 and 23 individuals displayed HR and 25 did not. DNA was isolated from each plant and tested for the presence of the pTG34 construct by the polymerase chain reaction (PCR) using $N$ gene-specific primers to amplify a 450 bp product. We observed that the 23 HR(+) individuals possessed the 450 bp PCR product, whereas the 25 HR(-) individuals lacked the product. This result demonstrated that the $N$ gene is directly responsible for the TMV-dependent HR in transgenic tomato. The segregation of pTG34(+) to pTG34(-) plants is also in agreement with the expected 1:1 segregation for a single dominant gene. Therefore, the multiple copies of pTG34 present in VTG34-1 are tightly linked and comprise a single T-DNA locus.

## SYSTEMIC TMV MOVEMENT IS BLOCKED IN TRANSGENIC TOMATO PLANTS

The formation of HR lesions in tobacco coincides with the inhibition of viral movement and localization of TMV to the inoculated leaf. We tested if systemic movement of TMV was abolished in HR(+) tomato plants. A probe from the 30 kDa movement protein (P30) gene of TMV-U1 was hybridized to RNA gel blots containing total RNA from inoculated and upper, uninoculated leaves of HR(-) VF36 and HR(+) VTG34-1 progeny. In VF36 plants, viral RNA was expected to be detected in both the inoculated and upper, uninoculated leaves of the same plant. However, in HR(+) progeny of VTG34-1, we expected that TMV-U1 RNA would be detected at relatively low levels in the inoculated leaf and would be absent in upper, uninoculated leaves. We observed no hybridization of the P30 probe to total RNA isolated from a mock inoculated VF36 plant (Fig. 2, lane 1), whereas copious amounts of TMV-U1 genomic and subgenomic RNAs were observed in RNA of inoculated and upper, uninoculated leaves (Fig. 2, lanes 2 and 3). In contrast, very little P30 hybridization was observed in RNA isolated from inoculated leaves of two transgenic plants (Fig. 2, lanes 4 and 6) and no hybridization was observed in upper, uninoculated leaves (Fig. 2, lanes 5 and 7). The results are indistinguishable from P30 hybridization to total RNA from inoculated and upper, uninoculated leaves of TMV-susceptible and TMV-resistant tobacco cultivars (Figure 2, lanes 8-12). These results demonstrate that expression of the $N$ gene in tomato effectively inhibits replication and systemic movement of TMV, as in tobacco bearing the wild type $N$ gene.

## N-MEDIATED RESISTANCE TO TMV IS REVERSIBLY INACTIVATED BY ELEVATED TEMPERATURES

A useful property of the *N*-mediated HR to TMV in tobacco is that it is reversibly inactivated at elevated temperatures. At temperatures of 28°C and above, *N*-mediated HR is suppressed and TMV moves systemically (Samuel 1931). The HR is restored when the temperature is reduced below 28°C and necrosis occurs throughout the plant, presumably due to systemic movement of TMV followed by massive cell death mediated by *N*. The property of temperature sensitivity of the *N*-mediated HR was exploited to identify *N* loss-of-function mutations in tobacco, which led to the cloning of *N* by transposon tagging. If this property is conserved in tomato, then it could be employed in a genetic selection scheme to isolate mutations in other loci that are necessary for *N*-mediated resistance to TMV. To test whether the *N*-mediated HR was reversibly inactivated at elevated temperatures in tomato, SR1, Samsun NN, VF36, and VTG34-1 plants were inoculated with TMV-U1 and placed at 30°C. At four days postinoculation, the plants were shifted to the permissible temperature of 23°C. The SR1 tobacco and VF36 tomato plants were not affected by this treatment because they lack a functional *N* gene. However, both the Samsun NN and VTG34-1 plants developed systemic necrosis within 12 to 18 hours following the temperature shift and were completely dead at 7 day following the shift. Following the temperature shift to 23°C, the HR was restored as evidenced by the formation of systemic necrosis on the Samsun NN and VTG34-1 plants. These results confirmed that the reversible, temperature sensitive nature of the *N*-mediated HR is conserved in tomato. Although the mechanism of the temperature sensitive block is not known, this result further demonstrates that the pathway leading to TMV resistance is highly conserved in tomato and tobacco.

## Discussion

The major finding of this study is that we have demonstrated that a member of the NBS/LRR class of disease resistance genes can be transformed into another crop plant species where it confers resistance to a significant pathogen. The *N* gene mediates resistance to the viral pathogen TMV by the formation HR lesions and inhibition of TMV movement in transgenic tomato, as it does in tobacco. The finding that N-mediated TMV resistance is reconstituted in transgenic tomato demonstrates that all of the components required by *N* for both TMV recognition and signal transduction are conserved in tomato. Tomato and tobacco are closely related by virtue of being members of the plant family, Solanaceae. There are a number of *N* homologues in tomato that might function in disease resistance which may explain why the signal transduction components required for *N*-mediated resistance are present in tomato. It will be interesting to test if N-mediated resistance to TMV can reconstituted in plant species more distantly related to tobacco, such as the crucifers which can also be hosts for tobamovirus infection. Significantly, *N* confers resistance to TMV in a genetically tractable species where genetic tools and a powerful selection for loss-of-function mutations using the temperature shift protocol can be utilized to identify genes encoding the components of the signal transduction pathway leading to HR and resistance to TMV.

Another resistance gene, *Pto,* the unique member of the serine-threonine kinase class of disease resistance genes, has also been demonstrated to confer resistance in heterologous plant species. *Pto* confers resistance to *Pseudomonas syringae* bacterial strains possessing the avirulence gene *avrPto* in tomato. *Pto* has been shown to encode resistance to *P. syringae* strains expressing avrPto in the heterologous species *Nicotiana benthamiana* and *N. tabacum.* The ability to engineer plant species with new pathogen resistance specificities indicates that within the Solanaceae the molecular constituents of pathogen induced signal transduction pathways are conserved. This work demonstrates the efficacy of exchanging resistance genes between members of this plant family for engineering genetic resistance to destructive pathogens.

## Acknowledgements

We thank B. Osborne for discussions and critical reading of the manuscript. We thank M. Zaitlin for the U1 strain of TMV, W. Dawson for ToMV, S. Naito for TMV-Cg, and F. Ausubel for the pOCA28 T-DNA vector. This work was supported by United States Department of Agriculture grants (5335-22000-004-00D and 5335-21000-008-00D).

## References

1.      Padgett, H. S., and R. N. Beachy. 1993 Analysis of a tobacco mosaic virus strain capable of overcoming N gene-mediated resistance. Plant Cell 5: 577-586.
2.      Staskawicz, B. J., F. M. Ausubel, B. J. Baker, J. G. Ellis, and J. D. Jones. 1995 Molecular genetics of plant disease resistance. Science 268: 661-667.
3.      Watterson, J. C. 1993, Development and breeding of resistance to pepper and tomato viruses. p. 278, In: Book. , edited by M. M. Kyle. Timber Press, Portland, Oregon.
4.      Whitham, S., S. P. Dinesh-Kumar, D. Choi, R. Hehl, C. Corr, and B. Baker. 1994 The product of the tobacco mosaic virus resistance gene N: similarity to toll and the interleukin-1 receptor. Cell 78: 1101-1115.

# GENETIC INTERACTIONS SPECIFYING DISEASE RESISTANCE IN THE BACTERIAL SPECK DISEASE OF TOMATO.

Christian Tobias, John Salmeron, Giles Oldroyd, Caius Rommens, Steven Scofield and Brian Staskawicz.

Department of Plant Biology, University of California, Berkeley, CA 94720 and The NSF Center of Engineering Plants for Resistance Against Pathogens (CEPRAP), University of California, Davis, CA 95616.

During the past couple of years several major discoveries in the field of molecular plant pathology have greatly advanced our knowledge concerning the molecular mechanisms by which plants recognize and express resistance to invading pathogens. First and foremost among these has been the cloning and characterization of genetically defined plant disease resistance genes that specify resistance to viral, bacterial and fungal pathogens. The molecular characterization of these resistance genes (R genes) has revealed the existence of several distinct classes of genes.

The first disease resistance gene to be cloned was the *Hml* gene of maize (Johal et al. 1992). *Hml* provides resistance to strains of the fungus *Cochliobolus carbonum* which produce HC-toxin. The *Hml* gene represents a unique class of resistance genes as the encoded gene product is a reductase that in activates the HC-toxin. The prototype of a second class of R gene is the *Pto* gene from tomato (Martin et al. 1993). *Pto*, which is required for resistance to *Pseudomonas syringae* pv. *tomato* (*Pst*) expressing the avirulence gene *avrPto*, has strong homology to the serine/threonine class of protein kinases. The largest class of disease resistance genes contains leucine rich repeat domains in their proteins. The *RPS2* and *RPMl* genes from *Arabidopsis* (Bent et al. 1994; Mindrinos et al. 1994; Grant et al. 1995), the *N* gene from tobacco (Whitham et al. 1994), the $L^6$ gene from flax (Lawrence et al. 1995) and the *Prf* gene (Salmeron et al. 1996) from tomato are all members of this class of resistance genes. This striking similarity between these five genes is especially interesting since they come from different plant families and are required for resistance to very different pathogens (bacteria, *RPS2*, *RPMl* and *Prf*; virus, *N*; fungi, *Cf-9*, *Cf-2*, and $L^6$). LRR regions have been implicated in protein-protein interactions (Kobe et al. 1994), raising the possibility that the LRR portion of these resistance gene products is required for specific recognition of a protein elicitor, or for interaction with other protein components of a signal transduction pathway. The LRR class of resistance genes can be divided into subclasses based on other conserved features (Dangl 1995; Staskawicz et al. 1995). *RPS2*, *RPM1*, *N*,*L6* and *Prf* all contain a potential nucleotide binding site (NBS) and a putative cytoplasmic LRR, whereas *Cf-9* and *Cf-2* do not have a nucleotide binding site and appear to contain an extracytoplasmic LRR region.

Interestingly, a recently cloned resistance gene from rice, *Xa-21*, encodes a protein with both an extracytoplasmic LRR region similar to *Cf-9* and a cytoplasmic serine/threonine kinase domain similar to *Pto* (Song et al. 1995) Thus, *Xa-21* has both a potential receptor domain and a potential signaling domain, suggesting that resistance gene products containing a putative receptor (such as those in the LRR class) might interact with a resistance gene product containing a putative signaling component (such as *Pto*, in the serine/threonine kinase class).

In addition to the resistance genes described above, the cloning and characterization of the corresponding pathogen avirulence genes has put researchers in a unique position to uncover the molecular events that specify disease resistance. More than 30 avirulence genes from phytopathogenic bacteria have been cloned and characterized (reviewed by (Keen 1990; Dangl 1994). While it is known that most bacterial avirulence genes appear to encode hydrophilic proteins, the predicted amino-acid sequences generally reveal very little about the molecular mechanism of avirulence. For the most part, avirulence genes bear no resemblance to other DNA sequences in available databases (Long et al. 1993; Dangl 1994).

Based on the one-to-one correspondence in gene-for-gene interactions, it has been predicted that the products of avirulence genes may interact directly with the products of specific resistance genes (Ellingboe 1981). In a few cases, an avirulence gene product itself has been definitively demonstrated to induce a specific response. For example, two avirulence genes from *C. fulvum*, *avr9* and *avr4*, were shown to encode precursors of peptides which are capable of specifically eliciting a necrotic response in tomato cultivars carrying the *Cf-9* or *Cf-4* resistance genes, respectively. The avirulence gene *avrBs3* from *Xanthomonas campestris* pv. *vesicatoria* also suggests a direct interaction of avirulence gene product with resistance gene product. *avrBs3* contains a region of 17.5 copies of a 34-amino acid repeat unit (Bonas et al. 1989); deletion of some of these repeats causes a change in host specificity, resulting in virulence on normally resistant hosts and in some cases avirulence in normally susceptible hosts (Herbers et al. 1992). The fact that structural changes in the avirulence gene can uncover new resistance specificities in the host plant suggests that the avirulence gene product itself may be acting as a specific elicitor in the host plant. In contrast, avirulence due to the *avrD* gene from *P. syringae* pv. *tomato* appears to operate in a different manner. Rather than interacting directly with a resistance gene product, *avrD* is required for the production of two homologous gamma lactones which elicit a specific response in resistant soybean cultivars (Keen et al. 1990; Kobayashi et al. 1990).

Elucidation of the molecular events controlling plant-pathogen specificity will require a complete understanding of the avirulence signal that is delivered to the host and the specific host molecules that interact with that signal. The availability of cloned corresponding pairs of genes from both the host and the pathogen allows experiments to be performed to test the hypothesis that proteins encoded by avirulence genes directly interact with resistance gene proteins to determine specificity and the expression of disease resistance in the plant.

Over the past several years, the bacterial speck disease of tomato has been developed as one of the best characterized model systems to study the molecular basis of pathogen recognition and the expression of bacterial plant disease resistance. In this system, several genes have been identified that are important in determining pathogen recognition and expression of plant disease resistance. Molecular and classical genetic analyses of both the host and the pathogen have revealed that the expression of resistance is genetically determined by the cloned avirulence gene, *avrPto,* in *Pseudomonas syringae* pv. *tomato* that specifically interacts with the semi-dominant *Pto* resistance gene in tomato. The most striking feature of the *Pto/avrPto* interaction is the *Pto* resistance gene itself which, unlike the majority of other cloned resistance genes, encodes a serine-threonine kinase with a single myristoylation site that possibly anchors *Pto* to the plasma membrane (Martin,et al. 1993). The LRR structural motif is conspicuously absent.

Another unique feature of this system, summarized in figure 1, involves the presence of a parallel signaling pathway that when activated by the organophosphate insecticide fenthion, produces necrotic lesions superficially similar to hypersensitive cell death in response to pathogens (Carland et al. 1993). The fenthion sensitivity also requires a functional *Fen* gene, one of several other *Pto*-like kinases at the *Pto* locus (Martin et al. 1994; Rommens et al. 1995). The 80% amino-acid identity that *Fen* and *Pto* share indicates that Fen is likely to be a functional counterpart to *Pto*. Both *avrPto*-specified elicitor and fenthion signaling pathways utilize a common step originally identified genetically by mutation as being tightly linked to the *Pto* locus. The gene, *Prf*, is required for both resistance to *P. syringae* expressing *avrPto* and sensitivity to fenthion (Salmeron et al. 1994).

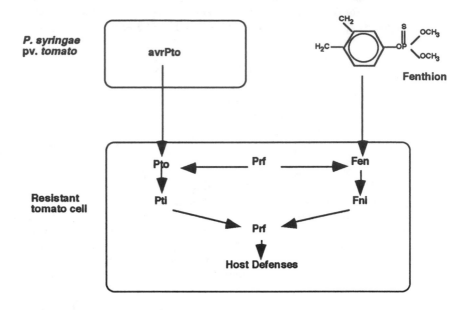

Fig 1.

The cloning and characterization of the predicted *Prf* gene product has revealed amino-acid similarities to a subset of other R gene products which are predicted to be cytoplasmic including *RPM1* and *RPS2* from *Arabidopsis*, the *N* gene of *Nicotiana glutinosa* and the *L6* gene of flax. These genes all share several kinase subdomains involved in nucleotide binding, a short conserved region of unknown function, and LRR's (the number of which varies). The identity of *Prf* to other resistance gene products indicates that resistance to *Pst* (*avrPto*) may have more in common with other examples of gene-for-gene interactions than had previously been assumed. Other unresolved aspects of the relationship between the parallel *Pto* and *Fen* pathways should become apparent when the bacterial elicitor produced by *P. syringae* strains expressing *avrPto* is identified. A structural similarity between fenthion and the *avrPto* elicitor in a form of biological mimicry is one possibility, it also may be the case that the AvrPto protein itself is the elicitor.

A downstream component of the plant signal transduction pathway has been isolated by the yeast two-hybrid system that interacts specifically with *Pto* (Zhou et al. 1995). This gene (*Pti1*) encodes a serine/threonine protein kinase and has been demonstrated to enhance resistance to *P.*

*syringae* when overexpressed in tobacco. Recombinant fusion proteins of *Pti1* are phosphorylated specifically by *Pto* and not by *Fen*.

Preliminary results from our lab have shown that when *avrPto* is expressed in plant cells containing the *Pto* gene via a PVX expression vector or an *Agrobacterium* transient assay, a resistance response is observed. These results are consistent with the notion that the AvrPto protein may be the elicitor. We also have constructed epitope-tagged *avrPto* genes with the 6x-His and 1x c-myc epitopes and have shown that these genes are still capable of inducing a race-specific hypersensitive reaction when expressed in bacteria. Future experiments are planned to use indirect immunofluoresence to determine if AvrPto can be found localized within the plant cell.

Literature Cited

Bent, A. F., et al. 1994. *RPS2* of *Arabidopsis thaliana*: a leucine-rich repeat class of plant disease resistance genes. Science 2651856-1860.
Bonas, U., et al. 1989. Genetic and structural characterization of the avirulence gene *avrBs3* from *Xanthomonas campestris* pv. *vesicatoria*. Mol. Gen. Genet. 218127-136.
Carland, F. and B. Staskawicz 1993. Genetic characterization of the *Pto* locus of tomato: semi-dominance and segregation of resistance to *Pseudomonas syringae* pathovar tomato and sensitivity to the insecticide Fenthion. Mol. Gen. Genet. 23917-27.
Dangl, J. L. (1994). The enigmatic avirulence genes of phytopathogenic bacteria. Bacterial Pathogenesis of Plants and Animals. J. L. Dangl, Springer-Verlag: 99-118.
Dangl, J. L. 1995. Pièce de résistance: novel classes of plant disease resistance genes. Cell 80363-366.
Ellingboe, A. H. 1981. Changing concepts in host-pathogen genetics. Ann. Rev. Phytopath. 19125-143.
Grant, M. R., et al. 1995. Structure of the *Arabidopsis RPM1* gene enabling dual specificity disease resistance. Science 269843-846.
Herbers, K., et al. 1992. Race-specificity of plant resistance to bacterial spot disease determined by repetitive motifs in a bacterial avirulence protein. Nature 356172-174.
Johal, G. S. and S. P. Briggs 1992. Reductase activity encoded by the Hm1 disease resistance gene in maize. Science 2585084: 985-987.
Keen, N. T. 1990. Gene-for-gene complementarity in plant-pthogen interactions. Annu. Rev. Genet. 24447-463.
Keen, N. T., et al. 1990. Bacteria expressing avirulence gene D produce a specific elicitor of the soybean hypersensitive reaction. Mol. Plant-Microbe Interact. 3112-121.

Kobayashi, D. Y., et al. 1990. Molecular characterization of avirulence gene D from *Pseudomonas syringae* pv. *tomato*. Mol. Plant-Microbe Interact. 32: 94-102.

Kobe, B. and J. Deisenhofer 1994. The leucine-rich repeat: a versatile binding motif. Trends Biochem. Sci. 1910: 415-421.

Lawrence, G. J., et al. 1995. The *L6* gene for flax rust resistance is related to the *Arabidopsis* bacterial resistance gene *RPS2* and the tobacco viral resistance gene *N*. Plant Cell 71195-1206.

Long, S. R. and B. J. Staskawicz 1993. Prokaryotic plant parasites. Cell 73921-935.

Martin, G. B., et al. 1993. Map-based cloning of a protein kinase gene conferring disease resistance in tomato. Science 2621432-1436.

Martin, G. B., et al. 1994. A member of the tomato Pto gene family confers sensitivity to fenthion resulting in rapid cell death. Plant Cell 611: 1543-1552.

Mindrinos, M., et al. 1994. The *A. thaliana* disease resistance gene *RPS2* encodes a protein containing a nucleotide-binding site and leucine-rich repeats. Cell 781089-1099.

Rommens, C. M. T., et al. 1995. Use of a gene expression system based on potato virus X to rapidly identify and characterize a tomato *Pto* homolog that controls fenthion sensitivity. Plant Cell

Salmeron, J. M., et al. 1994. Tomato mutants altered in bacterial disease resistance provide evidence for a new locus controlling pathogen recogniton. Plant Cell 6511-520.

Salmeron, J. M., et al. 1996. Tomato *Prf* is a member of the leucine-rich repeat class of plant disease resistance genes and lies embedded within the *Pto* kinase gene cluster. Cell in press

Song, W.-Y., et al. 1995. A receptor kinase-like protein encoded by the rice disease resistance gene, *Xa21*. Science 2701804-1806.

Staskawicz, B. J., et al. 1995. Molecular genetics of plant disease resistance. Science 268661-667.

Whitham, S., et al. 1994. The product of the tobacco mosaic virus resistance gene *N*: similarity to toll and the interleukin-1 receptor. Cell 781101-1115.

Zhou, J., et al. 1995. The tomato gene -Pti1 encodes a serine/threonine kinase that is phosphorylated by Pto and is involved in the hypersensitive response. Cell 83925-935.

# Signal Perception and Transduction in the Activation of Plant Defense by ß-Glucan Elicitors

Luis Antelo, Andrea Daxberger, Judith Fliegmann, Axel Mithöfer, Christel Schopfer, and Jürgen Ebel
*Botanisches Institut der Universität, D-80638 München, Germany*

ABSTRACT. The best characterized elicitors of the oomycete *Phytophthora sojae* for activating a multicomponent defense response, including the production of phytoalexins, in soybean (*Glycine max* L.) are the branched 1,3- and 1,6-linked ß-glucans that are structural polysaccharides of the hyphal walls of the pathogen. The soybean microsymbiont *Bradyrhizobium japonicum* synthesizes cyclic 1,3-1,6-ß-glucans that suppressed the fungal ß-glucan-induced phytoalexin response in soybean, indicating a novel mechanism by which the symbiotic bacteria might avoid a defense response that is normally activated in pathogenic interactions. A putative receptor ($M_r$ = 75 kDa) for the *P. sojae* ß-glucans was isolated from soybean membranes and partially characterized. Partial amino acid sequences were used to raise an anti-peptide antiserum and to generate two oligonucleotides which served as primers for PCR using soybean cDNA as template. Northern blot experiments with a PCR product indicated that the glucan-binding protein mRNA has a size of about 2.4 kb. Since several species of the plant family Fabaceae were shown to possess related high-affinity ß-glucan-binding sites, glucan-based perception mechanisms may be a more common feature of this plant family. Following plasma membrane binding of the ß-glucan elicitor, the subsequent signal transduction might involve a rapid, transient increase in the cytosolic $Ca^{2+}$ level and in permeability changes of the plasma membrane to $Ca^{2+}$, $H^+$, and $Cl^-$. Production of the isoflavonoid phytoalexins, glyceollins, in soybean involves a multi-step biosynthetic pathway. Among the pathway enzymes are several cytochrome P450s. A gene family-specific differential display approach was used in the identification of several elicitor-inducible P450 cDNAs.

# Introduction

Plants are resistant against most potential pathogens in their environment owing to the existence of effective defense mechanisms. For both nonhost and race/cultivar-specific resistance, models have been proposed which envision a ligand-receptor-like interaction between the product of a pathogen avirulence gene, called race-specific elicitor, or a more general elicitor and a corresponding plant receptor that triggers the initiation of a multicomponent defense response in the plant (Ebel and Cosio 1994). Elicitor perception and defense activation are thought to be linked by a signal transduction pathway. The confirmation of such a signaling process requires the identification and characterization of elicitor receptors, of components involved in signal transduction, and of the mechanisms underlying the regulation of the defense response.

Among the fungal compounds which elicit a phytoalexin defense response in soybean (*Glycine max* L.) are branched 1,3-1,6-ß-glucans (for review see Ebel and Cosio 1994). This type of glucan is a major component of the hyphal walls of the oomycete *Phytophthora sojae* (syn. *P. megasperma* f. sp. *glycinea*), a pathogen of soybean, and of other members of the genus *Phytophthora* which encompasses a large number of known pathogens for many legume species. Hyphal growth is usually accompanied by the release of appreciable quantities of these carbohydrates to the surrounding medium as demonstrated in vitro by germinating cysts of *P. sojae* (Waldmüller et al. 1992). During fungal growth on plant tissues, glucan release could be enhanced through the action of 1,3-ß-glucanases (Ham et al. 1991). These features may have resulted in an early selection of ß-glucans as markers for a potential pathogen in the activation of defense responses in plant families such as the Fabaceae (Cosio et al. 1996).

Work on the elicitor-active structures within the ß-glucans identified a characteristic hepta-1,3-1,6-ß-glucoside as being responsible for activity (Sharp et al. 1984). The hepta-ß-glucoside and ß-glucan fractions have been used as ligands for the identification of high-affinity ß-glucan-binding sites in soybean (Schmidt and Ebel 1987; Cosio et al. 1988, 1990; Cheong and Hahn 1991) and in other legume species (Cosio et al. 1996). Identical structural elements appeared to be required for membrane binding of the ß-glucan elicitors and for induction of the defense response (Schmidt and Ebel 1987; Cosio et al. 1988 1990, 1992; Cheong and Hahn 1991). This observation indicates that the ß-glucan-binding protein(s) in soybean might function as receptor in the transmission of the extracellular elicitor signal leading to the transcriptional activation of phytoalexin production.

## Glucan Elicitor Activity in Legumes

A screening for ß-glucan-binding sites in membranes of legumes other than soybean with a radioiodinated conjugate of a *P. sojae* ß-glucan fraction resulted in the identification of high-affinity binding in alfalfa, bean, lupin, and pea with ligand affinities very similar to those reported for soybean (Cosio et al. 1990, 1996; Cheong et al. 1991). The affinity of the binding sites of bean for the same ligand was significantly lower. Low binding activity for the ß-glucan fraction of relatively weak affinity was detected in membrane fractions of chickpea. Good correlation was found between the presence of high-affinity binding and the accumulation of isoflavonoid phytoalexins in roots of alfalfa, bean, chickpea, and pea seedlings after exposure to fungal ß-glucan. Lupin displayed a strong wound-induced accumulation of isoflavones making it impossible to determine phytoalexin induction in response to elicitor. No specific binding or phytoalexin accumulation in response to ß-glucans was observed in broadbean. The results indicate that the ß-glucan perception mechanism is probably not restricted to soybean but might be a more common feature among species of the plant family Fabaceae (Cosio et al. 1996).

In symbiotic interactions between *Rhizobium* or *Bradyrhizobium* and legumes, extracellular polysaccharides seem to play critical roles in the establishment of structural and metabolic cooperation. Some of the extracellular carbohydrates may be active as signal compounds to the plant and not merely as passive coat for protection (Long and Staskawicz 1993). Cyclic 1,3-1,6-ß-glucans synthesized by *B. japonicum* USDA 110 not only inhibited the binding of the elicitor-active ß-glucans from *P. sojae* to soybean membranes but they also prevented the stimulation of phytoalexin production in soybean when tested in combination with the fungal ß-glucans (Mithöfer et al. 1996a). When tested alone the bacterial cyclic ß-glucans were inactive as phytoalexin elicitors even at relatively high concentrations. These results may indicate a novel mechanism for the establishment of a successful plant-symbiont interaction. The production of the extracellular cyclic ß-glucans could enable *Bradyrhizobium* to suppress the plant's defense response (Mithöfer et al. 1996a). This model was supported by our finding that following inoculation of soybean roots with a *B. japonicum* mutant that was affected in the synthesis of the cyclic 1,3-1,6-ß-glucans or in the glucan transport to the extracellular medium not only ineffective nodules were formed (Bhagwat and Keister 1995) but also phytoalexin production was stimulated (Bhagwat, Kraus, Mithöfer, and Ebel, unpublished results).


## Characterization of ß-Glucan-binding Proteins

The candidate ß-glucan receptor protein(s) that are possibly involved in elicitor recognition in soybean plasma membranes are of low abundance. Purification of the elicitor-binding proteins is a prerequisite to further characterize them and to better define their role in elicitor signaling.

A low abundance ß-glucan elicitor-binding protein from soybean was recently isolated by a rapid, simple and one-step purification method yielding about 9000-fold enrichment (Mithöfer et al. 1996b). The final preparation consisted of one major protein with an apparent molecular mass of about 75 kDa (Mithöfer et al. 1996b) that appeared to be slightly larger than the mass determined earlier (Cosio et al. 1992; Frey et al. 1993). Electrophoretic analyses of the purified and photoaffinity-labeled (Cosio et al. 1992) binding protein showed that the native protein was an oligomer with apparent molecular mass of about 240 kDa. While the 75-kDa protein was the main component after photoaffinity labeling and separation of the products by SDS-PAGE, minor labeled protein bands of 100 and 150 kDa that had been detected earlier (Frey et al. 1993) were again visible (Mithöfer et al. 1996b). Of these, the 150-kDa component might represent an aggregate of the 75-kDa protein. A polyclonal anti-peptide antiserum that was raised against a synthetic 15-mer oligopeptide sequence derived from the 75-kDa protein recognized the purified binding protein in immunoblotting experiments.

Several internal peptides of the 75-kDa ß-glucan-binding protein were obtained after digestion with proteinase-Lys-C, sequenced, and used to design degenerate oligonucleotides. Two oligonucleotides were used as primers in PCR with soybean root cDNA as template. The PCR product generated by this approach was 1493 bp in size and included coding sequences corresponding to four of the sequenced peptides. Northern blot analysis indicated that the PCR product hybridized to one mRNA with a size of about 2.4 kb, compatible with a size of an mRNA required to encode a 75-kDa protein.

## Signal Transduction and Activation of Defense

In several plant cell systems, rapid permeability changes of the plasma membrane to ions including $Ca^{2+}$, $H^+$, $K^+$, and $Cl^-$ have been suggested to be involved in signal transduction leading to the activation of the defense (Ebel and Scheel 1992). In addition, the activation of the responses in several plant cell systems including soybean, depends on the presence of $Ca^{2+}$ in the cell culture medium (Ebel and Scheel 1992).

Experiments with soybean cell cultures demonstrated that a *P. sojae* ß-glucan fraction induced a rapid, transient increase in $^{45}Ca^{2+}$ uptake, an alkalinization of the cell culture medium, and an efflux of $^{36}Cl^-$ from the cells preceding the onset of phytoalexin synthesis (Ebel et al. 1995; Daxberger, unpublished results). Chloride channel antagonists were more effective than calcium channel antagonists in inhibiting both stimulation of the defense response and the inducible ion fluxes (Ebel et al. 1995; Daxberger, unpublished results). Even more rapid than the elicitor-inducible ion fluxes was the transient enhancement in the level of cytosolic $Ca^{2+}$ (Mithöfer, Boller, Ebel, and Neuhaus-Url, unpublished results). These observations indicate that a complex sequence of changes in fluxes and levels of ions may occur during the early events of elicitor signal transduction.

Production of the pterocarpanoid phytoalexins, glyceollins, in soybean comprises a complement of about 15 enzymes and is regulated by temporary gene activation. The biosynthetic pathway consists of a series of reactions which are catalyzed by enzymes of general phenylpropanoid metabolism, including cinnamate 4-hydroxylase and 4-coumarate:CoA ligase, and of flavonoid/isoflavonoid biosynthesis, as well as of enzymes specifically involved in later steps of pterocarpan phytoalexin biosynthesis (Ebel and Grisebach 1988). A remarkable feature of phytoalexin formation in soybean is the involvement of five cytochrome P450-dependent enzymes in the biosynthetic pathway. One of these, cinnamate 4-hydroxylase (C4H) has been extensively studied both at the biochemical and molecular level in several plant species including soybean. The differential display of soybean mRNA was used to identify transcripts encoding cytochrome P450s. Several corresponding full-length cDNAs were isolated. One of them encodes C4H. Other P450 cDNAs that are presumably related to phytoalexin biosynthesis will be further characterized by functional expression.

## Acknowledgements

This work was supported by the Deutsche Forschungsgemeinschaft (SFB 369) and Fonds der Chemischen Industrie.

## References

Bhagwat, A.A., and Keister, D.L. 1995. Cloning and mutagenesis of the ß-(1→3), ß-(1→6)-D-glucan synthesis locus of *Bradyrhizobium japonicum*. Mol. Plant Microbe Interact. 8:366-370.

Cheong, J.-J., and Hahn, M.G. 1991. A specific, high-affinity binding site for the hepta-ß-glucoside elicitor exists in soybean membranes. Plant Cell 3:137-147.

Cosio, E.G., Pöpperl, H., Schmidt, W.E., and Ebel, J. 1988. High-affinity binding of fungal ß-glucan fragments to soybean (*Glycine max* L.) microsomal fractions and protoplasts. Eur. J. Biochem. 175:309-315.

Cosio, E.G., Frey, T., Verduyn, R., van Boom, J., and Ebel, J. 1990. High-affinity binding of a synthetic heptaglucoside and fungal glucan phytoalexin elicitors to soybean membranes. FEBS Letters 271:223-226.

Cosio, E.G., Frey, T., and Ebel, J. 1992. Identification of a high-affinity binding protein for a hepta-ß-glucoside phytoalexin elicitor in soybean. Eur. J. Biochem. 204:1115-1123.

Cosio, E.G., Feger, M., Miller, C.J., Antelo, L., and Ebel, J. 1996. High-affinity binding of fungal ß-glucan elicitors to cell membranes of species of the plant family Fabaceae. Planta. in press.

Ebel, J., and Grisebach, H. 1988. Defense strategies of soybean against the fungus *Phytophthora megasperma* f. sp. *glycinea*: a molecular analysis. Trends Biochem. Sci. 13:23-27.

Ebel, J., and Scheel, D. 1992. Elicitor recognition and signal transduction. Pages 183-205 in: Plant gene research. Genes involved in plant defense. Vol. 8. T. Boller and F. Meins, Jr., eds. Springer-Verlag, Wien.

Ebel, J., and Cosio, E.G. 1994. Elicitors of plant defense responses. Int. Rev. Cytol. 148:1-36.

Ebel, J., Bhagwat, A.A., Cosio, E.G., Feger, M., Kissel, U., Mithöfer, A., and Waldmüller, T. 1995. Elicitor-binding proteins and signal transduction in the activation of a phytoalexin defense response. Can. J. Bot. 73 (Suppl. 1): S506-S510.

Frey, T., Cosio, E.G., and Ebel, J. 1993. Affinity purification and characterization of a binding protein for a hepta-ß-glucoside phytoalexin elicitor in soybean. Phytochemistry 32:543-550.

Ham, K.-S., Kauffmann, S., Albersheim, P., and Darvill, A.G. 1991. Host-pathogen interactions. XXXIX. A soybean pathogenesis-related protein with ß-1,3-glucanase acitivity releases phytoalexin elicitor-active heat-stable fragments from fungal walls. Mol. Plant Microbe Interact. 4:545-552.

Long, S.R., and Staskawicz, B.J. 1993. Prokaryotic plant parasites.Cell 73:921-935.

Mithöfer, A., Lottspeich, F., and Ebel, J. 1996a. One-step purification of the ß-glucan elicitor-binding protein from soybean (*Glycine max* L.) roots and characterization of an anti-peptide antiserum. FEBS Letters 381:203-207.

Mithöfer, A., Bhagwat, A.A., Feger, M., and Ebel, J. 1996b. Suppression of fungal ß-glucan-induced plant defence in soybean (*Glycine max* L.) by cyclic 1,3-1,6-ß-glucans from the symbiont *Bradyrhizobium japonicum*. Planta 199:270-275.

Schmidt, W.E., and Ebel, J. 1987. Specific binding of a fungal glucan phytoalexin elicitor to membrane fractions from soybean *Glycine max*. Proc. Natl. Acad. Sci. U.S.A. 84:4117-4121.

Sharp, J.K., McNeil, M., and Albersheim, P. 1984. The primary structures of one elicitor-active and seven elicitor-inactive hexa(ß-D-glucopyranosyl)-D-glucitols isolated from the mycelial walls of *Phytophthora megasperma* f. sp. *glycinea*. J. Biol. Chem. 259:11321-11336.

Waldmüller, T., Cosio, E.G., Grisebach, H., and Ebel, J. 1992. Release of highly elicitor-active glucans by germinating zoospores of *Phytophthora megasperma* f. sp. *glycinea*. Planta 188:498-505.

# Perception of Oligochitin (*N*-acetylchito-oligosaccharide) Elicitor Signal in Rice

Naoto Shibuya, Yuki Ito and Hanae Kaku

Department of Cell Biology, National Institute of Agrobiological Resources, Ministry of Agriculture, Forestry and Fisheries, Tsukuba, Ibaraki, Japan

Higher plants initiate various defense reactions when they are attacked by pathogens such as fungi, bacteria and viruses. These defense responses include the production of phytoalexins, proteinase inhibitors, hydrolases such as chitinase and $\beta$-glucanase, wall glycoproteins and lignin. These responses can be triggered by oligo-/polysaccharide elicitors derived from the cell surface of pathogenic microbes as well as host plants. Some of these elicitors could trigger these responses in plants at a very low concentration and their recognition by host plants seems to be very strict concerning to their structure. These results strongly indicate the presence of specific recognition systems in the host cells (Côté and Hahn, 1994).

To understand the whole process of signal perception and transduction leading to the activation of specific sets of genes on molecular basis, it is critically important to know the detailed properties of the receptor molecules which perceive the elicitor signal. However, the knowledge on such receptor molecules is still very limited and none of their genes have been cloned. Only the presence of high-affinity binding sites, putative receptor molecules, for some elicitors have been reported in the membrane preparations from several plants. A high-affinity binding protein for the elicitor active $\beta$-glucan fragments has been detected in microsomal/plasma membrane preparations from soybean cotyledon (Schmidt et al. 1987; Cheong et al. 1991). Mithöfer et al. recently reported the purification of this binding protein to apparent homogeneity (Mithöfer et al. 1996). The presence of high-affinity binding sites for a glycopeptide elicitor (Basse et al. 1993) and a peptide elicitor (Nürnberger et al. 1994) in the microsomal preparation from suspension-cultured tomato cells and parsley cells respectively, were also reported.

Chitin fragments (*N*-acetylchitooligosaccharides) are another class of important elicitor molecule derived from the cell surface of pathogenic fungi and have been reported to induce defense reactions in several monocots as well as some dicots. There have been only little information, however, on the putative receptor molecule for this elicitor. In this paper, we describe our recent results on the perception of *N*-acetylchitooligosaccharide elicitor signal in suspension-cultured rice cells and the characterization of the putative receptor molecule in the plasma membrane from the rice cells.

## Perception of *N*-Acetylchitooligosaccharide Elicitor Signal in Suspension-Cultured Rice Cells

Using the phytoalexin (momilactone A) production in suspension-cultured rice cells as an index of elicitor activity, we found that purified chitin fragments, *N*-acetylchitooligosaccharides, showed a very potent elicitor activity depending on their size and structure (Yamada et al. 1993). Namely, the larger oligosaccharides such as heptamer or octamer induced the phytoalexin biosynthesis even at nM concentration but the smaller fragments, dimer or trimer did not show detectable activity at the concentration tested. Deacetylated form of these oligosaccharides, chitosan fragments, also did not show elicitor activity. It was shown that the same oligosaccharides induced various cellular responses such as transient depolarization of membrane potential (Kuchitsu et al. 1993), ion flux (Kuchitsu et al. unpublished), transient generation of reactive oxygen species (Kuchitsu et al. 1995) and jasmonic acid (Nojiri et al. 1996), and transient expression of several unique "early responsive" genes (Minami et al. 1996) also at a very low concentration. All these responses showed the same specificity concerning to the structure of elicitor-active chitin fragments so far tested, indicating that they represent the reactions mediated through the same, single receptor molecule for this elicitor. Rapid and transient nature of most of these responses also suggested their possible involvement in the signal transduction cascade in rice cells. Some responses such as the depolarization of membrane potential, ion flux and transient generation of reactive oxygen species are thought to take place at the level of plasma membrane, indicating the site of perception of the elicitor signal also locates on the cell surface.

## Identification of A High-Affinity Binding Site for The
## Elicitor in The Membrane Preparation From Rice Cells

[125]I-Labeled tyramine conjugate of N-acetylchitooctaose was synthesized for the binding study with the membrane preparation from suspension-cultured rice cells. Using the microsomal membrane preparation and the radioactive ligand, the presence of a high-affinity binding site for the N-acetylchitooligosaccharide elicitor was detected (Shibuya et al. 1993). Scatchard plot analysis of the results of the binding assay indicated the presence of an independent, non-interactive binding site with the $K_d$ of 5.4 nM. Presence of the most binding activity in the plasma membrane preparation obtained by aquous two phase partitioning showed that most of the binding site is localized in this fraction (Shibuya et al. submitted). Detailed studies of the binding specificity of this site using various unlabeled oligosaccharides revealed that the binding specificity correlated well with the specificity for N-acetylchitooligosaccharide elicitor shown by rice cells. That is, larger sized N-acetylchitooligosaccharides such as heptamer/octamer showed several thousand to several ten thousand fold higher affinity to this binding site compared to trimer. Also, deacetylated form of these oligosaccharides did not show affinity to the binding site. Binding ability for the N-acetylchitooligosaccharide disappeared after the treatment of the membrane preparation with a protease, indicating the presence of corresponding binding protein(s) in the membrane preparation.

## Identification of The Binding Protein by Affinity Labeling

[125]I-Labeled photolabile 2-(4-azidophenyl)ethylamine conjugate and ?-(4-aminophenyl)ethylamine (APEA) conjugate of N-acetylchitooctaose were synthesized for photoaffinity labeling and affinity cross-linking, respectively. These radioactive probes were reacted with the plasma membrane preparation and cross-linked with the binding protein by the irradiation of UV light or the treatment with glutaraldehyde. Autoradiographic analysis of the SDS-PAGE gels revealed that a single, 70 kDa protein band was radioactively labeled by both affinity ligands (Ito et al. submitted). The incorporation of radioactivity was saturable concerning to the concentration of the ligand. The labeling of the 70 kDa protein was inhibited with unlabeled N-acetylchitooctaose with a half maximal concentration of 30 nM but not inhibited with N-acetylchitotriose nor deacetylated octamer, chitooctaose, at 25 μM. These characteristics of the 70 kDa protein corresponded well with that of the high-affinity binding site detected in the plasma membrane preparation, indicating that the 70 kDa protein identified by the affinity labeling is the actual molecule detected as

the "high-affinity binding site" in the previous binding studies. Furthermore, the corresponding binding specificity with the behavior of the rice cells and also its localization in the plasma membrane suggested that the 70 kDa protein might be a true, functional receptor molecule for this elicitor.

## Solubilization and Purification of The 70 kDa Protein

Using several detergents, the binding protein was successfully solubilized from the plasma membrane preparation in its active form. Among the detergents tested, Triton X-100 gave the best and reproducible results. Approximately one third of the binding activity was solubilized with this detergent. Solubilized fraction showed a saturable mode of binding with the $^{125}$I-labeled tyramine conjugate of $N$-acetylchitooctaose with an approximate $K_d$ of 90 nM. Affinity cross-linking with $^{125}$I-labeled APEA conjugate of $N$-acetylchitooctaose showed the presence of the same 70 kDa binding protein in the solubilized fraction. An affinity column, $N$-acetylchitoheptaosyl-lysil-agarose column, was designed for the affinity purification of this binding protein. Using this affinity column and specific elution with $N$-acetylchitooligosaccharides, we could purify the binding protein to apparent homogeneity. The purified fraction contained a binding affintiy toward the radioactive ligand which was specifically inhibited with unlabeled $N$-acetylchitooligosaccharides. SDS-PAGE of this fraction showed the presence of a single protein band corresponding to the 70 kDa protein detected by the affinity labeling.

## Summary and Future Perspectives

We showed that the $N$-acetylchitooligosaccharides of certain size could act as a potent elicitor signal in suspension-cultured rice cells and induce various cellular responses. This system gives an excellent model system for the study of signal perception and transduction in plants, because of the simplicity of the system, i.e., the use of a chemically well defined elicitor molecule and homogenous cultured cells. We identified the high-affinity binding site and the corresponding binding protein using binding assay with plasma membrane preparation as well as several affinity labeling techniques. A 70 kDa plasma membrane protein thus identified as a candidate of the receptor molecule for this elicitor, was successfully solubilized from the plasma membrane and purified to a single band on SDS-PAGE.

$N$-Acetylchitooligosaccharides have also been reported to induce the alkalinization of the reaction medium and the phosphorylation of some proteins in cultured tomato cells (Felix et al. 1993) and a corresponding high-affinity binding site was detected in the microsomal membrane preparation from the tomato cells (Baureithel et al. 1994) However, the preference to the size of the $N$-acetylchitooligosaccharides shown by the tomato cells seems to be pretty different from that of the rice cells. This might be explained from the different specificity of the binding proteins, as the specificity of both binding sites detected in each membrane preparation from these plants seem to reflect the characteristics of the cell themselves. Also, in the case of the tomato cells, there is no evidence that the perception of the $N$-acetylchitooligosaccharide elicitor signal is directly coupled to the expression of defense-related genes.

A high-affinity binding protein of the similar size was also identified by photoaffinity labeling in the case of the hepta-$\beta$-glucoside elicitor for soybean (Cosio *et al.* 1992). It would be very interesting to think the possibility of the presence of a family of receptor molecules which have a similar size but different specificity for carbohydrate elicitors.

Further characterization of these binding proteins and the cloning of the corresponding genes will answer such questions. These approaches will also supply critical information to answer whether these binding proteins really represent true, functional receptors.

## Acknowledgment

The authors thank to Mrs. Tomoko Yoshimura for her help in preparing plasma membrane preparation. This work was supported in part by a grant, Enhancement of Center-of-Excellence, the Special Coordination Funds for Promoting Science and Technology, from Science and Technology Agency, Japan.

## References

Basse, C. W., Fath, A. and Boller, T. 1993. High affinity binding of a glycopeptide elicitor to tomato cells and microsomal membranes and displacement by specific glycan suppressors. J. Biol. Chem., 268:14724-14731.

Baureithel, K., Felix, G. and Boller, T. 1994. Specific, high affinity binding of chitin fragments to tomato cells and membranes. J. Biol. Chem., 269:17931-17938.

Côté, F. and Hahn, M. G. 1994. Oligosaccharins: structure and signal transduction . Plant Mol. Biol., 26:1379-1411.

Cheong, J. J. and Hahn, M. G. 1991. A specific, high -affinity binding site for the hepta-β-glucoside elicitor exists in soybean membranes. Plant Cell, 3:137-147.

Cosio, E. G., Frey, T. and Ebel, J. 1992. Identification of a high-affinity binding protein for hepta-β-glucoside phytoalexin elicitor in soybean. Eur. J. Biochem., 204:1115-1123

Felix, G., Regenass, M. and Boller, T. 1993. Specific perception of subnanomolar concentrations of chitin fragments by tomato cells: induction of extracellular alkalinization, changes in protein phospholylation, and establishment of a refractory state. Plant *J*. 4:307-316.

Kuchitsu, K., Kikuyama, M. and Shibuya, N. 1993. *N*-acetylchitooligosccharides, biotic elicitor for phytoalexin production, induce transient membrane depolarization in suspension-cultured rice cells. Protoplasma. 174:79-81.

Kuchitsu, K., Kosaka, T. Shiga and Shibuya, N. 1995. EPR evidence for generation of hydroxyl radical triggered by *N*-acetylchitooligosaccharide elicitor and a protein phosphatase inhibitor in suspension-cultured rice cells. Protoplasma. 188:138-142.

Minami, E., Kuchitsu, K., He, D. Y., Kouchi, H., Midoh, N., Ohtsuki, Y., and Shibuya, N. 1996. Two novel genes rapidly and transiently activated in suspension-cultured rice cells by treatment with *N*-acetylchitoheptaose, a biotic elicitor for phytoalexin production. Plant Cell Physiol. 37:563-567.

Mithöfer, A., Lottspeich, F. and Ebel, J. 1996. One-step purification of the β-glucan elicitor binding protein from soybean (*Glycine max* L.) roots and characterization of an anti-peptide antiserum. FEBS Lett. 381:203-207.

Nojiri, H, Sugimori, M., Yamane, H., Nishimura, Y., Yamada, A., Shibuya, N., Kodama, O., Murofushi, N. and Omori, T. 1996. Involvement of jasmonic acid in elicitor-induced phytoalexin production in suspension-cultured rice cells. Plant Physiol. 110:387-392.

Nürnberger, T., Nennstiel, D., Jabs, T., Sacks, W. R., Hahlbrock, K. and Scheel, D. 1994. High affinity binding of a fungal oligopeptide elicitor to parsley plasma membranes triggers multiple defense responses. Cell. 78:449-460.

Schmidt, W. E. and Ebel, J. 1987. Specific binding of a fungal glucan phytoalexin elicitor to membrane fractions from soybean *Glycine max*. Proc. Natl. Acad. Sci. USA. 84:4117-4121.

Shibuya, N., Kaku, H. Kuchitsu, K. and Maliarik, M. J. 1993. Identificaton of a novel high-affinity binding site for *N*-acetylchitooligosaccharide elicitor in the microsomal membrane fraction from suspension-cultured rice cells. FEBS Lett. 329:75-78.

Yamada, A., Shibuya, N., Kodama, O. and Akatsuka, T. 1993. Induction of phytoalexin formation in suspension-cultured rice cells by *N*-acetylchitooligosaccharides. Biosci. Biotech. Biochem. 57:405-409.

# NIP1, a bifunctional signal molecule from the barley pathogen, *Rhynchosporium secalis*

Marion Fiegen, Angela Gierlich, Hanno Hermann, Volkhart Li, Matthias Rohe, and Wolfgang Knogge

Department of Biochemistry, Max-Planck-Institut für Züchtungsforschung, D-50829 Köln, Germany

Resistance of barley to the leaf scald fungus, *Rhynchosporium secalis*, is governed by a number of major resistance genes [1]. One of these genes, *Rrs1*, that was mapped to chromosome 3 [2], controls the interaction with fungal races carrying the *nip1* gene. Most fungal races virulent on Rrs1 cultivars lack the *nip1* gene while all avirulent races possess this gene [3]. Therefore, the *nip1* gene was presumed to be identical with the avirulence gene, *AvrRrs1*, that is complementary to resistance gene *Rrs1*. The gene product, NIP1, was originally identified as a member of a family of small, secreted necrosis inducing proteins that are toxic to leaves of barley and other cereals as well as of several dicot plants [4].

## NIP1 is the product of an avirulence gene

The *nip1* gene encodes an 82 amino acid pre-protein. Removal of a 22 amino acid putative secretory signal sequence yields a mature protein of 60 amino acids including 10 cysteines [3]. Resistance of barley to *R. secalis* does not involve a hypersensitive response [5]. However, treatment of leaves from Rrs1 barley, but not from rrs1 cultivars, with NIP1 results in the induction of PR protein synthesis in a very similar manner as upon inoculation with an avirulent fungal race [6]. Thus, NIP1 is a race-specific elicitor of plant defense reactions in host cultivars of a specific resistance genotype. This is exactly what was postulated for the product of an avirulence gene by the elicitor/receptor model of the gene-for-gene hypothesis [7] and had earlier been found for avirulence gene products of the tomato pathogen, *Cladosporium fulvum* [8].

Three experimental approaches were used to prove that NIP1 is the product of the fungal avirulence gene, *AvrRrs1*. The purified protein from an avirulent fungal race was mixed with spores of a virulent races prior to inoculation. This physiological complementation caused the inability of the originally virulent race to grow on Rrs1 plants. For genetic complementation and gene replacement studies, a transformation system was established for *R. secalis* using hygromycin B resistance as selective marker [9]. A vector was constructed that contains the bacterial hygromycin B phosphotransferase gene (*hph*) under the control of fungal promoter and terminator sequences [10] along with the *nip1* gene. Transformation of a virulent *R. secalis* race yielded transformants that were no longer capable of growing on Rrs1 plants. Both genetic and physiological complementation demonstrated that the *nip1* gene and its product are sufficient to determine avirulence in concert with resistance gene *Rrs1* [3]. Finally, another vector was constructed where the *nip1* gene was disrupted by inserting the selective marker gene into the *nip1* coding region. Upon transformation of an avirulent fungal race transformants were obtained in which the endogenous functional *nip1* gene was replaced by the non-functional gene through homologous recombination. These transformants were virulent on Rrs1 plants demonstrating that the *nip1* gene is necessary for avirulence determination.

### NIP1 is an unspecific toxin important for virulence expression

Careful analysis of the virulence phenotype obtained with the gene replacement mutants revealed that loss of the functional *nip1* gene does not only enable fungal growth on Rrs1 barley. Interestingly, there is also an impact on fungal development on susceptible plants (rrs1 barley). The mutants exhibited a reduced level of virulence, similar to the phenotype obtained with fungal wild-type races lacking the *nip1* gene. While nip1$^+$ races usually induce large necrotic lesions and little chlorosis on susceptible barley leaves, symptoms of nip1$^-$ wild-type races as well as of the mutants evolve slower and include more extensive chlorotic leaf areas. This indicates that the toxic activity exerted by the product of the *nip1* gene plays an important role in the expression of virulence that cannot be compensated for by other members of the NIP family. It is known that the toxic activity of NIP1is at least in part based on the stimulation of the plasma membrane-localized H$^+$-ATPase in barley, independent of the resistance genotype [11]. Apparently, this does not involve a direct interaction with the membrane enzyme. Therefore, NIP1 appears to be an unspecific toxin that contributes significantly to the killing of plant cells by stimulating the membrane enzyme. As a consequence, affected plant cells die and release nutrients required for pathogen development. However, the presence of the *Rrs1* gene enables barley to recognize the fungus *via* the virulence factor NIP1.

## Both functions of NIP1 appear to be mediated through a single receptor

The dual functions of the NIP1 molecule raise the question whether its activity is mediated through a single or two distinct plant receptors. The current evidence indicating the involvement of only one receptor in both functions is described in the following.

In different fungal races different *nip1* alleles were identified that encode proteins with single amino acid alterations [3]. These proteins were purified and analyzed for their elicitor activity and for their necrosis-inducing and ATPase-stimulating activity. The gene products from all analyzed avirulent fungal races exhibited elicitor and toxic activity while NIP1 from a virulent fungal race as well as from a race whose interaction with barley is not controlled by the resistance gene *Rrs1* were elicitor-inactive and non-toxic. Among the active proteins, those showing the highest elicitor activity were also the strongest necrosis inducers as well as ATPase stimulators, while isoforms with lower elicitor activity were also less toxic. The simplest explanation for this result is to assume a single receptor whose interaction with the NIP1 isoforms differs, thus leading to similar consequences for the observed functions.

In an approach to dissect elicitor and toxic activity of NIP1, 3 peptides were synthesized spanning the entire primary sequence of the molecule. When these peptides, alone or in all possible combinations, were applied onto barley leaves no elicitor activity could be detected 24 h post treatment. However, those combinations containing the central and the C-terminal peptide induced necrosis 4-5 days post treatment. A fourth peptide overlapping the sequence of these two peptides was also toxic without being elicitor-active. These data indicate that the structural constraints for elicitor activity are higher than those for toxicity. When barley leaves were treated with NIP1 in the presence of an excess of the toxic peptides, the detectable elicitor activity was drastically reduced. Again, the simplest explanation for this observation is that the peptides compete with NIP1 for the same binding site.

Currently, NIP1 is expressed in a heterologous system to enable the production of high protein amounts required for direct binding studies and the identification and isolation of the NIP1 receptor. If the working hypothesis of a single receptor is correct, the additional question arises whether this receptor is encoded by the resistance gene, *Rrs1*. If RRS1 is the receptor, different domains must be involved in triggering the signal transduction pathways leading to ATPase stimulation and defense gene activation, the latter being modified and, thus, inactivated in the product of the *rrs1* gene. Alternatively, *Rrs1* may encode a factor distinct from the receptor that couples the perception of the NIP1 signal with defense gene activation in resistant plants.

## Acknowledgements

This work was supported by grant 0136101 A from the Bundesministerium für Forschung und Technologie and by grant KN 225/3 from the Deutsche Forschungsgemeinschaft to W.K.

## References

1. Shipton, W.A., Boyd, W.J.R., and Ali, S.M. 1974. Scald of barley. Rev. Plant Pathol. 53:839-861.
2. Barua, U.M., Chalmers, K.J., Hackett, C.A., Thomas, W.T.B., Powell, W., and Waugh, R. 1993. Identification of RAPD markers linked to a *Rhynchosporium secalis* resistance locus in barley using near-isogenic lines and bulked segregant analysis. Heredity 71:177-184.
3. Rohe, M., Gierlich, A., Hermann, H., Hahn, M., Schmidt, B., Rosahl, S., and Knogge, W. 1995. The race-specific elicitor, NIP1, from the barley pathogen, *Rhynchosporium secalis*, determines avirulence on host plants of the *Rrs1* resistance genotype. EMBO J. 14:4168-4177.
4. Wevelsiep, L., Kogel, K.H., and Knogge, W. 1991. Purification and characterization of peptides from *Rhynchosporium secalis* inducing necrosis in barley. Physiol. Mol. Plant Pathol. 39:471-482.
5. Lehnackers, H. and Knogge, W. 1990. Cytological studies on  the infection of barley cultivars with known resistance genotypes by *Rhynchosporium secalis*. Can. J. Bot. 68:1953- 1961.
6. Hahn, M., Jüngling, S., and Knogge, W. 1993. Cultivar-specific elicitation of barley defense reactions by the phytotoxic peptid NIP1 from *Rhynchosporium secalis*. Molec. Plant-Microbe Interact. 6:745-754.
7. Keen, N.T. 1982. Specific recognition in gene-for-gene host-parasite systems. Adv. Plant Pathol. 1:35-82.
8. De Wit, P.J.G.M. 1995. Fungal avirulence genes and plant resistance genes: Unravelling the molecular basis of gene-for-gene interactions. Adv. Bot. Res. 21:147-185.
9. Rohe, M., Searle, J., Newton, A.C., and Knogge, W. 1996. Transformation of the plant pathogenic fungus, *Rhynchosporium secalis*. Curr. Genet. 29:587-590.
10. Punt, P.J., Oliver, R.P., Dingemanse, M.A., Pouwels, M.A., and Van den Hondel, C.A.M.J.J. 1987. Transformation of *Aspergillus* based on the hygromycin B resistance marker from *E. coli*. Gene 56:117-124.
11. Wevelsiep, L., Rüpping, E., and Knogge, W. 1993. Stimulation of barley plasmalemma $H^+$-ATPase by phytotoxic peptides from the fungal pathogen, *Rhynchosporium secalis*. Plant Physiol.101:297-301.

# The PGIP (Polygalacturonase-Inhibiting Protein) Family: Extracellular Proteins Specialized for Recognition

F. Cervone, G. De Lorenzo, B. Aracri, D. Bellincampi, C. Caprari, A. Devoto, F. Leckie, B. Mattei, L. Nuss, G. Salvi.

Dipartimento di Biologia Vegetale, Università "La Sapienza", Roma, Italy

Recognition of endopolygalacturonase (PG), a factor important for pathogenesis and therefore to be maintained by fungi during evolution for successful parasitism, may be exploited by plants as a clever strategy to establish incompatibility. Modular leucine-rich repeat (LRR) proteins, structurally related to some of the resistance gene products recently characterized, recognize and inhibit fungal PGs. These proteins, named PGIPs, have been found in the cell wall of many dicotyledonous plants including *Arabidopsis* and of some monocotyledonous plants. PGIPs are the only plant LRR proteins for which interaction with their ligands has been demonstrated; therefore the structural basis of the PG-PGIP may help to understand how plants recognize non-self molecules from pathogenic microorganisms. We are studying at the genetic and biochemical level the interaction between PGs from different fungi and PGIPs of *Phaseolus vulgaris* L. (De Lorenzo et al. 1994).

## Endopolygalacturonases and oligogalacturonides

PGs are among the first enzymes secreted when fungi are grown on plant cell wall material *in vitro*; their action on cell walls is sometimes a prerequisite for wall degradation by other enzymes. PGs exhibit a great variety of isoenzymatic forms; redundancy of these enzymes may allow to accomodate pathogenesis in a variety of different conditions and hosts, as well as protect the fungus from losses of pathogenicity. Different forms of PG may differ in terms of stability, specific activity, pH optimum, substrate preference or degradation kinetics, and types of oligosaccharides released (De Lorenzo and Cervone, 1996).

PGs hydrolize homogalacturonan in the plant cell wall to mono- di- and sometimes tri-galacturonic acid, assisting the colonization of the plant tissues and providing nutrients for the fungus during pathogenesis. PGs have also been shown to activate plant defense responses; evidence suggesting that this activation is due to oligogalacturonide fragments released from the plant cell wall has been obtained. Oligogalacturonides of chain length varying between 10 and 14, which are transiently produced by the action of PG on homogalacturonan, are elicitors of defense responses as well as regulators of several developmental processes; shorter oligomers have little or no elicitor activity (Bellincampi et al., 1993,1995,1996). PGIPs modulate the PG activity *in vitro*, and favour the formation of elicitor-active oligogalacturonides (De Lorenzo and Cervone, 1996).

### Modification of His-234 causes inactivation of PG while modification of other His residues causes inability of PG to be recognized by PGIP

We have cloned and characterized the gene encoding PG of *Fusarium moniliforme* (Caprari et al. 1993a). We have also shown that a single PG gene exists in *F. moniliforme* and that the four forms of PG secreted by the fungus into the medium derive from differential glycosylation of a single polypeptide (Caprari et al. 1993b). As a starting point for a systematic study of the structure-function relationships of the enzyme, we have developed a yeast-based expression system for the *F. moniliforme* PG gene and undertaken site-directed mutagenesis studies. Three residues, in a region highly conserved in all the sequences known to encode PGs and likely to represent the catalytic site, were separately mutated: His-234 was mutated into Lys, Ser 237 and Ser 240, into Gly. Replacement of His-234 with Lys abolished the enzymatic activity, indicating that this residue is involved in the catalytic activity of the enzyme. Replacement of either Ser-237 or Ser-240 with Gly reduced the enzymatic activity to 48% and 6%, respectively.

The interaction between the variant enzymes and PGIP of *P. vulgaris* was investigated using a biosensor based on surface plasmon resonance (SPR). The three variant enzymes were still able to interact and bind to PGIP with association constants comparable to that of the wild type enzyme (Caprari et al. 1996). However a photooxidative modification of all His residues caused both inactivation and inability of PG to interact with PGIP suggesting that one or more His residues different from His-234 are involved in the recognition by PGIP.

## PGIPs with distinct specificities are expressed in *P. vulgaris*.

A family of *pgip* genes, likely clustered on chromosome 10, is present in the genome of *P. vulgaris* and the composition of the family is different among the different cultivars of bean (Frediani et al. 1993). A major objective of our laboratory is to know how many PGIP proteins exist, under what circumstances they are expressed and their specificity. A gene encoding PGIP (*pgip-1*) was cloned from *P. vulgaris* (Toubart et al. 1992) Transgenic plants expressing the bean PGIP-1 have been obtained. While PGIP purified from bean, i.e. the sum of the individual PGIP proteins, inhibits PGs from different fungi, PGIP-1 transgenically expressed in tomato and tobacco plants inhibits some of them, but not others. For example, compared to PGIP purified from bean hypocotyls, the transgenically expressed bean PGIP-1 maintains the ability to inhibit PG from *Aspergillus niger*, but does not inhibit PG from *F. moniliforme*.

The discrepancy between the specificity of bean PGIP and of trans-genically expressed PGIP-1 suggests that PGIP of *P. vulgaris* is not a single protein but rather a mixture of different forms indistinguishable in terms of chromathographic and electrophoretic properties. By differential affinity chromathography we have been able to separate two forms of PGIP from *P. vulgaris* extracts, one of which interacted with PGs of both *F. moniliforme* and *A. niger* while the other interacted only with PG of *A. niger*. Moreover, two cDNAs clones, corresponding to *pgip* genes different from *pgip-1*, have now been isolated and sequenced. *Pgip-1, pgip-2* and *pgip-3* encode proteins of the same size and differing only for a limited number of amino acids. Expression of each cloned gene in heterologous systems is being carried out to clarify whether they exhibit distinct abilities to recognize PGs from different fungi.

In conclusion, three *pgip* cDNA clones have been isolated and two PGIPs have been separated, indicating that multiple members of the gene family, encoding PGIP proteins with distinct specificities, are expressed in *P. vulgaris*.

## Transcripts of *pgip* with distinct developmental and signal-induced expression are present in *P. vulgaris*

The *pgip-1* was used for Northern analysis in developing *P. vulgaris* tissues and organs as well as in cultured cells. *Pgip* transcript accumulation, taken as the sum of expression of the individual *pgip* genes, is regulated during development and in response to different environmental cues. For example, *pgip* transcripts are found all over the different organs and tissues of the plant, and accumulate in suspension-cultured cells following addition of elicitor-active oligogalacturonides or fungal glucan, and in bean hypocotyls in response to wounding or treatment with salicylic acid (Bergmann et al. 1994). Accumulation of *pgip* mRNA also occurs in different race-cultivar interactions (either compatible or incompatible) between *Colletotrichum lindemuthianum* and bean. Rapid accumulation of *pgip* mRNA correlates with the appearance of the hypersensitive response in incompatible interactions, while a more delayed increase, coincident with the onset of lesion formation, occurs in compatible interactions (Nuss et al. 1996). In incompatible interactions, the accumulation of the *pgip* mRNA is more intense in epidermal cells proximal to the site of infection and within a layer of parenchymal cells above the schlerenchymatic sheath tissue.

The 5' flanking sequences of the *pgip-1* gene have been cloned, sequenced and analyzed functionally. A *pgip-1*-β-glucuronidase (GUS) gene fusion has been expressed in *Nicotiana tabacum*. High levels of GUS activity have been detected specifically in the epidermal and subepidermal cells of the stigma; low level expression was also observed in vascular tissues and anthers. Activity was wound-inducible in leaf and stem tissues. In leaf or suspension-cultured cells, the promoter did not respond to treatment with oligogalacturonides, fungal glucan, salicylic acid or cryptogein. In intact plants the promoter did not respond to pathogen infection.

Using specific sense and antisense primers which differentially amplify *pgip-1, pgip-2 and pgip-3*, RT-PCR experiments have been performed to analyze the expression of the individual genes in response to several treatments and stimuli. Indeed, *pgip-1* transcripts accumulate in response to wounding, but not to salicylic acid and elicitors, while transcripts of the other genes accumulate in response to other stimuli.

These observations indicate that members of the *pgip* gene family are differentially regulated in *P. vulgaris*.


## PGIP interacts with plasma membrane proteins

The presence of a putative PGIP-like protein of about 100 kD in *P. vulgaris* plasma membrane preparations has been detected by immunoblots with a specific PGIP antibody. By Northern blot analysis, the presence of a 2.9 kb mRNA that hybridizes with a *pgip*-specific probe has also been detected. We have now solubilised and partially purified a plasma membrane protein fraction able to interact with the extracellular PGIP, as detected by SPR measurements. We have not been able yet to associate any activity (for example protein phosphorylation) with the plasma membrane PGIP-interacting protein.

## Conclusions

In plants, the LRR protein structure appears specialized for recognition of non-self molecules in race-cultivar interactions, and possibly in non-host interactions. PGIP, a widespread extracellular LRR protein which recognizes and binds fungal endopolygalacturonases, is evolutionary related to several resistance gene products. Therefore PGIP belongs to a super-family of proteins specialized for recognition of non-self molecules and rejection of pathogens.

The interaction between fungal endopolygalacturonases and PGIPs has the requisites for functioning not only in defense mechanisms, but also in the perception that leads to incompatibility. Both molecules are synthesized very early during an attempted infection and physically interact to give rise to the formation of oligogalacturonides that act as elicitors of several defense responses.

We have shown that a number of *pgip* genes encoding for proteins with distict specificities and differentially regulated during development and by treatment with elicitors are expressed in *P. vulgaris*. We have also evidence suggesting that the interaction PG-PGIP may depend on the presence of single aminoacid residues in one of the two proteins. Finally we have indication that PGIP may interact with plasma membrane proteins, probably to transmit signals inside the cell. All our data argue for an involvement of the PGIP protein family in the resistance of plants to fungi.

## Acknowledgements

Research in our laboratory is supported by the Ministero dello Risorse Agricole e Forestali (MIRAAF) and by the European Community Grant AIR3-CT94-2215.

## Literature cited

Bellincampi D., Salvi G., De Lorenzo G., Cervone F., Marfa' V., Eberhard S., Darvill A. and Albersheim P. 1993. Oligogalacturonides inhibit the formation of roots on tobacco explants. Plant J. 4: 207-213.

Bellincampi D., Forrest R., Nuss L., Salvi G., De Lorenzo G., and Cervone F. 1995. Extracellular accumulation of an auxin-regulated protein in *Phaseolus vulgaris* L. cells is inhibited by oligogalacturonides. J. Plant Physiol. 147: 367-370.

Bellincampi D., Cardarelli M., Zaghi D., Serino G., Salvi G., Gatz C., Cervone F., Altamura M.M., Costantino P. and De Lorenzo G. 1996.

Oligogalacturonides prevent rhizogenesis in *rolB*-transformed tobacco explants by inhibiting auxin-induced expression of the *rolB* gene. The Plant Cell 8: 477-488.

Bergmann C., Ito Y., Singer D., Albersheim P., Darvill A.G., Benhamou N., Nuss L., Salvi G., Cervone F. and De Lorenzo G. 1994. Polygalacturonase-inhibiting protein accumulates in *Phaseolus vulgaris* L. in response to wounding, elicitors and fungal infection. Plant J. 5: 625-634.

Caprari C., Richter A., Bergmann C., Lo Cicero S., Salvi G., Cervone F. and De Lorenzo G. 1993a. Cloning and characterization of the gene encoding the endopolygalacturonase of *Fusarium moniliforme*. Mycol. Res. 97: 497-505.

Caprari C., Bergmann C., Migheli Q., Salvi G., Albersheim P., Darvill A., Cervone F., and De Lorenzo G. 1993b. *Fusarium moniliforme* secretes four endopolygalacturonases derived from a single gene product. Physiol. Mol. Plant Path. 43: 453-462.

Caprari C., Mattei B., Basile M.L., Salvi G., Crescenzi V., De Lorenzo G. and Cervone F. 1996. Mutagenesis of endopolygalacturonase from *Fusarium moniliforme*: hystidine residue 234 is critical for enzymatic and macerating activity and not for PGIP (Polygalacturonase-Inhibiting Protein) binding. Mol. Plant Microbe Inter. (In Press)

De Lorenzo G., Cervone F., Bellincampi D., Caprari C., Clark A.J., Desiderio A., Devoto A., Forrest R., Leckie F., Nuss L. and Salvi G. 1994. Polygalacturonase, PGIP and oligogalacturonides in cell-cell communication. Bioch. Soc. Trans. 22: 394-397.

De Lorenzo G. and Cervone F. 1996. Polygalacturonase inhibiting proteins (PGIPs): their role in specificity and defense against pathogenic fungi. In: Plant-Microbe Interactions, vol. 3. G. Stacey and N. Keen, eds. Chapman & Hill, New York (In Press)

Frediani M., Cremonini R., Salvi G., Caprari C., Desiderio A., D'ovidio R., Cervone F. and De Lorenzo G. 1993. Cytological localization of the *pgip* genes in the embryo suspensor cells of *Phaseolus vulgaris* L. Theor. Appl. Genetics 87: 369-373.

Nuss L., Mahé A., Clark A., Grisvard J., Dron M., Cervone F. and De Lorenzo G. 1996. Differential accumulation of PGIP (Polygalacturonase-inhibiting protein) mRNA in two near-isogenic lines of *Phaseolus vulgaris* L. upon infection with *Colletotrichum lindemuthianum*. Physiol. Mol. Plant Path. 48: 83-89.

Toubart P., Desiderio A., Salvi G., Cervone F., Daroda L., De Lorenzo G., Bergmann C., Darvill A. and Albersheim P. 1992. Cloning and characterization of the gene encoding the endopolygalacturonase-inhibiting protein (PGIP) of *Phaseolus vulgaris* L. Plant J. 2: 367-373.

# Biochemical Characterization of Nod Factor Binding Sites in *Medicago* Roots and Cell Suspension Cultures

Bono J-J.[¶†], Gressent F.[¶], Niebel A.[†], Cullimore J.V.[†], and Ranjeva R.[¶]

[¶]Signaux et Messages Cellulaires chez les Végétaux. UMR CNRS-UPS n°5546 Toulouse, France. [†]Laboratoire de Biologie Moléculaire des Relations Plantes-Microorganismes. UMR CNRS-INRA n°215 Castanet-Tolosan, France.

The establishment of the Legume-*rhizobium* symbiosis involves a signal exchange between the eukaryotic and the prokaryotic partners. The plant root exudes compounds which activate the rhizobial *nod* genes, resulting in the synthesis of lipo-oligosaccharidic signals: the Nod factors. These compounds elicit, at subnanomolar concentrations, a number of biological responses on the host plant, which are similar to those induced by the bacteria (for a review see Dénarié and Cullimore 1993). Studies using bacterial mutants as well as purified Nod factors suggest that these signals may be perceived by specific and possibly multiple receptors (Ardourel et al. 1994) that remain to be identified. Recent progress in the synthesis of radiolabelled Nod factors has permitted experimentations to address directly the question of the existence of binding sites for these signals in plant extracts. We report here our results concerning the preparation of ligands suitable for these studies, their use in the characterization of two binding sites, and we will speculate on the putative roles of these sites in the specific perception of Nod factors.

## Preparation of radiolabelled Nod factors

Since Nod factors are active at very low concentrations it is anticipated that their putative binding sites (i) are not abundant

(ii) display high affinity and (iii) may be specific for precise chemical structures in order to discriminate between different members of the *Rhizobiaceae*. Consequently, the ligand has to be chemically and radiochemically pure with a high specific radioactivity.

Working on the *Medicago-R. meliloti* symbiosis we have employed two types of radiolabelled ligands. Firstly, the major Nod factor of *R. meliloti* NodRm-IV(Ac, S, C16:2) has been chemically synthesized (Nicolaou et al. 1992), and has been labelled in the fatty acid moiety with tritium (Bono et al. 1995). The $^3$H- NodRm factor has been shown to induce root hair deformation in concentrations similar to their biologically produced counterparts. However, the specific radioactivity of the $^3$H-NodRm factor used as ligand was only 56 Ci mmol$^{-1}$ preventing its use for the characterization of high affinity binding sites.

For such a purpose, we have developed an enzymatic method for radiolabelling non-sulphated NodRm factors with $^{35}$S to yield a highly labelled, biologically active ligand. The process involves a two-step procedure consisting of (i) preparing [$^{35}$S] PAPS (3'-phosphoadenosine 5'phosphosulphate), the active sulphate donor by taking advantage on the presence of ATP sulphurylase and APS (adenosine 5'-phosphosulphate) kinase in yeast extracts (ii) using recombinantly expressed rhizobial NodH protein to catalyze the sulphation of NodRm factors using [$^{35}$S] PAPS.

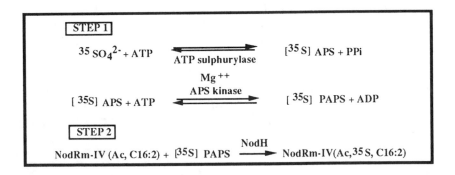

Fig. 1. Enzymatic two step labelling of NodRm-IV(Ac, S, C16:2) with $^{35}$S.

The final yield in our published procedure (Bourdineaud et al. 1995) was only 5-10 %. Recently we have improved the experimental system as follows. We have shown that formation of PAPS, the yield-limiting step, depends upon the relative concentrations of 4 ATP : 1 $Mg^{2+}$. Moreover, as described by Ehrhardt et al. (1995), addition of 5 mM GTP to the assay results in the stimulation of the reaction by a still unknown mechanism. After purification by reverse phase chromatography on a C18 column the specific radioactivity of the $^{35}S$-NodRm factor was 800 Ci mmol$^{-1}$, namely 14 times higher than for the tritium-labelled compound. With the optimized conditions for PAPS synthesis, up to 30 % of the $^{35}S$ from the initial inorganic sulphate was incorporated into the end product.

## Binding of Nod factors to particulate plant extracts

Characterization of NFBS1

$^{3}H$-NodRm-IV(Ac, S, C16:2) binds to crude-root extracts of nitrogen-starved *M. truncatula* and to subfractions of these extracts obtained by differential centrifugations. Most of the binding activity was associated with a "high-density root fraction" that sedimented at 3000 **g**. Binding of the $^{3}H$-NodRm factor to this fraction proceeded in a time-dependent manner and addition of excess unlabelled ligand led to the loss of specifically bound radioactivity. Binding of the ligand was saturable with an affinity of 86 nM and binding characteristics were consistent with a single class of binding sites.

Palmitate or tetra-N-acetyl-chitotetraose, respectively the fatty acyl and carbohydrate moieties of the Nod factors, were very poor competitors suggesting that the binding site was specific to the lipo(chito)oligosaccharidic structure. However, competition with modified Nod factors showed that the binding site was independent of both the O-acetyl and the sulphyl group and did not depend on the unsaturation of the fatty acid. Moreover, the binding activity was identical for roots of *M. truncatula* plants grown under conditions of high nitrogen, that inhibit nodulation, and similar binding sites have been characterized in a non-leguminous plant. Thus, the presence of this non-selective binding site, referred to as NFBS1 for Nod Factor Binding Site 1, does not correlate with the potential to nodulate. Its characteristics are summarized in Table 1.

Tab. 1. Characteristics of Nod Factor Binding Site 1 (NFBS1)

| Subcellular fraction | 3000 g |
|---|---|
| Thermodynamic binding constants | |
| $K_d$ | 86 nM |
| $B_{max}$ | 2 pmol/mg protein |
| Specificity | Specific for lipo-chito-oligo-saccharides but not for Nod factor decorations |
| Expression | Not influenced by nitrogen status<br>Similar binding site in tomato roots |

Characterization of NFBS2

From the above mentioned results, it appears that the thermodynamic characteristics and the expression pattern of NFBS1 are not consistent with what would be expected for a receptor specifically involved in the early events of symbiosis. Moreover, roots are complex organs with different cell types so that it is quite difficult to have large amounts or relatively homogenous material. For these reasons, we have used [35]S NodRm factors to look for binding sites on particulate fractions from *M. varia* cell suspension cultures. In contrast to the root extract from *M. truncatula*, most of the binding sites were associated with the 45 000 **g** fraction (microsomes). Binding was saturable and reversible and was lost when microsomes were pretreated by proteases suggesting that proteins are involved in the process.

Further analysis has shown that two classes of binding sites, differing by their affinity for Nod factors, coexist in microsomal fractions from *M. varia* cell suspension cultures. One of these sites is of the "low affinity" type ($K_d$=72 nM) similar to that of NFBS1 ($K_d$=86 nM). The other presents higher affinity for NodRm factor ($K_d$=1.9 nM, i.e more than one order of magnitude higher than for NFBS1) and is termed NFBS2.

At present, the selectivity of NFBS2 with regard to differently substituted Nod factors has not been established. However, in contrast to NFBS1 which has been shown to exist in non leguminous plants, attempts to characterize high affinity binding sites in carrot cell suspension cultures were unsuccessful, suggesting that their distribution among plants couldbe more limited.

## Discussion

By challenging particulate fractions from plant extracts with radiolabelled Nod factors we have been able to characterize binding sites specific to lipo chito-oligosaccharidic compounds. At least two classes of binding sites have been identified.

NFBS1, a low affinity binding site, was associated mainly with a "high density fraction" from roots of *M. truncatula* . A site with a similar affinity seems to occur in the microsomal fraction from cell suspension cultures of *M. varia*. The results obtained with root extracts suggest that NFBS1 is unlikely to be directly involved in symbiosis because of its low affinity and low selectivity. However, it might be a general lipo-chito-oligosaccharide binding site since several lines of evidence suggest that Nod-like factors may be perceived by non-leguminous plant and may play a more general role than symbiosis (De Jong et al., 1993; Röhrig et al., 1995).

The second class of binding sites has been characterized in cell suspension cultures only, and its existence in root extracts has to be confirmed. Regarding its higher affinity, this binding site termed NFBS2, may mediate the biological effects of Nod factors at subnanomolar concentrations. However, its specificity remains to be determined in order to assess its potential role in the early steps of symbiosis.

## Literature cited

Ardourel, M., Demont, N., Debellé, F., Maillet, F., De Billy, F., Promé, J-C., Dénarié, J., and Truchet, G. 1994.
*Rhizobium meliloti* lipooligosaccharide nodulation factors: different structural requirements for bacterial entry into target root hair cells and induction of plant symbiotic developmental

responses. Plant. Cell. 6: 1357-1374.

Bono, J-J., Riond, J., Nicolaou, K. C., Bockovich, N. J., Estevez, V.A., Cullimore, J. V., and Ranjeva, R. 1995. Characterization of a binding site for chemically synthetized lipo-oliligosaccharidic NodRm factors in particulate fraction prepared from roots. Plant. J. 7: 253-260.

Bourdineaud, J-P., Bono, J-J., Ranjeva, R. and Cullimore, J. V.1995.Enzymatic radiolabelling to a high specific activity of legume lipo-oligosaccharidic nodulation factors from *Rhizobium meliloti* . Biochem. J. 306: 259-264.

De Jong, A. J., Heidstra, R., and Spaink, H. P. 1993. *Rhizobium* lipooligosaccharides rescue a carrot somatic embryo mutant.Plant. Cell. 5: 615-620.

Dénarié, J. and Cullimore, J. V. 1993. Lipo-oligosaccharidic nodulation factors: a new class of signalling molecules mediating recognition and morphogenesis. Cell. 74: 951-954.

Ehrhardt, D. W., Atkinson, E. M., Faull, K. F., Freedberg, D. I., Sutherlin, D.P., Armstrong, R., and Long, S. R. 1995. *In vitro* sulfotransferase activity of NodH, a nodulation protein of *Rhizobium meliloti* required for host-specific nodulation. J. Bacteriol. 177: 6237-6245.

Nicolaou, K. C., Bockovitch, N. J., Carcanague, D. R., Hummel, C.W. and Even, L. F. 1992. Total synthesis of the NodRm-IV factors the *Rhizobium* nodulation signals. J. Am. Chem. Soc. 114: 8701-8702.

Röhrig, H., Schmidt, J., Walden, R., Czaja, I., Miklasevics, E., Wieneke, U., Schell, J. and John, M. 1995. Growth of tobacco protoplasts stimulated by synthetic lipo-chitooligosaccharides. Science. 269: 841-843.

# DO LEGUME VEGETATIVE TISSUE LECTINS PLAY ROLES IN PLANT-MICROBIAL INTERACTIONS?

Marilynn E. Etzler and Judith B. Murphy
Section of Molecular and Cellular Biology
University of California, Davis, CA, USA

The variety of carbohydrate specificities found among lectins of different plant species led to early speculation that these carbohydrate-binding proteins may function in specific plant-microbe interactions (for review see Etzler, 1986). A variety of roles have been proposed for these lectins, including their involvement in defense and as determinants of the host-strain specificity of rhizobium-legume. The possibility that lectins may play such roles was strengthened by the identification of specific oligosaccharide signals involved in the defense response (for review see Ryan, 1994) and the lipooligosaccharidic Nod factors produced by Rhizobia that elicit the formation of nodules (for review see Denairie and Cullimore, 1993). Should proteins serve as the receptors for these oligosaccharide signals, by definition they would be classified as lectins.

To date, most studies on the role lectins may play in the above processes have focused on the seed lectins of legumes. These carbohydrate-binding proteins have been found to constitute up to 10% of the total soluble protein in the cotyledons (Etzler, 1986), and a large amount of information is available on their carbohydrate specificities and structures (Goldstein and Poretz, 1986). Yet, in a number of legume species these particular lectins appear to be confined to the cotyledons and are not present at the time or place at which many plant-microbe interactions occur (Etzler, 1986).

Our studies on the legume, *Dolichos biflorus*, have shown that this plant contains a family of lectins, encoded by separate genes that are differentially expressed both spatially and temporally (Harada, et al., 1990; Etzler, 1992). A comparison of the properties of some of these lectins and their differential distribution suggest that they may be adapted to perform different role(s) in different tissues. The properties of some of these lectins are discussed below in context with the possibility of their involvement in plant-microbial interactions.

### Isolation and molecular properties of vegetative tissue lectins

In contrast to the high amounts of lectin found in the seeds of legumes, lectins normally constitute only a small percentage (< .05%) of the soluble protein extracted from the vegetative tissues (Etzler, 1992). We routinely isolate these lectins from the vegetative tissues of *Dolichos biflorus* by chromatography of tissue extracts on an affinity resin constructed by coupling hog gastric mucin blood group A + H substance to CNBr-activated Sepharose 4B. In recent years we have increased the capacity of this resin by digesting the blood group substance with pronase before coupling to the resin (Etzler, 1994a). The oligosaccharide heterogeneity of this resin enables the binding of lectins with a wide range of carbohydrate specificities. The vegetative tissue lectins show different salt sensitivities and can be eluted from the column by stepwise increases in NaCl concentration. This salt sensitivity of the vegetative tissue lectins distinguishes them from the blood group A specific seed lectin of this plant, which remains bound to the column at NaCl concentrations higher than 0.5 M, but can be eluted from the column with N-acetyl-D-galactosamine, the blood type A immunodominant sugar (Etzler and Kabat, 1970).

A comparison of properties of two of these *Dolichos biflorus* vegetative tissue lectins with the seed lectin from this plant is shown in Table 1. The names of these lectins are derived from their molecular weights. The stem and leaf lectin, DB58, is a heterodimer (Etzler, 1994b) that was first detected by its cross reaction with antibodies to the seed lectin (Talbot and Etzler, 1978). These lectins are encoded by genes located within 3 kilobases of one another; a high degree of homology in nucleotide sequence and the identical transcriptional orientation of these genes suggests that they evolved by gene duplication (Harada, et al., 1990). The amino acid sequences of these proteins show 88% identity; each of these proteins is synthesized as a single prepeptide, which undergoes further posttranslational alteration after the removal of the signal sequence (Etzler, 1992). One of these posttranslational alterations is the removal of a short segment of amino acids from the carboxyl termini; this modification occurs with only about half of the peptide chains and results in the creation of heteroligomers composed of approximately equal amounts of truncated and nontruncated subunits (Etzler, 1992, 1994b). Although the biological significance of this modification is not yet known, it is of interest that in the case of the seed lectin, carbohydrate binding has been found to be associated only with the nontruncated subunit (Subunit I), resulting in two carbohydrate binding sites per tetramer (Etzler, 1992).

Although neither the seed lectin nor DB58 have been detected in the roots of the plant, a lectin, named DB46, is present in this tissue (Quinn and Etzler, 1987). Small amounts of this lectin are also present in the stems and leaves, although the stem/leaf DB46 shows some differences from the root lectin and may be an altered molecular form of the lectin (Quinn and Etzler, 1987). A lectin, DB57, that cross reacts with antibodies against DB46 is present in the stems and leaves. Both DB46 and DB57 have been

Table 1. Comparison of *Dolichos biflorus* lectins.

|                         | SEED LECTIN          | DB58              | DB46                  |
|-------------------------|----------------------|-------------------|-----------------------|
| Affinity column eluant  | 10 mM GalNAc         | 0.15 M NaCl       | 0.5 M NaCl            |
| Molecular weight        | 110,000              | 58,000            | 46,000                |
| Subunit Structure       | Tetramer ($I_2II_2$) | Dimer ($\alpha\beta$) | Monomer           |
| Localization            | Cotyledons           | Stem/leaves       | Roots Stem/leaves     |

found to exist primarily as monomeric proteins in solution although some higher aggregates of these proteins are detected (Quinn and Etzler, 1987; Etzler, 1992). The complete cDNA sequence of DB46 has recently been determined, and this sequence shows no significant homology with the cDNA of any plant or animal lectin yet described (Etzler, et al., submitted for publication).

## Activities of vegetative tissue lectins

The vegetative lectins from *Dolichos biflorus* appear to be monovalent in that they do not agglutinate erythrocytes nor precipitate with multivalent ligands (Etzler, 1992). We have recently established solid phase assays to explore the carbohydrate binding properties of these lectins (Etzler, 1994a). Despite their similarity in amino acid sequence, the carbohydrate binding properties of DB58 and the seed lectin are quite different. Whereas the seed lectin recognizes terminal nonreducing $\alpha$-linked N-acetyl-D-galactosamine residues, DB58 shows no significant binding with any common monosaccharide and seems to prefer a larger oligosaccharide determinant of the blood group substance (Etzler, 1994a). These lectins also differ from one another in sensitivity to salt and to urea. Both lectins are metalloproteins and are inactive in chelating agents that remove divalent metal ions.

In addition to their carbohydrate binding properties, both the seed lectin and DB58 have hydrophobic sites that bind with high affinity to adenine and the cytokinin class of plant hormones (Gegg, et al., 1992). In contrast to the differences in carbohydrate specificity found between the seed lectin and DB58, the hydrophobic binding sites of these two lectins have been shown to have similar specificities (Gegg, et al., 1992). No interaction has as yet been found between the carbohydrate binding and hydrophobic binding sites of these lectins.

Although DB46 binds to hog blood group A + H substance, we have recently found that this lectin appears to recognize a site other than the blood group specific determinants and binds well with the chitin tetrasaccharide. This protein has also been found to be a nucleotide phosphatase (Etzler, et al., submitted for publication).

### Implications on studies of lectins in plant-microbe interactions

The tissue specific distribution and differences in properties among the above *Dolichos biflorus* lectins suggest that these individual carbohydrate-binding proteins may play different roles in the plant. It is thus important to focus on the appropriate lectin when examining its possible involvement in specific plant-microbe interactions.

For example, considerable effort has been expended on exploring the possibility that the conventional seed lectins may be involved in Rhizobium-legume symbiosis. Yet, only trace amounts of these lectins have been reported in the roots of a few legume species and no detectable levels have been found in others (Etzler, 1986). Although clover hairy roots transformed with cDNA encoding the pea seed lectin were nodulated with a rhizobial strain that normally nodulates peas (Diaz, et al., 1989), the properties of this lectin should be considered in interpreting these results. The pea seed lectin has never been shown to bind to the Nod factors produced by the rhizobia that nodulate pea nor clover, and all existing information on the pea seed lectin specificity (Goldstein and Poretz, 1986) predicts that it will not recognize this structure. Furthermore, the chromosomal location of the gene encoding this lectin has recently been mapped to a site different than the position of mutations that result in altered strain specificity (Lu, et al., 1994). It thus appears that the pea seed lectin is acting in some other role than as a Nod factor receptor. In addition to the *Dolichos biflorus* seed lectin described above, a number of other legume seed lectins have been found to bind to cytokinins (Roberts, et al., 1983), raising the possibility that the effects of the seed lectin on nodulation obtained in the above and other transgenic studies may be due to cytokinin binding properties of the lectins. Indeed, a cytokinin elevation has been found to induce at least one of the early plant response genes (Dehio and deBrujn, 1992) as well as to suppress the Nod⁻ phenotype of nodulation mutants (Cooper and Long, 1994).

Although we do not yet know the role of any of the *Dolichos biflorus* lectins described above, the properties of some of these proteins suggest the possibility that they may be involved in plant-microbe interactions. For example, the specificity of DB46 for chitin oligosaccharides makes it a possible candidate for a Nod factor receptor or for recognizing the chitooligosaccharide signals produced upon microbial infection. The nucleotide phosphatase activity associated with this lectin raises the possibility that it may function as a signal transducer. Recent studies on DB58 show that it is under similar regulatory controls as other defense proteins in the plant (Bunker and Etzler, work in progress), and the elevation of DB57 levels upon heat shock (Etzler, 1992) suggests the possible involvement of this lectin in defense. A combination of biochemical and transgenic studies on these lectins in progress in our laboratory should yield further information on the role of these vegetative tissue lectins in the plant.

## Acknowledgments

The research discussed in this paper was supported by NIH Grant GM21882 and NSF Grant DM82-15758.

## Literature Cited

Cooper, J.B. and Long, S.R. 1994. Morphogenetic rescue of *Rhizobium meliloti* nodulation mutants by *trans*-zeatin secretion. Plant Cell 6: 215-225.

Dehio, C. and de Bruijn, F.J. 1992. The early nodulin gene SrENOD2 from *Sebastiana rostrata* is inducible by cytokinin. Plant J. 2: 117-128.

Denairie, J. and Cullimore, J. 1993. Lipo-oligosaccharide nodulation factors: A new class of signalling molecules mediating recognition and morphogenesis. Cell 74: 951-954.

Diaz, C.L., L.S. Melchers, P.J.J. Hooykass, B.J.J. Lugtenberg and Kijne, J.W. 1989. Root lectin as a determinant of host-plant specificity in the *Rhizobium*-legume symbiosis. Nature 338: 579-581.

Etzler, M.E. 1986. Distribution and function of plant lectins. Pages 371-435 in: The Lectins. Properties, Functions and Applications in Biology and Medicine. I.E. Liener, N. Sharon and I.J. Goldstein, eds. Academic Press, Inc., New York.

Etzler, M.E. 1992. Plant lectins: Molecular biology, synthesis and function. Pages 521-539 in: Glycoconjugates: Composition, Structure, and Function. H.J. Allen and E.C. Kisailus, eds. Marcel-Dekker, Inc., NY.

Etzler, M.E. 1994a. A comparison of the carbohydrate binding properties of two *Dolichos biflorus* lectins. Glycoconj. J. 11: 395-399.

Etzler, M.E. 1994b. Isolation and characterization of subunits of DB58, a lectin from the stems and leaves of *Dolichos biflorus*. Biochemistry 33: 9778-9783.

Etzler, M.E. and Kabat, E.A. 1970. Purification and characterization of a lectin (plant hemagglutinin) with blood group A specificity. Biochemistry 9: 869-877.

Gegg, C.V., D.D. Roberts, I.H. Segel and Etzler, M.E. 1992. Characterization of the adenine binding sites of two *Dolichos biflorus* lectins. Biochemistry 31: 6938-6942.

Goldstein, I.J. and Poretz, R.D. 1986. Isolation, physicochemical characterization, and carbohydrate-binding specificity of lectins. Pages 33-247 in : The Lectins. Properties, Functions and Applications in Biology and Medicine. I.E. Liener, N. Sharon and I.J. Goldstein, eds. Academic Press, Inc., New York,

Harada, J.J., J.P. Spadoro-Tank, J.C. Maxwell, D.J. Schnell and Etzler, M.E. 1990. Two lectin genes differentially expressed in *Dolichos biflorus* differ primarily by a 116 base pair sequence in their 5' flanking regions. J. Biol. Chem. 265: 4997-5001.

Lu, J., N.F. Weeden and LaRue, T.A. 1994. Chromosomal location of lectin genes indicates they are not the basis of *Rhizobium* strain specificity mutations identified in pea (*Pisum sativum L.*). J. Hered. 85: 179-182.

Quinn, J.M. and Etzler , M.E. 1987. Isolation and characterization of a lectin from the roots of *Dolichos biflorus*. Arch. Biochem. Biophys. 258: 535-544.

Roberts, D.D. and Goldstein, I.J. 1983. Binding of adenine and cytokinins to lima bean and other legume lectins. Pages 131-141 in: Chemical Taxonomy, Molecular Biology and Function of Plant Lectins. Goldstein, I.J. and Etzler, M.E., eds. Alan R. Liss, NY.

Roberts, D.M. and Etzler, M.E. 1984. Development and distribution of a lectin from the stems and leaves of *Dolichos biflorus*. Plant Physiol. 76:879-884.

Ryan, C.A. 1994. Oligosaccharide signals: From plant defense to parasite offense. Proc. Natl. Acad. Sci. (USA) 91: 1-2.

Schnell, D.J. and Etzler, M.E. 1988. cDNA cloning, primary structure, and *in vitro* biosynthesis of the DB58 lectin from *Dolichos biflorus*. J. Biol. Chem. 263: 14648-14653.

Talbot, C.F. and Etzler, M.E. 1978. Development and distribution of *Dolichos biflorus* lectin as measured by radioimmunoassay. Plant Physiol. 61: 847-850.

# *Agrobacterium*: A natural genetic engineer exploited for plant biotechnology

Eugene W. Nester, John Kemner, Wanyin Deng, Yong-Woog Lee,
Karla Fullner, Xiaoyou Liang, Shen Pan, and Joe Don Heath

Department of Microbiology, University of Washington,
Seattle, Washington 98195-7242, USA

## Introduction

The interaction between *Agrobacterium* and plants represents the only documented case in nature in which prokaryotic DNA is transferred, integrated, and expressed in a eukaryotic genome. As such, studies on this case of natural genetic engineering have attracted considerable interest. In this review, we will focus on the major features of this gene transfer process but touch only lightly, if at all, on subject material that will be covered in more detail in succeeding papers of this volume. Further, we will discuss important new reports in the genetic engineering of certain members of the *Gramineae* which should allow *Agrobacterium* to become even more valuable as a vector for the genetic engineering of higher plants.

## Overview

The overall process of the gene transfer process from *Agrobacterium* to plants is shown in Fig. 1. In a soil environment *Agrobacterium* is attracted to wound sites on higher plants by plant metabolites such as sugar precursors of plant cell walls which act as chemoattractants. Phenolic compounds, such as acetosyringone (AS) which serve as inducers of the virulence (*vir*) genes, are inactive as chemoattractants. Another early step in the interaction of *Agrobacterium* with plant cells is the attachment of the bacteria to the plant cells, a process that is poorly understood. Mutants have been isolated which are defective in attachment and are avirulent (Douglas et al. 1985; Thomashow et al. 1987). All of these mutants involve the synthesis of β-1,2-glucan or its transfer into the periplasm.

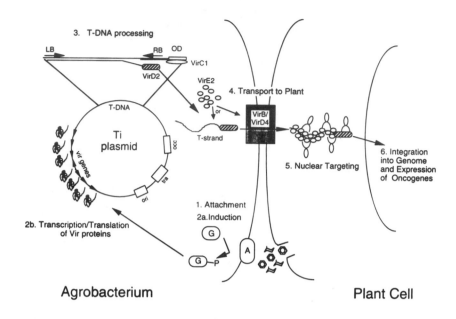

Fig. 1.  Gene trasnfer process from *Agrobacterium* to plants

However, it is not at all clear how this molecule is involved in the
attachment process.  It has not been possible to demonstrate that β-1,2
glucan or the supernatant of wild type cells can complement mutants
deficient in β-1,2 synthesis.  The plant receptor for *Agrobacterium* has been
reported to be a vitronectin-like molecule  (Wagner and Matthysse 1992).

TI-PLASMID ENCODED *VIR* GENES

The sugar monomers which are precursors of the plant cell wall and
phenolic compounds which are related to precursors of lignin, in an acidic
environment, serve to induce the virulence (*vir*) genes which are required for
the processing and transfer of the plasmid DNA (T-DNA).  Both molecules
as well as the acidic environment, are sensed by *Agrobacterium* by means of
a two component system consisting of the sensor molecule, a product of the
*virA* gene and VirG, the response regulator.  The VirA protein appears to
sense the phenolic compound directly (Lee et al. 1995) but the sugars
interact with a periplasmic binding protein which in turn interacts with
VirA (Cangelosi et al. 1990a; Shimoda et al. 1990).  (See later discussion.)
The phosphorylated form of VirG interacts with upstream regions of each of

the *vir* genes to transcriptionally activate each of them (Jin et al. 1990). The *virD* operon plays an important role in the early events in the processing of the T-DNA. VirD1 and VirD2 serve as an endonuclease which cleaves the bottom strand of the T-DNA at the right and left borders (Yanofsky aet al. 1986), and then remains covalently attached to the 5' end of the T-DNA, serving as a pilot protein (Young and Nester 1988). Following strand displacement of the single stranded T-DNA, the T-strand with its pilot protein associates with the transport mechanism by which the DNA is transferred into plant cells. The other members of the *virD* operon include *virD3* which apparently is not required for processing or transfer as well as *virD4*, whose protein product forms part of the transfer complex. The fact that knock out mutations in *virD3* have no effect on tumor formation, however, may mean only that another copy of this gene exists in the bacterial chromosome.

It was initially believed that inside the bacterial cell, the T- strand was coated by 60 molecules of the VirE2 protein, a single stranded DNA binding protein which nonspecifically interacts with single stranded DNA (Citovsky et al. 1988). However, this conclusion was brought into question when it was shown that a transgenic plant synthesizing the VirE2 could complement a *virE2 Agrobacterium* mutant (Citovsky et al. 1992). It has been further shown that the VirE2 protein in conjunction with functional protein VirE1 can be transferred into the plant cell independent of the T-strand through the same transport machinery as the T-strand (Sundberg et al. 1996). This transport machinery consists of the 11 proteins coded by the *virB* operon as well as *virD4*. These gene products are homologous to those used in the transport of wide host range plasmids in other Gram negative bacteria as well as the transfer of the protein toxin of *Bordetella pertussis* (Winans et al. 1996) and *Helicobacter pylori* (Tummuru et al. 1995). One of the *virB* genes, *virB2*, shows a weak similarity to the pilin subunit gene of the F plasmid of *E. coli*. Recently it has been shown that the *virB* operon and *virD4* are required for the synthesis of a pilus (Fullner et al. 1996). Although it seemed likely that this would be the case based upon sequence homology with other known pili systems, it had not been shown directly (Shirasu and Kado 1993). One of the secrets to demonstrating these pili was the recent observation that *Agrobacterium* grown at 28°C, a temperature which is optimal for the growth of the bacteria, transfers a broad host range plasmid very poorly into other bacterial cells, apparently because of a defective transfer apparatus (Fullner and Nester 1996). In contrast, cells grown at 19°C transfer the plasmid very efficiently. The transfer of the Ti plasmid into plant cells requires the same Ti plasmid encoded *vir* genes, as does the transfer of the broad host range plasmid between cells of *Agrobacterium*, suggesting that the mechanism of transfer is the same (Beijersbergen et al. 1992).

Once inside the plant cell, the T-strand is targeted to the plant cell nucleus. Both VirE2 and VirD2 contain nuclear targeting sequences and both seem to be necessary for optimal targeting into the plant cell nucleus

(Citovsky et al. 1994). The integration of the T-strand into the plant genome will be covered in more detail in other reports in this volume.

Different T-DNA's encode the synthesis of different opines, a class of compounds which *Agrobacterium* can use as a source of carbon, nitrogen, and energy. Strains are frequently referred to by the opine synthesized by the crown gall tumor they induce. Two common opines are octopine and nopaline and the Ti plasmids from octopine and nopaline strains differ in their *vir* gene composition to some extent. The Ti plasmids from both strains encode *virA, G, B, C, D,* and *E*. However, only the nopaline strain contains the *tzs* locus which encodes a cytokinin whose synthesis is under the control of the plant inducing molecules acetosyringone (AS) and sugar monomers. At a comparable position on the Ti plasmid, the octopine strains contain two copies of a gene concerned with the synthesis of a P450 enzyme. The function of these two different genes is not known but some evidence suggests that the *tzs* locus may extend the host range of *Agrobacterium* (Zhan et al. 1990) and the P450 enzyme may detoxify some inhibitory compounds of the plant cell environment. Another interesting gene is *virF*, a gene found in octopine but not nopaline strains. This gene appears to be a host range determinant since a *virF⁻* mutation alters the host range of the *Agrobacterium* strain. Like VirE2, the VirF protein apparently can enter and function within the plant cell. Another *vir* gene is also intriguing in terms of its location and function. *virJ* maps to the Ti plasmid of octopine but not nopaline strains. A similar gene, in terms of function, maps to the chromosome of both strains of *Agrobacterium*. In

**Table 1.** Ti plasmid encoded *vir* genes of an octopine strain

| virH | virA | virB | virG | virC | virD | virE | virF |
|------|------|------|------|------|------|------|------|
| ⟶ | ⟶ | ⟶ | ⟶ | ⟵ | ⟶ | ⟶ | ⟶ |

| vir | Induci-bility | Size kb | ORF's | Function |
|-----|---------------|---------|-------|----------|
| A | + | 2.8 | 1 | Plant signal sensor |
| G | + | 1.0 | 1 | Transcriptional activator |
| B | + | 9.5 | 11 | Transport of T-DNA |
| D | + | 4.5 | 4 | Processing of T-DNA, endonuclease, nuclear targeting |
| C | + | 1.5 | 2 | Processing of T-DNA |
| E | + | 2.2 | 2 | S.S. DNA binding protein, nuclear targeting |
| H | + | 3.4 | 2 | Cytochrome P450 enzyme |
| F | + | 0.6 | 1 | Host range determinant |

nopaline strains, avirulent mutants have been isolated by mutating the chromosomally encoded *virJ* gene (named *acvB*). The precise function of this gene is not certain, but a mutation in *virJ* in a strain with a defective *acvB* gene results in a transfer defective strain (Pan et al. 1995). The *vir* encoded genes with their functions are summarized in Table 1.

CHROMOSOMALLY ENCODED VIRULENCE GENES (*CHV* GENES)

Although investigators have focused on the role of the Ti plasmid in the *Agrobacterium*-plant interaction, numerous chromosomal genes have been shown to play important roles in the ability of *Agrobacterium* to transform plants. In general, the functions of these genes have not been elucidated and mutations in these genes are pleiotropic. It appears that whereas *vir* genes on the Ti plasmid are dedicated solely to the various steps in plant cell transformation, the chromosomal virulence genes play important roles in the physiology of *Agrobacterium* and only ancillary but important roles in the interaction of *Agrobacterium* with plant cells. An understanding of their function has contributed to our understanding of the steps required for processing and transfer of the T-DNA into plants.

*chvA, B and exoC.* A number of chromosomal loci are important in attachment of *Agrobacterium* to plant cells. Mutations in *chvA, B,* and *exoC* (*pscA*) (Marks et al. 1987) result in an avirulent phenotype which apparently results from the inability of the mutants to attach to plant cells. All three loci are concerned with either the synthesis or the transport of β-1,2 glucan. The mutations are pleiotropic and may also be involved in equalizing the osmotic pressure between the inside and outside of the cell. Cells defective in β-1,2 glucan synthesis or transport do not grow in a hypotonic medium (Cangelosi et al. 1990b). Other mutants (termed att⁻) are also defective in attachment of *Agrobacterium* to plant cells (Matthysse 1987).

*ChvG-I.* This is another two component regulatory system consisting of the sensor protein ChvG, and the response regulator ChvI. Null mutations in either of these genes result in the cells which are unable to induce the *vir* genes or to grow at a pH of 5.5. The molecular basis for this phenotype is not known. We do not know the signal to which the ChvG protein responds or the gene(s) which ChvI controls. It is of interest that a partial mutation in ChvG results in approximately 75% of normal *vir* gene induction. However, these mutants are avirulent which suggests that it is not the lack of *vir* gene induction which is responsible for the avirulent phenotype. Another mutation was isolated in which the *vir* genes are not inducible (Metts et al. 1991). This mutation, *ivr211*, was mapped in a gene concerned with sugar uptake. The molecular basis of this phenotype is not known.

*chvE.* This gene codes for a periplasmic protein which interacts with a wide variety of monosaccharides all of which are components of the plant cell wall. This periplasmic binding protein with its associated sugar interacts with the periplasmic domain of the VirA protein which is anchored in the cytoplasmic membrane. This interaction puts the VirA protein in a confirmation that allows it to interact with the phenolic inducer, AS. Mutations in *chvE* result in a strain which is avirulent on some, but not all, plants. Such mutants are also defective in chemotaxis towards these sugars thereby strongly suggesting that *chvE* can also interact with a membrane protein involved in chemotaxis to the various sugars. It has not been possible to demonstrate that the same mutants are defective in the growth on monosaccharides which induce the *vir* genes. However, it seems likely that the *chvE* gene product interacts with the uptake system required for the uptake of the sugars. The *chvE* gene provides a good example of a locus which plays dual roles, one in the physiology of *Agrobacterium* growing in the absence of plant cells, and the other in the functioning of a key step (signal transduction) required for plant cell transformation.

YEAST TRANSFORMATION

Two groups have now reported that *Agrobacterium* can transfer its T-DNA into yeast where appropriate constructs of the T-DNA can either replicate or integrate via homologous recombination into the yeast chromosome (Piers et al. 1996; Bundock et al. 1996). As discussed in greater detail in another paper in this volume, the data on integration strongly suggest that the enzymes involved in integration of the T-DNA are encoded by the host cells and not by the *vir* genes of *Agrobacterium*. The transfer of T-DNA into yeast cells requires *vir* gene activation, as well as the participation of the same *vir* genes that are required for the processing and transfer of T-DNA into plant cells. However, the *chv* genes concerned with the attachment of *Agrobacterium* to higher plant cells are not required (Piers et al. 1996). These data suggest that specificity exists in the attachment of *Agrobacterium* to plant cells which is not required in the attachment to yeast cells. There is no evidence that *Agrobacterium* can transfer its T-DNA into other bacteria such as *Agrobacterium* or *E. coli* (X. Liang, unpub. observ.). However, the entire Ti plasmid can be transferred via conjugation into other cells of *Agrobacterium*. This process is under the control of opines excreted by plant cell tumors which in turn activate the synthesis of an auto inducer, a homoserine lactone. This control system, termed quorum sensing (Fuqua et al. 1994), is found in a wide variety of different plant associated bacteria and is covered in more detail in a subsequent paper in this volume.

Independent of this mechanism of Ti plasmid transfer, which involves the transfer genes of the Ti plasmid, wide host range plasmids can be mobilized into *Agrobacterium* if the entire *virB* operon as well as *virD4* is functional (Binns et al. 1995; Fullner et al. 1996). This rapid assay for the study of certain *vir* gene functions should be useful since the transfer of wide host range plasmids into cells of *Agrobacterium* and the transfer of the T-DNA into higher plants have the same Ti plasmid encoded requirements. It should be possible to analyze this *Agrobacterium - Agrobacterium* transfer system more readily, since it may be possible to isolate suppresser mutations in the individual genes (*virB1-11* and *virD4*), all of which are necessary for the formation of the transfer pilus.

## TRANSFORMATION OF MONOCOTS - RICE, MAIZE, AND RYE

Within the past two years, a group of Japanese investigators has provided convincing data that *Agrobacterium* is capable of transforming rice and maize at a very high efficiency (Hiei et al. 1994; Ishida et al. 1996). In both cases, stable integration, expression and transfer to progeny of *trans* genes could be shown using molecular and genetic analysis of transformants. The boundaries of the T-DNA of transgenic rice plants as well as transgenic maize were similar to those in transgenic dicotyledonous plants. In the case of rice, the highest levels of transformation were obtained with rice varieties (*japonica*) that were most amenable to tissue culture manipulations and regeneration into whole plants. Stable integration required cocultivation of rice explants with *Agrobacterium* in the presence of acetosyringone and glucose. A strain of *Agrobacterium* containing the superbinary vector genes from a strain of *Agrobacterium*, Bo542, which is especially virulent on a wide variety of recalcitrant plants, proved to be a suitable vector. The use of the superbinary vector was especially critical for varieties of rice which grew poorly in tissue culture. A large number of different rice explants were used and the calli induced from scutella proved to be especially good as starting material. Both the *japonica* and *indica* rice varieties have been stably transformed although the former varieties are much more susceptible in general. Whether all *indica* varieties can be stably transformed and transgenic plants regenerated is not yet clear. Many of these varieties are especially difficult to regenerate and grow in tissue culture.

In the case of maize, the inbred line A188 was transformed from immature embryos cocultivated with *Agrobacterium* that carry the superbinary vector. The frequencies of transformation were between 5 and 30% and almost all transformants were normal in morphology and more than 70% were fertile. Attempts to transform other varieties of maize have not yet been possible. Whether this will require the construction of more efficient transforming vectors of *Agrobacterium* or better regeneration technology for the maize is unclear. What these two reports do prove however, is that the host range of *Agrobacterium* is far broader than many investigators believed. Indeed, at this meeting, a poster was presented describing experiments that showed that rye grasses (*Lolium perenne* and *L. multiflorum*) could be transformed using *Agrobacterium* (Poster M30). In this case the vector was based upon the Bo542 *vir* genes and embryo-derived cell suspension cultures were used as the starting material. Currently, these investigators are following the inheritance and segregation of the transgenes in subsequent generations. However, it certainly does appear that *Agrobacterium* will play an increasingly prominent role in gene transfer into plants which many investigators had previously considered resistant to infection by *Agrobacterium*. The surprises continue!

ACKNOWLEDGMENTS

The research conducted by the investigators in Seattle was supported by NIH and PMIF.

LITERATURE CITED

Beijersbergen, A., Dulk-Ras, A., Schilperoort, R., and Hooykaas, P. 1992. Conjugative transfer by the virulence system of *Agrobacterium tumefaciens*. Science 256:1324-1326.

Binns, A. N., Beaupré, C. E., and Dale, E. M. 1995. Inhibition of VirB-mediated transfer of diverse substrates from *Agrobacterium tumefaciens* by the IncQ plasmid RSF1010. J. Bacteriol 177:4890-4899.

Bundock, P., den Dulk-Ras, A., Beijersbergen, A., and Hooykaas, P. J. 1995. Trans-kingdom T-DNA transfer from *Agrobacterium tumefaciens* to *Saccharomyces cerevisiae*. EMBO J. 14:3206-3214.

Cangelosi, G. A., Ankenbauer, R. G., and Nester, E. W. 1990a. Sugars induce the *Agrobacterium* virulence genes through a periplasmic binding protein and a transmembrane signal protein. Proc. Natl. Acad Sci. USA 87:6708-6712.

Cangelosi, G. A., Martinetti, G., and Nester, E. W. 1990b. Osmosensitivity phenotypes of *Agrobacterium tumefaciens* mutants that lack periplasmic beta-1,2-glucan. J. Bacteriol. 172:2172-2174.

Citovsky, V., De Vos, G., and Zambryski, P. 1988. Single-stranded DNA binding protein encoded by the *virE* locus of *Agrobacterium tumefaciens*. Science 240:501-504.

Citovsky, V., Warnick, D., and Zambryski, P. 1994. Nuclear import of *Agrobacterium* VirD2 and VirE2 proteins in maize and tobacco. Proc. Natl. Acad. Sci. USA. 91:3210-3214.

Citovsky,V., Zupan, J., Warnick, D., and Zambryski, P. 1992. Nuclear localization of *Agrobacterium* VirE2 protein in plant cells. Science. 256:1802-1805.

Douglas, C. J., Staneloni, R. J., Rubin, R. A., and Nester, E. W. 1985. Identification and genetic analysis of an *Agrobacterium tumefaciens* chromosomal virulence region. J. Bacteriol. 161:850-860.

Fullner, K. J., Lara, J. C., and Nester, E. W. 1996. Pilus assembly by *Agrobacterium* T-DNA transfer genes. Science *In Press*.

Fullner, K. J.,and Nester, E. W. 1996. Temperature affects the T-DNA transfer machinery of *Agrobacterium tumefaciens*. J. Bacteriol. 178:1498-1504.

Fuqua, W. C., Winans, S. C., and Greenberg, E. P. 1994. Quorum sensing in bacteria: The LuxR-LuxI family of cell density-responsive transcriptional regulators. J. Bacteriol. 176:269-275.

Hiei, Y., Ohta, S., Komari, T., and Kumashiro, T. 1994. Efficient transformation of rice (*Oryzae sativa* L.) mediated by *Agrobacterium* and sequence analysis of the boundaries of the T-DNA. Plant J. 6:271-282.

Ishida, Y., Saito, H., Ohta, S., Hiei, Y., Komari, T., and Kumashiro, T. 1996. High efficency transformation of maize (*Zea mays* L.) mediated by *Agrobacterium tumefaciens*. Nature Biotech. 14:745-750.

Jin, S., Roitsch, T., Christie, P. J., and Nester, E. W. 1990. The regulatory VirG protein specifically binds to a *cis*-acting regulatory sequence involved in transcriptional activation of *Agrobacterium tumefaciens* virulence genes. J. Bacteriol. 172:531-537.

Lee, Y. W., Jin, S., Sim, W. S., and Nester, E. W. 1995. Genetic evidence for direct sensing of phenolic compounds by the VirA protein of *Agrobacterium tumefaciens*. Proc. Natl. Acad. Sci USA 92:12245-12249.

Marks, J. R., Lynch, T. J., Karlinsey, J. E., and Thomashow, M. F. 1987. *Agrobacterium tumefaciens* virulence locus *pscA* is related to the *Rhizobium meliloti exoC* locus. J. Bacteriol. 169:5835-5837.

Matthysse, A. G. 1987. Characterization of nonattaching mutants of *Agrobacterium tumefaciens*. J. Bacteriol. 169:313-323.

Metts, J., West, J., Doares, S. H., and Matthysse, A. G. 1991. Characterization of three *Agrobacterium tumefaciens* avirulent mutants with chromosomal mutations that affect induction of *vir* genes. J. Bacteriol. 173:1080-1087.

Pan, S.Q., Jin, S., Boulton, M.I., Hawes, M., Gordon, M.P., and Nester, E.W. 1995. An *Agrobacterium* virulence factor encoded by a Ti plasmid gene or a chromosomal gene is required for T-DNA transfer into plants. Mol-Microbiol. 17:259-69.

Piers, K., Heath, J. D., Liang, X., Stephens, K. M., and Nester, E. W. 1996. *Agrobacterium tumefaciens*-mediated transformation of yeast. Proc. Natl. Acad. Sci. USA 93:1613-1618.

Shimoda, N., Toyoda-Yamamoto, A., Nagamine, J., Usami, S., Katayama, M., Sakagami, Y., and Machida, Y. 1990. Control of expression of *Agrobacterium vir* genes by synergistic actions of phenolic signal molecules and monosaccharides. Proc. Natl. Acad. Sci USA 87:6684-6688.

Shirasu, K. and Kado, C. I. 1993. Membrane location of the Ti plasmid VirB proteins involved in the biosynthesis of a pilin-like conjugative structure on *Agrobacterium tumefaciens*. FEMS Microbiol. Lett. 111:287-294.

Sundberg, C., Meek, L., Carroll, K., Das, A., and Ream, W. 1996. VirE1 protein mediates export of the single-stranded DNA-binding protein VirE2 from *Agrobacterium tumefaciens* into plant cells. J. Bacteriol. 178:1207-1212.

Thomashow, M. F., Karlinsey, J. E., Marks, J. R., and Hurlbert, R. E. 1987. Identification of a new virulence locus in *Agrobacterium tumefaciens* that affects polysaccharide composition and plant cell attachment. J. Bacteriol. 169:3209-3216.

Tummuru, M. K. R., Sharma, S. A., and Blaser, M. J. 1995. *Helicobacter pylori picB*, a homologue of the *Bordetella pertussis* toxin secretion protein, is required for induction of IL-8 in gastric epithelial cells. Mol. Microbiol. 18:867-876.

Wagner, V. T. and Matthysse, A. G. 1992. Involvement of a vitronectin-like protein in attachment of *Agrobacterium tumefaciens* to carrot suspension culture cells. J. Bacteriol. 174:5999-6003.

Winans, S. C., Burns, D. L., and Christie, P. C. 1996. Adaptation of a conjugal transfer system for the export of pathogenic macromolecules. Trends Microbiol. 4:64-68.

Yanofsky, M. F., Porter, S. G., Young, C., Albright, L. A., Gordon, M. P., and Nester, E. W. 1986. The *virD* operon of *Agrobacterium tumefaciens* encodes a site-specific endonuclease. Cell 47:471-477.

Young, C., and Nester, E. W. 1988. Association of the VirD2 protein with the 5' end of T strands in Agrobacterium tumefaciens. J. bacteriol. 170:3367-3374.

Zhan, X.. C., Jones, D. A., and Kerr, A. 1990. The pTiC58 *tzs* gene promotes high-efficiency root induction by agropine strain 1855 of *Agrobacterium rhizogenes*. Plant Mol. Biol. 14:785-792.

# Biogenesis of the *Agrobacterium tumefaciens* T-complex Transport Apparatus

D. FERNANDEZ, G. M. SPUDICH, T. A. DANG, X.-R. ZHOU, S. RASHKOVA, AND P. J. CHRISTIE*

Department of Microbiology and Molecular Genetics, The University of Texas Health Science Center, Houston, TX  77030

Recently, the view has emerged that products of the *Agrobacterium tumefaciens virB* operon assemble into a cell surface machinery which is dedicated to the conjugal delivery of oncogenic DNA to susceptible plant cells.  Toward development of a structural model of this transport apparatus, we used a combination of approaches, including cellular fractionation, fusion to periplasmically-active alkaline phosphatase, and protease susceptibility, to characterize the localizations of the VirB proteins.  Based on a discovery that several mutants of strain A348 sustaining nonpolar *virB* gene deletions accumulate aberrantly low levels of VirB proteins, we developed an assay for identifying stabilizing interactions among the VirB proteins.  With this assay, we have shown: i) VirB7, an outer membrane lipoprotein, forms intermolecular disulfide bridges *in vivo* with itself and with VirB9, ii) formation of the disulfide cross-links is essential for stabilization of both proteins, and iii) formation of the cross-linked VirB7/VirB9 heterodimer is essential for stabilization of other VirB proteins.  Our findings are consistent with a model in which VirB7 maturation as a lipoprotein and its subsequent covalent cross-linking with VirB9 represent critical early events in the biogenesis of the T-complex transport apparatus.

## Introduction

*Agrobacterium tumefaciens* incites tumor formation in plants by processing and transferring a segment of its genome (T-DNA) to susceptible plant cells.  The translocation-competent form of T-DNA (T-complex) is thought consist of a single-stranded molecule (T-strand) covalently capped at its 5' end with the VirD2 protein and coated along its length by the VirE2 single-stranded DNA binding protein (SSB).  If this indeed is the form in which T-DNA is delivered to plant cells, its estimated size in excess of $50 \times 10^6$ daltons easily makes T-complexes one of the largest substrates thus far identified to be translocated across any biological membrane (see Zupan and Zambryski 1995).

Three lines of study have led to the suggestion that *A. tumefaciens* has evolved as a phytopathogen in part by coopting a set of conjugal *tra* genes encoded by the *virB* operon for use in delivering its oncogenes to susceptible plant cells. First, sequence analyses have demonstrated that very close evolutionary relationships exist between the *virB* genes and the *tra* genes responsible for interbacterial conjugal transmission of the IncN, IncW, and IncP broad-host-range (BHR) plasmids. Not only do all 11 of the *virB11* genes have IncN counterparts, but related genes are co-located in both operons. Ten of the *virB* genes are similar to 10 *trw* genes of the IncW plasmid R388, and 6 *virB* genes are similar to 6 *trb* genes of the IncP plasmid RP4 (see Winans et al. 1996). Second, genetic experiments have revealed that each of these transport machineries delivers a common substrate, mobilizable IncQ plasmids, to recipient cells (Lessl and Lanka 1994; Beijersbergen et al. 1992; Buchanon-Wollaston et al. 1987). Finally, there is recent evidence that the BHR Tra as well as the VirB proteins are involved in elaboration of a sex pilus (Lessl and Lanka 1994; Fullner and Nester 1996).

While the relatedness of the T-complex and BHR conjugation machineries is certainly significant, it is perhaps more intriguing that components of these systems are also highly similar to the Ptl proteins which are postulated to assemble into a transport machinery for delivery of the 6-subunit pertussis toxin across the *Bordetella pertussis* outer membrane (see Winans et al. 1996). Thus, a family of transporters has been identified that collectively exhibits the ability to export macromolecules as diverse as nucleoprotein particles and multicomponent proteins across the gram-negative bacterial envelope for delivery to bacterial, plant, or mammalian recipient cells.

## Results

LOCALIZATIONS OF VIRB PROTEINS

Table 1 summarizes results of cellular fractionation, protease susceptibility, and PhoA fusion studies of many of the VirB proteins. Taken together with available sequence information and data from other laboratories (Fernandez at al. 1996a & references therein), we suggest that the T-complex transport apparatus consists of a VirB protein complex whereby the VirB4 and VirB11 ATPases are located predominantly on the cytoplasmic face of the inner membrane , VirB4 and VirB6 are polytopic inner membrane proteins, VirB10 is a monotopic inner membrane protein with a large C-terminal periplasmic domain, and VirB1, VirB5, VirB7, VirB8, and VirB9 are membrane-associated proteins with large periplasmic domains that may form a structure spanning the inner and outer membranes. Our PhoA fusion data suggest that VirB2 and VirB3 also contain exported domains; these proteins may possess periplasmic or extacellular domains.

# Table 1. Properties of VirB proteins

| VirB[a] Protein | Signal Sequence | Subcellular[b] Localization | | Protease[c] Susceptibility Spheroplasts | PhoA[d] Junction | PhoA[e] Activity |
|---|---|---|---|---|---|---|
| | | IM | OM | | | |
| VirB1 | + | 50 | 50 | + | 164 | 901 |
| VirB2 | - | | | | 121 | 1298 |
| VirB3 | - | | | | 47 | 93 |
| VirB4 | - | 90 | 10 | +[f] | (I)[g] | 35 |
| | | | | | (II) | 20 |
| VirB5 | + | 60 | 40 | + | 93 | 1130 |
| VirB6 | - | ? | | ? | | ? |
| VirB7 | + | 40 | 60 | + | 41 | 3070 |
| VirB8 | - | 90 | 10 | + | 72 | 411 |
| VirB9 | + | 50 | 50 | + | 74 | 309 |
| VirB10 | - | 50 | 50 | + | 276 | 1126 |
| VirB11 | - | 70 | 30(C)[h] | - | 323 | 0 |

a. Data from Fernandez et al. 1996a; Dang and Christie 1996; and Rashkova and Christie 1996.
b. Percentage of protein in inner (IM) and outer (OM) membrane fractions of sucrose gradients.
c. Proteinase K treatment of A348 spheroplasts.
d. Defined by the amino acid immediately preceding the fusion junction.
e. Expressed as Units of AP activity per $OD_{600}$ in *A. tumefaciens*.
f. Protease treatment of spheroplasts resulted in appearance of two immunoreactive degradation products of ~45-kDa and ~35-kDa, suggesting that a protease susceptible domain is near the middle of the protein.
g. Fusion junctions of active VirB4::PhoA fusions mapped to two regions, Region I (Residues 58-84) and Region II (Residues 450-514).
h. Amount of VirB11 protein found in the cytoplasmic (C) fraction.

DEVELOPMENT OF AN ASSEMBLY PATHWAY FOR THE T-COMPLEX
TRANSPORT APPARATUS

*Effects of nonpolar* virB *gene deletions on steady-state levels of VirB proteins.* Each of the 11 ΔvirB mutations generated in a previous study was shown by genetic complementation analysis to be nonpolar for expression of downstream genes. Interestingly, however, several of the ΔvirB mutants possessed aberrantly low levels of several VirB proteins. Most notably, the ΔvirB7 mutant strain, PC1007, accumulated low levels of many VirB proteins, and, more significantly, undetectable levels of VirB9. These findings formed the basis of a hypothesis that certain VirB proteins such as VirB7 function as critical stabilizing elements during biogenesis of the T-complex transport apparatus (Berger and Christie 1994).

*Modulation of VirB9 protein abundance through the regulated expression of* virB7. To test whether the effect of the ΔvirB7 mutation on VirB protein levels was due to the absence of VirB7 protein, we introduced *virB7* in *trans* into strain PC1007. Immunoblot studies confirmed that steady-state levels of several VirB proteins, including VirB9, were enhanced in these cells. To test whether VirB7 has a stabilizing effect on VirB9, we introduced plasmid pPC970, which expresses *virB7* from an IPTG-inducible $P_{lac}$ promoter and *virB9* from an acetosyringone-inducible (AS) $P_{virB}$ promoter into strain PC1000, which carries a precise deletion of the entire *virB* operon. Immunoblot analyses showed that PC1000(pPCB970) cells induced with 50 μM AS did not accumulate detectable levels of VirB7 or VirB9. Conversely, cells induced with 1 mM IPTG accumulated high levels of VirB7 but undetectable levels of VirB9. Cells induced with both AS and IPTG accumulated VirB9 in rough proportion to the level of VirB7. These findings suggested that VirB9 accumulates specifically in response to VirB7 synthesis (Fernandez et al. 1996b).

*VirB7 forms two intermolecular disulfide bridges.* In the experiment described above, protein samples were electrophoresed under reducing conditions. When we electrophoresed corresponding protein samples under nonreducing conditions, we could not detect the monomeric forms of the 4.5-kDa VirB7 protein or the 32-kDa VirB9 protein. Instead, we visualized a ~36-kDa complex, designated C1, that was immunologically crossreactive with both the VirB7 and VirB9 antisera and increased in abundance upon induction with increasing concentrations of IPTG (Fernandez 1996b).

We next tested whether the VirB7 and VirB9 antisera would co-precipitate the presumptive VirB7/VirB9 heterodimer from cell extracts suspended in nonreducing buffers. SDS-PAGE using a glycine buffer system followed by immunoblot analysis revealed the presence of complex C1 in immunoprecipitates recovered with both antisera. Interestingly, complex C1 dissociated into the VirB7 and VirB9 monomers when immunoprecipitates suspended in nonreducing sample buffer were electrophoresed adjacent to a lane containing β-mercaptoethanol (Spudich et al. 1996). The dissociation of complex C1 into VirB7 and VirB9 monomers was further demonstrated using more conventional two-dimensional nonreducing/reducing gel electrophoresis.

Electrophoresis of the same protein samples through a tricine-SDS/polyacrylamide gel system for resolution of low-molecular-weight proteins showed that VirB7 antisera but not VirB9 antisera precipitated a ~9-kDa complex, designated C3. The size of this complex suggests that VirB7 also assembles as a β-mercaptoethanol-dissociable homodimer.

*Cys-24 of VirB7 and Cys-262 of VirB9 are reactive cysteine residues.* We examined the effects of substituting Ser for Cys-24 of VirB7 and Cys-262 of VirB9 - the only Cys residues in the mature forms of these proteins. The C24S and C262S derivatives did not accumulate to detectable levels, demonstrating that dimerization via these reactive Cys residues is essential for stabilization of both VirB7 and VirB9 (Spudich et al. 1996).

*Stabilizing effects of the VirB7/VirB9 heterodimer on other VirB proteins.*
As noted above, PC1007 cells accumulate enhanced levels of several VirB
proteins when carrying a *virB7 trans* expression plasmid. We similarly
found that PC1009 cells, which sustain a Δ*virB9* mutation, accumulate
enhanced levels of VirB proteins when transformed with a *virB9 trans*
expression plasmid. These observations suggest that VirB7 and VirB9 are
both needed for accumulation of other VirB proteins to wild-type levels.
Interestingly, PC1009 cells accumulate wild-type levels of VirB7 in the
form of a homodimer and yet accumulate low levels of other VirB proteins.
Thus, we postulate that it is the VirB7/VirB9 heterodimer, not the VirB7
homodimer, that contributes to stabilization of other VirB proteins
(Fernandez et al. 1996b).

## Discussion

Based on results of these investigations, we propose that the T-
complex transport system is assembled according to the sequence of events
depicted in Fig. 1. In stage I, VirB7 is processed as a lipoprotein, most
probably according to the lipoprotein biosynthetic pathway developed for
Lpp, the major outer membrane lipoprotein of *E. coli*. In stage II, newly-
processed VirB7 monomers form intermolecular disulfide cross-links which
are critical for protein stabilization. We postulate that the VirB7/VirB9
heterodimer is the physiologically relevant dimer for transporter assembly
and that the VirB7 homodimer may or may not assemble as a component of
the transporter. In stage III, the VirB7/VirB9 heterodimer is sorted by an
unknown mechanism to its final destination at the *A. tumefaciens* outer
membrane. In stage IV, the VirB7/VirB9 heterodimer is postulated to
function as a nucleation center for recruitment of other VirB proteins for
further assembly of the T-complex transport apparatus. In future studies,
we will test features of this model, and we will attempt to identify steps in
the assembly pathway that are upstream and downstream of those depicted.

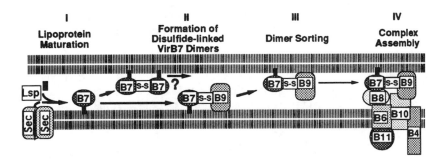

**Figure 1.** Proposed Stabilization Pathway for Assembly of the T-complex
Transport Apparatus.

## ACKNOWLEDGMENTS

We thank Drs. Anath Das and Christian Baron for sharing their simultaneous discoveries that VirB7 forms disulfide cross-linked dimers prior to publication. This work was supported by Public Health Service Grant GM48746 to PJC.

## LITERATURE CITED

Beijersbergen, A., Dulk-Ras, A. D., Schilperoort, R. A, and Hooykaas, P. J. J. 1992. Conjugative transfer by the virulence system of *Agrobacterium tumefaciens.* Science 256:1324-1327.

Berger, B. R., and Christie, P. J. 1994. Genetic complementation analysis of the *Agrobacterium tumefaciens virB* operon: *virB2* through *virB11* are essential virulence genes. J. Bacteriol. 176:3646-3660.

Buchanan-Wollaston, V., Passiatore, J. E., and Cannon, F. 1987. The *mob* and *oriT* mobilization functions of a bacterial plasmid promote its transfer to plants. Nature 328:172-175.

Christie, P. J., Ward, J. E., Gordon, M. P., and Nester, E. W. 1989. A gene required for transfer of T-DNA to plants encodes an ATPase with autophosphorylating activity. Proc. Natl. Acad. Sci. USA 86:9677-9681.

Dang, T. A., and Christie, P. J. 1996. Unpublished results.

Fernandez, D., Dang, T. A., Spudich, G. M., Zhou, X.-R., Berger, B. R., and Christie, P. J. 1996a. *Agrobacterium tumefaciens virB7* gene product, a proposed component of the T-complex transport apparatus, is a membrane-associated lipoprotein exposed at the periplasmic surface. J. Bacteriol. 178:3156-3167.

Fernandez, D., Spudich, G. M., Zhou, X.-R., and Christie, P. J. 1996b. The *Agrobacterium tumefaciens* VirB7 lipoprotein is required for stabilization of VirB proteins during assembly of the T-complex transport apparatus. J. Bacteriol. 178:3168-3176.

Fullner, K. J., and Nester, E. W. 1996. Science In press.

Lessl, M., and Lanka, E. 1994. Common mechanisms in bacterial conjugation and Ti-mediated T-DNA transfer to plant cells. Cell 77:321-324.

Rashkova, S., and Christie, P. J. 1996. Unpublished results.

Spudich, G. M., Fernandez, D., Zhou, X.-R., and Christie, P. J. 1996. Intermolecular disulfide bonds stabilize VirB7 homodimers and VirB7/VirB9 heterodimers during biogenesis of the *Agrobacterium tumefaciens* T-complex transport apparatus. Proc. Natl Acad. Sci. USA 93:XXX-XXX.

Winans, S. C., Burns, D. L., and Christie, P. J. 1996. Adaptation by bacterial pathogens of a conjugal transfer system for export of macromolecules. Trends Microbiol. 4:64-68.

Zupan, J. R., and Zambryski, P. 1995. Transfer of T-DNA from *Agrobacterium* to the plant cell. Plant Physiol. 107:1041-1047.

# Agrobacterium and Plant Genes Affecting T-DNA Transfer and Integration

Soma Narasimhulu, Jaesung Nam, Xiao-bing Deng, and Stanton Gelvin
Dept. Biological Sciences, Purdue University, West Lafayette, IN 47907

During the inception of crown gall tumorigenesis, *Agrobacterium tumefaciens* processes a region of DNA (T-DNA) from the tumor-inducing (Ti) plasmid and transfers this DNA to plant cells. Proteins encoded by the virulence (*vir*) region of the Ti plasmid regulate T-DNA processing and transfer. Nicking of 25-bp directly repeated T-DNA "border" sequences by the VirD2 endonuclease results in the generation of single-stranded T-DNA molecules (T-strands) with which VirD2 is tightly associated at the 5' end. These single-stranded DNA molecules are transferred to the plant cytoplasm (Yusibov et al., 1994). Targeting of the T-DNA to the plant nucleus may be mediated by nuclear localization sequences (NLS) within the associated VirD2 and VirE2 proteins (Citovsky et al., 1994). T-DNA molecules eventually integrate into the plant chromosomes, thereby stabilizing the oncogenes and opine biosynthesis genes encoded by the T-DNA. However, non-integrated copies of T-DNA may persist in the nucleus for a period of time.

Although we now have a fairly detailed understanding of the early events that occur in *A. tumefaciens*, little is known about the events that take place within the plant cell involving the targeting of T-DNA to the nucleus, its ultimate integration into plant nuclear DNA, and the early stages of T-DNA expression. Similarly, our knowledge of the role that plant genes play in these processes is scant. In an attempt to elucidate the kinetics of T-DNA transport to the nucleus as well as to determine which Vir proteins are necessary to effect this process, we developed a very sensitive reverse transcriptase-PCR assay to detect T-DNA transcription early after infection. In addition, we show that variation in susceptibility to tumorigenesis among *Arabidopsis thaliana* ecotypes is a heritable trait. One recalcitrant ecotype is deficient in the ability to integrate T-DNA into the plant genome.

## A novel system to detect the early transcription of T-DNA

To determine the kinetics of transport of the T-DNA from the cytoplasm to the nucleus and the importance of various Vir proteins in this process, we developed a sensitive assay to detect early transcription of the T-DNA. We constructed a T-DNA binary vector, pBISN1, that contains a *gusA* gene under the transcriptional control of a super- promoter. The *gusA* gene contains a 189 bp intron. We developed PCR primers that would amplify a 732 bp region containing this intron. However, if the intron were processed from a *gusA* transcript, we would amplify (using RT-PCR) a 543 bp fragment. RT-PCR analysis of *in vitro*-transcribed *gusA* RNA indicated that, by using this assay, we could readily detect 10 fg of *gusA* mRNA.

We infected rapidly dividing tobacco BY-2 suspension culture cells with three *A. tumefaciens* strains harboring pBISN1. *A. tumefaciens* At793 lacks a Ti plasmid, *A. tumefaciens* At789 contains the octopine-type Ti plasmid pTiA6, and *A. tumefaciens* At790 contains the agropine-type supervirulent Ti plasmid pTiBo542. We isolated total cellular RNA from infected plant cells at various times starting from the initiation of cocultivation, and subjected the RNA to RT-PCR.

Figure 1 shows that we could detect *gusA* transcripts beginning 20-24 hr after infection of tobacco BY-2 suspension culture cells by the octopine *A. tumefaciens* strain At789. The level of *gusA* mRNA increased and peaked at 36 hr, after which there was a slight decline in *gusA* mRNA. We could not detect *gusA* mRNA in noninfected plant cells and in plant cells from which RNA was extracted immediately after the start of cocultivation (0 hr time point). The synthesis of *gusA* mRNA in cocultivated plant cells

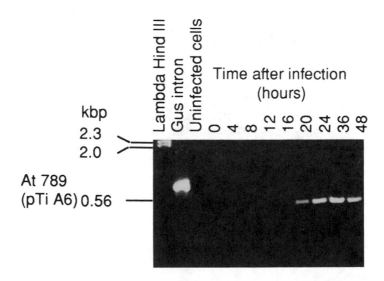

Figure 1. Kinetics of *gusA* gene expression in tobacco suspension culture cells. Tobacco cells were infected with *A. tumefaciens* At789 (an octopine-type strain) for various periods of time. RNA was extracted, subjected to RT-PCR, and analyzed by agarose gel electrophoresis. GUS intron, PCR amplification of a 732 bp *gusA*-intron gene region.

depended upon the presence of a Ti plasmid within the infecting *A. tumefaciens* strain (data not shown). These data suggest that *gusA* gene expression was *vir* gene-dependent.

## Early T-DNA transcription and *vir* gene function

To determine the importance of different virulence proteins in the early stages of T-DNA transfer, transcription, and perhaps integration, we introduced pBISN1 into *A. tumefaciens* containing mutations in various *vir* genes, incubated each mutant strain separately with tobacco BY-2 cells, and assayed for the presence of *gusA* mRNA by RT-PCR (data not shown). *A. tumefaciens* strains harboring mutations in *virB*, *virD1*, and *virD4* could not direct the synthesis of *gusA* mRNA in infected tobacco cells. These results were expected because mutations in *virD1* abolish T-DNA processing, and mutations in *virB* and *virD4* inhibit T-DNA transfer.

Cocultivation of tobacco cells with *A. tumefaciens* strains mutant in *virE* and *virC* result in the delayed and highly attenuated expression of *gusA* mRNA. Stachel and Nester (19869) previously showed that similar mutations in *virE* render the bacterium severely attenuated in virulence, despite the fact that normal levels of T-strands accumulate in induced cells. Thus, the presence of VirE2 single-stranded DNA binding protein is not necessary to protect the T-strand within the bacterium. Furthermore mutations in *virC* severely attenuate the virulence of *A. tumefaciens* on most plant species, including tobacco (Stachel and Nester, 1986).

*A. tumefaciens* containing the non-polar insertion mx304 in the 3' end of *virD2* are avirulent, although they accumulate T-strands following induction by acetosyringone. The Tn*3*-HoHo1 insertion occurs 73 amino acids from the C-terminus of VirD2 and translationally fuses β-galactosidase protein to the C-terminus of VirD2 (Koukolikova-Nicola et al., 1993). The resulting altered VirD2 protein lacks the C-terminal NLS and the ω region that is important for efficient tumorigenesis (Shurvinton et al., 1992). Therefore, we were surprised to detect relatively high levels of *gusA* transcripts in tobacco cells infected with this *A. tumefaciens* strain. Quantitative RT-PCR indicated that the level of *gusA* transcript resulting from infection of tobacco cells by this mutant strain was 20-30% that found in tobacco cells cocultivated with an *A. tumefaciens* strain containing a wild-type *virD2* gene. The presence of a high level of *gusA* mRNA in these cells was considerably more transient than that resulting from infection by the wild-type bacterium. These data indicate that deletion of the carboxy-terminal NLS and/or ω region of VirD2 results in high level transient expression of T-DNA-encoded genes. The high level of transient *gusA* gene expression following infection of tobacco cells by the mx304 mutant strain suggests that, despite the lack of a carboxy-terminal NLS in the VirD2 protein encoded by this strain, the T-DNA is efficiently directed to the plant nucleus.

We repeated the tobacco cell infections using an *A. tumefaciens* strain that encodes a mutant VirD2 protein containing two serine residues in place of four of the five amino acids in the ω domain. It therefore lacks the ω region while still retaining the NLS. Despite the fact that this strain only induces 2-3% of the tumors compared to a wild-type strain, the ω mutant directed a high level of transient transcription of the *gusA* gene. Quantitative RT-PCR indicated that *gusA* transcripts accumulated to 20-30% the level as when we infected tobacco cells with the wild-type strain. These data indicate that the ω region of VirD2 is not required for T-DNA transport to the plant cytoplasm or nucleus, replication to a double-stranded form, or transcription. The data suggest, however, that this region of VirD2 may be involved in the stabilization of T-DNA transcription, perhaps by mediating T-DNA integration into the plant genome.

## *Arabidopsis* ecotypes and susceptibility to crown gall tumors

Recently, a number of laboratories have used different ecotypes of *Arabidopsis thaliana* as a model plant system to investigate differential host response to various strains or races of bacteria, fungi, and viruses. We wished to determine whether differences also exist among *Arabidopsis* ecotypes with regard to tumorigenesis by *A. tumefaciens*, and what the biochemical and genetic basis of such differences were. We therefore developed an *in vitro* root inoculation assay and screened *Arabidopsis thaliana* ecotypes for susceptibility or resistance to crown gall disease. We have identified several ecotypes that are hyper-susceptible to crown gall tumorigenesis, as well as other ecotypes that are recalcitrant (less susceptible) to tumorigenesis. Interestingly, we have identified several ecotypes that appear to permit T-DNA transfer and nuclear transport, but are deficient in T-DNA integration.

## Response of *Arabidopsis* ecotypes to *Agrobacterium* strains

We characterized the response of 39 *Arabidopsis* ecotypes to four different *A. tumefaciens* strains using an *in vitro* root segment infection tumorigenesis assay. From among these ecotypes, we chose Aa-0 as highly susceptible (yielding large green teratomas), and UE-1 as highly recalcitrant (yielding at best small yellow calli). We tested four *A. tumefaciens* strains, all containing the same C58 chromosomal background. We chose the nopaline-type strain A208 as the best for this study because it caused more severe symptom development in most of the ecotypes we tested.

We initially characterized basic metabolic properties of these two ecotypes (ability to incorporate $^3$H-thymidine and $^{14}$C-amino acids into macromolecules, the ability of exudates of these ecotypes to induce *Agrobacterium vir* genes, or differences in the phytohormone sensitivity of these ecotypes. We could not detect any major differences (data not shown).

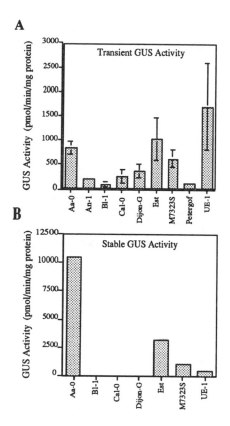

Figure 2. Transient and stable GUS activity in the roots of various *Arabidopsis* ecotypes. *A. tumefaciens* A208 (pCNL65) was used to infect sterile root segments, the bacteria were killed with antibiotics after 2 days, and GUS activity measured 5 days (**A**) or 30 days (**B**) later.

## Efficiency of T-DNA transfer and integration in ecotypes Aa-0 and UE-1

We investigated the transient and stable T-DNA-mediated transfer to and expression of a *gusA*-intron gene in several ecotypes. Figure 2 shows that transient GUS activity is actually higher in ecotype UE-1 than in Aa-0. However, stable GUS activity in UE-1 was much lower than in Aa-0. Because not only tumorigenesis, but stable GUS activity and the ability to be transformed to kanamycin-resistance (data not shown) was low in UE-1 but high in Aa-0, we investigated directly the ability of these two ecotypes to integrate T-DNA into high molecular plant DNA. Figure 3 shows that, in plant cells grown under non-selective conditions, ecotype Aa-0 integrated approximately 5-fold more T-DNA than did ecotype UE-1. This deficiency in T-DNA integration was paralleled by a 5-fold increase in γ-radiation sensitivity of UE-1 relative to Aa-0 (data not shown). This is, to our knowledge, the first example of natural variability in T-DNA integration within a plant species.

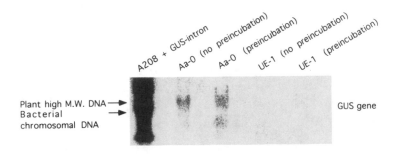

Figure 3. Integration of T-DNA into high molecular weight Arabidopsis DNA. Root segments of *Arabidopsis* ecotypes Aa-0 and UE-1 were infected with *A. tumefaciens* A208(pCNL65) and calli grown on CIM under non-selective conditions. High molecular weight plant DNA was blotted onto nitrocellulose and hybridized with a *gusA* gene probe.

## Literature Cited

Citovsky, V., D. Warnick, and P. Zambryski. 1994. Nuclear import of *Agrobacterium* VirD2 and VirE2 proteins in maize and tobacco. Proc. Natl. Acad. Sci. USA 91:3210-3214.

Koukolikova-Nicola, Z., D. Raineri, K. Stephens, C. Ramos, B. Tinland, E.W. Nester, and B. Hohn. 1993. Genetic analysis of the *virD* operon of *Agrobacterium tumefaciens*: A search for functions involved in transport of T-DNA into the plant cell nucleus and in T-DNA integration. J. Bacteriol. 175:723-731.

Shurvinton, C.E., L. Hodges, and W. Ream. 1992. A nuclear localization signal and the C-terminal omega sequence in the *Agrobacterium tumefaciens* VirD2 endonuclease are important for tumor formation. Proc. Natl. Acad. Sci. USA 89:11837-11841.

Stachel, S.E., and E.W. Nester. 1986. The genetic and transcriptional organization of the *vir* region of the A6 Ti plasmid of *Agrobacterium tumefaciens*. EMBO J. 5:1445-1454.

Yusibov, V.M., T.R. Steck, V. Gupta, and S.B. Gelvin. 1994. Association of single-stranded transferred DNA from *Agrobacterium tumefaciens* with tobacco cells. Proc. Natl. Acad. Sci. 91:2994-2998.

# THE ROLES OF THE VIRULENCE PROTEINS D2 AND E2 IN NUCLEAR TARGETING, PROTECTION AND INTEGRATION OF T- DNA

Barbara Hohn*, Bruno Tinland[1], Ana María Bravo Angel, Fabrice Schoumacher[2], Jesús Escudero[3] and Luca Rossi. Friedrich Miescher-Institut, CH-4002 Basel; present addresses: [1]ETH Zürich, Institut für Pflanzenwissenschaften, CH-8092 Zürich; [2]Zentrum für Lehre und Forschung, CH-4056 Basel, Switzerland; [3]Clusius Laboratory, Leiden University, The Netherlands

ABSTRACT
DNA artificially introduced into cells suffers severe degradations due to inefficient nuclear entry and damage during integration. In Agrobacterium tumefaciens mediated DNA transfer strategies were devised for efficient transfer and integration of only a few units of T-DNA. Ti-plasmid encoded virulence proteins, in particular VirD2 and VirE2 thereby play instrumental roles. The VirD2 protein mediates nuclear localisation sequence dependent nuclear targeting of the T-DNA, as shown by intracellular localisation studies and mutation analysis. The VirE2 protein is a sequence non-specific single-strand DNA binding protein, in the absence of which only severely truncated versions of the T-DNA become integrated. This points to a direct role of this protein in protection of the T-DNA, most likely by directly interacting with it. An indirect effect on nuclear import can not be excluded.
The involvement of VirD2 in integration of T-DNA was investigated by mutation analysis. The studied mutation reduced the efficiency of T-DNA transfer, most probably by decreasing the level of T-strand formation, but the efficiency of the integration step was unchanged. However, integration events obtained with the mutated VirD2 protein revealed an aberrant pattern, with truncations at the 5'terminus. This result shows the conclusion that the wildtype VirD2 protein directly or indirectly is responsible for integration of the 5'end of the T-DNA.

When *Agrobacterium tumefaciens* cells find a plant host for transformation, a series of events are triggered with the ultimate goal of introducing into the plant a complex of T-DNA (transferred DNA) and virulence proteins. The single-stranded T-DNA, called T-strand, is produced by the site-specific endonucleolytic activity coded for by the virulence proteins VirD1 and VirD2, whereby VirD2 remains covalently attached to the 5' end of the T-strand. Another protein, VirE2, a single-stranded DNA binding protein, has been proposed to bind to T-DNA at some step of the transfer (for reviews see Zupan and Zambryski, 1995; Rossi et al. 1996b). The functions that these virulence proteins play in the plant are only partially understood. They will be discussed in terms of the envisaged requirements of the T-DNA to survive in the intracellular environment of the plant, to cross the nuclear membrane and to integrate into nuclear DNA.

## Vir E2 protects T-DNA from degradation

In order to address the question of the function of VirE2, a complete deletion of the gene coding for this protein was produced and its phenotype analysed (Rossi et al. 1996a). As was found for previously tested virE2 mutants the efficiency of transfer was severely reduced but the accumulation of T-strands in induced bacterial cells was to wildtype levels. This confirmed the notion that the main activity of VirE2 protein must take place inside the plant cells, as also shown by trans-complementation to full virulence by cotransformation with a strain expressing VirE2 (Rossi, unpublished). This latter result matches the finding that T-DNA, derived from virE2 mutant strain, could be complemented to full virulence by VirE2 protein molecules expressed by a recipient plant transgenic for a constitutively expressed virE2 gene (Citovsky et al. 1992).

In contrast, when the pattern of integration in the few recovered tobacco transformants was analysed, severe truncations over the whole length of the T-DNA (except for the part containing the selectable marker, of course) were found (Rossi et al. 1996a) The function of the 5' terminus of T-DNA with the plant chromosomal DNA resembled wildtype junctions, implying proper VirD2 functions (see below). From this it was concluded that in the presence of VirE2 T-DNA along its entire length is preserved from nucleolytic degradation. This can most easily be envisaged as physical protection of the T-DNA by VirE2 protein molecules bound to the single-stranded DNA. This is consistent with the finding that *in vitro* assembled single-stranded DNA-VirE2 complexes are resistant to digestion by nucleases (Citovsky et al. 1989; Sen et al. 1989).

## Nuclear targeting of T-DNA

Entry of (mostly viral) nucleic acids into eukaryotic cell nuclei relies in the studied cases on nuclear localisation sequences, NLSs, contained in proteins binding to the nucleic acid. The small SV40 virus enters the nucleus in an apparently intact form, whereas adenovirus subcomplexes travel through the nuclear membrane pores (reviewed by Greber and Kasamatsu 1996). Also lentiviruses such as HIV travel through the nuclear pores as subviral preintegration complexes (reviewed by Stevenson 1996), while influenzavirus enters nuclei as viral ribonucleoprotein complexes (reviewed by Whittaker et al. 1996). In all these analysed cases karyophilic protein components of the viruses seem to mediate nuclear entry of the viral genomes, and to use the cellular machinery for NLS dependent transport.

Inspection of the sequences of VirD2 and VirE2 indeed revealed the presence of NLSs, and when these sequences were tested for transport of otherwise cytoplasmically localised markerproteins, nuclear targeting could be confirmed (reviewed in Zupan and Zambryski 1995; Rossi et al. 1996b). The significance of these sequences for T-DNA transfer was analysed for VirD2. It was shown that targeting to the plant cell nucleus was dependent on the C-terminal NLS (Koukolíková-Nicola et al. 1993; Rossi et al. 1993), confirming earlier observations (Shurvington et al. 1992). However, the quantitative aspect of the contribution of VirD2 to targeting seems to be dependent on the plant analysed and the assay used (Shurvington et al. 1992; Koukolíková-Nicola et al. 1993; Rossi et al. 1993; Narasimhulu et al. 1996).

It was tested whether the VirE2 contributes to nuclear targeting. A strain carrying a deletion of the NLS sequence of VirD2, in combination with a deletion of the virE2 gene transferred its T-DNA with an even further reduced efficiency (Rossi et al. 1996a). Thus, full tansfer activity requires both the NLS of VirD2 and the VirE2 protein. The contribution of the VirE2 NLSs cannot be directly evaluated since mutations in the NLS domain were found to affect the single-stranded DNA binding properties of VirE2 (Citovsky et al. 1992). This implies that the NLS domains of VirE2 may be masked in the interior of the VirE2-DNA complex and therefore may not be available to the NLS binding proteins. VirE2 may indirectly contribute to nuclear entry by enlongating the complex. Analysis of nuclear targeting of microinjected complexes (Zupan et al. 1996) are compatible with this interpretation (discussed in more detail in Rossi et al. 1996b).

### Integration of T-DNA

The process of integration of T-DNA occurs following a mode of illegitimate recombination (reviewed in Tinland and Hohn 1995; Tinland 1996). In contrast to plant transposable elements T-DNA does not code for functions required for integration. Consequently, integrated T-DNA is permanently immobilised. A conspicuous feature of integrated T-DNA is the finding that the right end of the T-DNA is frequently conserved up to the nucleotide which is cleaved in the bacterium, whereas the left end of the T-DNA is conserved much more poorly. These observations suggest that the twoT-DNA ends are joined to plant DNA by different mechanisms.

The VirD2 protein, which is covalently attached to the 5' terminus of the T-strand and presumably remains there until after the T-DNA has entered the nucleus, was suggested to be involved in the ligation of the T-DNA to plant DNA. This was tested by analysing integration of a T-DNA which was delivered from a bacterial strain carrying a mutation in the virD2 gene, located in a motif common to VirD2 and transfer proteins of conjugative plasmids (Tinland et al. 1995). As shown in the Table the mutant, designated virD2R129G, exhibited a reduced transfer efficiency, which is due to a reduced production of T-strands.

| Relevant mutation of bacterial strain used | Efficiency, in %, of | | |
| | a | b | |
| | T-DNA transfer (transient expression) | Plant transformation (antibiotic resistant calli) | T-DNA integration (ratio of a and b) |
|---|---|---|---|
| wt | 100 | 100 | 100 |
| virD2⁻ | 0 | 0 | |
| virD2 NLS⁻ | ~5 | ~5 | ~100 |
| virD2R129G | 0.4-1 | 0.4-1 | ~100 |
| virE2⁻ | 0.02-0.04 | 0.01-0.04 | ~100 |
| virE2⁻ virD2 NLS⁻ | ~0.0023 | | |

Data are from Rossi et al. 1993; Tinland et al. 1995; Rossi et al. 1996a

Unexpectedly, the integration efficiency, defined as ratio of frequences of transformation and transfer, was normal (see Table). Southern and sequence analyses of integration events obtained with the mutated VirD2 protein revealed an aberrant pattern of integration. Specifically, a loss of precision of integration of the 5' end of the T-DNA was noted. This led to the conclusion that in the wildtype situation VirD2 plays an essential role in preserving the precision of the integration while not contributing to the efficiency of this process.

As can be seen from the Table, also the VirE2 protein is not involved in the integration process *per se*, since the ratio of transformation- and transfer efficiencies is not changed in comparison to the wildtype situation. The main functions of VirE2 are therefore confined to protection of T-DNA and possibly to some step of nuclear targeting. The results on VirD2 and VirE2 taken together imply that the active steps of T-DNA integration are probably taken over by plant factors. In agreement with this is the isolation of radiation defective mutants of *Arabidopsis thaliana* which are deficient in T-DNA integration (Sonti et al. 1995).

The roles of the virulence proteins in the plants are mainly to protect the T-DNA, to lead it through the nuclear pore, and to allow precise junction of T-DNA to plant DNA (at least at the 5' T-DNA end). This is the basis for the efficiency with which *Agrobacterium tumefaciens* transforms plants.

## LITERATURE CITED

Citovsky, V., Wong, M.L. and Zambryski, P. 1989. Cooperative interaction of *Agrobacterium* virE2 protein with single-stranded DNA: implications for the T-DNA transfer process. Proc. Natl. Acad. Sci. 86:1193-1197.

Citovsky, V., Zupan, J., Warnick, D. and Zambryski, P. 1992. Nuclear localization of *Agrobacterium* VirE2 protein in plant cells. Science 256:1802 1805.

Greber, U.F. and Kasamatsu, H. 1996. Nuclear targeting of SV40 and adenovirus. Trends in Cell Biology 6:189-195.

Koukolíková-Nicola, Z., Raineri, D., Stephens, K., Ramos, C., Tinland, B., Nester E.W. and Hohn, B. 1993. Genetic analysis of the *vir*D operon of *Agrobacterium tumefaciens*: a search for functions involved in the transport of T-DNA into the plant cell nucleus and in T-DNA integration. J. Bacteriol. 175:723-731.

Rossi, L., Hohn, B. and Tinland, B. 1993. VirD2 protein carries nuclear localization signals important for transfer of T-DNA to plants. Mol. Gen. Genet. 239:345-353.

Rossi, L. Hohn, B. and Tinland, B. 1996a. Integration of complete T-DNA units is dependent on the activity of VirE2 protein of *Agrobacterium tumefaciens*. Proc. Natl. Acad. Sci. USA. 93: 126-130.

Rossi, L., Tinland, B. and Hohn, B. 1996b. Role of virulence proteins of *Agrobacterium* in the plant. In "The *Rhizobiaceae*", eds. H. Spaink, A. Konderosi and P. Hooykaas; Kluwer, in press.

Sen, P. Pazour, G.J., Anderson, D. and Das, A. 1989. Cooperative binding of *Agrobacterium tumefaciens* VirE2 protein to single-stranded DNA. J. Bacteriol. 171:2573-2580.

Shurvinton, C.E., Hodges, L. and Ream, W. 1992. A nuclear localization signal in the *Agrobacterium tumefaciens* VirD2 endonuclease is important for tumor formation. Proc. Natl. Sci. USA. 89:11837-11841.

Sonti, R. Chiurazzi, M., Wong, D., Davies, C., Harlow, G. and Mount, D. 1995. *Arabidopsis* mutants deficient in T-DNA integration. Proc. Natl. Acad. Sci. USA. 92:11786-11790.

Stevenson, M. 1996. Portals of entry: uncovering HIV nuclear transport pathways. Trends in Cell Biology 6: 9-15.

Tinland, B. and Hohn, B. 1995. Recombination between prokaryotic and eukaryotic DNA: integration of *Agrobacterium tumefaciens* T-DNA into the plant genome. Pages 209-229 in: Genetic Engineering, Principles and Methods, J.K. Setlow ed. Plenum New York 17:209-229.

Tinland B., Schoumacher, F., Gloeckler, V., Bravo Angel, A.M. and Hohn, B. 1995. The *Agrobacterium tumefaciens* Virulence D2 protein is responsible for precise integration of T-DNA into the plant genome. EMBO J. 13:5764-5771.

Tinland, B. 1996. The integration of T-DNA into plant genomes. Trends in Plant Sciences 1:178-184.

Whittaker, G., Bui, M. and Helenius, A. 1996.The role of nuclear import and export in influenza virus infection. Trends in Cell Biology 6:67-71.

Zupan, J.R. and Zambryski, P. 1995. Transfer of T-DNA from *Agrobacterium* to the Plant Cell. Plant Physiol. 107:1041-1047.

# Transfer and integration of *Agrobacterium tumefaciens* T-DNA in the *Saccharomyces cerevisiae* genome.

Paul Bundock, Amke den Dulk-Ras, Alice Beijersbergen[1], Eddy Risseeuw & Paul J.J.Hooykaas.

Institute of Molecular Plant Scieneces, Clusius Laboratory, Leiden University,Wassenaarseweg 64, 2333 AL, Leiden, The Netherlands.
[1]Present address: Unilever Research Laboratorium Vlaardingen, PO Box 114, 3130 AC,Vlaardingen, The Netherlands.

## Introduction

  *Agrobacterium tumefaciens* is a Gram-negative soil bacterium able to induce tumours, or crown galls, at plant wound sites. During tumorigenesis part of its tumour inducing (Ti) plasmid, the T-DNA, can be mobilized from the bacterium to the plant cell. *A.tumefaciens* induces tumours in a wide range of dicotyledonous plant species, but is unable to do so on monocotyledonous plant species even though T-DNA transfer to monocots has been demonstrated. The transfer of T-DNA from the bacterium to the plant cell is effected by the virulence (*vir*) genes also located on the Ti plasmid. The VirD2 protein covalently attached to the 5' end of the T-DNA contains nuclear localization sequences which target the T-DNA to the plant cell nucleus. Once in the nucleus the T-DNA integrates randomly via illegitimate recombination. Expression of the *onc* genes carried on the T-DNA leads to uncontrolled plant cell proliferation and tumour formation (Hooykaas & Beijersbergen,1994;Zambryski,1992)

  Although much is known about the processing of T-DNA and its transfer to the plant cell, little is known about the plant factors involved in tumorigenesis. Therefore, we were interested to see if *A.tumefaciens* could transfer T-DNA to the yeast *Saccharomyces cerevisiae*, an organism belonging to another kingdom (fungi) than that of plants. Not only would this by itself be interesting, but also this would allow us to develop this organism as model system for studying T-DNA transfer and integration. This yeast is known for its ease of handling, the availability of suitable vectors and its fast growth rate. Another asset is that sequencing of the complete *S.cerevisiae* genome was completed recently. Below we will show that T-DNA transfer and integration in *S.cerevisiae* is indeed possible (Bundock et al. 1995) and that this yeast can serve as a good model organism for studying T-DNA transfer and integration into the recipient genome.

T-DNA transfer from *A.tumefaciens* to *S.cerevisiae*.

Transformation of *S.cerevisiae* with a replicative vector is much more efficient than with an integrative vector. Therefore, in order to establish whether *A.tumefaciens* could transfer T-DNA to *S.cerevisiae*, we constructed the binary vector pRAL7101. Included between the border repeats of the binary vector pBIN19 (Bevan. 1984) were the yeast selection gene *URA3* (Rose et al. 1984) and the yeast 2μ origin of replication. pRAL7101 was electroporated to the *A.tumefaciens* strain LBA1100. LBA1100 contains a helper Ti plasmid pAL1100 carrying the *vir* genes necessary for T-DNA mobilization but lacks the T-DNA as well as genes involved in Ti plasmid conjugation. LBA1100(pRAL7101) was co-cultivated on solid medium with the Ura⁻ haploid *S.cerevisiae* strain M5-1a. As shown in Table 1, Ura⁺ yeast colonies were indeed obtained after co-cultivation of LBA1100(pRAL7101) with M5-1a if the *vir* gene inducing compound acetosyringone (AS) was present in the co-cultivation medium.

| *Agrobacterium* strain | Plasmid | Medium | Yeast strain | Yeast colonies on medium without uracil | Yeast colonies on medium with uracil | Freq.of Ura⁺ colonies per output recipient |
|---|---|---|---|---|---|---|
| LBA1100 | pRAL7101 | +AS | M51a | 272 | $1.6 \times 10^8$ | $1.7 \times 10^{-6}$ |
| | | -AS | | 0 | $0.2 \times 10^8$ | $<5 \times 10^{-8}$ |
| LBA1100 | pRAL7100 | +AS | M51a | 200 | $0.6 \times 10^8$ | $3.3 \times 10^{-6}$ |
| | | -AS | | 0 | $2.2 \times 10^8$ | $<4.5 \times 10^{-9}$ |
| LBA1126 | pRAL7102 | +AS | RSY12 | 12 | $1.7 \times 10^8$ | $7 \times 10^{-8}$ |
| | | -AS | | 0 | $1.5 \times 10^8$ | $<6.6 \times 10^{-9}$ |
| LBA1126 | pRAL7103 | +AS | RSY12 | 16 | $1.2 \times 10^8$ | $1.3 \times 10^{-7}$ |
| | | -AS | | 0 | $1.5 \times 10^8$ | $<6.6 \times 10^{-9}$ |

**Table 1.** T-DNA transfer from *Agrobacterium tumefaciens* to *Saccharomyces cerevisiae*.

Forty eight of these Ura⁺ colonies were further characterized for their T-DNA content. We suspected that a plasmid carrying the 2μ replicator was present in these strains. Therefore, a DNA preparation was made from these strains and transformed to *E.coli*, selecting for carbenicillin resistance (Cbʳ), the selectable marker present on the T-DNA of pRAL7101.

Restriction analysis of the plasmids rescued to *E.coli* showed that three of the forty eight Ura⁺ colonies contained a plasmid formed from circular T-DNA. We concluded that in these co-cultivations a T-DNA copy had been transferred to the yeast cell and the T-DNA had then circularized via a ligation step. Sequencing revealed that the plasmids in these three strains each contained one identical intact border repeat. These border repeats had been formed via a precise fusion of intact T-DNA ends. We believe that the VirD2 protein, which is covalently attached to the 5' end of the T-DNA during transfer, may have carried out this ligation reaction. *In vitro* this protein has nickase/ligase activity (Pansegrau et al, 1993).

Sixteen of the Ura⁺ colonies contained plasmids smaller than expected if T-DNA transfer and circularization had occurred. These plasmids were later found to be the result of homologous recombination between two 150bp direct repeats present on the T-DNA of pRAL7101.

Surprisingly, the remaining twenty nine Ura⁺ contained the entire pRAL7101 plasmid. Control experiments were done to exclude that these strains were generated by uptake of plasmid DNA by the yeast . We could never show uptake of DNA by this yeast, even when large amounts of plasmid DNA were added to control co-cultivations. Therefore, the presence of the entire pRAL7101 plasmid can be best explained by skipping of the LB during T-DNA processing in the bacterial cell. Aberrant T-DNA processing has been shown to occur *in vitro* (Stachel et al, 1987) and after T-DNA transfer to plants (Martineau et al. 1994). Skipping of the LB, transfer of the entire binary vector, and ligation at the RB would be expected to re-create the original pRAL7101 plasmid in *S.cerevisiae*.

To define the *vir* genes necessary for T-DNA transfer to yeast, we introduced pRAL7101 into a series of *A.tumefaciens vir* mutants and carried out co-cultivations with M5-1a (data not shown). We could show that transfer of T-DNA to *S.cerevisiae* required the same *vir* genes that are essential for T-DNA transfer to plants.

**Evidence for T-DNA integration into the *S.cerevsiae* genome.**

Homologous recombination

In plant species (T-)DNA integration occurs via illegitimate recombination (Offringa et al, 1990), but DNA introduced into *S.cerevisiae* integrates predominantly via homologous recombination. We were therefore interested to see how T-DNA would integrate in the yeast genome. Therefore,the integrative T-DNA pRAL7100 was constructed. The T-DNA of this binary vector does not carry a yeast origin of replication and therefore the T-DNA of pRAL7100 can only be maintained after integration into the yeast genome. pRAL7100 carries the the *URA3* gene surrounded by the flanking regions of the *S.cerevisiae PDA1* gene (Bundock et al. 1995).

This could promote homologous recombination between the T-DNA and the *PDA1* locus on chromosome V. The results of the co-cultivations between LBA1100(pRAL7100) and yeast are shown in Table 1. Twenty of these were purified for further characterization. Total yeast DNA was isolated and Southern blots were done to detect the presence of integrated T-DNA at the *PDA1* locus. For 12 of the Ura$^+$ colonies we could show that the T-DNA had recombined with the homologous wild type *PDA1* locus and had introduced the *URA3* gene by a double cross over or gene conversion. Three of the strains contained a integrated complete binary vector which had integrated at the *PDA1* locus via a single cross over.

   We were unable to demonstrate that the T-DNA had integrated in the genome of the remaining five Ura$^+$ colonies. These may have represented ligation of the T-DNA to nuclear located mitochondrial DNA fragments. This has been shown to occur at high frequency after transformation of yeast with dsDNA (Schiestl et al. 1993).

   In our lab we have also used homologous recombination of T-DNA in *S.cerevisiae* to test the suitability of T-DNA to act as a insertion type vector (Risseeuw et al, 1996). An artificial target locus was integrated in *S.cerevisiae* and insertion of the introduced T-DNA at this artificial locus was demonstrated to have occurred via a gap repair mechanism. Risseeuw et al therefore concluded that T-DNA can be used as an insertion type vector, even though this event could not be detected using the same T-DNA construct in plants.

Illegitimate recombination

   T-DNA integrates into the plant genome via illegitimate recombination (IR). IR is a mechanism that joins two DNA molecules that do not share extensive homology. In higher eukaryotic organisms such as plants IR is the predominant mechanism of DNA integration (Offringa et al. 1990; Paskowski et al. 1988). IR of T-DNA in the plant genome has been previously described (Gheysen et al. 1991;Matsumoto et al. 1990;Mayerhofer et al. 1991) but litle is known about the plant factors involved. Therefore, we were interested in whether it was possible to use *S.cerevisiae* as a model organism to also study IR of T-DNA. Although homologous recombination is the dominant form of recombination in *S.cerevisiae*, IR events can be selected for by transforming yeast with stretches of DNA lacking homology with the yeast genome (Schiestl et al. 1994). Therefore we constructed the binary vectors pRAL7102 and pRAL7103. The T-DNA of pRAL7102 carries the *S.cerevisae URA3* gene. In addition to the *URA3* gene, the T-DNA of pRAL7103 also carries a carbenicillin resistance gene and a pUC origin of replication for *E.coli*. Neither of these T-DNAs share extensive homology with the genome of *S.cerevisiae* strain (RSY12) which is deleted for *URA3*. In these co-cultivations we used LBA1126, a derivative of the *A.tumefaciens* helper strain LBA1100. The results of the co-cultivations between LBA1126(pRAL7102) and LBA1126(pRAL7103) and yeast are shown in Table 1. As expected, the frequency with which Ura$^+$ yeast colonies were obtained was very low. To establish if the T-DNA had indeed integrated via illegitimate recombination in these Ura$^+$ colonies, intact chromosomes were isolated,separated on a CHEF gel and then blotted onto a membrane. The blot was then probed using a labelled *URA3* gene. We could show that the T-DNA from LBA1126(pRAL7102) had indeed integrated in a different chromosome in each of the Ura$^+$ colonies (data not shown). This suggested that the T-DNA had indeed integrated via an illegitimate recombination mechanism. To find out whether this was the case, integrated pRAL7103 T-DNA together with flanking yeast sequences were cloned back to *E.coli*. Primers homologous to the T-DNA were then used to obtain the sequence of the genomic DNA flanking the T-DNA. The results showed that T-DNA integration results in deletion of the yeast target DNA, truncation of the borders and in some cases the formation of filler DNA.

Conclusions.

We have shown that T-DNA can be transferred from *Agrobacterium tumefaciens* to the yeast *Saccharomyces cerevisiae*. The frequency with which replacement T-DNA vectors were transferred was similar to that of integrative T-DNA vectors. This is different from yeast transformation or electroporation experiments in which replicative vectors transform yeast much more efficiently than integrative vectors. This difference may be due to the fact that T-DNA is introduced as a linear ssDNA-protein complex. Double strand formation and circularization of T-DNA may be inefficient steps, especially since this does not occur when the T-DNA enters the plant. There it integrates as a linear molecule.

T-DNA integrates into the plant genome by illegitimate recombination (IR). We have shown that T-DNA integrates preferentially into the yeast genome by homologous recombination (HR). The preferred mechanism of T-DNA integration is thus determined mainly by the recipient cell rather than by the incoming T-DNA. If homology with the yeast genome is omitted from the T-DNA, integration occurs by IR. Analysis of the integration sites reveals that IR of T-DNA produces similar target DNA deletions, T-DNA border truncations and filler DNA as seen after T-DNA integration into the plant genome. Therefore, we suggest that plant species and *S.cerevisiae* share a similar mechanism of illegitimate recombination. Genes involved in IR have been identified in *S.cerevisiae* (Schiestl et al. 1994) and it will be interesting to investigate if plant species contain homologues of such genes.

*S.cerevisiae* will prove invaluable as a model system to study the interaction of *Agrobacterium* with its recipient cell and will help identify as yet unknown plant proteins which interact with the T-DNA.

Acknowledgements.

Strain RSY12 was a generous gift from Dr R.Schiestl. We would like to thank Amke Den-Dulk Ras and Emma Scheeren-Groot for construction of LBA1126 and also Yde Steensma and Aloys Teunissen for invaluable help with the CHEF gels.

## References.

Bevan, M. (1984) Agrobacterium vectors for plant transformation. *Nucleic Acids Res.*, **22**, 8711-8721.

Gheysen, G., Villarroel, R. & Van Montagu, M. (1991) Illegitimate recombination in plants: a model for T-DNA integration. *Genes & Dev.*, **5**, 287-297.

Hooykaas, P.J.J. & Beijersbergen, A. (1994) The virulence system of Agrobacterium tumefaciens. *Annu.Rev.Phytopathol.*, **32**, 157-179.

Martineau, B., Voelker, T.A. & Sanders, R.A. (1994) On defining T-DNA. *Plant Cell*, **6**, 1032-1033.

Matsumoto, S., Ito, Y., Hosoi, T., Takahashi, Y. & Machida, Y. (1990) Integration of Agrobacterium T-DNA into a tobacco chromosome: Possible involvement of DNA homology between T-DNA and plant DNA. *Mol.Gen.Genet.*, **224**, 309-316.

Mayerhofer, R., Koncz-Kalman, Z., Nawrath, C., Bakkeren, G., Crameri, A., Angelis, K., Redei, G.P., Schell, J., Hohn, B. & Koncz, C. (1991) T-DNA integration: a mode of illegitimate recombination in plants. *EMBO J.*, **10**, 697-704.

Offringa, R., de Groot, M.J.A., Haagsman, H.J., Does, M.P., van den Elzen, P.J.M. & Hooykaas, P.J.J. (1990) Extrachromosomal homologous recombination and gene targetting in plant cells after Agrobacterium mediated transformation. *EMBO J.*, **9**, 3077-3084.

Pansegrau, W., Schoumacher, F., Hohn, B. & Lanka, E.(1993) Site specific cleavage and joining of single-strand DNA by VirD2 protein of Agrobacterium tumefaciens Ti plasmids: analogy to bacterial conjugation. *Proc.Natl.Acad.Sci.USA*, **90**, 11538-11542.

Paszkowski, J., Baur, M., Bogucki, A. & Potrykus, I. (1988) Gene targetting in plants. *EMBO J.*, **7**, 4021-4026.

Risseeuw, E., Franke-van Dijk, M.E.I. & Hooykaas, P.J.J. (1996) Integration of an insertion type T-DNA vector from Agrobacterium tumefaciens into the Saccharomyces cerevisiae genome by gap repair. *Mol.Cell.Biol.* (in press)

Rose, M., Grisafi, P. & Botstein, D. (1984) Structure and function of the yeast URA3 gene: expression in E.coli. *Gene*, **29**, 113-124.

Schiestl, R.H., Zhu, J. & Petes, T.D. (1994) Effect of mutations in genes affecting homologous recombination on restriction enzyme mediated and illegitimate recombination in Saccharomyces cerevisiae. *Mol.Cell.Biol.*, **14**, 4493-4500.

Schiestl, R.H., Dominska, M. & Petes, T.D. (1993) Transformation of Saccharomyces cerevisiae with non-homologous DNA:Illegitimate integration of transforming DNA into yeast chromosomes and in vivo ligation of transforming DNA to mitochondrial DNA sequences. *Mol.Cell.Biol.*, 13, 2697-2705.

Stachel, S.E., Timmerman, B. & Zambryski, P. (1987) Activation of Agrobacterium tumefaciens vir gene expression generates multiple single-strand molecules from the pTiA6 T-region: requirement for 5' virD products. *EMBO J.*, **6**, 857-863.

# Bacterial Determinants of Pathogenicity and Avirulence--An Overview

Noel T. Keen
Department of Plant Pathology, Univ. of California, Riverside, CA 92521, USA

There has been considerable recent progress toward understanding the role of toxins, enzymes and other bacterial pathogenicity determinants as well as the surveillance mechanisms utilized by plants to detect these pathogens. Indeed, a surprising relationship has emerged between the virulence of bacterial pathogens and their ability to elicit defense reactions in host plant species or cultivars carrying particular disease resistance genes. This was first suggested when Lindgren et al. (1986) showed that certain mutations in *Pseudomonas syringae* pv. *phaseolicola* eliminated both pathogenicity on bean plants and the ability to elicit the hypersensitive response (HR) on a non-host plant, tobacco. The mutated genes occurred in a large cluster, referred to as the *hrp* genes. There have been important recent developments in understanding how *hrp* gene products facilitate plant reactions as well as their interaction with avirulence gene proteins.

## Toxins and Polysaccharides

Progress has continued in understanding the biology of bacterial exotoxins and polysaccharides that are important virulence factors. For example, Liyanage et al. (1995) characterized the gene cluster for synthesis of coronatine by *P. syringae* pv. *glycinea* and defined the region required for biosynthesis of toxin components. Hutchison et al. (1995) studied the mechanism of toxicity by the biosurfactant toxin, syringomycin. These workers demonstrated the formation of ion channels in plant plasma membranes that in turn appear to initiate a cascade of subsequent intracellular signaling events. The groups of K. Geider and D. Coplin transferred amylovoran genes from *Erwinia amylovora* to a stewartan deficient *E. stewartii*, and stewartan genes from *E. stewartii* to an amylovoran deficient *E. amylovora* (Bernhard et al., 1996). The amylovoran genes conferred production of this polysaccharide to *E. stewartii* and restored virulence on corn seedlings. However, while the *E. amylovora* mutant produced stewartan when transformed with these biosynthetic genes, it did not regain full virulence on apple or pear. This suggests that amylovoran has special properties required for virulence on pear and apple but that *E. stewartii* has less stringent structural requirements for its extracellular polysaccharide on corn plants.

## Pectate Lyases

Pectate lyases have been regarded as important virulence factors of soft-rotting erwinias since early in this century. This role has been supported by gene knock-outs of pectate inducible *pel* genes, but these mutants retained the ability to produce limited maceration of plant tissue. Kelemu and Collmer (1993) showed that this was probably due to the presence of additional *pel* genes that are only induced at a high level during plant infection. Two of these genes, *pelL* and *pelZ*, have recently been cloned and characterized (Alfano et al., 1995; Lojkowska et al., 1995; N. Hugouvieux-Cotte-Pattat, personal communication). While *pelL* was not linked to other *pel* genes, *pelZ* occurred immediately 3' to the *pelBC* cluster but lacked significant homology. The increasing number of chain-splitting pectate lyase genes cloned from these bacteria underscores the complexity of the virulence system. Their regulation is also complex (see Surgey et al., 1996) and probably reflects the varied environments in which the bacteria occur as well as their relatively wide host range.

The crystal structures of pectate lyases C and E from *E. chrysanthemi* and a pectate lyase from *Bacillus subtilis* revealed a new structural fold, called the parallel β-helix (see Yoder and Jurnak, 1995). The helical backbone has 22 amino acids per turn and is stabilized by the interactions of similar or identical amino acids arrayed vertically in stacks. These include stacks of asparagines, hydrophobic amino acids and aromatic residues. Amino acids involved in the active site are provided by loop-outs from the parallel β-helix backbone. Subsequent identification of the parallel β-helix in other proteins involved in cell surface interactions raises the possibility that this structural fold has broader biological implications.

The array of pectate lyases produced by *Erwinia* species leads to the suspicion that the various proteins interact differently with plant cells. Indeed, these proteins differ widely in their ability to macerate plant tissue as well as their product spectra from polypectate. Significantly, *E. chrysanthemi* cells mutated in particular *pel* genes showed differential reductions in virulence on particular host plant species (Beaulieu et al., 1993). These observations indicate that considerable complexity exists in the interaction of pectate lyases and the cell walls of various plant species.

Mutation studies have identified several essential amino acids in PelC which appear to be involved in binding of the required cofactor, calcium, or for electron transfer or substrate binding (Kita et al., 1996). Some of the mutants (particularly of lysine 172) were greatly reduced in specific activity on sodium polypectate *in vitro*, but retained significant maceration activity. More surprising was the finding that the lysine 172 mutants retained virtually all of the wildtype activity as elicitors of phytoalexin production in soybean cotyledons. These results also suggest that the interactions between Pel proteins and plant cell walls are more complex than previously suspected.

In order to further investigate the interaction of Pel proteins and plant cell walls, we devised a screen for PelC resistant *Arabidopsis* mutants (C. Wattad and N. Keen, unpublished data). *Arabidopsis* seedlings in trays were sprayed with solutions of pectate lyase C sufficient to macerate and kill greater than 99.99% of the wildtype Columbia plants in 1-3 days. A few M2 plants from EMS treated seed, however, survived exposure to the enzyme. Several of these putative mutant plants have been retained for further characterization. Significantly, a high percentage of the putative mutants exhibit flowering abnormalities. We have also observed that the Mur-1 and Mur-2 mutant lines of ecotype Columbia, isolated in the lab of C. Somerville, are supra-sensitive to maceration by PelC. The Mur-1 mutant was recently shown to involve the replacement of L-fucose by L-galactose in the xyloglucan fraction of the cell wall (Zablackis et al., 1996). Further study of *Arabidopsis* mutants altered in sensitivity to maceration by pectate lyases should illuminate how these proteins interact with plant cell walls.

## The *hrp* Genes and their Relationship to Avirulence Genes

The *hrp* genes of several Gram negative pathogens of plants and animals are remarkably similar and many of these gene products are required for a type III secretion pathway (see Salmond, 1994). As noted earlier, *hrp* gene mutants are not only deficient in pathogenicity on the normal host plant, but also do not elicit the HR in non-host plants. In some cases, these dual properties may be accounted for by the inability of *hrp* mutant strains to deliver harpin and other proteins that are involved in production of the HR on non-host plants and contribute to virulence on host plants (e.g. Bauer et al., 1994). Recently, the *hrp* secretion system has also been suggested to deliver avirulence gene proteins to plant cells. This was first suggested when it was observed that the *hrp* gene regulons as well as all known avirulence genes in *P. syringae* pathovars are controlled by the same promoter, which utilizes a unique sigma factor encoded by *hrpL* (Xiao and

Hutcheson, 1994).  Mutations in certain *P. syringae* avirulence genes (e.g. Ritter and Dangl, 1995) reduce pathogen virulence.  It is therefore appealing to think that the Hrp system delivers proteins which contribute to bacterial virulence in susceptible hosts, but these same proteins may be recognized as elicitors of defense reactions in genetically resistant plants.

Functional alleles of avirulence gene *avrD* are present in many but not all isolates of *P. syringae*.  These genes direct the production of low molecular weight elicitors, called syringolides, which function only in soybean plants carrying the complementary disease resistance gene, *Rpg4* (Keen et al. 1990; Keen and Buzzell, 1991; Midland et al., 1993).  Searches for the production of high or low molecular weight elicitors from other cloned avirulence genes have thus far resulted in negative results. Furthermore, purified avirulence gene proteins have not been observed to elicit defense reactions when infiltrated into the intercellular spaces of resistant host plants.    Bacterial avirulence genes other than *avrD* therefore appear to function differently.   This situation was clarified when Pirhonen et al. (1996) demonstrated that *E. coli* cells carrying the *P. syringae hrp* gene cluster as well as one of several different *P. syringae* avirulence genes (*avrA, avrB, avrRpm1, avrPph3, avrRpm1* or *avrPto*) caused the HR on plant cultivars carrying the matching disease resistance genes.  This was in contrast to the total inactivity of *E. coli* cells expressing the same avirulence genes but lacking a functional *hrp* gene cluster.  The results of Pirhonen et al. also showed that mutations in genes conferring the secretion functions of the *hrp* gene cluster eliminated the ability to deliver avirulence gene signals to plant cells.  These results accordingly raised the possibility that the *hrp* gene system was directly delivering *avr* gene proteins to or into plant cells.

Recently, several laboratories have provided data suggesting that the *P. syringae hrp* genes deliver avirulence gene proteins into plant cells. Introduction of the cloned *avrB* gene into *Arabidopsis* plants carrying the matching disease resistance gene resulted in plant lethality, indicating that the *avrB* protein was an HR elicitor, but only when delivered inside the plant cell (S. He and A. Collmer, personal communication).  Similarly, introduction of the cloned *avrPto* gene into tomato cells with an *Agrobacterium* vector resulted in a necrotic response on *Pto* but not *pto* tomato leaves (B. Staskawicz, personal communication).   These results support the possibility that certain bacterial avirulence gene proteins function as elicitors *per se* when delivered to the plant cell.

### Delivery of avr Elicitor Signals to Plant Cells

Recent evidence also suggests that certain *hrp* gene proteins are involved in construction of a pilus-like bridge between bacteria and plant cells.   The groups of M. Romantschuk and S. He (personal communication) examined

*P. syringae* pv. *tomato* cells grown on *hrp*-inducing media and microscopically observed the production of pilus-like appendages (6 to 8 nm in diameter). These structures did not form on strains carrying *hrpH* or *hrpS* mutations or in wildtype bacteria grown on non-inducing media. The pilus-like structures were shown to be composed of 50 kDa and 10 kDa proteins. N-terminal sequencing of the latter protein disclosed that it was encoded by the *hrpA* gene, located immediately upstream from and in the same transcriptional unit as the *hrpZ* (harpin) gene. As predicted, a non-polar mutation in the *hrpA* gene eliminated production of the Hrp pili and eliminated pathogenicity and ability to elicit the HR. Additional results using the yeast two hybrid system indicate that *avrPto* physically interacts in tomato cells with the resistance gene protein, *Pto*, but not with mutant *Pto* proteins (S. He and B. Staskawicz, personal communications). These exciting results all support the idea that several avirulence gene proteins in *P. syringae* are delivered to the plant cell cytoplasm by a *hrp*-gene mediated pilus and directly function as elicitors.

The results with *P. syringae hrp* genes are consistent with findings involving certain avirulence genes in *Xanthomonas*. The laboratory of Dean Gabriel observed that the *avrBs3* family of avirulence genes in *Xanthomonas campestris* contains functional nuclear localization signals at their conserved carboxyl termini as well as sequences suggestive of eucaryotic transcriptional regulators (Yang and Gabriel, 1995). These observations are consistent with the idea that *hrp* gene products deliver avirulence gene proteins to the plant cell where they may localize to the plant nucleus and specifically affect gene transcription. This is all quite exciting because several of the *Xanthomonas* avirulence genes provoke specific plant responses such as water-soaking or canker formation in compatible host plants (Yang et al., 1996). The suggestion is therefore clear that avirulence gene proteins may function as highly specific virulence factors to pirate plant gene expression. Plants, on the other hand, have evolved plant disease resistance genes which intercede by recognizing the avirulence gene protein or an elicitor specified by it and initiating active defense reactions.

### The Syringolide Elicitors and Binding Sites in Soybean Cells

Results with *P. syringae avrD* also point to the possible occurrence of an intracellular receptor protein in soybean plants. Since the purified syringolides are sufficient to elicit the HR in *Rpg4* soybean cells in the absence of bacteria, the *hrp* gene cluster is not required. This may be due to the amphipathic properties of the syringolides, which permit them to enter and exit both bacterial and plant cells without active transport systems. However, it is also possible that the *avrD* protein may be targeted to plant cells by the Hrp system and generate syringolide production in the plant cell itself.

Recently, radiolabeled syringolides have been used in binding studies with cellular fractions from *Rpg4* and *rpg4* soybean leaves. These studies failed to demonstrate a ligand specific binding site associated with soybean membranes, including the plasma membrane. However, a ligand specific binding site was detected in the soluble fraction of soybean leaf extracts (Y. Okinaka, C. Ji, Y. Takeuchi, N. Yamaoka, M. Yoshikawa, R. Buzzell and N. Keen, unpublished data). The syringolide binding site was protease sensitive and was surprisingly present in both *Rpg4* and *rpg4* cultivars. The binding of radiolabeled syringolides was inhibited by several chemical derivatives in direct proportion to their activity as elicitors in *Rpg4* soybean leaves. Thus, the observed binding site may be physiologically important in the reception of syringolides. It is surprising that the binding site is intracellular and does not seem to be the product of the *Rpg4* gene.

Syringolide affinity columns permitted the isolation of a complex of three proteins (ca. 34, 32 and 29 kDa) from soybean leaf extracts that bound $^{125}$I-syringolide 1 (C. Ji and N. Keen, unpublished data). In crude extracts, syringolide binding occurs to a larger complex of proteins. The 32 and 34 kDa proteins isolated by affinity chromatography also bind ligand in the absence of the other proteins, but with lower affinity. It is therefore possible that the smaller two proteins are proteolytic products of the 34 kDa protein. These proteins may be part of a syringolide receptor in soybean.

Literature Cited

Alfano, J.R., Ham, J.H., and Collmer, A. 1995. Use of Tn5Tac1 to clone a *pel* gene encoding a highly alkaline, asparagine-rich pectate lyase isozyme from an *Erwinia chrysanthemi* EC16 mutant with deletions affecting the major pectate lyase isozymes. J. Bacteriol. 177:4553-4556.

Bauer, D.W., Bogdanove, A.J., Beer, S.V., and Collmer, A. 1994. *Erwinia chrysanthemi hrp* genes and their involvement in soft rot pathogenesis and elicitation of the hypersensitive response. Molec. Plant-Microbe Interact. 7:573-581.

Beaulieu, C., Boccara, M., and Van Gijsegem, F. 1993. Pathogenic behavior of pectinase-defective *Erwinia chrysanthemi* mutants on different plants. Molec. Plant-Microbe Interact. 6:197-202.

Bernhard, F., D. Schullerus, P. Bellemann, M. Nimtz, D.L. Coplin and K. Geider. 1996. Genetic transfer of amylovoran and stewartan synthesis between *Erwinia amylovora* and *Erwinia stewartii*. Microbiology 142:1087-1096.

Hutchison, M.L., Tester, M.A. and Gross, D.C. 1995. Role of biosurfactant and ion channel-forming activities of syringomycin in transmembrane ion flux: a model for the mechanism of action in the plant-pathogen interaction. Molec. Plant-Microbe Interact. 8:610-620.

Keen, N.T., and Buzzell, R.I. 1991. New disease resistance genes in soybean against *Pseudomonas syringae* pv. *glycinea*: evidence that one of them interacts with a bacterial elicitor. Theor. Appl. Genet. 81:133-138.

Keen, N.T., Tamaki, S., Kobayashi, D., Gerhold, D., Stayton, M., Shen, H., Gold, S., Lorang, J., Thordal-Christensen, H., Dahlbeck, D., and Staskawicz, B. 1990. Bacteria expressing avirulence gene D produce a specific elicitor of the soybean hypersensitive reaction. Molec. Plant-Microbe Interact. 3:112-121.

Kelemu, S. and Collmer, A. 1993. *Erwinia chrysanthemi* EC16 produces a second set of plant-inducible pectate lyase isozymes. Appl. Environ. Microbiol. 59:1756-1761.

Kita, N., Boyd, C.M., Garret, M.R., Jurnak, F. and Keen, N.T. 1996. Construction of site-directed mutations in *pelC* and their differential effect on pectate lyase activity, plant tissue maceration and elicitor activity. J. Biol. Chem. (in press).

Lindgren, P.B., Peet, R.C., and Panopoulos, N.J. 1986. Gene cluster of *Pseudomonas syringae* pv. *"phaseolicola"* controls pathogenicity on bean plants and hypersensitivity on nonhost plants. J. Bacteriol. 168: 512-522.

Liyanage, H., Palmer, D.A., Ullrich, M. and Bender, C.L. 1995. Characterization and transcriptional analysis of the gene cluster for coronafacic acid, the polyketide component of the phytotoxin coronatine. Appl. Environ. Microbiol. 61:3843-3848.

Lojkowska, E., Masclaux, C., Boccara, M., Robert-Baudouy, J. and Hugouvieux-Cotte-Pattat, N. 1995. Characterization of the *pelL* gene encoding a novel pectate lyase of *Erwinia chrysanthemi* 3937. Molec. Microbiol. 16:1183-1195.

Midland, S.L., Keen, N.T., Sims, J.J., Midland, M.M., Stayton, M.M., Burton, V., Smith, M.J., Mazzola, E.P., Graham, K.J. and Clardy, J. 1993. The structures of syringolides 1 and 2, novel C-glycosidic elicitors from *Pseudomonas syringae* pv. *tomato*. J. Org. Chem. 58: 2940-2945.

Pirhonen, M.U., Lidell, M.C., Rowley, D.L., Lee, S.W., Jin, S., Liang, Y., Silverstone, S., Keen, N.T., and Hutcheson, S.W. 1996. Phenotypic expression of *Pseudomonas syringae avr* genes in *E. coli* is linked to the activities of the *hrp*-encoded secretion system. Molec. Plant-Microbe Interact. 9: 252-260.

Ritter, C. and Dangl, J.L. 1995. The *avrRpm1* gene of *Pseudomonas syringae* pv. *maculicola* is required for virulence on *Arabidopsis*. Molec. Plant-Microbe Interact. 8:444-453.

Salmond, G.P.C. 1994. Secretion of extracellular virulence factors by plant pathogenic bacteria. Ann. Rev. Phytopathol. 32: 181-200.

Surgey, N., Robert-Baudouy, J. and Condemine, G. 1996. The *Erwinia chrysanthemi pecT* gene regulates pectinase gene expression. J. Bacteriol. 178:1593-1599.

Xiao, Y., and Hutcheson, S.W. 1994. A single promoter sequence recognized by a newly identified alternate sigma factor directs expression of pathogenicity and host range determinants in *Pseudomonas syringae*. J. Bacteriol. 176: 3089-3091.

Yang, Y. and Gabriel, D.W. 1995. *Xanthomonas* avirulence/pathogenicity gene family encodes functional plant nuclear targeting signals. Molec. Plant-Microbe Interact. 8: 627-631.

Yang, Y., Q. Yuan and Gabriel, D.W. 1996. Watersoaking function(s) of XcmH1005 are redundantly encoded by members of the *Xanthomonas avr/pth* gene family. Molec. Plant-Microbe Interact. 9:105-113.

Yoder, M.D. and Jurnak, F. 1995. The parallel β helix and other coiled folds. FASEB J. 9: 335-342.

Zablackis, E., York, W.S., Pauly, M., Hantus, S., Reiter, W-D., Chapple, C.C.S., Albersheim, P. and Darvill, A. 1996. Substitution of L-fucose by L-galactose in cell walls of *Arabidopsis mur1*. Science 272:1808-1810.

# *Pseudomonas syringae hrp* Genes:
# Regulation and Role in Avirulence Phenotypes

Hutcheson, Steven W., Songmu Jin, Michael C. Lidell, and Zhisheng Fu

Department of Plant Biology and the Center for Agricultural Biotechnology of the University of Maryland, College Park MD 20742

*Pseudomonas syringae* strains require the expression of the *hrp* gene cluster for pathogenicity in susceptible plant tissue and the elicitation of the hypersensitive response in resistant plants (see Hutcheson et al. 1996). In the strains characterized thus far, the major *P. syringae hrp* cluster consists of 26 genes organized as at least 8 transcriptional units (Huang et al. 1995). The majority of these *hrp* genes appear to encode for components of a Type III protein secretion system. At least 8 gene products, located in the *hrpJ*, *hrpU*, *hrpC* and *hrpZ* operons, share extensive similarity with components of other Type III protein secretion systems involved in mammalian pathogenesis or flagellar biosynthesis and have been given the designation "Hrc" to reflect this conservation (Bogdanove et al. 1996). Type III secretion systems in other bacteria have been associated with the injection of proteins into eukaryotic cells presumably through some kind of pilus structure (Rosqvist et al. 1994). This chapter will discuss the coordination and environmental regulation of *hrp* gene expression and the function of *hrp* genes in the phenotypic expression of *avr* genes.

## The Regulatory System Controlling *hrp* Expression

Three regulatory determinants have been identified in the *P. syringae hrp* cluster that are required for expression of the operons necessary to form the *hrp*-encoded secretion system (Grimm et al. 1995; Grimm and Panopoulos 1989; Xiao et al. 1994). HrpR and HrpS are unusual members of the enhancer-binding protein family. This protein family includes the effector components of two component regulatory systems that are positive transcriptional factors for $s^{54}$-dependent promoters (Shingler 1996). HrpR and HrpS, which exhibit 57% identity with each other, are much smaller than most other members of the protein family. Domains A and B are absent from HrpR and HrpS whereas domain C (including regions 1-7) and domain D are conserved. The absence of domain A suggests that HrpR and HrpS are not part of a two component regulatory system. The third component, HrpL, shares similarity with the ECF family of alternative sigma factors (Lonetto et al. 1994). The deduced HrpL gene product retains domains 2 and 4 that are broadly conserved among members of the $\sigma^{70}$ family of sigma factors.

A mechanism by which HrpR, HrpS and HrpL coordinate the expression of *hrp* genes was proposed by Xiao et al (1994). This mechanism invokes a regulatory cascade involving three sigma factors (see Hutcheson et al. 1996). At the top level of the cascade are *hrp*RS which appear to be expressed from a $\sigma^{70}$-dependent promoter. HrpR and HrpS in turn were shown to activate the *hrp*L promoter. The *hrp*L promoter includes a $\sigma^{54}$ promoter consensus sequence (14 of 17 nucleotides are conserved) and requires rpoN for activity (Hutcheson et al. 1994). A regulatory site necessary for activation of the *hrp*L promoter is located 100 bp upstream of the $\sigma^{54}$ promoter. Expression of HrpL, the putative ECF sigma factor, was sufficient to activate expression of the *hrpK, hrpJ, hrpU, hrpC and hrpZ* operons as well as most *avr* genes and *hrm*A that form the *hrp* regulon. A HrpL-linked promoter consensus sequence (GGAACCNA N14 CCACNNA) has been identified that is conserved in each of the HrpL-dependent promoters characterized thus far. This conserved sequence is a required cis-acting element for promoter activity.

Recently an alternative mechanism was proposed for regulation of the *P. syringae hrp* cluster in which HrpR activates a $\sigma^{54}$-dependent promoter internal to *hrp*R to induce expression of *hrp*S (Grimm et al. 1995). Expression of *hrp*S was reported to be sufficient to generate an HR positive phenotype in transformants. An important distinction in the experimental approaches used to build each model is while the former used specific promoter reporter gene fusions to monitor regulatory activity the latter monitored the ability of a strain to elicit the HR as an indication of gene expression. Although two different strains were used in these studies, the high sequence conservation in this region argues that these genes should be regulated in a similar manner. To test this alternative model, the *hrpS* coding region was amplified from strain Pss61 by PCR and cloned into pDSK600 such that *hrpS* was constitutively expressed from a strong *lac*UV5 promoter. *E. coli* transformants carrying a plasmid-borne *hrpL* promoter-lacZ fusion and this *hrpS* construct exhibited increased but low basal expression of the reporter gene (Jin and Hutcheson, unpublished results). This increased activity, however, was less than 1% that observed when a construct carrying the native *hrpRS* region was introduced into the strain and less than 5% the activity observed when a *hrpR* construct was transformed into the strain. Similar results were obtained in *P. syringae* strains. Since full activation of the *hrpL'-lacZ* reporter fusion was only observed when both *hrpR* and *hrpS* are expressed, these results are most consistent with the original model for HrpR and HrpS regulatory activity. It may be that the expression induced by engineered expression of *hrp*S under these conditions is sufficient to form a *hrp*-encoded secretion system. Previous studies have suggested that low level expression of the *hrp* cluster is sufficient to generate a HR positive phenotype in *P. syringae* strains.

Expression of the *P. syringae hrp* regulon has been shown to be affected by several environmental conditions, such as medium composition (carbon source and presence of broad spectrum amino acid sources), osmolarity, and pH (Huynh et al. 1989; Rahme et al. 1992; Xiao et al. 1992). An argument has also been made that plant factors may also influence expression of the *hrp* regulon (Rahme et al. 1992). In most strains, expression of the *hrp* regulon can be induced in culture by growth in a minimal salts medium with low osmolarity and a slightly acidic pH. Recently we have shown that expression of the *hrp* regulon is also affected by temperature. Temperatures less that 20° C were found to specifically suppress the activity of several *hrp* promoters in *P. syringae* transformants (Fu and Hutcheson, manuscript in preparation).

The mechanism by which medium composition and temperature affect expression of the *hrp* regulon genes in *P. syringae* strains has not been established. Transduction of these environmental signals is thought to involve additional regulatory determinants (see Hutcheson et al. 1996). Promoters at each level of the regulatory cascade appear to be regulated independently by medium composition whereas temperature-dependent regulation of the *hrp* cluster occurs at the level of the *hrp*L promoter (Fu and Hutcheson, manuscript in preparation). At least part of the media-dependent regulation of the *hrp*L promoter may be due to exponential silencing of $\sigma^{54}$ (Cases et al. 1996). There is also some evidence for an indirect auto-regulation of *hrp*L (Rowley and Hutcheson, unpublished results).

In other Type III secretion systems, regulatory systems coordinate secretion activities and transcription of secretion components (see Hughes et al. 1993; Plano et al. 1991). In *P. syringae* strains, there is also evidence for an interaction between the activities of the *hrp*-encoded secretion system and expression of *hrp* regulon genes (Rowley, Fu and Hutcheson, unpublished results). Polar mutations in *hrp*J and *hrc*V inhibit the activity of a plasmid-borne *hrp*Z promoter fusion in *P. syringae* transformants. Plasmid-directed expression of *hrp*AZ strongly suppressed *hrp*Z promoter activity in a *P. syringae* transformant. These results suggest that a regulatory system similar to that controlling flagellar biosynthesis may affect expression of *hrp* regulon genes.

## Interactions between *hrp* and *avr* Genes

The *avr* genes encode for host range determinants whose phenotype is only evident in plant tissue expressing a corresponding resistance gene (Staskawicz et al. 1995). Lack of phenotype in bacterial *hrp* mutants (Huynh et al. 1989; Knoop et al. 1991; Lindgren et al. 1988) and inclusion of *avr* genes in the *hrp* regulon as discussed above suggest that the Avr phenotype of bacterial strains may be dependent upon the activities of the *hrp* cluster (see Fenselau et al. 1992). Phenotypic expression of several *P. syringae avr* genes in *E. coli* was recently shown to be dependent upon the secretion activities encoded by the *P. syringae hrp* cluster (Pirhonen et al. 1996). *E. coli* transformants expressing both the *P. syringae hrp* cluster and one of several *P. syringae avr* genes exhibited the expected Avr phenotype in indicator soybean and *Arabidopsis thaliana* plants (i.e., elicited the HR only in plants expressing a corresponding resistance gene). This activity was specifically dependent upon several *hrp* genes whose deduced gene products are conserved and essential components of Type III protein secretion systems but did not require harpin production. Subsequent studies showed that *hrpA* was also essential for this activity (Jin and Hutcheson, unpublished results). Since most plant resistance gene products that interact with *P. syringae avr* gene products appear to be cytoplasmic proteins and Type III secretion systems have been proposed to function as protein injection systems, it seems likely that Avr phenotype results from Hrp-mediated injection of an Avr gene product into plant cells.

Our attempts to directly detect *hrp*-mediated secretion of AvrB by monitoring localization of several AvrB' translational fusions with green fluorescent protein or b-glucuronidase during interactions of *E. coli* and *P. syringae* transformants with cultured *A. thaliana* cells have been unsuccessful thus far. Instead a TMV-based expression system, developed by J. Culver, was used to show that the AvrB gene product appears to be phenotypically active when produced in planta (Lidell, Culver and Hutcheson, manuscript in preparation). This viral expression system expresses a gene cloned into the TMV coat protein region of the TMV genome. TMV has been shown to replicate asymptomatically in *A. thaliana* (see Urban et al. 1990). By monitoring immunologically for presence of the 126 kDa TMV replicase, TMV-*avr*B construct was found to replicate in a *rpm1* mutant of A. thalina Col-0 but not in Col-0 which expresses *RPM1*. In contrast, a TMV-*avr*B derivative carrying a frameshift deletion in the cloned *avr*B gene was able to replicate in Col-0. These results then are consistent with the model that the *hrp*-encoded secretion system functions to inject Avr gene products into plant cells. The inability to directly detect secretion of the Avr gene product by using fusions or immunological approaches may be due to low level secretion of the Avr gene product, posttranslational processing of the fusion or inability of the secretion system to secrete these fusions. This observations predicts that *avr* and R gene products will directly interact during the recognition process that initiates a defense response during gene-for-gene interactions.

Finally, a key step in this process appears to be adhesion of the bacteria to the plant cells. *P. syringae* strains adhere strongly to cultured tobacco and *A. thaliana* cells whereas *E. coli* transformants carrying an expressed *hrp* gene cluster are only weakly adsorbed. The weak adsorption of *E. coli* transformants to plant cells may account for the higher inoculum levels required by these bacteria to elicit plant responses. Hutcheson et al. (1989) reported that the magnitude of the hypersensitive response of cultured tobacco cells was correlated with the number of adsorbed bacteria. It may be that phytopathogenic bacteria produce adhesins similar to those produced by mammalian pathogens.

LITERATURE CITED

Bogdanove, A. J., Beer, S. V., Bonas, U., Boucher, C. A., Collmer, A., Coplin, D. L., Cornelis, G. R., Huang, H. C., Hutcheson, S. W., Panopoulos, N. J., and VanGijsegem, F. 1996. Unified nomenclature for broadly conserved *hrp* genes of phytopathogenic bacteria. Mol. Microbiol. 20: 681-683.

Cases, I., deLorenzo, V., and J., P.-M. 1996. Involvement of $\sigma^{54}$ in exponential silencing of the *Pseudomonas putida* TOL *Pu* promoter. Mol. Microbiol. 19: 7-17.

Fenselau, S., Balbo, I., and Bonas, U. 1992. Determinants of pathogenicity in *Xanthomonas campestris* pv. vesicatoria are related to proteins involved in secretion in bacterial pathogens of animals. Mol. Plant-Microbe Interact. 5: 390-396.

Grimm, C., Aufsatz, W., and Panopoulos, N. J. 1995. The *hrpRS* locus of *Pseudomonas syringae* pv. phaseolicola constitutes a complex regulatory unit. Mol. Microbiol. 15: 155-165.

Grimm, C., and Panopoulos, N. J. 1989. The predicted protein product of a pathogenicity locus from *Pseudomonas syringae* pv. phaseolicola is homologous to a highly conserved domain of several prokaryotic regulatory proteins. J. Bacteriol. 171: 5031-5038.

Huang, H. C., Lin, R. H., Chang, C. J., Collmer, A., and Deng, W. L. 1995. The complete *hrp* gene cluster of *Pseudomonas syringae* pv. *syringae* 61 includes two blocks of genes required for harpin$_{Pss}$ secretion that are arranged colinearly with *Yersinia ysc* homologs. Mol. Plant-Microbe Interact. 8: 733-746.

Hughes, K. T., Gillen, K. L., Semon, M. J., and Karlinsky, J. E. 1993. Sensing structural intermediates in bacterial flagellar assembly by export of a negative regulator. Science 262: 1277-1280.

Hutcheson, S. W., Collmer, A., and Baker, C. J. 1989. Elicitation of the hypersensitive response by *Pseudomonas syringae*. Physiol. Plant. 76: 155-163.

Hutcheson, S. W., Heu, S., Jin, S., Lidell, M. C., Pirhonen, M. U., and Rowley, D. L. 1996. Function and regulation of *Pseudomonas syringae hrp* genes. Pages 512-521 in: *Pseudomonas*: Molecular Biology and Biotechnology. T. Nakazawa, K. Furukawa, S. Harayama and S. Silvers, eds. American Society for Microbiology, Washington, D.C.

Hutcheson, S. W., Heu, S., and Xiao, Y. 1994. Mechanism for environmental regulation of *Pseudomonas syringae* pathogenicity and host range determinants. Pages 33-36 in: Advances in the molecular genetics of plant-microbe interactions. Vol. 3. M. J. Daniels, J. A. Downie and A. E. Osbornes, eds. Kluwer Academic Publishers, Dordrecht.

Huynh, T., Dahlbeck, D., and Staskawicz, B. J. 1989. Bacterial blight of soybean: Regulation of a pathogen gene determining host cultivar specificity. Science 245: 1374-1377.

Knoop, V., Staskawicz, B. J., and Bonas, U. 1991. The expression of the avirulence gene *avr*Bs3 from *Xanthomonas campestris* pv. vesicatoria is not under the control of *hrp* genes and is independent of plant factors. J. Bacteriol. 173: 7142-7150.

Lindgren, P. B., Panopoulos, N. J., Staskawicz, B. J., and Dahlbeck, D. 1988. Genes required for pathogenicity and hypersensitivity are conserved and interchangable among pathovars of *Pseudomonas syringae*. Mol. Gen. Genet. 21: 499-506.

Lonetto, M. A., Brown, K. L., Rudd, K. E., and Bittner, M. J. 1994. Analysis of the *Streptomyces coelicolor sig*E gene reveals the existence of a subfamily of eubacterial RNA polymerase σ factors involved in the regulation of extracytoplasmic functions. Proc. Natl. Acad. Sci. USA 91: 7573-7577.

Pirhonen, M. U., Lidell, M. C., Rowley, D., Lee, S. W., Silverstone, S., Liang, Y., Keen, N. T., and Hutcheson, S. W. 1996. Phenotypic expression of *Pseudomonas syringae avr* genes in *E. coli* is linked to the activities of the *hrp*-encoded secretion system. Mol. Plant-Microbe Interact. 9: 252-260.

Plano, G. V., Barve, S. S., and Straley, S. C. 1991. LcrD, a membrane-bound regulator of the *Yersinia pestis* low calcium response. J. Bacteriol. 173: 729-7303.

Rahme, L. G., Mindronos, M. N., and Panopoulos, N. J. 1992. Plant and environmental sensory signals control the expression of *hrp* genes in *Pseudomonas syringae* pv. phaseolicola. J. Bacteriol. 174: 3499-3507.

Rosqvist, R., Magnusson, K. E., and Wolf-Watz, H. 1994. Target cell contact triggers expression and polarized transfer of *Yersinia* YopE cytotoxin into mammalian cells. EMBO J. 13: 964-972.

Shingler, V. 1996. Signal sensing by σ[54]-dependent regulators: derepression as a control mechanism. Mol. Microbiol. 19: 409-416.

Staskawicz, B. J., Ausubel, F. M., Baker, B. J., Ellis, J. G., and Jones, J. D. G. 1995. Molecular genetics of plant disease resistance. Science 268: 661-667.

Urban, L. A., Sherwood, J. L., Rezende, J. A. M., and Melcher, U. 1990. Examination of mechanisms of cross protection with non-transgenic plants. Pages 415-426 in: Recognition and response in plant-virus interactions. Vol. H41. R. S. S. Frasers, eds. Springer-Verlag, Berlin.

Xiao, Y., Heu, S., Yi, J., Lu, Y., and Hutcheson, S. W. 1994. Identification of a putative alternate sigma factor and characterization of a multicomponent regulatory cascade controlling the expression of *Pseudomonas syringae* pv. *syringae* Pss61 *hrp* and *hrmA* genes. J. Bacteriol. 176: 1025-1036.

Xiao, Y., Lu, Y., Heu, S., and Hutcheson, S. W. 1992. Organization and environmental regulation of the *Pseudomonas syringae* pv. *syringae* 61 *hrp* cluster. J. Bacteriol. 174: 1734-1741.

# Secreted Proteins, Secretion Pathways, and the Plant Pathogenicity of *Erwinia chrysanthemi* and *Pseudomonas syringae*

Alan Collmer[a], James R. Alfano[a], David W. Bauer[a], Gail M. Preston[a], Amy O. Loniello[a], Alison Conlin[a], Jong Hyung Ham[a], Hsiou-Chen Huang[b], Suresh Gopalan[c], and Sheng Yang He[c].
[a]Department of Plant Pathology, Cornell University, Ithaca, NY 14853-4203 USA. [b]Agricultural Biotechnology Laboratories, National Chung-Hsing University, Taichung, Taiwan 40227 ROC. [c]Department of Energy Plant Research Laboratory, Michigan State University, East Lansing, Michigan 48824-1312 USA.

The type III protein secretion pathway is important for the virulence of several plant and animal pathogens (Salmond 1994). We are exploring the involvement of type III (Hrp) protein traffic in the plant interactions of two contrasting plant pathogens, *Erwinia chrysanthemi* and *Pseudomonas syringae*, which typically cause host-promiscuous soft rots and host-specific leaf spots, respectively (Collmer and Bauer 1994). Both bacteria share the ability to elicit the hypersensitive response (HR), a rapid, localized, apparent programmed cell death that is associated with the defense of plants against "incompatible" host-specific pathogens (Goodman and Novacky 1994). Although the ability of *E. chrysanthemi* to elicit the HR is masked by the killing and maceration effects of its abundant extracellular pectate lyase (Pel) isozymes (secreted by the type II pathway), Pel-deficient mutants elicit a typical HR in tobacco (Bauer *et al.* 1994). The ability to elicit the HR is correlated with the ability to be pathogenic in (nontumorigenic) Gram-negative bacteria, and by exploring HR elicitation in these two contrasting pathogens, we hope to ultimately learn more about pathogenesis.

## *hrp* Genes and the Type III Secretion System

The ability of incompatible *P. syringae* pathovars and nonpectolytic *E. chrysanthemi* mutants to elicit the HR resides in *hrp* (*h*ypersensitive response and *p*athogenicity) genes, which may be universal to nontumorigenic phytopathogenic bacteria (Bonas 1994). *hrp* genes are carried in ca. 25-kb clusters, and the cluster from *P.s. syringae* 61, carried on cosmid pHIR11, permits nonpathogens to elicit the HR (but not cause disease) in tobacco and several other plants (Huang *et al.* 1988). As shown in Figure 1, the *P.s. syringae* 61 *hrp* genes includes nine that have been redesignated as *hrc* because they are conserved components of the type III protein secretion system in plant pathogenic *Pseudomonas*, *Xanthomonas*, and *Erwinia* spp. and animal pathogenic *Yersinia*, *Shigella*, and *Salmonella* spp.

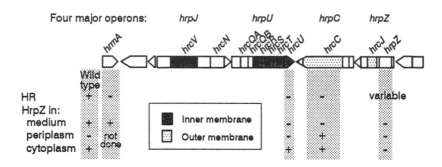

**Fig. 1.** Arrangement of *hrc* genes predicted to encode inner and outer membrane proteins in the four major *hrp* operons in pHIR11 (directions of transcription indicated). The phenotypes of key mutations are shown in the shaded areas below the respective genes. The HR phenotypes given are for *P. fluorescens* carrying mutated derivatives of pHIR11, as discussed in the text.

Little is known about mechanisms of protein secretion by type III systems, but the grouping of genes encoding putative inner and outer membrane proteins in pHIR11 suggests that HrpZ (discussed below) translocation across the two bacterial membranes may be differentially directed by the *hrpJ/U* and *hrpC/Z* operons, respectively. To test this hypothesis, we used polyclonal anti-HrpZ antibodies and immunoblot analysis to determine the distribution of HrpZ in the cytoplasmic, periplasmic, and extracellular fractions of wild-type and mutant *P.s. syringae* 61 cultures. The results summarized in Figure 1 indicate that HrpZ secretion through the Hrp pathway can be genetically dissected into two steps, although a free periplasmic intermediate may not occur during native secretion. A functionally nonpolar *hrcC::nptII* (terminatorless cartridge) mutation was used in the experiment depicted in Figure 1, but identical HrpZ distribution patterns were observed with *hrcC*::Tn*phoA* and *hrcJ*::Tn*phoA* mutants. Similarly, $\Omega Sp^R$ insertions affecting all *hrc* genes in the *hrpJ* and *hrpU* operons produced the same phenotype as indicated for *hrcU*::Tn*phoA*. All of the Hrc proteins encoded in the *hrpJ/U* operons have flagellar secretion/biogenesis homologs, and flagellar system components would be particularly suitable for a secretion system that is known in *Yersinia*, *Shigella*, and *Salmonella* spp. to be capable of delivering virulence proteins into host cells (reviewed in Galan 1996).

## Relative Roles of Harpins and the Hrp Secretion System in Elicitation of the HR by Different Bacteria

The *E. chrysanthemi* EC16 *hrpN_{Ech}* and *P.s. syringae* 61 *hrpZ* genes encode harpins, which are glycine-rich, cysteine-lacking proteins that are secreted in culture by the Hrp system and possess heat-stable HR elicitor activity in tobacco and other plants (He *et al.* 1993; Bauer *et al.* 1995). *hrp* secretion mutants and *hrpN_{Ech}* mutants in *E. chrysanthemi* no longer elicit the HR (as observable in *pel* mutants), and they have a reduced infectivity in witloof chicory leaves but a wild-type capacity to cause maceration (Bauer *et al.* 1994; Bauer *et al.* 1995). This suggests that *HrpN_{Ech}* is a key protein travelling the Hrp pathway, and it may contribute to a "pre-pectolytic" phase of pathogenesis (Collmer and Bauer 1994). In contrast, *hrp* secretion mutations in *P.s. syringae* 61 completely abolish HR and pathogenicity phenotypes, but functionally nonpolar Δ*hrpZ* mutations only partially reduce HR elicitation activity in tobacco (Alfano *et al.* 1996). Recently constructed, functionally nonpolar Δ*hrpZ* mutations in *P.s. syringae* B728a, *P. s. tomato* DC3000, and *P.s. glycinea* race 4 similarly retain HR elicitor activity but are subtly altered in their pathogenicity on compatible hosts. These observations highlight the importance of the Hrp secretion pathway in *P. syringae* pathogenesis, but diminish the importance of HrpZ in that traffic. Furthermore, deletion of *hrpZ* from pHIR11 leaves *P. fluorescens* cells carrying the derivative still able to elicit a variable and spotty necrosis (Alfano *et al.* 1996).

## Evidence that HrmA is an Avr Protein

As indicated in Figure 1, *P. fluorescens*(pHIR11) *hrmA* mutants secrete wild-type levels of HrpZ but fail to elicit the HR (Alfano *et al.* 1996). *hrmA*, which encodes a 41.5 kD protein of unknown biochemical function (Heu and Hutcheson 1993), has several characteristics of an *avr* gene that interacts with an undefined resistance (*R*) gene in tobacco (Alfano *et al*. 1996). Most importantly, DNA gel blots revealed that *P. s. tabaci* lacks *hrmA*, but transformants expressing the *P.s. syringae* 61 *hrmA* gene elicited the HR rather than disease in tobacco. This suggested that HrmA is directly involved in the elicitation of the HR in tobacco by *P. fluorescens*(pHIR11). Given the potential of the type III system to deliver proteins directly into host cells, it raised the possibility that HrmA acts as an elicitor inside rather than outside tobacco cells. Evidence for this hypothesis was gained by adapting the biolistic cobombardment assay of Mindrinos *et al*. (1994): suspension-cultured tobacco cells cobombarded with plasmids expressing a *GUS* reporter and *hrmA* failed to produce β-glucuronidase activity, presumably because of HR elicitation, whereas cells similarly cobombarded with the expression vector instead of *hrmA* showed strong β-glucuronidase activity.

### Evidence that Hrp-Delivered AvrB Acts within Plant Cells

More compelling evidence for Avr protein action in plant cells was found with *P.s. glycinea* AvrB (Gopalan *et al*. 1996). *avrB* interacts with the *RPG1* and *RPM1 R* genes in soybean and Arabidopsis, respectively, to elicit a genotype-specific or "gene-for-gene" HR (Staskawicz *et al*. 1995). *P. fluorescens*(pHIR11) does not effectively elicit the HR in these two plants, but cells carrying pAVRB1 or expressing AvrB tagged with the FLAG epitope (IBI, New Haven, CT) elicited a strong HR in all plants carrying the cognate *R* genes. *P. fluorescens*(pHIR11) carrying *hrcC*::Tn*phoA* or nonpolar Δ*hrpZ* mutations failed to elicit an AvrB-dependent HR, suggesting that the Hrp secretion system and HrpZ were both involved in delivering the AvrB signal. Pirhonen *et al*. (1996) similarly observed that *E. coli* MC4100(pHIR11) could deliver the signals for AvrB and several other *P. syringae* Avr proteins. Interestingly, the requirement for HrpZ was variable and not obligate, possibly because the *hrp* genes were hyperexpressed through HrpL alternate sigma factor manipulation. In *P. fluorescens*(pHIR11), the requirement for a complete Hrp system could not be obviated by supplying exogenous HrpZ or AvrB, by overexpressing *avrB* within mutants, or by using other independently constructed (Alfano *et al*. 1996) nonpolar *hrpZ* mutants. The results suggested that AvrB does not act within plant cells or the apoplast and left the interior of the plant cell as the default site of action.

Direct evidence for this site of AvrB action was found by transforming an Arabidopsis *rpm1* mutant with constructs expressing

*avrB*, and then crossing *avrB*⁺ transgenic plants with wild-type *RPM1*
plants (Gopalan *et al.* 1996). F1 seedlings carrying both genes
exhibited extensive necrosis on cotyledon leaves 10 days post-
germination. Interestingly, the only symptomless transformants
obtained in the *rpm1* mutant were those in which AvrB was produced
at a low level and with a signal peptide fusion such that the protein
would be present in the plant cytoplasm only transiently. To express
*avrB* without a signal peptide in Arabidopsis and soybean leaves, the
previously described biolistic cobombardment method was used. With
both plants, no $\beta$-glucuronidase activity developed in leaves containing
AvrB and either *RPM1* or *RPG1*. Thus, both stable and transient
expression experiments support the hypothesis that AvrB (and
probably other Avr proteins) act in plants following delivery by the
Hrp secretion system. This further suggests that Avr proteins have an
underlying function (perhaps subtle) in promoting bacterial parasitism
in plants.

## Acknowledgments

This research was supported in part by NSF Grant No. MCB
9305178 and NRI Competitive Grants Program/USDA Grant No. 94-
37303-0734. We thank Noel T. Keen for pAVRB1.

## Literature Cited

Alfano, J. R., Bauer, D. W., Milos, T. M., and Collmer, A. 1996.
Analysis of the role of the *Pseudomonas syringae* pv. *syringae* HrpZ
harpin in elicitation of the hypersensitive response in tobacco using
functionally nonpolar deletion mutations, truncated HrpZ
fragments, and *hrmA* mutations. Mol. Microbiol. 19:715-728.
Bauer, D. W., Bogdanove, A. J., Beer, S. V., and Collmer, A. 1994.
*Erwinia chrysanthemi hrp* genes and their involvement in soft rot
pathogenesis and elicitation of the hypersensitive response. Mol.
Plant-Microbe Interact. 7:573-581.
Bauer, D. W., Wei, Z.-M., Beer, S. V., and Collmer, A. 1995. *Erwinia
chrysanthemi* harpin$_{Ech}$: an elicitor of the hypersensitive response
that contributes to soft-rot pathogenesis. Mol. Plant-Microbe
Interact. 8:484-491.
Bonas, U. 1994. *hrp* genes of phytopathogenic bacteria. Pages 79-98 in:
Current Topics in Microbiology and Immunology, Vol. 192:
Bacterial Pathogenesis of Plants and Animals - Molecular and
Cellular Mechanisms. J. L. Dangl, ed. Springer-Verlag, Berlin.

Collmer, A., and Bauer, D. W. 1994. *Erwinia chrysanthemi* and *Pseudomonas syringae*: Plant pathogens trafficking in virulence proteins. Pages 43-78 in: Current Topics in Microbiology and Immunology, Vol. 192: Bacterial Pathogenesis of Plants and Animals - Molecular and Cellular Mechanisms. J. L. Dangl, ed. Springer-Verlag, Berlin.

Galan, J. E. 1996. Molecular genetic bases of *Salmonella* entry into host cells. Mol. Microbiol. 1996:263-271.

Goodman, R. N., and Novacky, A. J. 1994. The Hypersensitive Reaction of Plants to Pathogens. A Resistance Phenomenon. APS Press, St. Paul.

Gopalan, S., Bauer, D. W., Alfano, J. R., Loniello, A. O., He, S. Y., and Collmer, A. 1996. Expression of the *Pseudomonas syringae* avirulence protein AvrB in plant cells alleviates its dependence on the hypersensitive response and pathogenicity (Hrp) secretion system in eliciting genotype-specific hypersensitive cell death. Plant Cell (in press).

He, S. Y., Huang, H.-C., and Collmer, A. 1993. Pseudomonas syringae pv. syringae harpin$_{Pss}$: a protein that is secreted via the Hrp pathway and elicits the hypersensitive response in plants. Cell 73:1255-1266.

Heu, S., and Hutcheson, S. W. 1993. Nucleotide sequence and properties of the *hrmA* locus associated with the *Pseudomonas syringae* pv. *syringae* 61 *hrp* gene cluster. Mol. Plant-Microbe Interact. 6:553-564.

Huang, H.-C., Schuurink, R., Denny, T. P., Atkinson, M. M., Baker, C. J., Yucel, I., Hutcheson, S. W., and Collmer, A. 1988. Molecular cloning of a *Pseudomonas syringae* pv. *syringae* gene cluster that enables *Pseudomonas fluorescens* to elicit the hypersensitive response in tobacco. J. Bacteriol. 170:4748-4756.

Mindrinos, M., Katagiri, F., Yu, G.-L., and Ausubel, F. M. 1994. The A. thaliana disease resistance gene *RPS2* encodes a protein containing a nucleotide-binding site and leucine-rich repeats. Cell 78:1089-1099.

Pirhonen, M. U., Lidell, M. C., Rowley, D. L., Lee, S. W., Jin, S., Liang, Y., Silverstone, S., Keen, N. T., and Hutcheson, S. W. 1996. Phenotypic expression of *Pseudomonas syringae avr* genes in *E. coli* is linked to the activities of the *hrp*-encoded secretion system. Mol. Plant-Microbe Interact. 9:252-260.

Salmond, G. P. C. 1994. Secretion of extracellular virulence factors by plant pathogenic bacteria. Annu. Rev. Phytopathol. 32:181-200.

Staskawicz, B. J., Ausubel, F. M., Baker, B. J., Ellis, J. G., and Jones, D. G. 1995. Molecular genetics of plant disease resistance. Science 268:661-667.

# Genetic and molecular dissection of the *hrp* regulon of *Ralstonia (Pseudomonas) solanacearum*

Marenda, M. ; Van Gijsegem, F. ; Arlat, M. ; Zischek, C.; Barberis, P.; Camus, J. C. ; Castello, P. and Boucher, C.A.

Laboratoire de Biologie Moléculaire des Relations Plantes-Microorganismes, CNRS-INRA, Castanet Tolosan, France.

Like most phytopathogenic gram-negative bacteria, *Ralstonia solanacearum* [previously refered to as *Pseudomonas solanacearum* (Yabbuchi *et al.*, 1995)] possesses an *hrp* gene cluster of about 25 kb governing both the ability to elicit a hypersensitive response (HR) on non-host or resistant plants and to produce disease on susceptible plants. The *R. solanacearum hrp* genes are organized in 7 transcriptional units, 5 of which are indispensable for interactions with plants (units 1 to 4 and 7), while mutants in transcription units 5 and 6 produce leaky phenotypes (Fig. 1) (Van Gijsegem *et al.*, 1995).

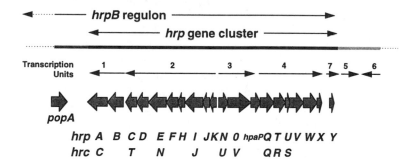

Fig 1 Genetic organization of the *hrp* gene cluster of *R. solanacearum.*

Transcription units 1 to 4 have been entirely sequenced revealing a coding capacity for 19 proteins. However due to the polar effect of Tn5 induced mutations, it was not possible to ascertain the role of each individual *hrp* gene in the control of plant bacteria interactions. It was known from previous experiments that mutations in *hpaP,* the 3' distal gene of transcription unit 3, had no effect on HR inducing ability and on pathogenicity of the bacteria. This indicates that certain *hrp* genes might, at least under certain conditions, be dispensable for normal interactions with plants (Gough *et al.*, 1993).

*R. solanacearum hrp* genes have been shown to control the biogenesis of a secretion system which is required for the transit of the PopA1 protein towards the extracellular space (Arlat *et al.*, 1994). This protein which acts as an HR elicitor on tobacco and on certain petunia genotypes, is encoded by a gene located to the left hand end of the *hrp* gene cluster. As for *hrp* genes belonging to transcription units 1-4, expression of the *popA* gene as well as expression of other surrounding genes is dependent on the *hrpB* regulatory protein. This defines a *hrp* regulon which includes *hrp* transcription units 1-4 and 7 and whose left border remains to be defined.

In this paper we present the strategies which have been used to establish the role of individual *hrp* genes in the process of HR elicitation and disease development and to define the role of the left hand end of the *hrp* regulon in the control of plant-bacteria interactions. This has led to the identification of proteins which are good candidates for secretion through the *hrp* secretion machinery.

### Non-polar mutagenesis of the *R. solanacearum hrp* gene cluster

Transcription units 1 to 4 encode 19 ORFs including the *hrpB* regulatory gene. Among the 18 other *hrp*-encoded ORFs, nine are conserved in all *hrp* systems studied so far. Homologues of these proteins are also present in several animal pathogenic enterobacteria such as *Yersinia*, *Salmonella* or *Shigella* and they were shown to be components of a new so called type III secretion machinery which delivers active proteins to host cells (Bonas, 1994, Van Gijsegem *et al.*, 1995). The *hrp* genes encoding these conserved proteins have been recently renamed *hrc* (Bogdanove *et al.*, 1996). Based on sequence and biochemical data, the corresponding Hrc proteins consist of one outer membrane protein (HrcC), one membrane-associated lipoprotein (HrcJ), an ATP-binding protein (HrcN) and 5 polytopic inner membrane proteins (HrcR to V). The HrcQ protein does not have characteristics allowing us to predict a precise localization in the cell.

Two other Hrp proteins, HrpF and HrpW, share similarities with proteins present in some but not all type III secretion systems. No similarities were found in data bases for the 5 remaining Hrp proteins, except for the highly related Hrp system of *Xanthomonas* species.

To assess the role of each Hrp protein in interactions with plants, non-polar mutants in each (except *hrcU* and *hrcQ* to date) of the *hrp* genes were constructed using the *apha3* cassette devised by Ménard et al.. (1993). This cassette is designed to split the studied gene in two parts, generating two truncated proteins corresponding to the parts of the gene located upstream and downstream of the point of insertion of the cassette.

The non-polar mutants constructed either using the *apha3* cassette, or by Tn5B20 insertions for the genes located at the end of a transcription unit, were tested for their ability to elicit the HR on the non-host plant tobacco and for their virulence on the susceptible plant tomato.

The *hrpB*, *hrcN* and *hrcR* to *V* as well as the *hrpF* and *hrpW* mutants did not elicit any reaction on either plant. The *hrcJ* mutant provoked a partial HR when highly concentrated bacterial suspensions were infiltrated into tobacco leaves and this mutant was also able to partially wilt tomato plants. Insertion of the *apha3* cassette in this gene produces a truncated protein lacking only the 70 carboxy-terminal amino acids raising the possibility that such a protein may retain partial activity. Other constructions in which the cassette will be inserted in other positions in the gene or with partial internal deletions are underway to test this hypothesis.

For the *hrp* genes encoding proteins with no similarities in data banks, the *hrpK* and *hrpX* mutants were also totally Hrp⁻. The *hrpH* and the *hpaP* mutants behaved like the wild type parental strain on both plants, the *hrpD* mutant elicited a normal HR on tobacco but was hypovirulent on tomato and the *hrpJ* and *hrpV* mutants elicited partial responses on both plants.

The null phenotype found for the *hrc* (with the exception of *hrcJ*) and *hrpF, K, W, X* mutants strongly suggests that the encoded proteins are parts of a secretion machinery. This has of course to be confirmed by analyzing the localization of the PopA1 protein in both cell lysates and in the external medium, after growth of cells under *hrp* inducing conditions.

On the other hand, the proteins encoded by *hrp* genes whose mutants have altered but still significant effects on plants are good candidates for transiting through the *hrp* secretion pathway.


### Analysis of the left hand end of the *hrp* regulon

Previous data had established that the region between the left hand end of the *hrp* gene cluster and *popA* belongs to the *hrp* regulon. This suggested that additional genes controlling plant-bacteria interactions might be present in this region. In order to test this hypothesis, a deletion mutant was constructed in strain GMI1000. This deletion extended over the 12 kb immediately adjacent to the left hand end of the *hrp* gene cluster. The resulting mutant, called GMI1554, was hypoaggressive on tomato plants and induced partial and delayed HR on tobacco.

Tn5-B20 mutagenesis was performed in this region in order to further define its genetic organization. This transposon generates transcriptional fusions. Individual mutants were tested for their ability to induce HR on tobacco and to cause disease on tomato. In addition, the insertion mutants were tested for *hrpB* dependent expression. The nucleotide sequence of the 9 kb DNA stretch located to the left hand end of *hrcC* was also determined. The results of these experiments are summarized in figure 2

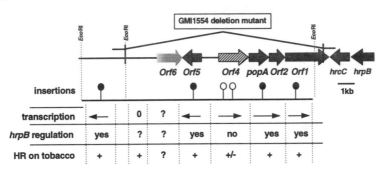

Fig. 2   Genetic organisation of the 16 kb region flanking the left-hand side of the *hrp* gene cluster of *R. solanacearum.*

## ENCODED PROTEINS, HOMOLOGIES AND STRUCTURAL FEATURES

Sequence analysis of the 9 kb DNA stretch led to the identification of 6 ORFs which have a high coding probability as deduced from codon usage. Search for homologies with sequences present in data bases and search for structural or functional motifs revealed the following features for the different ORFs.

*ORF 1* codes for a putative protein of 1024 aa which contains 19 tandem copies of a 24-amino acid leucine-rich repeat (LRR)-(Kobe and Deisenhofer, 1995). This domain presents significant homologies with LRR domains of proteins encoded by several plant resistance genes and various adenylate cyclase genes. Interestingly, LRR-containing proteins have been shown to be secreted by the type III secretion pathway in the animal pathogens *Yersinia pestis* (Leung *et al.*, 1989) and *Shigella flexneri* (Hartman *et al.*, 1990).

*ORF 2* codes for a protein of 174 aa which appears to contain a putative bipartite nuclear localization signal (NLS) (KRDDETDPNAETEGGKKK)-(Chelsky *et al.*, 1989) in its C-terminal region. In plant pathogenic bacteria, functional NLSs, have already been characterized in the VirD2 and VirE2 proteins of *Agrobacterium tumefaciens* (Zupan and Zambryski, 1995) and in the PthA protein encoded by *Xanthomonas citri* (Yang and Gabriel, 1995).

*ORF 3* encodes the PopA1 protein which has already been described (Arlat *et al.*, 1994).

*ORF 4* codes for a putative protein of 771 aa which shares homology over its entire length with FptA, the outer membrane Fe(III)-pyochelin receptor of *Pseudomonas aeruginosa* and with several other siderophore receptors. These outer membrane proteins are believed to interact with TonB to facilitate the movement of the outer membrane-bound ligand into the periplasmic space. The three conserved domains found in these TonB-dependent receptors were partially conserved in the putative protein encoded by ORF4 (Bitter *et al.*, 1991). Analysis of a multiple alignment of members of this family suggests that the protein encoded by ORF4 might be the first member of a new sub-family. This is also supported by the fact that, as opposed to other members of this gene family, transcription of the gene corresponding to ORF4 is not induced during growth under iron limiting conditions.

*ORF6.* Although only partialy sequenced, this ORF most probably corresponds to the previously characterized *pglA* gene which codes for the major endopolygalacturonase secreted by *R. solanacearum* (Huang and Schell, 1990*)*. This gene was already located to the left of the *hrp* gene cluster in strain K60 (Allen *et al.*, 1991).

REGULATION

Most of this region appears to be part of the *hrpB* regulon, the left border of which remains to be defined. This regulon is interrupted by ORF4 which shows a low level of transcription irrespective of the growth conditions tested and possibly by OFR6, corresponding to the putative *pglA* gene.

PLANT INTERACTIONS

None of the mutants carrying insertions within *hrpB*-regulated genes were found to be affected in their interactions with tobacco or tomato plants. This somewhat unexpected result is reminiscent of the situation found for PopA-deficient mutants. On the the other hand mutants in which ORF4 had been disrupted showed a reduced HR-inducing activity comparable to the phenotype of the deletion mutant GMI1554. However as opposed to the deletion mutant, an ORF4-deficient mutant was not significantly affected in its ability to elicit disease on tomato.

## Discussion

We have shown that the *hrp* gene cluster is a mosaic of genes whose integrity may or may not be essential for the ability of the bacteria to interact with plants and that genes belonging to these two categories are present in the same operon. A similar situation has also been reported for

*Pseudomonas syringae*, for which a single *hrp* transcription unit has been shown to encode both the secreted HrpZ protein and components of the secretion machinery (Huang *et al.*, 1995).

In *R. solanacearum* essential genes include all (possibly with a single exception) of the *hrc* genes which have been tested. These genes are assumed to code for components of the secretion machinery. Four additionnal *hrp* genes are also essential and could also participate in the biogenesis of this apparatus. Alternatively they might encode proteins (secreted or not) which play a major role in plant interactions. Monitoring the ability of the corresponding mutants to secrete PopA1 will permit us to discriminate between these two hypotheses.

Non essential *hrp* genes are good candidates to encode secreted proteins since for certain of the corresponding mutants PopA1 activity as already been found in culture supernatants indicating that the corresponding gene products are not required for secretion.

Two other good secretion candidates are the proteins encoded by ORF1 and ORF2 in the region flanking the left border of the *hrp* gene cluster. These proteins have characteristics of proteins normally found in eucaryotic cells. Moreover, the presence of a NLS in ORF2 is an indication that the corresponding protein might be injected into host cell similarly to what has been described for some of the Yop proteins of *Yersinia* . An intriguing feature of ORF1 is the presence of LRR motifs which are also present in certain plant resistance genes. An attractive hypothesis is that the ORF1 product interferes with molecular events taking place in the process of plant-bacteria recognition.

In *R. solanacearum*, as in animal pathogens, a significant number of proteins might transit through the *hrp* secretion pathway. This may be related to the unusually large host range of this bacterium, which could require different factors for different hosts. This would explain the lack of phenotype for mutants generated in these genes, considering that they have only been tested on the two species, tomato and tobacco. Alternatively the various secreted proteins might act synergistically and the absence of one of them might be compensated by the presence of the others.

One gene (*orf4*) has been identified in the left flanking region which partly controls the HR-eliciting ability. This gene does not belong to the *hrp* regulon and encodes a protein which has similarity with a group of outer membrane proteins which govern trafficking of iron and small molecules towards the periplasm of bacteria. We hypothesise that this gene is involved in the perception of a plant signal which modulates *hrp* gene expression.

ACKNOWLEDGMENTS :

We thank Stephane Genin and Belen Brito for stimulating discussions. This work was supported by the European Economic Community

(CHRX• CT93-0171), by the Ministère de l'Enseignement Supérieur et de la Recherche (ACC-SV6), and by the Region Midi-Pyrénées (RECH/9300249).

LITTERATURE CITED

Allen, C., Huang, Y. and Sequeira, L. 1991. Cloning of genes affecting polygalacturonase production in *Pseudomonas solanacearum*. Mol. Plant Microbe Interact. 4 : 147-154.

Arlat, M., Van Gijsegem, F., Huet, J.C., Pernollet, J.C., and Boucher, C. 1994. PopA1, a protein which induces a hypersensitivity-like response on specific petunia genotypes, is secreted via the Hrp pathway of *Pseudomonas solanacearum*. EMBO J. 13 : 543-553.

Bitter W., Marugg, J. D., . de Weger, L. A., Tommassen, J. and Weisbeek, P. J. 1991. The ferric-pseudobactin receptor PupA of Pseudomonas putida WCS358: Homology to TonB-dependent *Escherichia coli* receptors and specificity of the protein. Mol. Microbiol. 5: 647-655.

Bogdanove, A, *et al.*, 1996. A unified nomenclature for broadly conserved *hrp* genes of phytopathogenic bacteria. Mol. Microbiol. 20 : 681-683.

Bonas, U. 1994. *hrp* genes of phytopathogenic bacteria. Curr Top Microbiol Immunol 192: 79.

Chelsky, D., Ralph, R. and Jonak, G.1989. Sequence requirements for synthetic peptide-mediated translocation to the nucleus. Mol. Cell. Biol. 9: 2487-2492

Gough, C.L., Genin, S., Lopes, V., Zischek, C. and Boucher, C.A. 1992. Homology between the HrpO protein of *Pseudomonas solanacearum* and bacterial proteins implicated in a signal peptide-independent secretion mechanism. Mol. Gen. Genct. 239 : 378-392.

Hartman, A.B., Venkatesan, M., Oaks, E.V. and Buysse, J.M. 1990. Sequence and molecular characterization of a multicopy invasion plasmid antigen gene, ipaH, of Shigella flexneri. J. Bacteriol. 172 : 1905-1915.

Huang , J. and . Schell, M. A1990. DNA sequence analysis of *pglA* and mechanism of export of its polygalacturonidase product from *Pseudomonas solanacearum*. J. Bacteriol. 172 : 3879-3887.

Huang, H.C., . Lin, R. H, Chang, C. J., Collmer, A. and Deng, W. L. 1995. The complete *hrp* gene cluster of *Pseudomonas syringae* pv. *syringae* 61 includes two blocks of genes required for harpin Pss secretion that are arranged colinearly with Yersinia *ysc* homologs. Mol. Plant Microbe Interact.. 8: 733-746 .

Kobe, B. and Deisenhofer, J. 1995. A structural basis of the interactions between leucine-rich repeats and protein ligands. Nature 374: 183-186.

Leung, K. J. and Straley, S.C. 1989. The *yopM* gene of *Yersinia pestis* encodes a released protein having homology with the human platelet surface protein GPIbα. J. Bacteriol. 171 : 4623-4632.

Ménard, R., Sansonetti, P. J. and Parsot, C. 1993. Nonpolar mutagenesis of the *ipa* genes defines IpaB, IpaC, and IpaD as effectors of *Shigella flexneri* entry into epithelial cells. J. Bacteriol 175 : 5899-906

Van Gijsegem, F., Zischek, C., Gough, C., Niqueux, E., Arlat, M., Genin, S., Barberis, P., German, S., Castello, P. and Boucher, C. 1995. The *hrp* gene locus of *Pseudomonas solanacarum* which controls a type III secretion system, encodes eight proteins related to components of the bacterial flagellar biogenesis complex. Mol. Microbiol. 15 : 1095-1114

Yabbuchi, E., Kosako, Y. , Yano, I., Hotta, H.,and Nishluchi Y. 1995. Transfer of two *Burkholderia* and an *Alcaligenes* species to *Ralstonia* Gen. Nov.: Proposal for *Ralstonia picketii, Ralstonia solanacearum* and *Ralstonia eutropha*. Microbiol. Immunol. 39 : 897-904.

Yang, Y. and Gabriel, D.W. 1995. *Xanthomonas* avirulence/pathogenicity gene family encodes functioal plant nuclear targeting signals. Mol. Plant Microbe Interact. 8 : 627-631.

Zupan, J.R., and Zambryski, P. 1995. Transfer of T-DNA from *Agrobacterium* to plant cell. Plant Physiol. 107 : 1041-1047.

# Homoserine Lactone-Mediated Microbial Signaling: A Communication System Common to Plant-Associated Bacteria

Stephen K. Farrand[1,2], Kevin R. Piper[2], Rebecca Sackett[1], Gao Ping[1], Paul D. Shaw[1], and Kun-Soo Kim[2].

Departments of Crop Sciences[1] and Microbiology[2], University of Illinois at Urbana-Champaign, Urbana, Illinois 61801, USA.

Signaling systems play crucial roles in the specific interactions between plants and their host-associated microbes. In the two most well characterized systems, the *Agrobacterium*-crown gall interaction, and the *Rhizobium (Bradyrhizobium)*-legume interaction, chemical signals are produced by both microbe and host. The perception of and the response to these signals dictates the outcome of the interaction. With respect to the bacterium, signals produced by the plant result in the activation of gene systems, the products of which are required for the interaction between the bacterium and that host plant.

More recently, it has become apparent that bacteria within a population communicate with each other. This is a particularly intriguing concept; it has long been assumed that, except for some social microorganisms such as the myxobacteria, eubacteria function at the unicellular level and that any cooperativity is a strictly gratuitous result of a large number of independent cells all doing the same thing at the same time. Furthermore, evidence from several systems indicates that interbacterial communication is important to the ability of that population of microorganisms to succeed in certain, specialized niches.

While there may be many mechanisms by which bacteria communicate with each other, autoinduction mediated by acyl-homoserine lactone (HSL-AI) signal molecules is the best characterized and most intensively studied of such systems. First described for the bioluminescent marine symbiont, *Vibrio (Photobacterium) fischeri*, autoinduction constitutes a mechanism by which the bacterium induces expression of one or more gene systems only when it has reached a critical population density. This so-called Quorum Sensing phenomenon (Fuqua, *et al.*, 1994) results from the ability of the bacteria to sense the levels of a signal molecule it itself produces. Produced during growth this acyl-homoserine lactone signal molecule is freely diffusible through cell membranes. When it has reached a critical concentration throughout the population, it is believed to form a ligand with a specific transcriptional regulator. This interaction is required in order for the regulator to affect expression of the target genes.

## The production of acyl-HSL signals is common among Gram-negative plant-associated bacteria.

We have developed simple techniques for screening bacteria for the production of molecules with acyl-HSL activity. The fundamental assay is based on the activation of a *lacZ* fusion to a gene from *Agrobacterium tumefaciens* that requires an active acyl-HSL for expression. The detector system responds to nM to µM levels of a broad range of acyl-substituted homoserine lactones. Using this bioassay, we surveyed a large number of isolates for the production of acyl-HSLs. Individual isolates representing the genera *Pseudomonas*, *Erwinia*, *Agrobacterium*, *Rhizobium*, and *Bradyrhizobium* could be found that produced active compounds. However, not all isolates of a given genus, or even of a given species, produced such molecules under the conditions tested. For example, of two isolates of *P. fluorescens* tested, only one clearly produced one or more acyl-HSL signals. The second did not produce any detectable signal molecules. Interestingly, *P. fluorescens* 2-74, the isolate that produces acyl-HSLs also produces phenazine antibiotics. Furthermore, *P. aureofaciens* strain 30-84 gives a positive reaction in our assay. This strain regulates production of phenazine antibiotics by acyl-HSL-mediated autoinduction (Pierson, *et al.*, 1994). This suggests that strain 2-74 also may regulate phenazine biosynthesis by autoinduction. Similarly, three of four isolates of *R. meliloti* produced detectable amounts of active signals. Within the genus *Pseudomonas*, most isolates of biovars *tabaci*, *angulata*, *coronofaciens*, *glycinea*, *savastanoi*, and *syringae* produced detectable signals. However none of the pathovar *phaseolicola* or *tomato* isolates produced detectable active species. Similarly, within the erwinia, most, if not all isolates of *E. carotovora*, *E. stewartii*, and *E. chrysanthemi* produced one or more acyl-HSLs, while none of the *E. amylovora* or *E. rhaponitici* isolates tested showed any activity. Finally, among the biovar 1 and biovar 2 isolates of *Agrobacterium tumefaciens* and *A. radiobacter* tested, most produced detectable signal molecule activity.

Notably, however, none of the *A. rhizogenes* strains tested produced detectable acyl-HSLs. The genus *Xanthomonas* presents a special case. While the majority of isolates representing species *campestris* pathovars *glycines, cerealis, holicola, malvacearum,* and *vesicatoria* produced active compounds, the amounts, as judged by the reaction of the bioindicator system, were very low. It is possible that the signals produced by *Xanthomonas* species are not among the acyl-HSLs that are efficiently recognized by the bioindicator. Alternatively, the activity that we observe may not result from true acyl-HSL signals.

## *Agrobacterium tumefaciens* regulates Ti plasmid conjugal transfer by acyl-HSL-mediated autoinduction

Although autoinduction appears to be a common regulatory paradigm in plant-associated bacteria, the individual systems regulated by this mechanism have been characterized in only a few cases. These include antibiotic and exoenzyme production by *E. carotovora* (Bainton, *et al.*, 1992; Pirhonen, *et al.*, 1993), capsular polysaccharide production by *E. stewartii* (Beck von Bodman and Farrand, 1995) and the production of phenazine antibiotics by *P. aureofaciens* (Pierson, *et al.*, 1994). However, the regulation of conjugal transfer of *A. tumefaciens* Ti plasmids is among the best characterized of the autoinduction systems present in any microorganism. The well-studied octopine- and nopaline-type Ti plasmids of *A. tumefaciens* both are conjugal elements. Transfer from bacterium to bacterium is dependent upon three sets of genes, the *traAFB* operon, the *traCDG* operon, and the *trb* operon (Alt-Mörbe, *et al.*, 1996; Farrand, *et al.*, 1996). Expression of the genes within these operons is regulated by autoinduction through the transcriptional activator, TraR, and the acyl-HSL *N*-(3-oxo-octanoyl)-L-homoserine lactone, also called AAI (Piper, *et al.*, 1993; Zhang, *et al.*, 1993; Fuqua and Winans, 1994). In addition, an antiactivator, TraM, modulates TraR-mediated *tra* gene activation, apparently by interacting directly with the transcriptional activator protein (Hwang, *et al.*, 1995). Conjugal transfer occurs only when the donor cells reach high population densities, but this can be overcome by providing exogenous AAI to the donor culture (Fuqua and Winans, 1996; Piper and Farrand, in preparation).

### REGULATION OF CONJUGATION BY OPINES

Opines are low molecular weight carbon compounds synthesized in the crown gall tumors by enzymes encoded by the integrated T-DNAs (Dessaux, *et al.* 1993). These compounds, which are exuded to the rhizosphere, can be utilized by the inducing agrobacteria as carbon

175

sources by virtue of catabolic gene systems encoded on the Ti plasmids. Although expression of the *tra* and *trb* operons is activated directly by TraR, conjugation as a phenomenon actually is regulated by a subset of opines produced by the tumors induced by that particular *A. tumefaciens* strain. For classic octopine-type Ti plasmids such as pTi15955, pTiA6 and pTiR10, octopine is the conjugal opine. For nopaline-type Ti plasmids such as pTiC58 and pTiT37, the agrocinopine opines serve to induce conjugation. These requirements are highly specific; octopine will not induce conjugal transfer of pTiC58, and agrocinopines will not induce transfer of the octopine-type Ti plasmids.

How, then, is TraR-mediated autoinduction of conjugation super-regulated by the opines? The regulation of conjugal transfer of pTiC58 by agrocinopines is mediated by AccR, a transcriptional repressor that also controls expression of the *acc* operon encoding catabolism of the conjugal opine (Beck von Bodman, *et al.*, 1992). Mutations in *accR* result in constitutive expression of Acc and Tra phenotypes. Our sequence analysis indicates that *traR* is part of an operon that is adjacent to, and divergently transcribed from *acc* (Fig. 1).

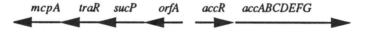

Fig. 1. Organization of the *acc-traR* region of pTiC58

Using a *lacZ* fusion to *traR* we show that expression of this gene is negatively regulated by AccR and agrocinopines. In addition, deletion analysis indicates that this regulated expression of *traR* is dependent upon a promoter located just upstream of *orfA* (see Fig. 1). When the interval between *orfA* and *accR* was cloned into a promoter probe,both orientations were active. This suggests that the region contains a divergent promoter system with transcription in both directions being regulated by AccR.

OPINE REGULATION RESULTS FROM SERENDIPTIOUS GENE PLACEMENT: A COMMON THEME

These results indicate that the regulation of conjugal transfer by the agrocinopines results from the placement of *traR* in an operon that is regulated by AccR, the opine-responsive transcriptional repressor. Fuqua and Winans (1996) recently reported that a similar strategy is responsible for the octopine-inducible regulation of conjugal transfer in the octopine-type Ti plasmid, pTiR10. In this case, *traR* is the distal gene of a seven-gene operon encoding a transport system, *oph*, for an unknown substrate (Fig. 2). This operon is located directly downstream from *occ*. Induction by *occR* in the presence of octopine leads to the expression of a 14 kb transcript which includes, in addition to *occ*, the seven-gene *oph* operon containing *traR*. Intriguingly, the operon in which *traR* of pTiR10 is located is not related to the operon in which *traR* of pTiC58 is resident (Fig. 2).

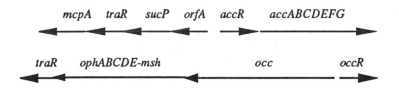

Fig. 2. Relationship between the opine-regulated *traR* gene sets from nopaline- (upper) and octopine-type (lower) Ti plasmids.

More recently, in our analysis of a region which encodes transport of mannopine, a member of the mannityl opine family, from the same octopine-type Ti plasmid, we discovered another *traR*-like gene. The gene is the penultimate orf of a six-gene operon which includes *motABCD* and an orf which could encode an mcp-like protein (Fig. 3).

Fig. 3. Organization of the mannopine transport operon containing a *traR*-like pseudogene present in octopine-type Ti plasmids.

The four *mot* genes encode the periplasmic protein-type MOP transport system. The amino two-thirds of the *traR*-like gene, called *trlR* is greater than 90% identical to the octopine-type *traR* gene. However, *trlR* contains a frameshift mutation just before the putative helix-turn-helix motif in the carboxy-region of the translation product. Consistent with this, mannopine does not induce conjugal transfer. However, as assessed by *lacZ* fusions, the opine does induce the expression of *trlR*, as well as the upstream *mot* genes.

The *mot* genes are not related to those associated with the *traR* genes of the agrocinopine or the octopine regulons described above. Thus this represents a third, independent gene arrangement in which *traR* is located in an operon that is regulated by an opine. Clearly, this theme by which conjugation is placed under control of an opine is important in the biology of the Ti plasmid and its host bacterium.

## Mannopine Does Induce Conjugal Transfer of Other *Agrobacterium* Plasmids

In the octopine/mannityl opine-type Ti plasmids the *traR* associated with the octopine regulon appears to have become dominant over that linked to the mannopine system. However, the existence of the mannopine-associated *traR*-like gene suggested that there may be plasmids in *Agrobacterium* in which conjugation is regulated by the mannityl opine. We examined a series of wild-type isolates of *Agrobacterium* spp. known to catabolize mannopine. Some of these isolates are tumorigenic, while others are avirulent. In 10 of these isolates, some of which are tumorigenic, mannopine induced the production of AAI, and also conjugal transfer of a plasmid encoding mannopine catabolism. Transconjugants derived from these matings showed the virulence phenotype of the parental donor. These transconjugants also became effective donors, transferring mannopine utilization to recipients in subsequent matings. This conjugal transfer, as well as production of AAI, required induction by mannopine. We suggest that these plasmids contain a *traR* associated with and regulated by the mannopine regulon, perhaps in a gene arrangement ancestral to that present in the octopine-mannityl opine Ti plasmids.

## Literature Cited

Alt-Mörbe, J., Stryker, J. L., Fuqua, C., Li, P.-L., Farrand, S. K., and Winans, S. C. 1996. The conjugal transfer system of *Agrobacterium tumefaciens* octopine-type Ti plasmids is closely related to the transfer system of an IncP plasmid and distantly related to Ti plasmid *vir* genes. J. Bacteriol. In press.

Bainton, N. J., Stead, P., Chhabra, S. R., Bycroft, B. W., Salmond, G. P. C., Stewart, S. A. B., and Williams, P. 1992. N-(3-oxo-hexanoyl)-L-homoserine lactone regulates carbapenem antibiotic production in *Erwinia carotovora*. Biochem J. 288:997-1004.

Beck von Bodman, S., and Farrand, S. K. 1995. Capsular polysaccharide biosynthesis and pathogenicity in *Erwinia stewartii* require induction by an N-acylhomoserine lactone autoinducer. J. Bacteriol. 177:5000-5008.

Dessaux, Y., Petit, A., and Tempé, J. 1992. Opines in *Agrobacterium* biology. Pages 109-136 in Molecular Signals in Plant-Microbe Communications. D. P. S. Verma, ed. CRC Press, Boca Raton, FL.

Beck von Bodman, S., Hayman, G. T., and Farrand, S. K. 1992. Opine catabolism and conjugal transfer of the nopaline Ti plasmid pTiC58 are coordinately regulated by a single repressor. Proc. Natl. Acad. Sci. (USA) 83:643-647.

Farrand, S. K., Hwang, I., and Cook, D. M. 1996. The *tra* region of the nopaline-type Ti plasmid is a chimera with elements related to the transfer systems of RSF1010, RP4, and F. J. Bacteriol. In press.

Fuqua, C., and Winans, S. C. 1994. A LuxR-LuxI type regulatory system activates *Agrobacterium* Ti plasmid conjugal transfer in the presence of a plant tumor metabolite. J. Bacteriol. 176:2796-2806.

Fuqua, C., and Winans, S. C. 1996. Localization of OccR-activated and TraR-activated promoters that express two ABC-type permeases and the *traR* gene of Ti plasmid pTiR10. Mol. Microbiol. 20:1199-1210.

Fuqua, C., Winans, S. C., and Greenberg, E. P. 1994. Quorum sensing in bacteria: the LuxR-LuxI family of cell density-responsive transcriptional regulators. J. Bacteriol. 176:269-275.

Hwang, I., Cook, D. M., and Farrand, S. K. 1995. A new regulatory element modulates homoserine lactone-mediated autoinduction of Ti plasmid conjugal transfer. J. Bacteriol. 177:449-458.

Pierson, L. S., Keppenne, V. D., and Wood, D. W. 1994. Phenazine antibiotic biosynthesis in *Pseudomonas aureofaciens* 30-84 is regulated by PhzR in response to cell density. J. Bacteriol. 176:3966-3974.

Piper, K. R., Beck von Bodman, S., and Farrand, S. K. 1993. Conjugation factor of *Agrobacterium tumefaciens* regulates Ti plasmid transfer by autoinduction. Nature (London) 362:448-450.

Pirhonen, M., Flego, D., Heikinheimo, R., and Palva, E. T. 1993. A small diffusible signal molecule is responsible for the global control of virulence and exoenzyme production in the plant pathogen *Erwinia carotovora*. EMBO J. 12:2467-2476.

Zhang, L., Murphy, P. J., Kerr, A., and Tate, M. E., 1993. *Agrobacterium* conjugation and gene regulation by N-acyl-L-homoserine lactones. Nature (London) 362:446-448.

# Molecular analysis of protein secretion systems involved in *Erwinia carotovora* virulence

Thomas, J., Wharam, S., Vincent-Sealey, L., Harris, S., Shih, Y.-L. and Salmond, G.P.C.[1]

Department of Biological Sciences, University of Warwick, Coventry, CV4 7AL, UK & [1]Department of Biochemistry, University of Cambridge, Cambridge, CB2 1QW, UK

**Introduction.**

*Erwinia carotovora* subspecies *carotovora* (Ecc), *E. carotovora* subspecies *atroseptica* (Eca) and *Erwinia chrysanthemi* (Echr) can cause soft rot and/or blackleg disease of potato (Barras et al, 1994). The virulence of these bacteria is largely due to secreted plant cell wall-degrading enzymes, including pectate lyases, pectin lyases, polygalacturonases, pectin methyl esterases, cellulases (endoglucanases) and proteases (Barras et al, 1994). Mutants defective in the secretion of these enzymes show impaired virulence. There are at least three functionally-distinct protein secretion pathways in *Erwinia* spp., classified as Types I, II and III (Salmond, 1994; Salmond and Reeves, 1993; Van Gijsegem et al, 1993;). Only Type I and II pathways are discussed here as they are involved in exoenzyme targeting. But homologues of the proteinaceous components of all three secretory machineries are widely distributed throughout Gram-negative bacteria, including other plant pathogens and various animal pathogens (Salmond, 1994; Van Gijsegem et al, 1993; Wharam et al, 1995).

THE TYPE I PATHWAY.

The proteases of Echr are targeted through the Type I pathway involving a secretory "apparatus" of three membrane proteins thought to form a transmembrane channel (Fig 1). Although the "signal" information for protease targeting is in the C-terminal domain, it is not cleaved. The genes for the Type I machinery and the corresponding enzymes are clustered (Delepelaire and Wandersman, 1991). Do Ecc and Eca target their proteases via an analogous pathway? Despite repeated attempts in various labs, there are no positive reports of the reconstitution of the equivalent Type I system from Ecc/Eca in *E. coli*. Ecc does have genes encoding Type I secretion machinery homologues (S. Harris, unpublished) but either Ecc proteases are secreted by a different system from that of Echr, or the enzyme genes and secretion genes are unlinked in Ecc and so are never co-cloned. The protease secretion system in Ecc is clearly worth further study.

## THE TYPE II PATHWAY

Soft rot *Erwinia* mutants (Out⁻) that do not secrete the pectinases (Pel) and cellulases (Cel) show reduced virulence (e.g. see Barras et al, 1995) and Pel and Cel enzymes accumulate in the periplasm while proteases are secreted. The Out proteins define the Type II secretory machinery or "general secretory pathway" (GSP) which also has homologues in many plant and animal pathogens (Pugsley, 1993; Salmond and Reeves, 1993; Wharam et al, 1995). The Type II mechanism is two step; step 1 involving export from the cytoplasm to the periplasm and step 2 involving true secretion across the outer membrane (Barras et al, 1994; Pugsley, 1993; Salmond, 1994;)

In Ecc, the Type II secretory apparatus (Wharam et al, 1995) comprises 15 Out proteins encoded by a large gene cluster. This clustering of Type II secretion genes is common but there are differences in the transcriptional organisation of this locus between bacterial genera - implying different modes of regulation of the GSP system genes. In Echr some *out* genes are regulated via the pectinolysis repressor, KdgR, temperature and growth phase (e.g. see Barras et al, 1994). There is a putative KdgR "box" some distance 5' of the *outC* gene of Ecc suggesting that KdgR might also influence *out* gene expression in Ecc (S. Wharam, unpublished) and we are currently testing this hypothesis. Any role for Ecc *out* gene regulation via the signaling pheromone, OHHL, is unclear (S. Wharam, unpublished).

In Out mutants Pel and Cel accumulate in the periplasm i.e. they complete the first step of the two step pathway. This export step is probably identical to the Sec-dependent translocation step of *E. coli* K12 (e.g. see Barras et al, 1994) - since *E. coli* K12, carrying either *pel* or *cel* genes of Ecc or Echr, exports the enzymes to the periplasmic space. In *E. coli* these enzymes cannot traverse the outer membrane as there is no appropriately functional GSP, despite the recently-identified presence of "cryptic" GSP genes (Francetic and Pugsley, 1996).

Of the 15 GSP protein homologues, only the D and S proteins are outer membrane-associated (Fig 1). Some S homologues have lipoprotein N-terminal signal sequences, whereas D homologues have classical signals (Pugsley, 1993; Wharam et al, 1995). The pIV protein of filamentous bacteriophage is an OutD homologue and this protein multimerises in the outer membrane (10-12) in the process of phage "secretion" (Kazmierczak et al, 1994). The homologues of OutD are widely distributed and are also found as component parts of the Type III secretory system (Salmond, 1994; Van Gijsegem et al, 1993; Wharam et al, 1995). The Type II pathway OutD homologues may be critical for determining the specificity of secretion, perhaps acting as "gatekeepers" controlling access of the periplasmic versions of the secreted proteins to the final translocation step across the outer membrane (Fig 1) (Lindeberg et al, 1996). OutD homologues have been predicted to be embedded in the outer membrane via their C-terminal domains, with N-terminal domains in the periplasm (Kazmierczak et al, 1994) but this view may now be controversial (Hardy et al, 1996).

Most of the Out proteins are embedded in the inner membrane (Fig 1) as determined by sequence and Bla fusion analysis. OutF and OutO are polytopic cytoplasmic membrane proteins; the former having three transmembrane stretches (J. Thomas, unpublished) and the latter having eight (Reeves et al, 1994). Most of the OutF protein is cytoplasmic. Similarly, most of OutO is exposed to the cytoplasm. OutC,G,H,I,J,K,L,M and Out N have a single transmembrane "anchor" but with varying amounts extending into the periplasm and cytoplasm. In each, except OutL, the vast majority of the protein is periplasmic (J. Thomas, unpublished).

OutGHI and OutJ have N-terminal homology with type IV pilin monomers. This homology covers the target site for cleavage by the N-MePhe peptidase and a hydrophobic stretch acting as transmembrane anchor, and so the GHI and J components have been called "pseudopilins". The OutO polytopic membrane protein is also homologous to the N-MePhe peptidase (PilD/XcpA) that cleaves (and methylates) the type IV pilins. OutO processes the OutGHIJ pseudopilins by "signal" cleavage on the cytoplasmic face of the inner membrane, but leaving the transmembrane anchor in position (Fig 1). The overall topology of the OutO peptidase is critical for function, because, although the catalytic site is in the cytoplasmic N-terminal domain of the protein, the last C-terminal transmembrane stretch of OutO is essential for biological activity (Reeves et al, 1994).

There is evidence that an OutS homologue (PulS) has a chaperone-like function by assisting the insertion of PulD (the OutD homologue) into the outer membrane (Fig 1) (Hardy et al, 1996). OutB is probably cytoplasmic. Although in-frame Bla fusions can be generated in OutE, they do not gain access to the periplasm - confirming that OutE is essentially cytoplasmic.

Apart from the peptidase activity of OutO, there is little data on biochemical functions of the Out proteins. Although the pseudopilins are substrates for the peptidase cleavage reaction, the structural purpose of this is unknown as there is no evidence that the cleavage promotes the formation of a transmembrane channel equivalent to a type IV pilus. Based on sequence conservations and biochemical studies on homologues, it is clear that OutE must be a "traffic ATPase"(Pugsley, 1993; Salmond, 1994) but the function of this in the Type II system is unclear. We have identified mutations within *outE*, including a conditional (Ts) mutation, which block secretion of the Pel and Cel proteins across the outer membrane. Therefore, we assume that the "activity" of OutE is somehow relayed across the inner membrane, directly or indirectly. There is evidence that the N-terminal domain of an OutE homologue (EpsE from *Vibrio cholerae*) interacts with the OutL homologue (EpsL) allowing a transient inner membrane interaction (Sandkvist et al, 1995). In this way the L protein could act to "transduce" a signal across the inner membrane which ultimately leads to protein secretion across the outer membrane. Interestingly, the N-terminal 2/3 of OutL extends into the cytoplasm with the C-terminal 1/3 exposed in the periplasm (Fig 1). We have found mutations mapping to the cytoplasmic face of OutL which also abolish secretion across the outer membrane and these genetic data agree with the notion that OutE and OutL could interact. Does the periplasmic domain of OutL have important interactions with other GSP proteins (or the enzymes) in the periplasm?

The OutD and OutC proteins have been identified as candidate "gatekeepers" of the Type II pathway (Lindeberg et al, 1996). However, although the OutD outer membrane protein might act as part of a gated channel across the outer membrane (Fig 1) (Lindeberg et al, 1996; Salmond, 1994), there is still no biochemical evidence for this.

In the periplasm, the Pel and Cel enzymes may reach a folded state before secretion across the outer membrane. Similarly, cholera toxin, which is secreted via the *Vibrio cholerae* Type II system, folds into a complex of 1A:5B subunits, in the periplasm (e.g. see Pugsley, 1993). So it is likely that all proteins moving through the GSP will require some form of folding in the periplasm. The folding may be essential for the formation of a "patch motif" of residues acting as a molecular recognition "signal" enabling access to the Type II machinery in the periplasm. The Pels and CelZ of Echr are disulfide bonded in the periplasm. Dsb mutants of Echr fail to secrete either Pel or Cel - the unfolded Pels are proteolytically degraded in the Echr periplasm (Schevchik et al, 1995). However, the CelV protein of Ecc has no Cys residues and so cannot disulfide bond. DsbA and DsbC mutants of Ecc have been used to study the secretion of CelV and Pel in Ecc (L. Vincent-Sealey, unpublished). As in Echr, the Pels are not secreted in Ecc DsbA mutants and are probably degraded in the periplasm, but contrary to expectations, it seems that Dsb mutants of Ecc are affected in CelV production and targeting efficiency. These preliminary results suggest an indirect, but measurable, effect on CelV due to Dsb defects. We are investigating a possible feedback regulation system, imposed on the exoenzyme genes, from an unknown periplasmic signal.

After Sec-dependent cleavage in export, there is no second cleavage of the periplasmic Pel or Cel enzymes. The second step must be specific as only a subset of periplasmic proteins is secreted. The "secretion signal" in these proteins is not defined but the specificity of this signal is highlighted by the observation that neither Echr nor Ecc will secrete each other's Pel or Cel - despite high homology between the respective enzymes and despite high homology between the corresponding Out proteins. This curious specificity is further highlighted by the paradoxical observation that, within either species the Type II machinery must "recognise" apparently structurally-diverse proteins such as Pel and Cel with no obvious common motifs. This species-specificity paradox may imply a co-evolution of the GSP and exoenzymes within species. The functional exchangeability studies (Lindeberg et al, 1996) suggest that the mystery of species-specificity may lie, at least in part, in the "gatekeeper" functions of OutC and OutD.

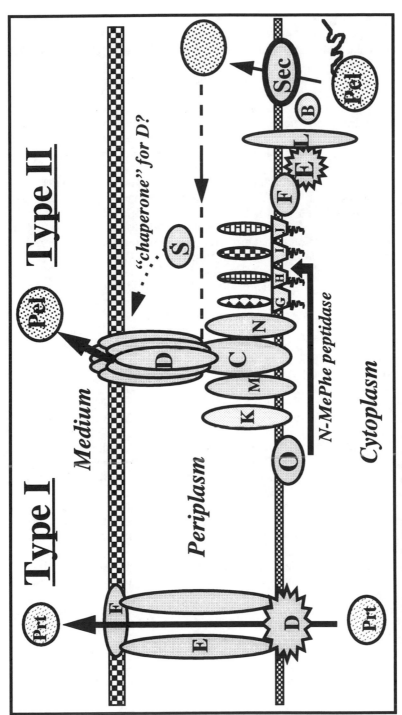

**Fig 1. The Type I & II secretion systems**

LITERATURE CITED

Barras, F., van Gijsegem, F. and Chatterjee, A.K. (1994) Extracellular enzymes and pathogenesis of soft rot *Erwinia*. Ann. Rev. Phytopathol. 32:201-134.

Delepelaire, P. and Wandersman, C. (1991) Characterisation, localization and transmembrane organisation of the three proteins PrtD, PrtE and PrtF necessary for protease secretion by the Gram-negative bacterium, *Erwinia chrysanthemi*. Mol. Microbiol. 5:2427-2343.

Francetic, O. and Pugsley, A.P. (1996) The cryptic general secretory pathway (gsp) operon of *Escherichia coli* K-12 encodes functional proteins. J. Bacteriol. 178:3544-3549.

Hardy, K.R., Lory, S. and Pugsley, A.P. (1996) Insertion of an outer membrane protein in *Escherichia coli* requires a chaperone-like protein. EMBO J. 15:978-988

Kazmierczak, B.I., Mielke, D.L., Russel, M. and Model, P. (1994) pIV, a filamentous phage protein that mediates phage export across the bacterial envelope, forms a multimer. J. Mol. Biol. 238:187-198.

Lindeberg, M., Salmond, G.P.C. and Collmer, A. (1996) Complementation of deletion mutations in a cloned functional cluster of *Erwinia chrysanthemi out* genes with *Erwinia carotovora out* homologues reveals OutC and OutD as candidate gatekeepers of species-specific secretion of proteins via the type II pathway. Mol. Microbiol. 20:175-190.

Pugsley, A.P. (1993) The complete general secretory pathway in Gram-negative bacteria. Microbiol. Revs. 57:50-108.

Salmond, G.P.C. (1994) Secretion of extracellular virulence factors by plant pathogenic bacteria. Ann. Rev. Phytopathol. 32:181-200.

Salmond, G.P.C. and Reeves, P.J. (1993) Membrane traffic wardens and protein secretion in Gram-negative bacteria. Trends in Biochem. Sci. 18:7-12.

Sandkvist, M., Bagdasarian, M. and Dirita, V.J. (1995) Interactions between the autokinase EpsE and EpsL in the cytoplasmic membrane is required for extracellular secretion in *Vibrio cholerae*. EMBO J. 14:1664-1669

Schevchik, V.E., Bortoli-German, I., Robert-Baudouy, J., Robinet, S., Barras, F. and Condemine, G. (1995) Differential effect of *dsbA* and *dsbC* mutations on extracellular enzyme secretion in *Erwinia chrysanthemi*. Mol. Microbiol. 16:745-754

van Gijsegem, F., Genin, S. and Boucher, C. (1993) Conservation of secretion pathways for pathogenicity determinants of plant and animal bacteria. Trends in Microbiol. 1:175-180.

Wharam, S., Mulholland, V. and Salmond, G.P.C. (1995) Conserved virulence factor regulation and secretion systems in bacterial pathogens of plants and animals. Eur. J. Plant Pathol. 101:1-13.

ACKNOWLEDGMENTS
This work was funded by the BBSRC, UK (Award PO1587).

# Regulation of $hrpN_{Ecc}$ and genes for other exoproteins in soft-rotting *Erwinia carotovora* by RsmA, a putative RNA-binding protein

Asita Mukherjee, Yaya Cui and Arun K. Chatterjee, Department of Plant Pathology, University of Missouri, Columbia, Missouri 65211 U. S. A.

Corresponding author: Arun Chatterjee. Telephone: 573 882 1892, Fax: 573 882 0588, E-mail: achatterjee@psu.missouri.edu

In addition to a large array of small diffusible metabolites, many plant pathogenic and plant associated bacteria produce extracellular proteins, polysaccharides, lipoproteins and lipopolysaccharides, collectively designated as exoproducts. While some of the macromolecules comprise integral components of bacterial cell wall and glycocalyx, others are secreted and released into the milieu. It is becoming increasingly apparent that such secreted exoproducts perform functions important in cellular physiology and host interaction. Recognition of such potentially significant ramifications has prompted studies of: (i) export mechanisms; (ii) molecular properties of the structural or biosynthetic genes; (iii) regulation of metabolite production; and (iv) biological functions. For various reasons we and others have used soft-rotting *E. carotovora* as a model system for the analysis of exoproducts (Barras et al., 1994; Wharam et al., 1995). Consequently, the following generalizations can be made: (i) These bacteria produce exoenzymes and diffusible metabolites in a growth phase dependent manner. (ii) Exoenzymes are required for pathogenicity; some may also elicit host defense responses. (iii) Production of exoenzymes is regulated negatively by the *rsmA* gene and positively by the *aep* (= *exp* or *car*) genes in conjunction with quorum sensing signals as well as plant signals. (iv) Protein export is mediated via highly evolved secretion systems. (v) Production of pectin lyase, one of the plant cell-wall-degrading enzymes, is activated along with the SOS genes upon DNA damage. (vi) The expression of $hrpN_{Ecc}$, a gene encoding the elicitor of the HR, is stringently regulated. Because of the space limitation, we focus on two aspects: (i) negative regulation of exoenzymes and secondary metabolism by *rsmA*; and (ii) regulation of expression of $hrpN_{Ecc}$ by RsmA and other factors.

## Post-transcriptional regulation of gene expression by RsmA

Recently, in *E. carotovora* subsp. *carotovora* strain 71 we discovered a gene, designated as *rsmA* for *r*egulator of *s*econdary *m*etabolism. The phenotypes of RsmA⁻ mutants and the trans-dominant effects of the cloned *rsmA*⁺ DNA have shown that this gene controls the production of exoenzymes, antibiotics, pigments and polysaccharides, the synthesis of flagella, and levels of the cell density (quorum) sensing signal, *N*-(3-oxohexanoyl)-L-homoserine lactone in various *Erwinia* species. In addition, it affects factors that determine plant pathogenicity and the ability to elicit the hypersensitive reaction (Cui et al., 1995; Mukherjee et al., 1996). *rsmA* has extensive homology with the *E. coli* regulatory gene, *csrA* (Liu et al., 1995 ). We have found that *rsmA* genes occur in all enterobacterial species tested, and that these genes are expressed under normal growth conditions. Furthermore, the *E. carotovora rsmA* gene affects the expression of various traits in several enterobacterial species. However, the mechanism by which *rsmA* controls gene expression is not yet fully understood. Our working hypothesis, based upon the findings of Liu et al. (1995), is that RsmA (or CsrA) binds to transcripts of the target genes and promotes their decay, thereby lowering the half life of the mRNA species. Indeed, our observations (i) that the levels of mRNAs of various *E. carotovora* genes are suppressed by RsmA, and (ii) that RsmA possesses RNA-binding motif, but no DNA-binding motif, are certainly consistent with this hypothesis. We propose that in enterobacteria, RsmA performs a key housekeeping function during exponential growth by preventing the accumulation of transcripts of genes for "non-essential" functions including virulence factors and secondary metabolic pathways. This hypothesis raises an important question: How do cells control RsmA action, *rsmA* expression or both during late exponential growth so as to initiate the production of secondary metabolites including virulence factors? As a prelude to resolving this issue, we have determined that *aepH* DNA, previously discovered in strain 71 and several other soft-rotting *Erwinia*, can neutralize the negative effect of *rsmA*. This neutralizing effect was seen with multiple copies of *aepH* in heterologous bacterial species. Experiments are in progress to determine if RsmA avidly binds to *aepH* transcripts and if this binding then depletes the pool of RsmA thereby lengthening the half lives of various transcripts that are otherwise susceptible to the RsmA action.

## The elicitation of the HR by pectolytic *E. carotovora* subspecies depends upon derepression of *hrpN*$_{Ecc}$ expression.

Many gram-negative phytopathogenic bacteria, when infiltrated into a nonhost plant such as tobacco, cause localized necrosis, generally known as the hypersensitive reaction (HR). *E. c.* subsp. *carotovora* and most other soft-rotting bacteria are unusual in that they do not elicit a typical HR when infiltrated into tobacco leaves.    However, Southern blot hybridizations by Laby and Beer (1992) revealed that several soft-rotting *Erwinia* possess *hrp* genes. Subsequently, a mutant strain of *E. chrysanthemi* deficient in the synthesis of the major pectate lyase isozymes, but not the pectolytic parent, was found to elicit the HR (Bauer et al. 1994). Bauer et al. (1995) also demonstrated that *E. chrysanthemi* possesses *hrp* genes including $hrpN_{Ech}$ which encodes an elicitor of the HR. These observations prompted us to critically analyze the occurrence and expression of *hrpN* alleles in *E. carotovora* subspecies. Our results are summarized below.

(1) A locus designated as $hrpN_{Ecc}$ has been cloned from *E. c.* subsp. *carotovora* strain 71. Homologs of $hrpN_{Ecc}$ occur in all strains of *E. carotovora* subspecies tested.

(2) $hrpN_{Ecc}$ is predicted to encode a glycine-rich protein of ca. 36 kDa possessing significant homology with the HR elicitors of *E. chrysanthemi* (Harpin$_{Ech}$) and *E. amylovora* (Harpin$_{Ea}$). Like the other harpins, the deduced 36 kDa protein does not possess a typical signal sequence, but it contains a putative membrane spanning domain. The 36 kDa protein has been identified as the $hrpN_{Ecc}$ product by an immunoblot analysis using anti-Harpin$_{Ech}$ antibodies.

(3) By mini-Tn5-Km and chemical mutagenesis we have isolated RsmA⁻ mutants of Ecc71 that elicit the HR in tobacco leaves. The inactivation of $hrpN_{Ecc}$ resulted in the HR⁻ phenotype and a loss of the ability to produce the 36 kDa protein. These observations collectively demonstrate that $hrpN_{Ecc}$ encodes an elicitor of the HR, designated as Harpin$_{Ecc}$.

(4) The expression of $hrpN_{Ecc}$ is markedly affected by media composition, carbon source and the *rsmA* allele. While the carbon source dependent regulation of $hrpN_{Ecc}$ expression occurs in both RsmA⁺ and RsmA⁻ bacteria, the mRNA levels are uniformly higher in the RsmA⁻ strains.

(5) The DNA region upstream of the $hrpN_{Ecc}$ transcriptional start site contains a consensus $\sigma^{54}$ promoter sequence and a *hrp* box. Thus, the structure of the putative $hrpN_{Ecc}$ promoter closely resembles those of *hrp* and *avr* genes of *Pseudomonas syringae* pathovars.

Based upon those observations we conclude that the inability of wild type pectolytic *E. carotovora* strains to elicit the HR is due to the lack of a significant level of *hrpN* expression.


ACKNOWLEDGMENTS

This research was supported by the National Science Foundation (grant

DMB-9018733) and the Food for 21st Century Program of the University of Missouri. This article is journal series 12,510 of the Missouri Agricultural Experiment Station.

LITERATURE CITED

Barras, F., van Gijsegem, F., and Chatterjee, A. K. 1994. Extracellular enzymes and pathogenesis of soft-rot *Erwinia*. Ann. Rev. Phytopathol. 32:201-234.

Bauer, D. W., Bogdanove, A. J., Beer, S. V., and Collmer, A. 1994. *Erwinia chrysanthemi hrp* genes and their involvement in soft rot pathogenesis and elicitation of the hypersensitive response. Mol. Plant-Microbe Interaction 7:573-581.

Bauer, D. W., Wei, Z. M., Beer, S. V., and Collmer, A. 1995. *Erwinia chrysanthemi* Harpin$_{Ech}$ : an elicitor of the hypersensitive response that contributes to soft-rot pathogenesis. Mol. Plant-Microbe Interaction 8:484-491.

Cui, Y., Chatterjee, A., Liu, Y., Dumenyo, C. K., and Chatterjee, A. K. 1995. Identification of a global repressor gene, *rsmA*, of *Erwinia carotovora* subsp. *carotovora* that controls extracellular enzymes, $N$-(3-Oxohexanoyl)-L-Homoserine lactone, and pathogenicity in soft-rotting *Erwinia* spp. J. Bacteriol. 177:5108-5115.

Laby, R. J., and Beer, S. V. 1992. Hybridization and functional complementation of the *hrp* gene cluster from *Erwinia amylovora* strain Ea321 with DNA of other bacteria. Mol. Plant-Microbe Interaction 5:412-419.

Liu, M. Y., Yang, H., and Romeo, T. 1995. The product of the pleiotropic *Escherichia coli* gene *csrA* modulates glycogen biosynthesis via effects of mRNA stability. J. Bacteriol. 177:2663-2672.

Mukherjee, A., Cui, Y., Liu, Y., Dumenyo, C. K., and Chatterjee, A. K. 1996. Global regulation in *Erwinia* species by *Erwinia carotovora rsmA*, a homologue of *Escherichia coli csrA*: repression of secondary metabolites, pathogenicity and hypersensitive reaction. Microbiology 142:427-434.

Wharam, S. D., Mulholland, V., and Salmond, G. P. C. 1995. Conserved virulence factor regulation and secretion systems in bacterial pathogens of plants and animals. Eur. J. Plant Path. 101:1-13.

# Genes and Proteins Involved in Aggressiveness and Avirulence of *Xanthomonas oryzae* pv. *oryzae* to Rice

J. E. Leach, W. Zhu, J. M. Chittoor, G. Ponciano, S. A. Young, and F. F. White

Department of Plant Pathology, Kansas State University, Manhattan, KS 66506-5502 USA

*Xanthomonas oryzae* pv. oryzae (Ishiyama) Dye is a vascular pathogen which causes bacterial blight of rice. In interactions between *X. o.* pv. *oryzae* and rice, resistance is governed by an interaction between single, dominant resistance gencs in rice and corresponding pathogen genes called avirulence (*avr*) genes (for review, see Mew 1987, Leach and White 1996). Although several *avr* genes have been cloned from *X. o.* pv. *oryzae* (Hopkins et al. 1992), how the products of these genes function to elicit remains a mystery. Here, we summarize structural and functional features of the *avr* genes from *X. o.* pv. oryzae, describe potential for interactions of the *avr* genes or their products with other bacterial genes or proteins, and present models for *avr/R* gene interactions.

## *X. o.* pv. *oryzae* avirulence genes

*X. o.* pv. *oryzae avr* genes *(avrxa5, avrXa7* and *avrXa10*) are homologs (Hopkins et al. 1992), and are members of a gene family from *Xanthomonas* that are typified by the first cloned member of the family, *avrBs3* from *X. campestris* pv. *vesicatoria* (Bonas et al. 1989). Strains of *X. o.* pv. *oryzae* contain twelve to fourteen copies of genes that hybridize to members of the family (Hopkins et al. 1992). To date, members of this gene family which function as *avr* genes have been cloned from pathogens of such diverse hosts as pepper (Bonas et al. 1989, Canteros et al. 1991, Bonas et al. 1993), cotton (De Feyter et al. 1991, 1993), and citrus (Swarup et al. 1991, 1992). The gene family has not been identified in other genera.

The most striking feature of the *avr* genes from *X. o.* pv. *oryzae* and other family members is the structure (Fig. 1). Within the middle third of the encoded protein of each gene is a repeated sequence of 34 amino acids; *avrXa10* and *avrXa7* contain 15.5 and 25.5 copies of the repeated sequence (Hopkins et al. 1992, White et al. unpublished). The repeated units are highly conserved with the exceptions of amino acids at positions 12 and 13, which vary (Fig. 1). We have exchanged the repeat region of *avrXa10* with the region of *avrXa7* and *avrBs3*, and specificity was found in all cases to correspond to the source of the repeat domain. The *avrBs3* repeat domain in *avrXa10* conferred a resistant phenotype to *X. c.* pv. *vesicatoria* on pepper containing the *Bs3* gene, and the *avrXa7* repeat in *avrXa10* conferred specificity for the resistance gene *Xa7*, and lost specificity for *Xa10*.

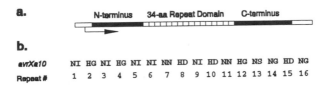

Figure 1. A. Schematic map of members of the *avr* gene family from *Xanthomonas*. The filled bars indicate the coding sequence on either side of the repeat domain, which consists of varying numbers of 102-bp directly repeated sequences. B. The order of the two-amino acid variable region from within each 34 amino acid repeat of *avrXa10* (Hopkins et al. 1992) from *X. o.* pv. oryzae.

The arrangements of the variable regions of the characterized *Xanthomonas avr* genes vary, and appear to be critical features for the race-specificity of the proteins. Some deletions in repeats of *avrBs3* resulted in the loss of avirulence activity on pepper plants containing the *Bs3* gene for resistance and, at the same time, gained new avirulence activity on tomato (Herbers et al. 1992). Site-directed mutagenesis altering amino acid 13 in some but not all repeats of *avrXa10* resulted in loss of avirulence function, indicating that the variable regions within the repeats play a role in specificity (Zhu and White, unpublished).

As intriguing as the repeat domain structure of the *avrXa10*-like genes are the phenotypic effects that some of the homologs have on pathogenesis. If the *avrXa7* gene, which normally confers avirulence activity on strains of *X. o.* pv. *oryzae* when inoculated to rice containing the *Xa7* gene, is inactivated, the mutant strains are less aggressive to a susceptible cultivar of rice (Fig. 2, Choi et al. unpublished). This potential role in aggressiveness or fitness also has been described for some, but not all, other members of the gene family (Swarup et al. 1991, De Feyter et al. 1993, Yang et al. 1996). These results may help explain the retention of avirulence genes, in general, and more specifically, the presence of many copies of the genes in *X. o.* pv. *oryzae*. The introduction of new plant genotypes to control the pathogen may select for new versions of the gene family that avoid recognition for resistance and retain functions for aggressiveness.

## Location of *avrXa10*

The simplest model for *avr/R* gene interactions predicts that the primary avirulence gene product is the elicitor of resistance. If this is true in the *X. o.* pv. *oryzae/* rice interactions, the products of the *avr* genes would likely be located outside the bacterial cells, either in the plant extracellular spaces (Fig. 3a) or inside the plant cell (Fig. 3b). However, the product of the *avrXa10* gene, like that of *avrBs3* (Brown et al. 1993), has

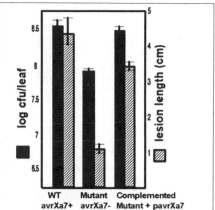

Fig. 2. Bacterial numbers and lesion lengths after clip inoculation of the susceptible rice cultivar IR24 with a wild type strain of *X. o.* pv. *oryzae* (*avrXa7*[+]), a marker exchange mutant (*avrXa7*) and the complemented mutant (*avrXa7*[+]). Bacterial numbers and lesion lengths are the average of three replicates; bar indicates standard deviation.

been shown by biochemical fractionation and immunoelectron microscopy experiments to be located mainly in the cytoplasm of the bacterial cells (Young et al. 1994). The protein was not detected in plant cells or in the extracellular spaces. In addition, neither the *avrXa10* gene product, which was purified from *E. coli* or *X. o.* pv. *oryzae* cells, nor the extracellular fluids of rice leaves undergoing an incompatible response elicited resistance in rice plants containing *Xa10* (Young et al. 1994).

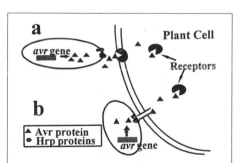

Fig. 3. Models for direct interactions between Avr proteins and plant receptors (possibly resistance gene products).

Although the *avr* gene products (a) do not contain secretion signals, (b) were not detected outside of the bacterial cells and (c) when purified and infiltrated into rice extracellular spaces did not elicit the HR, they may yet function outside the bacterial cell. Bacterial proteins can be exported through secretion pathways that do not require signal sequences

(Salmond and Reeves 1994), and some exported proteins require modification for activity (Wagner et al. 1988, Hughes et al. 1992). Thus, the *avr* gene products may function as elicitors and may not have been detected outside the bacterial cells because they are: (a) modified such that they are not recognized by the antibodies; (b) rapidly degraded by host proteases; (c) exported in very low amounts and/or at very specific times during infection, or (d) introduced by some mechanism directly into plant cells.

### Interaction of *avr* and *hrp* genes.

Like *avr* genes from other plant pathogenic bacteria, the activities of the *X. o.* pv. *oryzae avr* genes are dependent on functional *hrp* (*h*ypersensitive *r*eaction/*p*athogenicity) genes. *X. o.* pv. *oryzae hrp* genes also are induced to higher levels after culture in minimal medium than in rich medium (Zhu et al. unpublished). Expression of the *X. o.* pv. *oryzae avr* genes, however, is independent of *hrp* gene expression.

Similarity of the *hrp* genes of plant pathogenic bacteria to genes coding for type III secretory systems in other bacteria has lead to the speculation that the *hrp* genes might function in the secretion of *avr* proteins. Inactivation of $hrpCI_{Xoo}$, which is similar to *hrpN* in *Pseudomonas solanacearum* and *yscU* in *Yersinia pseudotuberculosis,* by marker exchange mutagenesis renders *X. o.* pv. *oryzae* non-pathogenic to rice and unable to elicit the HR. When leaves of rice containing the *Xa10* gene for resistance were co-inoculated with the mutant ($hrpCI_{Xoo}^{-}$ /$avrXa10^{+}$; interaction phenotype = no response) and a $hrpCI_{Xoo}^{+}$ /$avrXa10^{-}$ strain (phenotype = susceptible), no elicitation of resistance occurred, both strains multiplied (Zhu et al. unpublished). Instead, a susceptible response was observed. These results indicate that *avr* and *hrp* genes must function in the same bacterial cell for elicitor activity.

One intriguing feature of the YOP secretion system from *Yersinia* spp. is the use of cytoplasmic chaperones that are specific for individual secreted proteins called Yops (for review, see Cornelius 1994). In preliminary studies to identify proteins (possibly chaperones) that interact with *X. o.* pv. *oryzae avr* gene products, cytoplasmic proteins extracted from *X. o.* pv. *oryzae* cells were used in binding reactions (Carr et al. 1991) with purified, tagged-AVRXa10 (AVRXa10.F1). After incubation with protein extracts from *X. o.* pv. *oryzae* grown in *hrp* inducing conditions, migration of AVRXa10.F1 was retarded in native gels. Migration of AVRXa10.F1 was not retarded after incubation with proteins from *X. o.* pv. *oryzae* grown in *hrp* non-inducing conditions. A protein with apparent $M_r$ 27 kD co-purified with AVRXa10.F1.

## Concluding Remarks

The cloning of avirulence genes and the development of cultivars near-isogenic for resistance to *X. o.* pv. oryzae has added a new dimension to our understanding of resistant interactions. The fact that the so-far identified avirulence genes from *X. o.* pv. oryzae are multi-copy and members of a gene family common to other xanthomonads suggests that (a) the *avr* genes may have similar functions, (b) rice resistance genes may be functionally related to one another, and (c) rice resistance genes may be similar to resistance genes from other host plants for xanthomonads. The recently cloned *Xa21* gene from rice is in fact similar in structure to resistance genes cloned from other crop plants (Song et al. 1995). The structure of the *Xanthomonas avr* genes with the intriguing repeat domain that regulates specificity suggests a mechanism for creation of new specificities, or the loss of avirulence function, i.e., through recombination between the genes. The predominant location of the *avr* gene products inside the *X. o.* pv. oryzae cells would seem to suggest they do not function directly as elicitors. However, the structural features and requirements of the *avr* genes, the similarity of *hrp* genes to secretory genes in other bacteria and the requirements for functional *hrp* genes for the *avr* phenotype present the possibility that the *avr* gene products may serve as direct elicitors of resistance.

## Acknowledgements

This is contribution No. 97-12-A from the Kansas Agricultural Experiment Station.

## References

Bonas, U., Stall, R.E. and Staskawicz, B.J. 1989. Genetic and structural characterization of the avirulence gene *avrBs3* from *Xanthomonas campestris* pv. *vesicatoria*. Mol. Gen. Genet. 218:127-136.

Bonas, U., Conrads-Strauch, J. and Balbo, I. 1993. Resistance in tomato to *Xanthomonas campestris* pv. *vesicatoria* is determined by alleles of the pepper-specific avirulence gene *avrBs3*. Mol. Gen. Genetics 238:261-269.

Brown, I., Mansfield, J., Irlam, I., Conrads-Strauch, J.and Bonas, U. 1993. Ultrastructure of interactions between *Xanthomonas campestris* pv. *vesicatoria* and pepper, including immunocytochemical localization of extracellular polysaccharides and the *avrBs3* protein. Mol. Plant-Microbe Interact. 6:376-386.

Canteros, B., Minsavage, G., Bonas, U., Pring, D. and Stall, R. 1991. A gene from *Xanthomonas campestris* pv. *vesicatoria* that determines avirulence in tomato is related to *avrBs3*. Mol. Plant-Microbe Interact. 4:628-632.

Carr, D.W., Stofko-Hahn, R.E. Fraser, I.D.C., Bishop, S.M., Acott, T.S., Brennan, R.G., and Scott, J.D. 1991. Interaction of the regulatory subunit (RII) of cAMP-dependent protein kinase with RII-anchoring proteins occurs through an amphipathic helix binding motif. J. Biol. Chem. 266:14188-14192.

Cornelius, G.R. 1994. *Yersinia* pathogenicity factors. Pages 243-263 in: Bacterial Pathogenesis of Plants and Animals--Molecular and Cellular Mechanisms. J.L. Dangl ed. Springer Verlag, Berlin.

De Feyter, R. and Gabriel, D.W. 1991. At least six avirulence genes are clustered on a 90-kilobase plasmid in *Xanthomonas campestris* pv. *malvacearum*. Mol. Plant-Microbe Interact. 4:423-432.

De Feyter, R., Yang,Y. and Gabriel, D.W. 1993. Gene-for-genes interactions between cotton R genes and *Xanthomonas campestris* pv. *malvacearum avr* genes. Mol. Plant-Microbe Interact. 6:225-237.

Herbers, K., Conrads-Strauch, J. and Bonas, U. 1992. Race-specificity of plant resistance to bacterial spot disease determined by repetitive motifs in a bacterial avirulence protein. Nature 356:172-174.

Hopkins, C.M., White, F.W., Choi, S.H., Guo, A. and Leach, J.E. 1992. A family of avirulence genes from *Xanthomonas oryzae* pv. *oryzae* Mol. Plant-Microbe Interact. 5:451-459.

Hughes, C., Issartel, J.P., Hardie, K., Stanley, P., Koronakis, E. and Koronakis, V. 1992. Activation of *Escherichia coli* prohemolysin to the membrane-targeted toxin by HlyC-directed ACP-dependence fatty acylation. FEMS Microbiol. and Immunol. 5:37-43.

Leach, J.E. and White, F.F. 1996. Avirulence genes. (in press) in: Plant Microbe Interactions, Vol. 2. Stacey, G. and Keen, N. eds., Kluwer Acad Publishers.

Mew, T.W. 1987. Current status and future prospects of research on bacterial blight of rice. Annu. Rev. Phytopathol 25:359-382.

Salmond, G.P.C. and Reeves, P.J. 1994. Secretion of extracellular virulence factors by plant pathogenic bacteria. Annu. Rev. Phytopathol 32:181-200.

Song, W.-Y., Wang, G.-L., Chen, L.-L., Kim, H.-S., Pi, L.-Y., Holsten, T., gardner, J., Wang, B., Zhai, W.-Y., Zhu, L.-H., Fauquet, C., and Ronald, P. 1995. A receptor kinase-like protein encoded by the rice disease resistance gene, *Xa21*. Science 270:1804-1806.

Swarup, S., De Feyter, R., Brlansky, R. and Gabriel, D.W. 1991. A pathogenicity locus from *Xanthomonas citri* enables strains from several pathovars of *X. campestris* to elicit cankerlike lesions on citrus. Phytopathol 81:802-809.

Swarup, S., Yang, Y., Kingsley, M.T. and Gabriel, D.W. 1992. A *Xanthomonas citri* pathogenicity gene, *pthA*, pleiotropically encodes gratuitous avirulence on nonhosts. Mol. Plant-Microbe Interact. 5:204-213.

Wagner, W., Kuhn, M., Goebel, W. 1988. Active and inactive forms of hemolysin (Hy1A) from *Escherichia coli*. J. Biol. Chem. 369:39-46.

Yang, Y., Yuan, Q., and Gabriel, D.W. 1996. Watersoaking function(s) of XcmH1005 are redundantly encoded by members of the *Xanthomonas avr/pth* gene family. Mol. Plant-Microbe Interact. 9:105-113.

Young, S.A., White, F.F., Hopkins, C.M. and Leach, J.E. 1994. AVRXa10 protein is in the cytoplasm of *Xanthomonas oryzae* pv. oryzae. Mol. Plant-Microbe Interact. 7:799-804.

# ROLE OF NUCLEAR LOCALIZING SIGNAL SEQUENCES IN THREE DISEASE PHENOTYPES DETERMINED BY THE *XANTHOMONAS avr/pth* GENE FAMILY.

D.W. Gabriel, Q. Yuan, Y. Yang, and P.K. Chakrabarty.

Plant Molecular and Cellular Biology Program and Plant Pathology Department, University of Florida, Gainesville, FL 32611-0680, USA

The bacterial signals required to induce at least three different plant response phenotypes [hyperplastic cankers, water soaking and the hypersensitive responses (HRs)] are determined by different members of the *Xanthomonas avr/pth* gene family. Fusion experiments using *pthA, pthN* and *avrb6* revealed that both regions flanking the leucine-rich, direct tandem repeats are required for all three phenotypes, and are interchangeable. Members of this *avr/pth* gene family are known or suspected to be required in all forms of citrus canker, cotton blight, common bean blight and bacterial blight of rice diseases. DNA sequence analyses revealed that three nuclear localization signals (NLSs) are conserved in at least 10 members of the family (*avrBs3, avrBs3-2, avrXa10, avrb6, pthA, pthB, pthC, pthN, pthN2* and *avrB4*). Gus reporter fusions carrying the NLS regions of Avrb6 and PthA localize to onion cell nuclei, while site-directed mutations in any of the three NLSs abolished localization to onion cell nuclei. Site-directed mutations affecting the middle NLS also abolished the ability of *pthA* to elicit cankers on citrus and *avrb6* to elicit the gene-for-gene HR on cotton isoline Acb6, but had no effect on the ability of either gene to elicit a nonhost HR. Mutations of any two NLSs abolished the ability of *pthA* to elicit cankers on citrus, but did not abolish its ability to elicit a nonhost HR. Since the NLSs are needed for pathogenic phenotypes and some, but not all, HR reactions, these results indicate that there may be at least two signal pathways: one directly affecting programmed cell responses in the plant nucleus, and another that leads to an HR, but may not involve direct Avr/Pth nuclear transport.

## *Xanthomonas* and pathogenicity

Most mutations that abolish pathogenicity in plant pathogenic xanthomonads involve the *hrp* (hypersensitive response and pathogenicity) genes (1). The major plant pathogenic prokaryotic genera---*Erwinia, Pseudomonas* and *Xanthomonas*---all have highly similar *hrp* genes, and they are similarly organized (2,3). Mutations of any of the *hrp* genes abolish or reduce plant symptoms induced by *Erwinia, Pseudomonas* or *Xanthomonas*: pathogenicity (water soaking, wilts, blights, cankers) on susceptible hosts, a hypersensitive response (HR) on resistant hosts, and an HR on nonhosts.

The *hrp* genes of *P. syringae* are coordinately regulated (4) and there are striking DNA sequence and gene organization similarities when *hrp* genes are compared to the *Yersinia* virulence genes (*vir, lcr* or *ysc*), which strongly suggests a similar biochemical function (2). The *Yersinia* virulence genes are known to encode a Type III protein export system that enables the export of at least two large virulence or antihost proteins called Yops (5). The Type III export system differs from the General Secretory Pathway (Type II export system) in that a canonical export signal sequence is not defined for Type III export, and it differs from the Type I ($\alpha$ hemolysin) pathway in that only 3 genes are required for the latter (3).

Pathogenic symptoms are a host response phenotype. With both *Pseudomonas* and *Erwinia*, the effector molecules of the host response phenotype have been identified as proteins (harpins, HrpZ and PopA). These protein elicitors are (among) the exported product(s) of *hrp* genes (6,7,8,). These proteins are therefore functionally analogous to the *Yersinia* Yops proteins both in terms of export and in terms of being pathogenic effectors. Harpins, HrpZ and PopA directly elicit plant symptoms on both host and nonhost plants. In *Xanthomonas,* no such exported products have been identified. We propose that proteins encoded by the *Xanthomonas avr/pth* gene family, discussed in the next section, are functionally analogous to the *Yersinia* Yops and are among the primary products exported.

### The *Xanthomonas avr/pth* gene family

Race variation is determined by avirulence (*avr*) genes, which act as negative factors to limit the growth of strains within a pathovar or species to a subset of **hosts** within the range of the pathovar or species. The hosts on which avirulence is observed always carry resistance (*R*) genes that are genetically specific for particular *avr* genes; this genetic requirement is often termed gene-for-gene (*avr*-for-*R*) specificity (for reviews, refer 9,10,). Resistance in hosts or avirulence in bacterial pathogens is usually assayed phenotypically by a plant hypersensitive response (HR) following inoculation of the pathogen at high (>10$^7$ cfu/ml) levels.

. Surprisingly, most of the *X. campestris* pv. malvacearum *avr* genes are also pathogenicity (*pth*) genes and members of an *avr/pth* gene family, redundantly present in every strain of the pathovar, and additively responsible for the water soaked symptoms of angular leaf spot disease of cotton (11). Even more surprisingly, these genes are members of the same *avr/pth* gene family as the genes of *X. citri* that are responsible for inducing citrus canker disease (11,12,13 and some unpublished results). Most of the members of this *avr/pth* gene family were first isolated as *avr* genes, and without evidence of *pth* function. However, all of the *X. citri* and X.c. pv. aurantifolii genes (*pthA, pthB* and *pthC*) and *pthN* and *pthN2* of X.c. pv. malvacearum were first isolated as *pth* genes, and without evidence of *avr* function.

Members of this *Xanthomonas avr/pth* gene family are widely distributed in the genus *Xanthomonas*, although not all xanthomonads carry members of the family. Currently, at least 26 members have been cloned: from X.c. pv. malvacearum, genes *avrBn* (14), *avrB4, avrb6, avrb7, avrBIn, avrB101, avrB102* (15), *avrB103, avrB104, avrB5* (11), *pthN* and *pthN2* (Chakrabarty and Gabriel, submitted); from *X. oryzae*, genes *avrxa5, avrXa7* and *avrXa10* (16); from X.c. pv. vesicatoria, genes avrBs3 and avrBs3-2 (17,18); from *X. citri*, genes *pthA, avrXc1, avrXc2* and *avrXc3* (14); from X.c. pv. aurantifolii, genes *pthB, pthB2, pthC* and *pthC2* (Yuan and Gabriel, unpublished), and from *X. phaseoli, avrXp1* (Swarup and Gabriel, unpublished). Members of the gene family are required for pathogenicity of X.c. pv. malvacearum (cotton blight), *X. citri* (Asiatic citrus canker), X.c. pv. aurantifolii (false citrus canker and Mexican lime cancrosis), X. oryzae (bacterial blight of rice) and X. phaseoli (common bean blight).

The host-specific pathology may help condition host range by affecting dispersal. For example, the formation of cankers ruptures the citrus epidermis and releases abundant bacteria to the leaf surface (14). Similarly, at least seven out of ten *avr/pth* genes tested from a single X.c. pv. malvacearum strain confer ability to water soak cotton (15). When all seven of these *avr/pth* genes are destroyed in a single strain, HM2.2S, the strain grows *in planta* to the same levels as the wild type, but is completely asymptomatic, and 1,600 times less bacteria are released to the leaf surface. Since both citrus canker disease and bacterial blight of cotton are spread primarily by rain splash, the presence of larger numbers of bacteria on the leaf surface undoubtedly contributes to pathogen dispersal and host range.

There is no evidence that PthA and Avrb6 are direct elicitors. Unlike harpin, PopA or HrpZ, protein extracts containing PthA or Avrb6 elicited no symptoms when inoculated onto plants; *hrp* genes were required for *Xanthomonas* carrying *pthA* to elicit hyperplastic cankers on citrus (19).

## NLS signals in the *Xanthomonas* Avr/Pth predicted proteins

Our analyses (19) of the published, predicted amino acid sequences encoded by *avrb6, pthA, avrXa10, avrBs3* and *avrBs3-2* revealed the presence of three stretches of basic residues with complete homology with the nuclear localization consensus sequences (K-R/K-X-R/K) found in many characterized nuclear localized proteins (20). These three putative nuclear localization sequences (NLSs) are located near the C-terminus of the proteins, at positions

1020-1024 (K-R-A-K-P), 1065-1069 (R-K-R-S-R), and 1101-1106 (R-V-K-R-P-R) in PthA. The first and third putative NLSs also contain proline residues, which are often found in NLSs. The presence of the same three putative NLSs was also found in 5 additional genes that we have now partially sequenced: *pthB* and *pthC*, required for citrus canker disease caused by X.c. pv. aurantifolii strains, *pthN* and *pthN2*, which contribute to angular leaf spot disease caused by strain XcmN of X.c. pv. malvacearum, and *avrB4*, an avirulence gene from strain XcmH of X.c. pv. malvacearum. In short, NLS signals are found in the predicted peptide sequences of all functional *avr/pth* gene family members of which we are aware.

The NLS signals of PthA and Avrb6 are functional. The DNA coding sequences for the C-terminal regions of Avrb6 and PthA were independently fused to a β-glucuronidase (GUS) reporter gene. When introduced into onion cells, both of these translational fusions were transiently expressed, and GUS activity was specifically localized in the nuclei of transformed cells (19). When the resulting plasmid DNAs were delivered into onion epidermal cell layers, transient expression and accumulation of GUS activity was observed in both control and experimental treatments. In control experiments, in which plasmid pBI221 (encoding GUS) was introduced into onion epidermal cells, GUS activity was found only in the cytoplasm.

### Site-directed mutations.

PCR primers were used to create site-directed mutations that eliminated each of the three NLSs individually in both *pthA* and *avrb6*. The three NLSs, named in order towards the C terminus, are nlsA, nlsB and nlsC

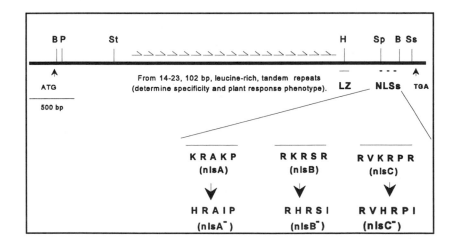

(refer figure). nlsA was changed from K-R-A-K-P to H-R-A-I-P; nlsB was changed from R-K-R-S-R to R-H-R-S-I, and nlsC was changed from R-V-K-R-P-R to R-V-H-R-P-I. In each case, lysine was changed to histidine, and a second lysine or arginine was changed to isoleucine (from a basic amino acid to a neutral one) on both genes. The mutant genes were sequenced to confirm that only the desired mutations were present and were then moved into appropriate *Xanthomonas* strains. For canker assays, citrus plants were inoculated with citrus-compatible xanthomonads carrying the *pthA* constructs. For host and nonhost resistance assays, resistant and susceptible cotton plants were inoculated with XcmN carrying both the *avrb6* and *pthA* constructs. For water soaking assays, susceptible cotton plants were inoculated with HM2.2S carrying the *avrb6* construct. The results were repeated many times and are unequivocal: the nlsB mutation completely negated normal *pthA* (hyperplastic canker) activity on citrus, but did not affect the HR conferred by the gene to

## NLS Knockout Mutations in *pthA*

|  | In any xanthomonad tested on citrus | In XcmN on resistant cotton line AcBln3 | In compatible xanthomonads on bean |
|---|---|---|---|
| Vector Only | No Canker | Water soaking | Water soaking |
| *pthA* | Canker | HR | HR |
| *pthA* (nlsA⁻) | Canker | HR | HR |
| *pthA* (nlsB⁻) | **No Canker** | **HR** | **HR** |
| *pthA* (nlsC⁻) | Canker | HR | HR |
| *pthA* (nlsA⁻B⁻) | **No Canker** | **HR** | **HR** |
| *pthA* (nlsA⁻C⁻) | **No Canker** | **HR** | **HR** |
| *pthA* (nlsB⁻C⁻) | **No Canker** | **HR** | **HR** |

## NLS Knockout Mutations in *avrb6*

|  | In HM2.2S on susceptible cotton Ac44 | In XcmN on resistant cotton Acb6 | In compatible xanthomonad on bean |
|---|---|---|---|
| Vector only | Asymptomatic | Water soaking | Water soaking |
| avrb6 | Water soaking | HR | HR |
| avrb6 (nlsA⁻) | Water soaking | HR | HR |
| avrb6 (nlsB⁻) | **Water soaking** | **Water soaking** | **HR** |
| avrb6 (nlsC⁻) | **Water soaking** | **Weak HR** | **HR** |

XcmN on cotton or to compatible xanthomonads on bean. The same nlsB mutation had no effect on water soaking of cotton caused by HM2.2S/*avrb6*, but it did abolish the gene-for-gene HR normally elicited by XcmN/avrb6 on resistant cotton line Acb6. The individual nlsA mutations had no effect on the normal functions of *pthA* or *avrb6*, but the nlsC mutation affected the gene-for-gene HR response of cotton Acb6 with *avrb6* (assayed in XcmN; see chart). Double NLS mutations were also created in *pthA*, and any combination of these were found to completely negate normal *pthA* (hyperplastic canker) activity in citrus-compatible xanthomonads on citrus. Again, however, the non-host HR on cotton and bean were unaffected.

We recreated the GUS gene fusions in pBI221 using the site-directed NLS mutations for a repeat of published the nuclear localization experiments in onion (19). Site-directed mutations of any single NLS coding region (fused with *gus*) did not localize to onion cell nuclei.

## Literature Cited

1. Willis, D.K. et al. 1991.Molec. Plant-Microbe Interact. 4:132-138.

2. Van Gijsegem et al. 1993. Trends Microbiol. 1:175-180.

3. Salmond, G.P.C. 1994.Annu. Rev. Phytopathol. 32:181-200.

4. Xiao, Y. et al. 1994. J. Bacteriol. 176:1025-1036.

5. Wattiau, P.et al. 1994. Proc. Natl. Acad. Sci. USA 91:10493-10497.

6. Arlat, M. et al. 1994. EMBO J. 13:543-553.

7. He, S.Y. et al. 1993. Cell 73: 1255-1266.

8. Preston, G., et al. 1995. Molec. Plant-Microbe Interact. 8:717-732.

9. Gabriel, D.W. et al. 1990.Annu. Rev. Phytopathol. 28:365-391

10. Keen, N.T. 1992. Plant Mol. Biol. 19:109-122.

11. Yang, Y. et al. 1996. Molec. Plant-Microbe Interact. 9:105-113.

12. Swarup, S. et al. 1992.Molec. Plant-Microbe Interact. 5:204-213.

13. Yang, Y., et al. 1994. Molec. Plant-Microbe Interact. 7:345-355.

14. Gabriel, D.W. et al. 1986. Proc. Natl. Acad. Sci. USA. 83:6415-6419.

15. De Feyter, R. et al. 1991. Mol. Plant-Microbe Interact. 4:423-432.

16. Hopkins, C.M. et al. 1992. Mol. Plant-Microbe Interact. 5:451-459.

17. Bonas, U. et al. 1993. Mol. Gen. Genet. 238:261-269.

18. Bonas, U. et al. 1989. Mol. Gen. Genet. 218:127-136.

19. Yang, Y. et al. 1995. Molec. Plant-Microbe Interact. 8:627-631.

20. Chelsky, D. et al. 1989. Mol. Cell. Biol. 9: 2487-2492.

# Xanthomonas campestris pv. vesicatoria hrp gene regulation and avirulence gene avrBs3 recognition

U. Bonas, E. Huguet, L. Noël, M. Pierre, O. Rossier, K. Wengelnik, and G. Van den Ackerveken.

Institut des Sciences Végétales, CNRS, 91198 Gif-sur-Yvette, France

*Xanthomonas campestris* pathovar *vesicatoria* (*Xcv*), a gram-negative bacterium, is the causal agent of bacterial spot, a non-systemic disease of pepper and tomato which is of great economic importance in growing regions with a warm and humid climate. Bacteria enter the plant tissue via stomata or wounds and multiply in the intercellular space. After infection of a plant with *Xcv* two different types of reactions can be observed depending on the genotype of both organisms. In the susceptible plant, infection with a virulent *Xcv* strain gives rise to watersoaked lesions that later on become necrotic (compatible interaction). If the plant is resistant and the bacterium avirulent, a hypersensitive reaction (HR) is induced (incompatible interaction). We study bacterial and plant genes involved in the interaction: *hrp*, *avrBs3*, *avrBs3-2*, and resistance genes *Bs3* and *Bs3-2*.

## Genetic and Molecular Analysis of *hrp* Genes

### UPDATE ON THE ANALYSIS OF THE LARGE *HRP* CLUSTER

Basic pathogenicity of *Xcv* is determined by *hrp* genes (hypersensitive reaction and pathogenicity), which are organized in six loci, *hrpA* to *hrpF*, in a 23-kb region (Bonas et al. 1991). No new *hrp* loci were identified when 70 new Tn*3-gus* insertions to the right and left of the known *hrp* cluster were marker exchanged into the wild-type strain. Based on DNA sequence analysis of the 23-kb region we predict 21 genes to be expressed, ten of which are predicted to encode proteins associated with or localized in the bacterial inner or outer membrane. Most of these putative membrane proteins show similarity with components of type III secretion systems (Bonas 1994), e.g., the HrpA1 protein, a homologue of the *Klebsiella oxytoca* PulD and the *Yersinia* spp. YscC proteins. Localization of HrpA1, which contains a classical signal peptide sequence at the N-terminus, was found to be mainly in the outer membrane of *Xcv*. Interestingly, expression of the HrpA1 protein in *Escherichia coli* induced the *psp* operon (phage shock protein; in collaboration with Dr. M. Russel, Rockefeller University, NY), suggesting that HrpA1 forms pores in the outer membrane by multimerization as has been proposed for some of its homologs, e.g., filamentous phage pIV protein (Wengelnik et al. 1996a).

Plant cell responses to *hrp* mutants were analyzed by electron microscopy. As expected, there were no macroscopic disease symptoms visible, but papillae were induced. Formation of papillae, a local plant defense reaction, was not observed in the interaction of wild-type bacteria with both susceptible and resistant plants. We believe therefore that the wild type produces a suppressor of papillae formation (Brown *et al.*, 1995), which might be a protein that is secreted via the Hrp secretion apparatus.

## *HRP* REGULATORY GENES

Expression of the *hrp* loci is regulated by environmental conditions. No or weak *hrp* gene expression is detected in bacteria grown in minimal or complex medium. However, expression of all six *hrp* loci is induced during growth of *Xcv* in the plant, in TCM (Schulte and Bonas 1992), and in the synthetic medium XVM2 (Wengelnik et al. 1996a). Which genes control *hrp* gene expression? We isolated the *hrpX* and *hrpG* regulatory genes, which are localized next to each other outside of the *hrp* cluster and are divergently transcribed (see Fig. 1). *hrpG* was found to be at the top of the regulatory cascade. HrpG is homologous to two-component response regulators.

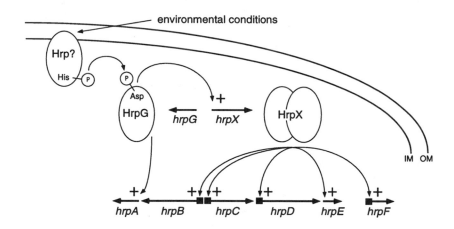

Fig. 1: Model of *hrp* gene regulation in *Xcv*. We propose that a so far unidentified sensor activates HrpG by phosphorylation (P) on a conserved Asp residue. HrpG then activates *hrpX* and *hrpA*. HrpX is postulated to activate the loci *hrpB* to *hrpF* as a homodimer. PIP-boxes in the promoters of *hrpB, C, D,* and *hrpF* are indicated as square blocks. IM: inner membrane; OM: outer membrane.

Expression of *hrpG* is low in complex medium and slightly increased in XVM2 (Wengelnik et al. 1996b). *hrpX* encodes an AraC-type regulator (Wengelnik and Bonas 1996). Expression of *hrpX* gene is induced in XVM2 and is required for induction of the five loci *hrpB* to *hrpF*. The HrpX protein is related to HrpB of *Burholderia solanacearum* which regulates expression of several transcription units within the *hrp* cluster and of the *popA* gene (Genin et al. 1992). Although the gene products only share significant similarity in their C-terminal domains, a *Xcv hrpX* mutant could be partially complemented by a p*lac-hrpB* construct (Wengelnik and Bonas 1996).

Promoter analysis revealed the presence of a conserved sequence motif, TTCGC-N15-TTCGC, the PIP-box (Plant Inducible Promoter), 44 bp upstream of the *hrpB* transcription start site as well as in the promoters of *hrpC*, *hrpD*, and *hrpF*. Furthermore, we found identical or highly similar motifs in the putative promoter regions of the *Xcv* avirulence gene *avrRxv* and of the transcription units regulated by the *hrpB* gene in *B. solanacearum* (Fenselau and Bonas 1995). No PIP-box was found in the *hrpX* and *hrpA* promoters which are expressed *hrpX*-independently (Wengelnik et al. 1996a). Taken together, these data suggest an important role for the PIP-box in transcriptional control of *hrp* genes through HrpX. A more general conclusion from our results is that *hrp* gene regulation in *Xcv* appears to be fundamentally different from that in *Pseudomonas* spp.

## Avirulence Gene *avrBs3*-mediated Recognition

HOW DOES RECOGNITION WORK?

The *avrBs3* gene (Bonas et al. 1989) represents a growing family of highly related avirulence genes in *Xanthomonas*, which is characterized by the presence of nearly identical 102 bp repeats in the internal coding region. Pepper lines containing the resistance gene *Bs3* specifically recognize *Xcv* strains expressing the avirulence gene *avrBs3*. A natural allele, the *avrBs3-2* gene (97% identical to *avrBs3*) has a different specificity: it governs the induction of an HR in almost all tomato lines tested but not in pepper (Bonas et al. 1993). The most fascinating feature of both *avrBs3* and *avrBs3-2* is their internal region which contains 17.5 copies of the 102 bp repeat motif. We have previously shown that new avirulence specificities could be generated by deleting repeat units in *avrBs3*, indicating that the repeats determine the specificity of the protein (Herbers et al. 1992).

Recognition between *Xcv* expressing *avrBs3* and *Bs3* pepper plants requires exchange of specific signals between the two organisms. How does this work? The most simple assumption was that the AvrBs3 protein is secreted and recognized by the plant. Since the polyclonal antibody available only reacts with the repeat region of AvrBs3 the FLAG epitope was introduced at the C-terminus of AvrBs3 to allow sensitive detection of the full size protein. Analysis of concentrated culture supernatants of bacteria grown in *hrp* inducing XVM2, or of intercellular washing fluid of infected pepper plants did not reveal the AvrBs3 protein outside of the bacteria, nor was any elicitor activity found. Furthermore, purified AvrBs3 protein does not induce the HR when injected into the leaf intercellular space of *Bs3* plants.

## ADDITIONAL COMPONENTS NEEDED FOR *AVRBS3* FUNCTION?

We wondered whether additional components besides the known *hrp* genes, e.g., a specific chaperone, are needed for *avrBs3* function. Therefore, new *Xcv* mutants were generated to look for additional genes involved in HR induction. After a severe NTG mutagenesis 1000 single bacterial colonies were inoculated into the resistant plant and screened for loss of HR-induction. 11 mutants in already known avirulence genes and the *hrp* cluster, as well as 3 mutants in *hrp* regulatory genes were obtained: two in *hrpX*, and one affected in the key regulatory gene *hrpG* (see above). No mutant other than in the *avrBs3* gene itself was affected in *avrBs3*-mediated HR induction on *Bs3* pepper plants

### *AVRBS3*-MEDIATED PHENOTYPE IN SUSCEPTIBLE PLANTS

Under normal conditions *Xcv* bacteria expressing *avrBs3* induce a distinct reaction only in the resistant plant. Recently we found *avrBs3* dependent formation of pustules on the lower side of the inoculated leaf of susceptible pepper and tomato plants. This phenotype is due to hypertrophy of mesophyll cells that become up to 10-fold larger as compared to cells in non-infected tissue. The pustules are only observed when the appearance of watersoaking symptoms is delayed and depend on *avrBs3* and its 17.5 repeats.

## IS AVRBS3 TRANSLOCATED INTO THE PLANT CELL?

The AvrBs3 protein is hydrophilic, does not contain a signal peptide sequence, is mainly localized in the bacterial cytoplasm, and is constitutively expressed in *Xcv* (Brown et al. 1993; Knoop et al. 1991) As *hrp* mutants expressing the AvrBs3 protein do not induce the HR, *hrp* genes might be involved in the formation or export of the HR elicitor. The relative ease with which new *avr* specificities are obtained (Herbers et al. 1992) suggests that the AvrBs3 protein is the molecule that is recognized by the resistant plant. Since we never found an extracellular elicitor we think that AvrBs3 (or part of it) might be translocated directly into the plant cell via the Hrp transport system (the so-called injection model). This idea was supported by the recent finding of three putative nuclear localization signals (NLSs) in AvrBs3. To test for functionality of the *avrBs3* NLSs in nuclear targeting, the C-terminal region of the *avrBs3* gene was fused to the *uidA* reporter gene in a plant expression vector. Onion epidermal cells were transiently transformed using particle bombardment. The AvrBs3 C-terminus clearly targeted β-glucuronidase (GUS) to the plant nucleus while in controls GUS activity was cytoplasmic. In a recent publication, D. Gabriel and colleagues reported that the C-terminus of the *avrBs3* homologous gene *pthA* of *Xanthomonas citri* also targets GUS to the plant nucleus (Yang and Gabriel 1995). The next step will be to test for the biological significance of this finding.

## Mapping of Pepper and Tomato Resistance Genes

For obvious reasons, we are interested in the plant genes that govern specific recognition of *Xcv* bacteria expressing the avirulence genes *avrBs3* or *avrBs3-2*. In order to fully understand the basis of specific recognition we want to isolate the corresponding disease resistance genes, *Bs3* and *Bs3-2*. A prerequisite has been the genetic characterization of the tomato resistance, i. e., identification of a susceptible line. Among 70 lines tested, only one susceptible *Lycopersicon pennellii* line was appropriate for crosses. Analysis of the F2 progeny showed that a single dominant R-locus, designated *Bs3-2*, is involved in the *avrBs3-2*-mediated resistance in tomato. We have chosen to isolate the *Bs3* and *Bs3-2* genes by a map-based cloning approach. Bulked segregant analysis of both pepper and tomato F2 plants using AFLP resulted in the identification of closely linked markers for both genes. Further analysis is in progress.

## Perspectives

Understanding the mechanism by which bacterial pathogens deliver pathogenicity and avirulence factors to the plant is a major goal of our laboratory. The cloning of the *hrp* regulatory genes will allow us to set up the *in vitro* conditions to demonstrate the postulated *hrp* function in protein secretion. Knowing that the AvrBs3 NLSs are functional in nuclear targeting key questions to be answered are: is AvrBs3 transported into the plant cell, and if so, how does it interact with the product of the resistance gene *Bs3*?

LITERATURE CITED

Bonas, U. 1994. *hrp* genes of phytopathogenic bacteria. Curr. Top. Microbiol. Immunol. 192:79-98.

Bonas, U., Stall, R. E., and Staskawicz, B. 1989. Genetic and structural characterization of the avirulence gene *avrBs3* from *Xanthomonas campestris* pv. *vesicatoria*. Mol. Gen. Genet. 218:127-136.

Bonas, U., Schulte, R., Fenselau, S., Minsavage, G. V., and Staskawicz, B. J. 1991. Isolation of a gene cluster from *Xanthomonas campestris* pv. *vesicatoria* that determines pathogenicity and the hypersensitive response on pepper and tomato. Mol. Plant-Microbe Interact. 4:81-88.

Bonas, U., Conrads-Strauch, J., and Balbo, I. 1993. Resistance in tomato to *Xanthomonas campestris* pv. *vesicatoria* is determined by alleles of the pepper-specific avirulence gene *avrBs3*. Mol. Gen. Genet. 238:261-269.

Brown, I., Mansfield, J., Irlam, I., Conrads-Strauch, and Bonas, U. 1993. Ultrastructure of interactions between *Xanthomonas campestris* pv. *vesicatoria* and pepper, including immunocytochemical localization of extracellular polysaccharides and the AvrBs3 protein. Mol. Plant-Microbe Interact. 6:376-386.

Brown, I., Mansfield, J., and Bonas, U. 1995. *hrp* genes in *Xanthomonas campestris* pv. *vesicatoria* determine ability to suppress papillae deposition in pepper mesophyll cells. Mol. Plant-Microbe Interact. 8:825-836.

Fenselau, S., and Bonas, U. 1995. Sequence and expression analysis of the *hrpB* pathogenicity operon of *Xanthomonas campestris* pv. *vesicatoria* which encodes eight proteins with similarity to components of the Hrp, Ysc, Spa, and Fli secretion systems. Mol. Plant-Microbe Interact. 8:845-854.

Genin, S., Gough, C. L., Zischek, C., and Boucher, C. A. 1992. Evidence that the *hrpB* gene encodes a positive regulator of pathogenicity genes from *Pseudomonas solanacearum*. Mol. Microbiol. 6:3065-3076.

Herbers, K., Conrads-Strauch, J., and Bonas, U. 1992. Race-specificity of plant resistance to bacterial spot disease determined by repetitive motifs in a bacterial avirulence protein. Nature 356:172-174.

Knoop, V., Staskawicz, B.J., and Bonas, U. 1991. The expression of the avirulence gene *avrBs3* from *Xanthomonas campestris* pv. *vesicatoria* is not under the control of *hrp* genes and is independent of plant factors. J. Bacteriol. 173:7142-7150.

Schulte, R., and Bonas, U. 1992. Expression of the *Xanthomonas campestris* pv. *vesicatoria hrp* gene cluster, which determines pathogenicity and hypersensitivity on pepper and tomato, is plant inducible. J. Bacteriol. 174:815-823.

Wengelnik, K., and Bonas, U. 1996. HrpXv, an AraC-type regulator, activates expression of five out of six loci in the *hrp* cluster of *Xanthomonas campestris* pv. *vesicatoria*. J. Bacteriol. 178:3462-3469.

Wengelnik, K., Marie, C., Russel, M., and Bonas, U. 1996a. Expression and localization of HrpA1, a protein of *Xanthomonas campestris* pv. *vesicatoria* essential for pathogenicity and induction of the hypersensitive reaction. J. Bacteriol. 178:1061-1069.

Wengelnik, K., Van den Ackerveken, G., and Bonas, U. 1996b. HrpG, a key *hrp* regulatory protein of *Xanthomonas campestris* pv. *vesicatoria* is homologous to two-component response regulators. Mol. Plant Microbe Interact. (in press)

Yang, Y., and Gabriel, D.W. 1995. *Xanthomonas* avirulence/pathogenicity gene family encodes functional plant nuclear targeting signals. Mol. Plant-Microbe Interact. 8:627-631.

# Some Novel Factors Required for Pathogenicity of *Xanthomonas campestris* pv. *campestris*

C.E. Barber, T.J.G. Wilson, H. Slater, J.M Dow and M.J. Daniels.

The Sainsbury Laboratory, John Innes Centre, Norwich NR4 7UH, UK

Many plant pathogenic bacteria produce extracellular products such as polysaccharides and enzymes which degrade plant cell components. There is general agreement that these factors are required for normal pathogenicity, but the precise role of individual components remains obscure. For example, mutations in structural genes for single enzymes may have little observable effect on disease symptoms, whereas mutations in pleiotropic regulatory or secretion genes affecting the set of factors give clearly reduced pathogenicity. However, such mutants retain the ability to induce a hypersensitive response on non-host plants.

*Xanthomonas campestris* pv. *campestris* (*Xcc*) is a vascular pathogen of cruciferous plants. The later stages of the disease (black rot) are usually characterized by extensive tissue necrosis, probably caused by bacterial enzymes, including proteases, pectate lyases, endoglucanase, lipase etc. Synthesis of enzymes and polysaccharide is regulated in a complex manner by several independent sets of genes (Dow and Daniels 1994). Presumably the several regulators respond to different input signals to allow the bacteria to adapt to changing conditions encountered during pathogenesis or saprophytic existence. One cluster of genes which we have studied extensively contains at least seven *rpf* genes which are required for synthesis of enzymes and polysaccharide. Here we describe the properties of three adjacent genes, *rpfA*, *rpfB* and *rpfF*, which are involved in two novel types of regulation.

## Aconitase Encoded by the *rpfA* Gene

Transposon mutations in the *rpfA* gene cause a reduction of *ca.* 90%

in synthesis of enzymes, measured either by appearance of enzyme
activity or by gene fusion techniques. Bacterial growth is not affected
by the mutations. Figure 1 shows that the expression of the *prtA*
protease gene, measured with a transcriptional fusion to Tn*5lac*, is much
reduced in *rpfA* compared with wild type backgrounds.
Complementation with the subcloned *rpfA* gene shows that the
phenotype is caused by the mutation.

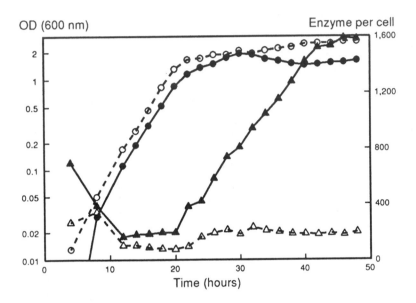

Fig. 1. pLAFR3 containing the *prtA* gene (Liu et al. 1990) interrupted
by Tn*5lac* was introduced into wild type and *rpfA* strains of *Xcc*.
Growth (circles) and β-galactosidase (triangles) were measured. Solid
symbols: wild type, open symbols: mutant.

Sequencing of the *rpfA* gene indicates that it encodes an aconitase,
an enzyme of the tricarboxylic acid cycle. *Xcc* is an aerobe and
aconitase is presumed to be essential for growth; the viability of the
mutants therefore implies the existence of a second aconitase. This was
confirmed by non-denaturing gel electrophoresis followed by staining
for aconitase activity. One of the two aconitase bands was missing
from *rpfA* mutants and was restored by complementation of the mutants.
    Eukaryotic cells also contain two aconitases. The mitochondrial
form functions in respiration. However the cytoplasmic protein is

believed to be an iron-responsive-element-binding protein (IRP) which regulates gene expression in response to changes in iron levels. This is mediated by partial dissociation of an iron-sulphur cluster with concomitant changes in protein folding. Aconitases or IRPs have not previously been implicated in iron regulation in prokaryotes. *Escherichia coli* contains two aconitase genes, but neither appears to be involved in regulation (Bradbury et al. 1996). When wild type *Xcc* was subjected to iron deprivation by addition of 2,2' dipyridyl or diethylenetriaminepentaacetic acid, protease synthesis was sharply reduced. Our data raise the possibility that *Xcc* senses iron and regulates synthesis of pathogenicity factors by a pathway involving aconitase. Iron has been shown to play a role in the pathogenicity of *Erwinia chrysanthemi* (Expert et al. 1996), but the pathways are not fully understood.

### *rpfB*, *rpfF* and a Diffusible Regulatory Molecule

Enzyme production by *rpfF* mutants can be restored by cultivation adjacent to *rpfF*⁺ strains, indicative of cross-feeding by a diffusible factor (DF). This cross-feeding can be used to detect and assay DF. Mutations in *rpfF* and the adjacent gene *rpfB* result in loss of DF production. However, in contrast to *rpfF* mutants, *rpfB* mutants cannot be cross-fed. Mutations in all other genes tested do not affect production of, or response to DF.

Many examples have been found of small molecules which mediate regulation or intercellular communication in bacteria. In most Gram-negative bacteria the effector molecules are N-acyl homoserine lactones (AHSL), and the genes most commonly found to be involved in synthesis and action of these substances belong to the *luxI* and *luxR* families (Swift et al. 1996). *rpfB* and *rpfF* do not belong to these classes. Sequence comparisons suggest that *rpfB* encodes a long-chain fatty acyl CoA ligase. The function of the *rpfF* product is less clear, although there is some relationship to enzymes of fatty acid metabolism.

The *Xcc* DF has no activity in bioassays which detect a range of AHSLs, and has some different chemical properties. We have been unable to detect any AHSL in *Xcc* cultures. A range of phytopathogenic and plant-associated xanthomonads, pseudomonads and erwinias were tested for DF production (defined by cross-feeding of *rpfF* mutants). Only certain xanthomonads produced active material.

*Xcc* enzymes accumulate predominantly in the stationary phase in laboratory cultures (Fig. 1), and DF is also not detectable until this time.

However DF accumulation cannot be the sole determinant of the timing of enzyme synthesis because addition of purified DF to exponentially growing cultures does not result in precocious enzyme synthesis. It has been postulated that the delay in production of AHSL-regulated virulence determinants (such as enzymes in *Erwinia*) until bacteria have reached a high density is a strategy to minimize early induction of plant defenses by elicitor-active products of enzyme action (Swift et al. 1996). An alternative interpretation for *Xcc* is that DF production and action are responses to physiological changes such as starvation which the bacteria undergo in plant tissues, rather than population density *per se*. *Xcc* has apparently evolved a novel regulatory system, probably acting through pathways of lipid metabolism, to solve the problems of parasitism.

## Acknowledgments

The Sainsbury Laboratory is supported by the Gatsby Charitable Foundation. TJGW is also supported by the Biotechnology and Biological Sciences Research Council.

## Literature Cited

Bradbury, A.J., Gruer, M.J., Rudd, K.E., and Guest, J.R. 1996. The second aconitase (AcnB) of *Escherichia coli*. Microbiology 142:389-400.

Dow, J.M., and Daniels, M.J. 1994. Pathogenicity determinants and global regulation of pathogenicity of *Xanthomonas campestris* pv, *campestris*. Pages 29-41 in: Bacterial Pathogenesis of Plants and Animals. J.L. Dangl, ed. Springer Verlag, Heidelberg.

Expert, D., Enard, C., and Masclaux, C. 1996. The role of iron in plant host-pathogen interactions. Trends Microbiol. 4:232-237.

Liu, Y.-N., Tang, J.-L., Clarke, B.R., Dow, J.M., and Daniels, M.J. 1990. A multipurpose broad host range cloning vector and its use to characterise an extracellular protease gene of *Xanthomonas campestris* pv. *campestris*. Mol. Gen. Genet. 220:433-440.

Swift, S., Throup, J.P., Williams, P., Salmond, G.P.C., and Stewart, G.S.A.B. 1996. Quorum sensing: a population-density component in the determination of bacterial phenotype. Trends Biochem. Sci. 21:214-219.

# Coronatine, a Plasmid-Encoded Virulence Factor Produced by *Pseudomonas syringae*

C. Bender[1], D. Palmer[2], A. Peñaloza-Vázquez[1], V. Rangaswamy[1], and M. Ullrich[3]

[1]110 Noble Research Center, Department of Plant Pathology, Oklahoma State University, Stillwater, OK 74078-3032, USA; [2]Zeneca Ag Products, Eastern Regional Technical Center, Route #1, Box 117, Whitakers, NC 27891 USA; [3]Max-Planck-Insitut für terrestrische Mikrobiologie, Karl-von-Frisch-Strasse, D-35043 Marburg, Germany

The phytotoxin coronatine (COR) is a virulence factor which operates in several diseases caused by the phytopathogen *Pseudomonas syringae*. The structure of coronatine (Fig. 1) is unusual and consists of two distinct parts: coronafacic acid is of polyketide origin, whereas coronamic acid is an ethylcyclopropyl amino acid derived from isoleucine. In addition to eliciting chlorosis, COR is known to induce hypertrophy, inhibit root elongation, and stimulate ethylene biosynthesis. Several reports have noted the striking structural and functional similarities between COR, jasmonic acid, and 12-oxo-phytodienoic acid, suggesting that COR may function as a molecular mimic of the octadecanoid signalling molecules of higher plants (Feys et al. 1994; Weiler et al. 1994). However, we recently identified several biological activities which are specifically induced by COR and not by methyl jasmonate in tomato, suggesting that the CMA moiety, or perhaps the amide linkage between CFA and CMA may impart additional biological activities to COR (Palmer and Bender, 1995). The mode of action for coronatine may remain obscure until a target site for the toxin is located in the plant.

Production of the phytotoxin coronatine has been demonstrated in five pathovars of *P. syringae* including *atropurpurea*, *glycinea*, *maculicola*, *morsprunorum*, and *tomato* which infect ryegrass, soybean, crucifers,

*Prunus* spp., and tomato, respectively. In several host-pathogen interactions, COR was shown to have a distinct role in virulence (Bender et al. 1987; Mittal and Davis 1995). However, it is interesting to note that production of COR is dispensable in many pathovars, since pathogenic strains of pvs. *glycinea, maculicola, morsprunorum,* and *tomato* have been isolated which are unable to produce the toxin (Mitchell 1982; Ullrich et al. 1993). This variability in COR production might be caused by the occurrence of the COR biosynthetic cluster on large transmissible plasmids (Bender et al. 1991; Ullrich et al. 1993). The tendency of the COR biosynthetic cluster to occur as a plasmid-encoded trait leads to the speculation that these genes have been horizontally transferred in *P. syringae* (Bender et al. 1991).

### Biosynthetic Pathway to Coronatine in *P. syringae*

Coronafacic acid (CFA) is a polyketide derived from three acetate units, one butyrate unit, and one unit of pyruvate (Fig. 1). Coronamic acid (CMA) originates from isoleucine via alloisoleucine and is cyclized by an unknown mechanism. Both CFA and CMA function as discrete intermediates and are secreted by COR-producing strains at low levels. The final step in the pathway to COR is presumed to be the coupling (CPL) or ligation of CFA and CMA via amide bond formation (Fig. 1).

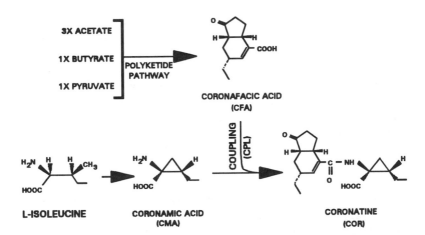

Fig. 1. Biochemical pathways involved in the synthesis of coronatine (COR). COR consists of a polyketide component (CFA) which is coupled (CPL) via amide bond formation to an ethylcyclopropyl amino acid (CMA).

The COR biosynthetic cluster in *P. syringae* pv. *glycinea* PG4180 is borne on a 90-kb plasmid designated p4180A. Various approaches have been used to characterize the COR biosynthetic cluster encoded by p4180A including: (1) saturation Tn5 mutagenesis; (2) feeding studies using exogenously supplied CFA and CMA; and (3) nucleotide sequence analysis (Bender et al. 1993; Liyanage et al. 1995b; Ullrich et al. 1994; Ullrich and Bender, 1994; Young et al. 1992). Fig. 2 shows the functional organization of the COR biosynthetic gene cluster in p4180A.

The region designated CPL was shown to be required for the coupling of CFA and CMA via amide bond formation (Bender et al. 1993). Liyanage et al. (1995b) sequenced this region and identified a 1.4-kb gene designated *cfl* (*corona*facate *l*igase). Cfl shows relatedness to enzymes which activate carboxylic acids by adenylation, and we suspect that this enzyme adenylates CFA and the CFA-adenylate is then ligated to CMA. The CMA biosynthetic gene cluster is contained within a 6.9 kb region, and the nucleotide sequence of this DNA revealed the presence of three genes designated *cmaA* (2.7 kb), *cmaT* (1.2 kb), and *cmaU* (0.9 kb) (Ullrich and Bender 1994). The deduced amino acid sequence of *cmaA* indicates that the enzyme contains an amino acid-activating domain and a putative iron-binding region. Furthermore, the deduced amino acid sequence of *cmaT* suggests that it functions as a thioesterase, providing further support to the role of a thiotemplate mechanism for CMA biosynthesis (Ullrich and Bender 1994). Sequence analysis of *cmaU* failed to reveal anything about its function in CMA biosynthesis. Complementation analyses, nucleotide sequencing, and exogenous feeding studies using blocked mutants were used to define the location of the CFA biosynthetic gene cluster. In *Sst*I fragment #7, Penfold (1995) identified three open reading frames (ORFs) which showed relatedness to genes involved in the synthesis of polyketide compounds. ORFs were identified with relatedness to acyl carrier proteins (ACP), acetyl dehydratases (AD), and β-keto synthases (KS). These homologies are consistent with a role for this region in the synthesis of CFA, which is a product of the polyketide pathway.

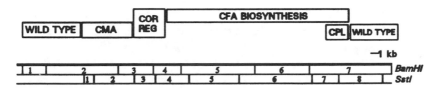

Fig. 2. Map of the PG4810 coronatine biosynthetic gene cluster.

Primer extension, complementation analyses, and nucleotide sequencing were used to define the two transcripts containing structural genes for CMA and CFA biosynthesis (Liyanage et al. 1995a,b; Ullrich and Bender 1994). These approaches indicated that *cmaA* and *cmaT*, the two major genes required for CMA biosynthesis, are co-transcribed as a single transcript with *cmaA* at the promoter proximal end. The CFA biosynthetic gene cluster is also encoded on a single transcript and transcribed with the coronafacate ligase gene (*cfl*) from a common promoter.

## Regulation of the Coronatine Biosynthetic Gene Cluster

One of the most striking effects on COR production is temperature. COR production is optimal at 18°C and negligible at 30°C, even though the bacterium grows well at both temperatures (Palmer and Bender 1993). CFA and CMA are also subject to the same pattern of temperature control, with optimal production at 18°C (Palmer 1995; Ullrich and Bender 1994). Recent data indicate that the production of both intermediates is regulated at the transcriptional level by temperature (Liyanage et al. 1995a; Ullrich and Bender 1994). The higher level of transcriptional activity for the CMA and CFA biosynthetic promoters at 18°C helps explain why COR production is optimal at this temperature.

The nucleotide sequence of the regulatory region in the COR biosynthetic cluster revealed the presence of three genes, *corP*, *corS*, and *corR* (Ullrich et al. 1995). The deduced amino acid sequence of *corP* and *corR* indicated relatedness to response regulators which function as members of two-component regulatory systems, whereas *corS* shows similarity to histidine protein kinases which function as environmental sensors. The COR regulatory system is modified from the two component paradigm since it contains two response regulator proteins together with a single sensor protein (Ullrich et al. 1995). The function of each regulatory protein is currently under investigation.

## Summary

We have shown that the production of COR requires two distinct biosynthetic gene clusters. Perhaps one of the most novel evolutionary events in COR biosynthesis is the joining of the CFA and CMA pathways by a centrally located regulatory region which is temperature-responsive (Ullrich et al. 1995). It is intriguing to ask why the COR regulatory region contains two response regulators (CorP and CorR) and how these proteins function with the histidine protein kinase (CorS) to achieve temperature-sensitive production of coronatine. In time, we anticipate that this and many other interesting questions regarding COR biosynthesis will be answered using a multidisciplinary approach which integrates biochemistry, genetics, and organic chemistry.

## Acknowledgements

C. Bender acknowledges financial support for this work from the Oklahoma Agricultural Experiment Station and National Science Foundation grants DMB-8902561, EHR-9108771, INT-9220628 and MCB-9316488.

## References

Bender, C.L., Stone, H.E., Sims, J.J., and Cooksey, D.A. 1987. Reduced pathogen fitness of *Pseudomonas syringae* pv. *tomato* Tn5 mutants defective in coronatine production. Physiol. Mol. Plant. Pathol. 30:273-283.

Bender, C.L., Young, S.A., and Mitchell, R.E. 1991. Conservation of plasmid DNA sequences in coronatine-producing pathovars of *Pseudomonas syringae*. Appl. Environ. Microbiol 57:993-999.

Bender, C.L., Liyanage, H., Palmer, D., Ullrich, M., Young, S., and Mitchell, R. 1993. Characterization of the genes controlling biosynthesis of the polyketide phytotoxin coronatine including conjugation between coronafacic and coronamic acid. Gene 133:31 38.

Feys, B.J.F., Benedetti, C.E., Penfold, C.N., and Turner, J.G. 1994. *Arabidopsis* mutants selected for resistance to the phytotoxin coronatine are male sterile, insensitive to methyl jasmonate, and resistant to a bacterial pathogen. The Plant Cell 6:751-759.

Liyanage, H., Palmer, D.A., Ullrich, M., and Bender, C.L. 1995a. Characterization and transcriptional analysis of the gene cluster for coronafacic acid, the polyketide component of the phytotoxin coronatine. Appl. Environ. Microbiol. 61:3843-3848.

Liyanage, H., Penfold, C., Turner, J., and Bender, C.L. 1995b. Sequence, expression and transcriptional analysis of the coronafacate ligase-encoding gene required for coronatine biosynthesis by *Pseudomonas syringae*. Gene 153:17-23.

Mitchell, R.E. 1982. Coronatine production by some phytopathogenic pseudomonads. Physiol. Plant. Pathol. 20:83-89.

Mittal, S.M., and Davis, K.R. 1995. Role of the phytotoxin coronatine in the infection of *Arabidopsis thaliana* by *Pseudomonas syringae* pv. *tomato*. Mol. Plant-Microbe Interact. 8:165-171.

Palmer, D.A. 1995. PhD dissertation, Oklahoma State University.

Palmer, D.A., and Bender, C.L. 1993. Effects of environmental and nutritional factors on production of the polyketide phytotoxin coronatine by *Pseudomonas syringae* pv. glycinea. Appl. Environ. Microbiol. 59:1619-1626.

Palmer, D.A., and Bender, C.L. 1995. Ultrastructure of tomato leaf tissue treated with the Pseudomonad phytotoxin coronatine and comparison with methyl jasmonate. Mol. Plant-Microbe Interact. 8:683-692.

Penfold, C. 1995. PhD dissertation, University of East Anglia.

Ullrich, M., Bereswill, S., Völksch, B., Fritsche, W., and Geider, K. 1993. Molecular characterization of field isolates of *Pseudomonas syringae* pv. *glycinea* differing in coronatine production. J. Gen. Microbiol. 139:1927-1937.

Ullrich, M., and Bender, C.L. 1994. The biosynthetic gene cluster for coronamic acid, an ethylcyclopropyl amino acid, contains genes homologous to amino acid activating enzymes and thioesterases. J. Bacteriol. 176:7574-7586.

Ullrich, M., Guenzi, A.C., Mitchell, R.E., and Bender, C.L. 1994. Cloning and expression of genes required for coronamic acid (2-ethyl-1-aminocyclopropane 1-carboxylic acid), an intermediate in the biosynthesis of the phytotoxin coronatine. Appl. Environ. Microbiol. 60:2890-2897.

Ullrich, M., Peñaloza-Vázquez, A., Bailey, A.M., and Bender, C.L. 1995. A modified two-component regulatory system is involved in temperature-dependent biosynthesis of the *Pseudomonas syringae* phytotoxin coronatine. J. Bacteriol. 177:6160-6169.

Weiler, E.W., Kutchan, T.M., Gorba, T., Brodschelm, W., Niesel, U., and Bublitz, F. 1994. The *Pseudomonas* phytotoxin coronatine mimics octadecanoid signalling molecules of higher plants. FEBS Lett. 345:9-13.

Young, S.A., Park, S.K., Rodgers, C., Mitchell, R.E., and Bender, C.L. 1992. Physical and functional characterization of the gene cluster encoding the polyketide phytotoxin coronatine in *Pseudomonas syringae* pv. *glycinea*. J. Bacteriol. 174:1837-1843.

# Molecular Genetic Approaches to the Study of Fungal Pathogenesis Revisited

Sally A. Leong

USDA-ARS Plant Disease Resistance Research Unit and Department of Plant Pathology, University of Wisconsin, Madison, 53706 U.S.A.

## The Threat of Fungal Disease

Fungal pathogens account for the greatest overall loses associated with plant disease. This year in the United States, high priority has been given to control of two emerging fungal epidemics, late blight on potato caused by the Oomycete *Phytophthora infestans*, and Karnal bunt of wheat caused by the Basidiomycete *Telletia indica*. In 1994, a major epidemic of late blight occurred in Wisconsin and in states along the Eastern coast of the U. S. Most of the potato crop in Northern Maine was lost. In 1995, cool wet weather in Idaho led to the incidence of late blight on potato for the first time in history. This year, cool wet weather is again favoring late blight in many potato growing areas of the United States. Control of this disease is becoming increasing difficult due to the emergence of fungal isolates with resistance to the fungicide ridomyl and the recent migration of the highly virulent A2 mating type into the United States. Last year late blight on potato also led to loss of 30% of the potato crop in China where it is an important food staple thus threatening China's food supply. The discovery of Karnal bunt on wheat in the U. S. for the first time this year may severely affect seed quality and will limit the exportation of wheat from the United States through quarantine. Clearly timely research is needed in order to design new and improved strategies to control these and other fungal diseases worldwide.

# Molecular Biology in the Real World

　　　　Several years ago I co-authored a paper entitled "Molecular Genetic Approaches to the Study of Fungal Pathogenesis" (Leong and Holden, 1989).  While this paper is still essentially current from a technological standpoint, I would like to use this opportunity to discuss the need to study the molecular biology of plant disease in a real world setting beyond the artificial laboratory environment.

　　　　In considering how fungi cause disease, we must understand the fungus' life strategy.  What does it take to survive and reproduce in nature?  The disease cycle provides a useful way in which to think about these questions.  In order to be a successful pathogen, the fungus must find its host, perhaps attach to the host surface, gain entry into the plant through natural openings or by mechanical or enzymatic means, colonize the host, produce new inoculum, and finally disperse to a new host often after an extended period of quiescence in the absence of the host.

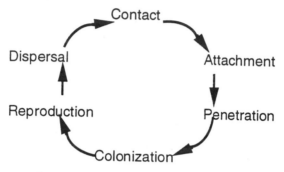

Figure 1.  The Disease Cycle

　　　　Historically, knowledge gained through the study of fungal pathogenesis came largely from physiological experiments, cytological observation and correlation.  With the advent of DNA-mediated transformation in phytopathogenic fungi over the last decade we have seen the cloning of a plethora of pathogenicity-related genes through complementation and reversed genetic approaches, the disruption of specific pathogenicity genes through targeted gene disruption, the random mutation of pathogenicity genes through REMI, Restriction-Enzyme-Mediated Integration of transforming DNA (Sweigard, 1996), and the fine structure analysis of pathogenicity gene organization, structure, expression and function.  Great strides have been made during this period to understand many aspects of fungal pathobiology (Leong and Holden, 1989; Oliver and Osbourn, 1995).  However, in virtually all cases, this work has been limited to the assay of pathogenicity in the laboratory.

　　　　I would like to call upon all of us to consider the need to begin to analyze how the genes we are studying affect survival and reproduction of fungal pathogens in the real world of the field, garden, park or forest.  There are very few studies which have been carried to this level and which attempt to reproduce an infection process that is natural. Two examples from the world of bacterial phytopathogens bear on this concern. The *lemA* gene of *Pseudomonas syringae* pv. *syringae* was isolated by Kyle Willis and associates as a mutation defective in lesion formation on bean plants grown under laboratory conditions (Willis *et al.*, 1990). Growth of the mutant was not impaired relative to wild type under these conditions. This factor was used as a criterion for the isolation of this pathogenicity gene. However, recent work by Susan Hirano and Chris Upper at the University of Wisconsin has shown that *lemA* not only affects lesion formation but also fitness in a field setting (Hirano *et al.*, 1994). The *lemA* mutant grows less vigorously in the bean phyllosphere when compared to wild type. In a second study conducted by Hirano and Upper in collaboration with Amy Loniello and Alan Collmer from Cornell University (Hirano *et al.*, 1996), *hrpZ* and *hrpK* mutants of *P. syringae* pv. *syringae* were examined for growth/survival and disease causation on snap bean in the field. A *hrpZ* mutant showed no difference from wild type in any of these parameters while a *hrpK* mutant showed decreased ability to cause disease and decreased fitness on leaves but not on germinating seeds. The results with *hrpZ* contrast those found by Loniello *et al* (1995) with laboratory assays in which *hrpZ* was found to affect virulence of surface inoculated bean leaves. Finally, Upper and Peterson have found that the momentum of simulated rainfall on laboratory grown bean plants causes severe physical damage to aerial plant tissues by tearing leaves (S. Hirano and C. Upper, personal communication). This implies that the physical structure of the leaf grown in the laboratory differs markedly from that of plants in the field. The natural microflora associated with plants grown in nature is also absent on growth chamber plants. In short, it would behoove those studying pathogenesis of fungi to evaluate laboratory findings in the field. We may find that the genes we are studying are less important than previously thought or perhaps more important. Ultimately our work must have relevance to the real world. The controversial role of cutinases in fungal infection of plants may be in part the result of the context in which the pathogenicity assay was done (Rogers *et al.*, 1994; Schafer, 1993).

In my own work on corn smut disease I would like to know definitively what role siderophores play in the disease cycle of *Ustilago maydis* (Leong *et al.*, 1987). Disruption of the *sid1* gene which encodes ornithine-$N^5$-oxygenase blocks siderophore production in culture (Mei *et al.*, 1993). *sid1* mutants are fully pathogenic on laboratory grown plants (Mei *et al.*, 1993). Does this mean that siderophores play no role in pathogenesis of maize by *U. maydis*? Our infection assay requires the injection of $10^5$ sporidial cells per plant in order to insure infection of all plants. Is this representative of what occurs in nature? I also question the use of sporidia as the inoculum; in nature the primary inoculum is likely to be the teliospore. What is the primary mode of infection by *U. maydis* in nature? Does the teliospore undergo reduction devision or does it infect directly after germination as a solopathogenic diploid? How frequently does inbreeding and outcrossing of sporidia take place in nature? These are questions which bear not only on questions one may ask in relation to mating type control during pathogenesis but also on the design of field inoculation experiments. For example, if a *sid1* homozygous teliospore was used as a spray inoculum in the field, what chance would there be for the spore to germinate, undergo meiosis and the progeny outcross with *sid1* prototrophic sporidia? These are key questions which must be answered before a valid test of the role of *sid1* in pathogenesis and survival in nature can be addressed.

ACKNOWLEDGMENT

I thank Susan Hirano and Chris Upper for sharing unpublished results.

LITERATURE CITED

Hirano, S., Ostertag, E. M., Savage, S. A., Willis, D. K. and Upper, C. D. 1994. Contribution of the regulatory gene *lema* to fitness of *Pseudomonas syringae* pv. *syringae* in the phyllosphere and spermosphere under field conditions. Microbial Ecology 3:607.

Hirano, S., Loniello, A. O., Collmer, A. and Upper, C. D. 1996. Behavior of *Pseudomonas syringae* pv. *syringae* B728A *HRPZ* and *HRPK* mutants on snap bean plants under field conditions. Abstract 8th International Congress on Molecular Plant-Microbe Interactions, Knoxville, TN.

Leong, S. A. and Holden, D. W. 1989. Molecular genetic approaches to the study of fungal pathogenesis. Ann. Rev. Phytopathol. 27:463-481.

Leong, S. A., Wang, J., Budde, A., Holden, D., Kinscherf, T. and Smith, T. 1987. Molecular strategies for the analysis of the interacction of *Ustilago maydis* and maize. Pages 95-106 in: Molecular Strategies for Crop Protection, C. J. Arntzen and C. Ryan, eds. Alan R. Liss, New York.

Loniello, A. O., Alfano, J. R., Bauer, D. W. and Collmer, A. 1995. Analysis of pathogenicity of a *Pseudomonas syringae* pv. *syringae* B728A *ΔHRPZ::NPII* mutant on bean. Phytopathology 85:267.

Oliver, R. and Osbourn, A. 1995. Molecular dissection of fungal phytopathogenicity. Microbiology 141:1-9.

Rogers, L. M., Flaishman, M. A. and Kolattukudy. 1994. Cutinase gene disruption in *Fusarium solani* f. sp. *pisi* decreases its virulence on pea. Plant Cell 6:935-945.

Schafer, W. 1993. The role of cutinase in fungal pathogenicity. Trends Microbiol. 1:69-71.

Sweigard, J. 1996. A REMI primer for filamentous fungi. IS-MPMI Reporter Spring:3-5.

Willis, D. K., Hrabak, E. M., Rich, J. J., Barta, T. M., Lindow, S. E. and Panopoulos, N. J. 1990. Isolation and Characterization of a *Pseudomonas syringae* pv. *syringae* mutant deficient in lesion formation on bean. Mol. Plant-Microbe Interaction 3:149-156.

# THE MOLECULAR BASIS OF COMPATIBILITY: LESSONS FROM THE HOST-SELECTIVE TOXIN OF *Cochliobolus carbonum*

Virginia Crane, Nasser Yalpani and Steve Briggs

Pioneer Hi-Bred, Int'l., Inc. Johnston, IA U.S.A.

The molecular interactions between plants and microbes that lead to compatibility, those that cause disease, are much less studied than those interactions providing resistance. An exception is what is known about the events that take place during the development of *Helminthosporium* leaf spot and ear mold in maize. Caused by *Cochliobolus carbonum* race 1, this disease occurs only on a few inbred lines, all of which carry a mutation in the *Hm1* gene. Infection is dependent upon the presence of a cyclic tetrapeptide, HC-toxin. Pathogenicity of *C. carbonum* is determined by a single locus, *Tox2*, which encodes the biosynthetic enzymes responsible for producing HC-toxin. Resistance in most maize lines, and probably in most plants, is accomplished by an NADPH-dependent reduction of the toxin to an inactive form. *Hm1* encodes the reductase.

## HC-toxin action

Precisely how the host-selective toxin of *C. carbonum* permits infection of susceptible plants is unknown. It is clear, however, that the pure toxin can cause a number of physiological effects in the host. These effects are striking in that they are not directly lethal to cells, but often result in cytostasis. We tested the hypothesis that *C. carbonum* infections may be successful because HC-toxin actively suppresses the induction of host defense responses (Brosch et al., 1995).

We investigated the action of HC-toxin in susceptible (*hm1/hm1*) maize plants in which markers of the defense response to fungal infections had been induced by pretreatment with UV light or fungal infection. We found that HC-toxin inhibits the UV-induced accumulation of salicylic acid (SA) and pathogenesis-related (PR) proteins in a dose-dependent manner, at physiologically significant concentrations. UV-C light treatment resulted in the accumulation of SA to levels at least 167% of control, as did *C. carbonum* spores from a toxin deficient strain of race 1. UV light and inoculation with a toxin-deficient *C. carbonum* isolate acted synergistically, eliciting accumulation of 2.5X more SA than control. As little as 0.1 μg HC-toxin/ml reduced the level of induction of SA, regardless of the eliciting treatment, and 0.25 μg HC-toxin/ml or greater held SA to background levels.

UV light also induced accumulation of maize proteinase inhibitor (MPI), PRm2 (recognized by antiserum against tobacco PR-1b) and a β-glucanase. Accumulation of MPI and PRm2 was additionally enhanced in UV-treated material by inoculation with spores. This UV- or fungus-induced accumulation was suppressed in a dose-dependent manner by HC-toxin, evident at 0.5 μg/ml. The effect of HC-toxin on glucanase expression differed in that the regulation by UV light overrode suppression by HC-toxin. However, seedlings that were inoculated with a toxin-deficient *C. carbonum* isolate experienced a dramatic suppression in accumulation of glucanase with increasing doses of toxin. We also observed that HC-toxin inhibits the heat-shock induction of cpn60.

A tantalizing clue to the molecular foundation for these observations was provided recently by the discovery that HC-toxin specifically inhibits the activity of histone deacetylase in a dose-dependent manner, at physiologically significant concentrations (Brosch et al, 1995). Acetylation of specific lysine residues near the N-terminus of histones has been associated with transcriptional activation (e.g., Durrin et al., 1991). Acetylation may reduce the charge-charge interactions between these basic lysine residues and the negatively-charged DNA backbone, thus allowing enough nucleosome unwinding to permit transcription-complex formation. The Tetrahymena histone acetyltransferase A, recently cloned by Brownell and coworkers (1996), is a homologue of GCN5p, a yeast transcription factor that directly interacts with the SWI/SNF transcription complex. Also, a human histone deacetylase has been cloned, and sequence comparisons reveal that it is 60% identical to another yeast transcription factor, RPD3, which is required for the proper expression of inducible genes (Taunton et al., 1996). Thus, both the acetylase and the deacetylase are specific transcription factors.

Exactly how histone acetylation regulates transcription is a mystery. It has been proposed that histone deacetylase may facilitate transcription by stabilizing chromatin structure through the constraining effects of deacetylation *after* transcription complexes form, thereby maintaining the induced, regulated transcription of particular genes. It may be that HC-toxin acts as a determining factor in pathogenesis by blocking the induction of the plant's defense response, possibly by preventing the inactivation of genes which suppress the defense response.

## Host-selective toxins and the gene-for-gene model

No Avr/R system has been defined in detail for maize pathology. As our understanding of disease, especially in dicots, increases, complex- or multi-locus relationships are becoming the rule rather than the exception. Expanding the genetic quadratic to include consideration of host-selective toxins may be useful (see figure). The *C. carbonum*/maize interactions are described by Mendelian genetics. The reductase encoded by *Hm1*, HCTR, is the single factor required by the plant for resistance. HC-toxin, encoded by *Tox2*, is the single necessary factor for pathogenesis. That these interactions may lie within the Avr/R quadrant is indicated by the presence of defenseless mutants; mutants that, in sterile culture, are susceptible to a toxin-deficient *C. carbonum* isolate. This class may contain r mutants; fungal avr mutants have yet to be discovered.

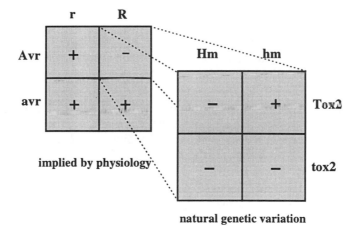

Figure. Possible relationships between host-selective toxins and the gene-for-gene model.

# The molecular basis of compatibility

The example provided by the *C. carbonum*/maize system may have anticipated the emerging story of bacterial pathogens. Topical application of the host-selective toxin on susceptible hosts results in disease by amending the action of toxin-deficient *C. carbonum* isolates. Compatibility factors have not been as readily recognized in bacterial pathogens, possibly because they rely on the type III secretion system for direct introduction of compounds into the host cell. Fungi, on the other hand, seem to use small, diffusible peptides and secondary compounds as compatibility factors. Host-selective toxin research has shown us that compatibility requires more than the absence of R/avr interactions.

## Acknowledgments

We thank Drs. Bob Meeley and Doug Rice for helpful comments and advice. We are especially grateful to Drs. Ray White and Tottempudi Prasad for the gifts of anti-PR1 and anti-cpn60 sera, respectively.

## References

Brosch, G., Ransom, R., Lechner, T., Walton, J.D., and Loidl, P. 1995. Inhibition of maize histone deacetylases by HC toxin, the host-selective toxin of Cochliobolus carbonum. The Plant Cell 7:1941-1950.

Brownell, J.E., Zhou, J., Ranalli, T., Kobayashi, R., Edmondson, D.G., Roth, S.Y., and Allis, C.D. 1996. Tetrahymena histone acetyltransferase A: a homolog to yeast Gcn5p linking histone acetylation to gene activation. Cell 84:843-851.

Durrin, L.K., Mann, R.K., Kayne, P.S., and Grunstein, M. 1991. Yeast histone H4 N-terminal sequence is required for promoter activation in vivo. Cell 65:1023-1031.

Ritter, C. and Dangl, J.L. 1995. The *avrRpm1* gene of *Pseudomonas syringae* pv. *maculicola* is required for virulence of Arabidopsis. MPMI 8:444-453.

Tauton, J., Hassig, C.H., and Schreiber, S.L. 1996. A mammalian histone deacetylase related to the yeast transcriptional regulator Rpd3p. Sci. 272:408-411.

# Review of Evidence Linking Hypovirus-mediated Disruption of Cellular G-Protein Signal Transduction and Attenuation of Fungal Virulence.

Baoshan Chen, Shaojian Gao, Lynn M. Geletka, Shin Kasahara, Ping Wang and Donald L. Nuss

Center for Agricultural Biotechnology, University of Maryland Biotechnology Institute, University of Maryland, College Park, MD, USA

Research interest in Hypoviruses stems primarily from the observations that members of this virus group persistently and stably alter a number of interesting phenotypic traits in their natural fungal host, the chestnut blight fungus *Cryphonectria parasitica*. Alterations can range from reduced orange pigment production (a convenient experimental marker), to reduced asexual and sexual reproduction (vital host functions), to reduced virulence, termed hypovirulence (a complex process of practical importance). An understanding of how a hypovirus modifies such a diverse range of host traits, functions and processes has been a long-standing research goal. This chapter reviews a growing body of evidence linking hypovirus-mediated disruption of cellular G-protein-linked signal transduction and altered fungal phenotype, including virulence attenuation.

REDUCED ACCUMULATION OF A HETEROTRIMERIC GUANINE NUCLEOTIDE-BINDING PROTEIN α SUBUNIT ASSOCIATED WITH HYPOVIRUS INFECTION.

The first indication for an involvement of G-protein-linked signal transduction in hypovirus-mediated alteration of fungal phenotype and virulence derived from the cloning and characterization by Choi et al. (1995) of two *C. parasitica* G-protein α subunit genes, designated *cpg-1* and *cpg-2*. The deduced protein product of *cpg-1*, CPG-1, was found to be 98% identical to a G-protein α subunit of the $G_i$ class previously cloned from *Neurospora crassa* (Turner and Borkovich 1993). The predicted protein product of *cpg-2*, CPG-2, failed to conform to any currently defined Gα family. An effect of hypovirus infection on CPG-1 accumulation was subsequently identified when Western analysis of protein extracts from isogenic virus-free (EP155) and virus-infected (EP713) *C. parasitica* strains revealed significantly less (essentially undetectable levels) CPG-1 accumulation in extracts of the latter strain. A transgenic approach to reduce CPG-1 levels in the absence of virus resulted in a related, but quite unexpected, result. Transformation of virus-free strain EP155 with a control sense orientation copy of *cpg-1* resulted in altered colony morphology and a corresponding reduction in CPG-1 accumulation in over 60% of transformants that contained the ectopic sense orientation *cpg-1* copy. In contrast, transformation with the antisense orientation copy of *cpg-1* failed to yield transformants with altered phenotype or reduced levels of CPG-1 accumulation. Choi et al. (1995) compared this observation to the well established phenomenon of transgenic co-suppression in plants (Napoli et al. 1990; Lindbo and Dougherty 1992; Flavell 1994). Although in both cases the introduction of a sense orientation transgene can result in suppressed accumulation of the protein products of both the endogenous gene homologue and the introduced transgene, there is no indication whether similar mechanisms are operating in co-suppressed plants and in *C. parasitica cpg-1* sense transformants. Irrespective of the mechanism involved, transformants exhibiting CPG-1 co-suppression were found to be attenuated in fungal virulence. In combination, these results i) suggested a role for G-protein-linked signal transduction in fungal virulence and ii) indicated that one important way in which a hypovirus alters the interaction between *C. parasitica* and its plant host is by suppressing the accumulation of a key fungal molecular signaling component.

CELLULASE INDUCTION IN *C. PARASITICA*: SUPPRESSION BY HYPOVIRUS INFECTION AND REQUIREMENT FOR AN INTACT CPG-1 SIGNALING PATHWAY.

A relationship between the production of *C. parasitica* encoded plant cell wall-degrading enzymes, putative virulence factors, and CPG-1 linked signaling was recently demonstrated by results of Wang and Nuss (1995). Consistent with earlier reports (McCarroll and Thor 1985), cellulose-induced cellulase activity was prevented by hypovirus infection. Significantly, EP155-transformants rendered hypovirulent due to CPG-1 co-suppression also failed to induce cellulase in the presence of cellulose as the sole carbon source. These results were extended to a specific member of the *C. parasitica* cellulase complex by examining transcript accumulation for the cloned cellobiohydrolase I gene, *cbh-1* . Thus, cellulase induction in *C. parasitic* appears to be reduced as a result of hypovirus infection and to require an intact CPG-1-linked signal transduction pathway.

HYPOVIRUS-MEDIATED ALTERATION OF FUNGAL GENE TRANSCRIPT ACCUMULATION AND ELEVATION OF G-PROTEIN REGULATED cAMP LEVELS.

More extensive similarities in the effect of hypovirus infection and CPG-1 co-suppression on fungal transcript accumulation were also recently provided by Chen et al. (1996) using the technique of mRNA differential display. Displays generated from mRNA preparations isolated from parallel grown virus-free *C. parasitica* strain EP155, hypovirus-infected isogenic strain EP713 and CPG-1 co-suppressed strain G1310 derived from strain EP155 revealed the following information. First, more than 400 PCR products were identified that changed in intensity as a result of hypovirus infection, 139 of which exhibited an intensity change of more than four fold. Second, nearly two thirds of the differential PCR products increased in

228

intensity, indicating that hypovirus infection can increase as well as decrease the accumulation of host transcripts. Third, 65% of the changes that occurred in PCR product intensity as a result of hypovirus infection were duplicated as a result of CPG-1 co-suppression in the absence of virus infection. These results were consistent with the view that hypovirus infection causes an extensive and persistent alteration of host gene expression/transcript accumulation and indicated that many of these alterations are mediated through modification of the CPG-1 signaling pathway.

Reports that mammalian $G_i\alpha$ subunits negatively regulate adenylyl cyclase activity (Kaziro et al. 1991) stimulated Chen et al. (1996) to also examine the effect of reduced CPG-1 accumulation on cAMP levels. They found a three to five fold elevation in cAMP levels in both hypovirus infected and CPG-1 co-suppressed strains, relative to the levels found in parallel grown control strain EP155. Moreover, it was possible to mimic the effect of virus infection and CPG-1 co-suppression on transcript accumulation for representative *C. parasitica* genes by drug-induced elevation of cAMP levels. Combined, these observations supported the prediction that, similar to mammalian $G_i\alpha$ subunits, CPG-1 functions as a negative modulator of adenylyl cyclase and suggested a role for G-protein-regulated cAMP accumulation in hypovirus-mediated alteration of fungal gene expression.

DISRUPTION OF GENES ENCODING CPG-1 AND CPG-2.

We recently disrupted the genes *cpg-1* and *cpg-2*, encoding CPG-1 and CPG-2, respectively (S. Gao and D. L. Nuss, Submitted). Disruption of *cpg-1* resulted in a set of phenotypic traits similar to, but more severe than, those associated with hypovirus infection. In contrast, *cpg-2* disruption resulted in only slight reductions in growth rate and asexual sporulation and no significant reduction in virulence or other hypovirulence-associated traits. It was concluded that CPG-1 is required for fungal virulence and reproduction, while CPG-2 is dispensable, ie., CPG-1 and CPG-2 serve distinct roles in fungal virulence, morphology and reproduction. The observation that *cpg-1* disruptants are deficient in the same set of traits and processes that are altered by hypovirus infection also provides additional evidence that host phenotypic alterations associated with hypovirus infection are primarily mediated through modification of the CPG-1 signaling pathway.

SUMMARY AND FUTURE DIRECTIONS.

A role for CPG-1-linked signaling in *C. parasitica* virulence and the impact of hypovirus infection can be partially envisioned in the minimal working hypothesis depicted in Fig. 1. The level of activated CPG-1 in a virus-free strain is determined throughout the infection process by events at the cell surface involving as yet undefined G-protein-linked receptors and unidentified ligands. By negatively regulating adenylyl cyclase, activated

CPG-1 modulates cAMP levels which, in turn, modulates gene expression to elicit appropriate adaptive responses, eg., cellulase induction. By suppressing CPG-1 accumulation, hypovirus infection constitutively elevates cAMP levels by relieving the negative regulation of adenylyl cyclase. This effectively compromises the ability of the invading fungus to respond appropriately to events at the fungus-plant interface, thereby impeding penetration and canker formation. The Gα subunit CPG-2 appears to be dispensable for virulence.

Much work remains to adequately test and expand this minimal model. Issues of central focus will include the mechanism(s) involved in the apparent hypovirus-mediated reduction in CPG-1 accumulation, the identity of hypovirus encoded protein(s) responsible for CPG-1 suppression, the role of the other G-protein subunits in the regulation of key fungal processes and the systematic identification of the range of environmental cues that are sensed by, and functions that are regulated through, CPG-1-linked signaling pathways. Finally, it is important to note that virus-mediated reduction in CPG-1 accumulation has to date only been reported for the prototypic member of the hypovirus group, CHV1-713. A survey of the effect of other members of the family *Hypoviridae* on cellular G-protein signal transduction is likely to be instructive.

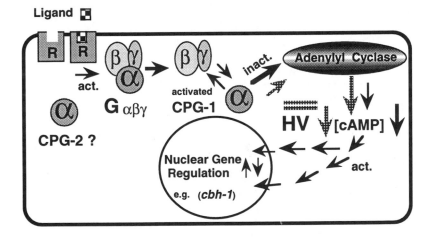

Fig. 1 Cartoon depicting a minimal working hypothesis of the role of G-protein alpha subunit CPG-1 in *C. parasitica* virulence and the impact of hypovirus infection. Abbreviations include: act., activation; *cbh-1*, the gene encoding a *C. parasitica* cellobiohydrolase I enzyme designated CPCBH-1; HV, hypovirus; inact., inactivation; R, receptor; Gαβγ, heterotrimeric guanine nucleotide-binding protein complex consisting of the α, β and γ subunits. The proposed impact of hypovirus infection in suppressing CPG-1 accumulation and constitutively elevating cAMP levels is indicated by the stippled arrows. This figure was adapted in part from Jans (1994).

ACKNOWLEDGMENT

This chapter was adapted in part from the following publications listed under "Literature Cited": Choi et al. 1995; Chen et al. 1996; Wang and Nuss 1995; Nuss 1996, and Gao and Nuss, Submitted.

LITERATURE CITED

Chen, B., S. Gao, G. H. Choi and D. L. Nuss. 1996. Extensive alteration of fungal gene transcript accumulation and elevation of G-protein-regulated cAMP levels by a virulence-attenuating hypovirus. Proc. Natl. Acad. Sci. USA. In Press.

Choi, G. H., B. Chen, and D. L. Nuss. 1995. Virus-mediated or transgenic suppression of a G-protein α subunit and attenuation of fungal virulence. Proc. Natl. Acad. Sci. USA 92:305-309.

Flavell, R. B. 1994. Inactivation of gene expression in plants as a consequence of specific sequence duplication. Proc. Natl. Acad. Sci. USA 91:3490-3496.

Gao, S. and D. L. Nuss. Distinct roles for two G-protein α subunits in fungal virulence, morphology and reproduction revealed by targeted gene disruption. Submitted.

Jans, D. A. 1994. Nuclear signaling pathways for polypeptide ligands and their membrane receptors? FASEB J. 8:841-847.

Kaziro, Y., Itoh, H., Kozasa, T., Nakafuku, M. and T. Satoh. 1991. Structure and function of signal-transducing GTP-binding proteins. Ann. Rev. Biochem. 60:349-400.

Lindbo, J. A. and W. G. Dougherty. 1992. Untranslatable transcripts of the tobacco etch virus coat protein gene sequence can interfere with tobacco etch virus replication in transgenic plants and protoplasts. Virology 189: 725-733.

McCarroll, D. R. and E. Thor. 1985. Pectolytic, cellulytic and proteolytic activity expressed by cultures of *Endothia parasitica* and inhibition of these activities by components extracted from Chinese and American chestnut inner bark. Physiol. Plant Pathol. 26:367-378.

Napoli, C., Lemieux, C. and R. Jorgensen. 1990. Introduction of a chimeric chalcone synthase gene into petunia results in reversible co-suppression of homologous genes *in trans*. Plant Cell 2:279-289.

Nuss, D. L. 1996.  Using hypoviruses to probe and perturb signal transduction processes underlying fungal pathogenesis.  Plant Cell. In Press.

Turner, G. E. and Borkovich, K. A. 1993.  Identification of a G protein $\alpha$ subunit from *Neurospora crassa* that is a member of the $G_i$ family.  J. Biol. Chem. 268:14805-14811.

Wang, P. and D. L. Nuss. 1995.  Induction of a *Cryphonectria parasitica* cellobiohydrolase I gene is suppressed by hypovirus infection and regulated by a GTP-binding protein-linked signal transduction pathway involved in fungal pathogenesis. Proc. Natl. Acad. Sci. USA 92:11529-11533.

# Saponins and Plant Disease

Jos P. Wubben, Rachel E. Melton, Michael J. Daniels, and Anne E. Osbourn.
The Sainsbury Laboratory, John Innes Centre, Norwich Research Park, Colney, Norwich NR4 7UH, UK.

Saponins are steroidal or triterpenoid glycosides which have been found in many different plant species (Hostettmann and Marston 1995; Price et al. 1987). Many saponins have potent antifungal properties, and hence it has been suggested that these compounds may play a role in protecting plants against attack by saponin-sensitive fungal pathogens (Osbourn 1996). The antifungal action of saponins is likely to result from their membraneolytic action (Osbourn 1996). Some fungi have intrinsic resistance to saponins because of their membrane composition, while others produce specific saponin-detoxifying enzymes that remove sugar molecules from the glycosyl chain at the C-3 carbon position, to give products which are less toxic to fungal growth (Osbourn 1996). Several saponin-detoxifying enzymes have been described so far from fungal pathogens of oat and tomato (Osbourn 1996).

Previously we have shown that the saponin-detoxifying enzyme avenacinase (which detoxifies avenacin A-1 [Fig. 1]), produced by the cereal pathogen *Gaeumannomyces graminis* var. *avenae*, is essential for pathogenicity of this fungus to oats (Bowyer et al. 1995). Recently, an unexpected relatedness between avenacinase of *G. graminis* var. *avenae*, and the tomatinase enzyme produced by the tomato leaf-infecting fungus *Septoria lycopersici* (which acts on the steroidal glycoalkaloid α-tomatine [Fig. 1]) has been described (Osbourn et al. 1995). These two enzymes share common physicochemical properties and are immunologically cross-reactive; however, there are critical differences in their substrate specificities which reflect the host preference of the fungus from which the enzymes were purified. Comparison of the predicted amino acid sequences of avenacinase and tomatinase revealed that the enzymes are clearly similar and that they belong to family 3 of the glycosyl hydrolases as defined by Henrissat (1991).

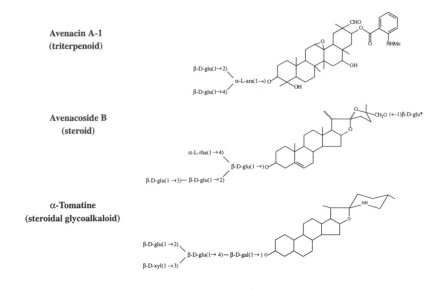

**Avenacin A-1**
**(triterpenoid)**

β-D-glu(1→2)
α-L-ara(1→)O
β-D-glu(1→4)

**Avenacoside B**
**(steroid)**

CH₂O (←1)β-D-glu*

α-L-rha(1→4)
β-D-glu(1→)O
β-D-glu(1→3)— β-D-glu(1→2)

**α-Tomatine**
**(steroidal glycoalkaloid)**

β-D-glu(1→2)
β-D-glu(1→4)— β-D-gal(1→)O
β-D-xyl(1→3)

Fig. 1. Structural representations of avenacin A-1, avenacoside B and α-tomatine.

This chapter will describe recent results of our research related to saponins and plant disease which is divided into three areas.

Whereas triterpenoid avenacin saponins occur only in oat roots, oat leaves contain an different class of saponins known as avenacosides. Here we describe detoxification of avenacosides by the oat leaf pathogen *Septoria avenae* and the possible significance of this in the infection process (i). In parallel with studies of saponin detoxification by fungi, the isolation of avenacin-minus mutants of oat has allowed us to directly investigate the contribution of avenacin to disease resistance (ii). Expression of saponin-detoxifying enzymes in heterologous fungi can give additional information on the role of these enzymes in fungal pathogenesis. The effects of expressing the *S. lycopersici* tomatinase gene in the biotrophic tomato pathogen *Cladosporium fulvum* (which is unable to detoxify α-tomatine) is currently being analysed and will be described (iii).

# The avenacosidase enzyme of *Septoria avenae* f. sp. *avenae*: a third saponin-detoxifying enzyme belonging to family three glycosyl hydrolases

The oat leaf saponins avenacosides A and B (Fig. 1) are steroidal saponins, that show no antifungal activity in their native form (Tschesche et al. 1969; Tschesche and Lauven 1971). Activation of the avenacosides requires removal of a glucose molecule from position C-26 of the saponin by a plant avenacosidase enzyme, resulting in the formation of antifungal 26-desglucoavenacosides A and B (26-DGAs A and B [Lüning and Schlösser, 1975; Gus-Mayer et al. 1994]).

Fungal isolates of different *Septoria* species were analysed for pathogenicity to oats and wheat and found to be either oat-attacking or wheat-attacking (Wubben et al. 1996). Mass spectrometry analysis of the major two antifungal compounds in oat leaves identified them as 26-DGAs A and B. Oat- but not wheat-attacking isolates of *Septoria* were able to detoxify these saponins by enzymatic hydrolysis of the sugar chain at C-3. A 26-DGA-hydrolysing enzyme was purified from culture filtrate of *Septoria avenae* f.sp. *avenae*. This enzyme (avenacosidase) was capable of removing both L-rhamnose and D-glucose molecules from the C-3 sugar chain of the saponins. The enzyme had a molecular weight of 110 kD, an isoelectric point of between 3.8 and 4.1, and optimal β-glucosidase activity at pH 5.4 (Wubben et al. 1996). Amino acid sequence analysis of two peptide fragments obtained after cyanogen bromide cleavage of the avenacosidase enzyme revealed homology with family 3 glucosyl hydrolases, among which are the two previously characterized saponin-detoxifying enzymes avenacinase of *G. graminis* (Bowyer et al. 1995) and tomatinase of *S. lycopersici* (Osbourn et al. 1995; Sandrock et al. 1995). Degenerate PCR primers were made based on conserved regions of amino acid sequences of glycosyl hydrolases family 3 and the avenacosidase peptide fragments. Several PCR fragments of the expected size were cloned and sequenced. The two peptide sequences derived from the purified avenacosidase were represented in the predicted amino acid sequence of the cloned PCR fragments. Southern analysis revealed that homologous DNA was present as a single copy in both oat- and wheat-attacking isolates of *S. avenae*, although clear polymorphisms did exist.

Transformation-mediated targeted gene disruption experiments will now allow specific mutants of *S. avenae* to be generated which are unable to make avenacosidase. The effects of these mutations on the ability to tolerate 26-DGAs and on pathogenicity to oats will be tested in near future.

## Generation of avenacin A-1-minus mutants of the diploid oat species *Avena strigosa*

The isolation of plant mutants defective in the biosynthesis of saponins allows a direct genetic test of the importance of these compounds in plant defence.  The oat root saponin avenacin A-1 is unusual amongst saponins in that it autofluoresces under ultra violet light.  This property was used to isolate avenacin A-1 mutants of the diploid oat *Avena strigosa* following sodium azide mutagenesis.  Of 1500 M2 families screened, 7 independent mutants lacking avenacin A-1 and one with reduced amounts of the saponin were isolated.  HPLC analysis of root extracts revealed that the mutants were also affected in the synthesis of the three other related compounds avenacins A-2, B-1 and B-2, indicating that these mutations must affect the early stages, or the regulation, of avenacin biosynthesis. Progeny of inter-mutant crosses are currently being screened for root autofluorescence to determine the number of complementation groups within this mutant collection.  The mutants show increased susceptibility to both *G. graminis* var. *avenae* and var. *tritici*, with some mutants being more susceptible than others.  Furthermore, they are more susceptible to infection by other root-infecting fungi such as *Fusarium avenaceum*, *F. culmorum* and *F. graminearum*.  Analysis of the F2 progeny of the crosses of each mutant back to the wild type should allow us to establish whether the absence of avenacins and increased disease susceptibility are causally related.  The mutants retain the ability to synthesise the foliar avenacoside saponins, and are unaffected in their interactions with the oat leaf pathogen *Septoria avenae* f. sp. *avenae*.

## Expression of *Septoria lycopersici* tomatinase in *Cladosporium fulvum*

A number of necrotrophic pathogens of tomato in addition to *S. lycopersici* produce α-tomatine-degrading enzymes (tomatinases) which may be required for successful infection of tomato plants.  The biotrophic fungus *Cladosporium fulvum* causes little damage to tomato leaf tissue during infection and so may avoid encountering antifungal levels of α-tomatine, which is believed to be localised primarily in the vacuole. This fungus does not appear to be able to enzymatically detoxify α-tomatine.  Inhibition of germ tube growth of *C. fulvum* does occur at α-tomatine levels as low as 8 μM (Dow and Callow 1978).  In order to test whether expression of *S. lycopersici* tomatinase in *C. fulvum* had any effect on interactions with tomato plants, the *S. lycopersici* tomatinase cDNA under the control of an appropriate constitutive promoter was transformed into *C. fulvum* race 4.  A number of transformants which expressed and secreted tomatinase were identified. For the compatible interaction between tomato and a tomatinase-producing *C. fulvum* race 4, sporulation appears at an earlier time point after inoculation and is heavier than sporulation with wild type race 4. For the incompatible interaction, microscopic analysis showed that fungal invasion was more advanced for the tomatinase producing race 4 when compared to the wild type race 4, although expression of tomatinase did not allow transformants to progress to sporulation.

## Conclusions

Additional information has been obtained on the effect of saponins on interactions between plants and fungal pathogens. Previously, it was shown that oat root saponin detoxification was required for pathogenicity of *G. graminis* var. *avenae* on oats.

Biological data presented here show a correlation between detoxification of oat leaf saponins by *S. avenae* and pathogenicity on oats. Molecular genetic evidence will be obtained to establish the importance of avenacosides in this plant fungus interaction.

Avenacin A-1-minus mutants of *A. strigosa* show increased disease susceptibility towards several root-infecting fungi. Although further genetic analysis of these mutants is required, the data suggest that avenacin saponins are factors which add to increased disease resistance in oat roots against various fungal pathogens.

## Acknowledgements

The authors would like to thank Richard Oliver and Linda Flegg of the University of East Anglia, Norwich, for performing the *C. fulvum* transformations and their help with the analysis of the transformants. The Sainsbury Laboratory is supported by the Gatsby Charitable Foundation. Jos Wubben is supported by the European Union Human Capital and Mobility Network grant, CHRX-CT-0244.

## Literature cited

Bowyer, P., Clarke, B.R., Lunness, P., Daniels, M.J. and Osbourn, A.E. 1995. Host range of a plant pathogenic fungus determined by a saponin-detoxifying enzyme. Science 267:371-374.

Dow, J.M., and Callow, J.A. 1978. A possible role for α-tomatine in the varietal-specific resistance of tomato to *Cladosporium fulvum*. Phytopath. Z. 92:211-216.

Gus-Mayer, S., Brunner, H., Schneider-Poetsch, H.A.W. and Rüdiger, W. 1994. Avenacosidase from oat: purification, sequence analysis and biochemical characterization of a new member of the BGA family of β-glucosidases. Plant Mol. Biol. 26:909-921.

Henrissat, B. 1991. A classification of glycosyl hydrolases based on amino acid sequence similarities. Biochem. J. 280:309-316.

Hostettmann, K. A. and Marston, A. 1995. Saponins. Chemistry and pharmacology of natural products. Cambridge University Press. Cambridge, UK.

Lüning, H.U and Schlösser, E. 1975. Role of saponins in antifungal resistance V. Enzymatic activation of avenacosides, Z. Pflanzenkr. Pflanzenschutz 82:699-703.

Osbourn, A. 1996. Saponins and plant defence - a soap story. Trends Plant Sci. 1:4-9.

Osbourn, A., Bowyer, P., Lunness, P., Clarke, B. and Daniels, M. 1995. Fungal pathogens of oat roots and tomato leaves employ closely related enzymes to detoxify host plant saponins. Mol. Plant-Microbe Interact. 8:971-978.

Price, K.R., Johnson, I.T. and Fenwick, G.R. 1987. The chemistry and biological significance of saponins in food and feedingstuffs. CRC Crit. Rev. Food Sci. Nutr. 26:27-133.

Sandrock, R.W., DellaPenna, D. and VanEtten, H.D. Purification and characterization of $\beta_2$-tomatinase, an enzyme involved in the degradation of α-tomatine and isolation of the gene encoding $\beta_2$-tomatinase from *Septoria lycopersici*. 1995. Mol. Plant-Microbe Interact. 8:960-970.

Tschesche, R. and Lauven, P. 1971. Avenacosid B, ein zweites bisdesmosidisches Steroidsaponin aus *Avena sativa*. Chem. Ber. 104:3549-3555.

Tschesche, R., Tauscher, M., Fehlhaber, H.W. and Wulff, G. 1969. Avenacosid A, ein bisdesmosidisches Steroidsaponin aus *Avena sativa*. Chem. Ber. 102:2072-2082.

Wubben, J.P., Price, K.R., Daniels, M.J., and Osbourn, A.E. 1996. Detoxification of oat leaf saponins by *Septoria avenae*. Phytopathology, *in press*.

# Control of mating, filamentous growth and pathogenicity in *Ustilago maydis*

Regine Kahmann, Tina Romeis, H. Andreas Hartmann, Heidi U. Böhnert, Michael Bölker and Jörg Kämper

Institut für Genetik und Mikrobiologie der Universität München, Germany

The maize pathogenic fungus *Ustilago maydis* is able to infect corn plants only in its dikaryotic stage which results from fusion of two compatible haploid cells. Fusion and subsequent pathogenic development are genetically controlled by the *a* and *b* mating-type loci (see Banuett, 1995). Haploid cells which grow yeast-like respond to pheromone secreted by cells of opposite *a* mating type by forming conjugation tubes that fuse at their tips (Snetselaar, 1993; Spellig et al., 1994). The resulting dikaryon can continue towards sexual and pathogenic development only if both of its nuclei carry different alleles of the multiallelic *b* locus. A switch to filamentous growth is the visible indication for the successful entry into this developmental pathway. The filamentous dikaryon needs the host plant for growth and sexual development (see Banuett, 1995). Genetic studies have revealed different contributions of the *a* and *b* mating type loci to these distinct stages of development. While the *a* locus alone controls cell fusion, and the *b* locus alone is responsible for regulating pathogenic development, both loci are required for maintenance of filamentous growth (see Banuett, 1995).

Molecular studies have revealed that the biallelic *a* locus contains structural genes for a pheromone based cell recognition system. Each allele encodes a lipopeptide pheromone precursor (*mfa*) and a receptor (*pra*) that recognizes pheromone secreted by cells of opposite mating type (Bölker et al., 1992; Spellig et al., 1994; Urban et al., 1996). The pheromone receptors belong to the family of the seven transmembrane domain receptors that are coupled to heterotrimeric G proteins. The pheromone signal is believed to be transmitted by a MAP kinase cascade (Banuett and Herskowitz, 1994). Transcription of all genes located in the *a* and *b* loci is induced upon pheromone stimulation (Urban et al., 1996). This leads to amplification of the pheromone signal during mating and guarantees that expression of the *b* genes is increased prior to fusion. A short DNA sequence, termed pheromone response element, is found in the vicinity of all pheromone inducible genes and is both necessary and sufficient for pheromone induction (Urban et al., 1996).

The multiallelic *b* mating type locus encodes the bE and bW homeodomain proteins that are involved in intracellular recognition through combinatorial interactions (see Kahmann et al., 1995). It has recently been shown that the *bE* and *bW* polypeptides from the same allele are unable to

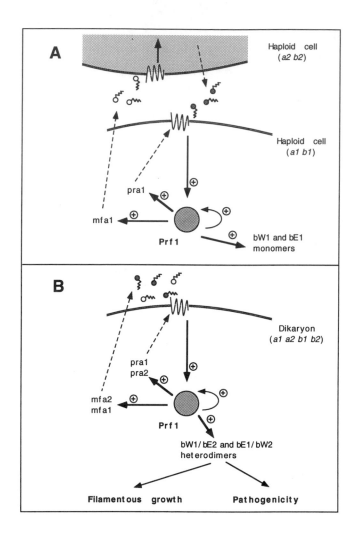

FIGURE 1   The regulatory connection between *a* and *b*.

In A) the consequences of pheromone stimulation of two compatible strains are shown. *a2* cells secrete a2 pheromone (filled circles with wavy tails) that are recognized by the Pra1 receptor (**MW**) present on a1 cells. The pheromone signal activates Prf1 which in turn induces transcription of the receptor (*pra*), pheromone (*mfa*) and *b* genes as well as of its own, resulting in fusion competence. B) represents the situation in the dikaryon. Pheromone signaling is on due to autocrine stimulation. This maintains a high level of *b* gene expression and due to the presence of two *b* alleles bE/bW heterodimers are formed. These heterodimers are thought to switch on genes for filamentous growth and pathogenicity.

interact, whereas the *bE* and *bW* gene products from different alleles can form heterodimers (Kämper et al., 1995). These heterodimers are thought to be transcription factors that switch on genes required for sexual development. The switch from yeast-like to filamentous growth associated with dikaryon formation requires autocrine stimulation of the pheromone response pathway (Bölker et al., 1992; Spellig et al., 1994). Haploid strains engineered to express an active bE-bW heterodimer are pathogenic, illustrating that the active *b* gene complex is sufficient to initiate pathogenic development (Gillissen et al., 1992; Bölker et al., 1995a). Such solopathogenic haploid strains grow yeast-like but can be induced to form filaments by pheromone of opposite mating type. This illustrates that activation of the pheromone response pathway is necessary to observe this morphological transition on plates (Gillissen et al., 1992; Spellig et al., 1994).

Recently, the complex interplay between the *a* and *b* loci during development has been untangled by discovering the molecular mechanism through which pheromone signalling is connected with the control exerted by the homeodomain proteins (Figure 1). This opens new avenues for studying the genetic switch to pathogenic development at the molecular level.

COUPLING THE PHEROMONE RESPONSE AND *b* GENE EXPRESSION

The products of the *b* genes are the key regulators for pathogenicity and central to their mechanism of action is the formation of bE/bW heterodimers that allows self/nonself recognition. Recently we have learned that the *b* genes themselves are subject to intricate transcriptional control. The transcription of the *bE* and *bW* genes is induced by the pheromone receptor system that operates during and after cell fusion (Fig. 1A). The pheromone signal is likely to be transduced via a heterotrimeric G protein and a MAP kinase cascade to the pheromone response factor Prf1 (Herskowitz, 1995; Hartmann et al., 1996). Prf1 is an HMG-domain protein that activates a set of pheromone responsive genes through binding to their cis-acting regulatory elements. Among these genes is the *prf1* gene itself as well as the pheromone and receptor genes (Fig. 1A). This creates an autoregulatory feedback loop and results in strong amplification of the initial pheromone signal prior to fusion. Since transcription of the *b* genes is also controled by Prf1 the amount of *b* gene products rises sharply upon pheromone stimulation. By linking *b* gene expression to the pheromone response pathway it may be ensured that the homeodomain proteins are synthesized in sufficient quantities only when the active heterodimer is needed. Interestingly, Prf1 not only controls the induced level of transcription of the pheromone-inducible genes but is also required for their basal transcription (Hartmann et al., 1996). It is presently not clear how Prf1 regulates basal transcription of these genes. As a consequence, deletion of *prf1* in a solopathogenic haploid strain leads to a decrease in *b* gene expression below a threshold level necessary to trigger pathogenic development (Hartmann et al., 1996). This pathogenicity defect can be fully suppressed by overproduction of bE1 and bW2. This illustrates that with respect to pathogenic development Prf1 exerts its crucial role solely by regulating the transcription of the *b* genes (Hartmann et al., 1996). The filamentous phenotype of such haploid *b* overproducers is likely to reflect the presence of high amounts of active bE/bW heterodimers. In the natural situation this is achieved after cell fusion when the autocrine pheromone stimulation activates Prf1 which in turn stimulates *b* gene expression (Fig. 1B).

## STRATEGIES FOR THE IDENTIFICATION OF PATHOGENICITY GENES IN *USTILAGO MAYDIS*

### A) REMI MUTAGENESIS

The construction of haploid solopathogenic strains of *U.maydis* (Bölker et al., 1995a) allowed for the first time a systematic approach to screen for pathogenicity mutants. After establishing the REMI mutagenesis technique for *U.maydis* (Bölker et al., 1995b) it has become feasible to conduct a saturation mutagenesis. In contrast to other DNA tagging techniques in *U.maydis* the REMI procedure leads to preferential single copy integrations of the transforming plasmid. This facilitates recloning of the integrated plasmid together with flanking sequences. About 1-2% of all mutants generated by the REMI procedure were completely defective in inducing disease symptoms while another 1-2% were affected in severity of disease symptoms. This suggests that a large number of genes participate in this process. Among these we would expect to find genes with direct functions in establishing an infection. In addition, we expect to find genes whose products play an essential role in fungal metabolism during growth in planta. Such genes we would consider as only indirectly required for fungal disease induction. Furthermore, all mutations that interfere with the regulation of the *prf1* gene could reduce the level of b heterodimers below what is needed to turn on pathogenicity genes. This class of mutants can be detected by introducing a plasmid overexpressing bE1 and bW2. Interestingly we were able to show that in about 50 % of all nonpathogenic REMI mutants the pathogenicity defect could be suppressed by overexpressing the *b* genes (H.Böhnert, K.H.Braun, J.Görl and R.Kahmann, unpublished). This allows for the first time to classify the pathogenicity mutants and to concentrate on the ones that interfere with *b*-regulated development.

### B) FINDING THE TARGETS FOR THE *b* GENE PRODUCTS

The *b* locus is the central regulator for pathogenic development. Since both bE and bW are homeodomain proteins, it is expected that the active bE/bW heterodimer acts as a transcriptional regulator by binding to specific target sequences upstream of pathogenicity genes. However, such direct targets have not yet been identified. Thus it remains open whether the bE/bW heterodimer functions as a positive regulator or as a negative regulator of a repressor for pathogenicity genes as suggested by Banuett

(1991). By differential techniques a number of genes have been isolated which are genetically controled by the *b* locus (Bohlmann et al., 1994; Schauwecker et al., 1995; Urban et al., 1996; Wösten et al., 1996). Unfortunately, the deletion of none of these genes has any affect on pathogenic development of the fungus, indicating that neither of these genes represent the much sought central targets of the bE/bW heterodimer. The analysis of their regulation and the identification of regulatory proteins necessary for their stage-specific induction may, however, provide a valuable tool to fill in the missing links that connect the expression of these genes to the *b* locus. Another strategy to identify direct targets for the bE/bW heterodimer makes use of inducible promoters recently identified in *U.maydis* (Banks et al., 1993; Bottin et al., 1996; A. Brachmann, J. Kämper and R. Kahmann, unpublished). The inducible expression of the bE and bW proteins should facilitate the identification of early induced genes and thus potential direct targets by differential display (Liang and Pardee, 1992). A third and most direct means would be the identification of targets by exploiting the presumed ability of the bE/bW heterodimer to bind specifically to DNA. However, all approaches to select for target sequences were hampered by the fact that two proteins had to be used which, in addition, had to form heterodimers. To circumvent this problem we have now constructed single-chain fusions in which the bE1 protein is linked via a flexible kink region to the bW2 protein. Haploid *U. maydis* strains expressing this synthetic *b*-fusion undergo a morphological switch from yeast-like to filamentous growth and became solopathogenic for corn (T.Romeis, J.Kämper and R.Kahmann, unpublished). Thus the synthetic single-chain *b*-fusion is able to substitute for an active bE/bW heterodimer consisting of two peptides. We expect that such fusion peptides will be instrumental in target site identification.

REFERENCES

Banks, G.R., Shelton, P.A., Kanuga, N., Holden, D.W. and Spanos, A. 1993. The *Ustilago maydis nar1* gene encoding nitrate reductase activity: sequence and transcriptional regulation. Gene 131: 69-78.

Banuett, F. 1991. Identification of genes governing filamentous growth and tumor induction by the plant pathogen *Ustilago maydis*. Proc. Natl. Acad. Sci. USA 88: 3922-3926.

Banuett, F. 1995. Genetics of *Ustilago maydis*, a fungal pathogen that induces tumors in maize. 1995. Annu. Rev. Genet. 29: 179-208.

Banuett, F., and Herskowitz, I. 1994. Identification of Fuz7, a *Ustilago maydis* MEK/MAPKK homolog required for *a*-locus-dependent and -independent steps in the fungal life cycle. Genes. Dev. 8: 1367-1378.

Bohlmann, R., Schauwecker, F., Basse, C. and Kahmann, R. 1994. Genetic regulation of mating and dimorphism in *Ustilago maydis*. Pages 239-245 in: Advances in Molecular Genetics of Plant Microbe Interactions, Vol. 3. Daniels, M.J., Downie, J.A. and Osbourn, A.E., eds. Kluwer Acad. Publ. Dordrecht, Netherlands.

Bölker, M., Genin, S., Lehmler, C., and Kahmann, R. 1995a. Genetic regulation of mating and dimorphism in *Ustilago maydis*. Can. J. Bot. 73: 320-325

Bölker, M., Böhnert, H.U., Braun, K.H., Görl, J., and Kahmann, R. 1995b. Tagging pathogenicity genes in *Ustilago maydis* by restriction enzyme mediated integration (REMI). Mol.Gen.Genet. 248, 547-552.

Bölker, M., Urban, M., and Kahmann, R. 1992. The *a* mating type locus of *U. maydis* specifies cell signaling components. Cell 68: 441-450.

Bottin, A., Kämper, J. and Kahmann R. 1996. Characterization of a carbon source-regulated gene from *Ustilago maydis*. Mol. Gen. Genetics, submitted

Gillissen, B., Bergemann, J., Sandmann, C., Schroeer, B., Bölker, M. and Kahmann, R. 1992. A two-component regulatory system for self/non-self recognition in *Ustilago maydis*. Cell 68: 1-20.

Hartmann, H.A., Kahmann, R. and Bölker, M. 1996. The pheromone response factor coordinates filamentous growth and pathogenicity in *Ustilago maydis*. EMBO J. 15: 1632-1641.

Herskowitz, I. 1995. MAP kinase pathways in yeast: for mating and more. Cell 80: 187-197.

Kahmann, R., Romeis, T., Bölker, M., and Kämper, J. 1995. Control of mating and development in *Ustilago maydis*. Curr. Opinion in Genet. & Dev. 5: 559-564.

Kämper, J., Bölker, M., and Kahmann, R. 1994. Mating-type genes in heterobasidiomycetes. Pages 323-332 in: The Mycota, Vol 1. J. Wessels and F. Meinhardt, eds. Springer Verlag, Heidelberg, Germany.

Kämper, J., Reichmann, M., Romeis, T., Bölker, M., and Kahmann, R. 1995. Multiallelic recognition: nonself-dependent dimerization of the bE and bW homeodomain proteins in *Ustilago maydis*. Cell 81, 73-83.

Liang, P. and Pardee, A.B. 1992. Differential display of eucaryotic messenger RNA by means of the polymerase chain reaction. Science 257: 967-971.

Schauwecker, F., Wanner, G., and Kahmann, R. 1995. Filament-specific expression of a cellulase gene in the dimorphic fungus *Ustilago maydis*. Biol. Chem. Hoppe-Seyler 376: 617-625.

Snetselaar, K. M. 1993. Microscopic observation of *Ustilago maydis* mating interactions. Exp. Mycol. 17: 345-455.

Spellig, T., Bölker, M., Lottspeich, F., Frank, R.W. and Kahmann, R. 1994. Pheromones trigger filamentous growth in *Ustilago maydis*. EMBO J. 13: 1620-1627.

Urban, M., Kahmann, R., and Bölker, M. 1996. Identification of the pheromone response element in *Ustilago maydis*. Mol. Gen. Genet. 251: 31-37.

Wösten, H.A.B., Bohlmann, R., Eckerskorn, C., Lottspeich, F., Bölker, M. and Kahmann, R. 1996. A novel class of small amphipatic peptides affect aerial hyphal growth and surface hydrophobicity in *Ustilago maydis*. EMBO J., in press.

# Genomic Organization of the *TOX2* Locus
## of *Cochliobolus carbonum*

Jonathan D. Walton, Joong-Hoon Ahn, John W. Pitkin, and Anastasia N. Nikolskaya

Department of Energy Plant Research Laboratory, Michigan State University, E. Lansing MI 48824 U.S.A.

Fungi in the genus *Cochliobolus* (*Helminthosporium*) are major pathogens of cereal crops throughout the world. Many of the most destructive pathogens in this genus owe their virulence to the production of host-selective toxins (HSTs). Race 1 isolates of *Cochliobolus carbonum* produce an HST known as HC-toxin, a cyclic tetrapeptide of structure cyclo(D-prolyl-L-alanyl-D-alanyl-L-Aeo), where Aeo is 2-amino-9,10-epoxi-8-oxodecanoic acid. Maize that is homozygous recessive at the *Hm* locus is sensitive to HC-toxin and highly susceptible to race 1 of *C. carbonum*. Races 2, 3, and 4 of the fungus do not make HC-toxin and are only weakly pathogenic on *hm/hm* maize.

It has been shown that the maize *Hm* resistance gene encodes an enzyme, HC-toxin reductase, that detoxifies HC-toxin by reduction of the carbonyl group of Aeo (Meeley et al. 1992; Johal and Briggs 1992). HC-toxin inhibits all three isoforms of histone deacetylase in a variety of eukaryotic organisms, and it has been proposed that HC-toxin induces susceptibility by preventing the expression of essential defense genes during infection (Brosch et al. 1995).

## Biosynthesis of HC-toxin

Scheffer et al. (1967) showed that toxin production in the progeny of crosses between HC-toxin producing (Tox2$^+$) and non-producing (Tox2$^-$) isolates of *C. carbonum* segregates 1:1, indicating that a single gene, *TOX2*, controls HC-toxin production. Tox2$^+$ isolates but not Tox2$^-$ isolates have a multifunctional enzyme, HC-toxin synthetase (HTS), that catalyzes ATP/PP$_i$ exchange activity dependent on L-proline, D-alanine, and L-alanine and epimerizes L-proline and L-alanine (Walton and Holden 1988). The gene encoding HTS, called *HTS1*, contains a 15.7-kb open reading frame. HTS, which has a predicted MW of 570 kDa, has four homologous "domains" that are 35 to 40% identical to each other and to similar domains from other cyclic peptide synthetases. In addition to amino acid activation and epimerization, HTS is presumed to polymerize and cyclize the peptide (Scott-Craig et al. 1992).

Molecular studies of *HTS1* revealed that it is found naturally only in strains that make HC-toxin and is completely lacking in Tox2$^-$ isolates. Most Tox2$^+$ strains have two functional copies of *HTS1*. When both copies, but not either copy alone, are mutated by targeted gene disruption, HC-toxin production, HTS activity, and pathogenicity are lost (Panaccione et al. 1992).

Although HTS is the central enzyme in HC-toxin biosynthesis, it is not sufficient. First, nothing in the primary amino acid sequence of HTS suggests participation in Aeo biosynthesis. Second, when transformed into a Tox2$^-$ strain, *HTS1* is not expressed and HC-toxin is not produced.

## OTHER GENES REQUIRED FOR HC-TOXIN BIOSYNTHESIS

*Auto-resistance to HC-toxin.* Flanking both copies of *HTS1* and transcribed in the opposite direction is a gene, *TOXA*, which, like *HTS1*, is present only in natural Tox2$^+$ isolates. The transcriptional start sites of the two genes are 386 bp apart. *TOXA* is predicted to encode an integral membrane protein related to various antibiotic transport pumps of the major facilitator superfamily. Gene disruption of either copy of *TOXA* is feasible and gives no phenotype, but it has proven impossible to construct a strain with both copies disrupted. We hypothesize that *TOXA* is essential for self-protection against HC-toxin (Pitkin et al. 1996).

*Biosynthesis of Aeo.* Because *HTS1* and *TOXA* are present only in Tox2$^+$ isolates of *C. carbonum*, we reasoned that other genes necessary for HC-toxin production might also be Tox2$^+$-unique. We have thereby discovered two additional Tox2$^+$-unique genes, *TOXC* and *TOXD*. *TOXC* encodes a protein that is highly similar in size (220 kDa) and sequence to the $\beta$ subunit of fatty acid synthase from yeast and other fungi. Most Tox2$^+$ isolates have three copies of *TOXC*. When all copies are disrupted, HC-toxin production and pathogenicity are lost. A logical function of the *TOXC* product is to contribute to the synthesis of the decanoic acid backbone of Aeo. It is not known if the *TOXC* gene product makes decanoic acid or a shorter fatty acid. The analysis of *TOXC* predicts that *C. carbonum* Tox2$^+$ isolates should also have a gene encoding an $\alpha$ subunit homolog that is dedicated to HC-toxin biosynthesis (J.-H. Ahn and J.D. Walton, submitted). The function, if any, of *TOXD* is unknown.

## Genomic Organization of *TOX2*

Analysis of individual genes indicates that in most strains *TOX2* contains two copies of *HTS1*, two copies of *TOXA*, and three copies of *TOXC*, in addition to other postulated but still unknown genes (see below). This raises the question of why toxin production appears to segregate genetically as a single Mendelian trait. This was addressed by physical mapping combined with pulsed field gel electrophoretic analysis. In the standard lab strain SB111, all copies of *HTS1*, *TOXA*, *TOXC*, and *TOXD* are found on the same chromosome. This chromosome is the largest (3.5 Mb) and SB114, a related Tox2⁻ isolate, lacks a chromosome of this size (Ahn and Walton, 1996). This result gives two possible, not mutually exclusive, explanations for monogenic inheritance of toxin production: linkage (or clustering) and lack of homologous chromosomes in the parents. In regard to the latter possibility, SB114 does contain a chromosome (of 2.0 Mb) that contains at least one single-copy sequence that is also present on the 3.5-Mb chromosome of SB111. Whether there is sufficient homology between these two chromosomes to allow pairing during meiosis, however, is not known (Ahn and Walton 1996).

Chromosome walking indicated that the two copies of *HTS1* are not tightly clustered. The physical relationship between the multiple copies of *HTS1*, *TOXA*, and *TOXC* was established by physical mapping using endogenous sites for 8-base pair recognition enzymes and by introducing PacI sites into various locations. This revealed that the two copies of *HTS1/TOXA* are 270 kb apart (Fig. 1). One of the copies of *TOXC* (*TOXC*-2) is between the two copies of *HTS1*, and the other two copies flank them. *TOXC*-3 is 220 kb from *HTS1*-1 and *TOXC*-1 is 50 kb from HTS1-2. Together, the copies of *TOXC* and *HTS1* are spread over 540 kb (Fig. 1). The conclusion is that the "*TOX2*" locus is not only complex but is very large and has no particular pattern of organization (Ahn and Walton 1996). It is not a typical gene cluster and did not evolve by a simple one-step mechanism such as horizontal gene transfer from one of the other filamentous fungi that makes Aeo-containing cyclic tetrapeptides (Nikolskaya et al. 1995).

The absolute positions of the copies of *HTS1* were determined by introducing the cos site of bacteriophage lambda into *HTS1*-1 and then cutting with lambda terminase (Ahn and Walton 1996).

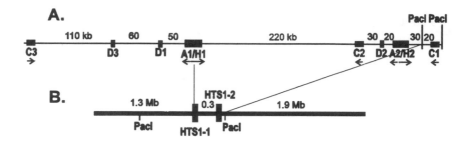

3.5-Mb "TOX2" chromosome of C. carbonum SB111

Figure 1. Map of the *TOX2* chromosome of SB111. A: Map of the region containing *HTS1*, *TOXA*, *TOXC*, and *TOXD* (distances in kb). B: Map of the entire chromosome (distances in Mb). Thin lines connect corresponding sites in A and B. Directions of transcription are indicated by arrows. Modified from Ahn and Walton (1996).

Naturally occuring Tox2$^+$ isolates typically have two copies of *HTS1* on the same chromosome. However, in a small sample of natural isolates, *HTS1* was found to be either on a 3.5-Mb chromosome (as in SB111) or on a 2.2-Mb chromosome. Using several probes, this particular chromosomal size polymorphism can be rationalized on the basis of a simple translocation event (Ahn and Walton 1996).

## Regulation of HC-toxin biosynthesis

Although HC-toxin is produced constitutively by the fungus growing on simple medium containing yeast extract, there is emerging evidence that it also under genetic regulation. First, when transformed into a Tox2$^-$ strain, *HTS1* is not expressed. Second, Weiergang et al. (1996) have shown that HC-toxin synthesis by germinating spores is linked to appressorium formation. Third, certain strains that contain at least one copy of all known genes for HC-toxin production (*HTS1*, *TOXA*, and *TOXC*) produce no detectable toxin in culture. One of these strains, 243-7, has a truncated *TOX2* chromosome of 2.1 Mb instead of the normal 3.5 Mb. It lacks *TOXC*-3, *TOXD*-3, *TOXD*-1, *TOXA*-1, and *HTS1*-1 but has *HTS1*-2, *TOXA*-2, *TOXC*-1, *TOXC*-2, and *TOXD*-2 (see Fig. 1). By physical mapping and pulsed-field electrophoresis, the *TOX2* chromosome of 243-7 has apparently undergone a break between *HTS1*-1 and *HTS*-2 ca. 20 kb from *HTS1*-1 (Fig.1). Strain 243-7 has a novel disease phenotype that is intermediate between race 1 and race 2: it causes lesions but the lesions develop more slowly and inoculated maize seedlings are never killed. We call this phenotype "race 1.5". Strain 243-7 makes no detectable HC-toxin in culture (detection limit ca. 10% of wild type level), does not have detectable HTS (ATP/PP$_i$ exchange) activity, and *TOXA* mRNA expression is strongly suppressed (Table 1). We conclude, however, that 243-7 still makes a low level of HC-toxin, at least *in planta*, and that this accounts for its intermediate pathogenic phenotype. This conclusion is based on the fact that when the remaining copy of *HTS1* (*HTS1*-2) is disrupted, 243-7 becomes completely non-pathogenic, i.e., it becomes race 2 (Table 1). We propose that the low level of toxin that 243-7 is producing is due to basal expression of the HC-toxin biosynthetic genes. This leads to the conclusion that Tox2$^+$ isolates of *C. carbonum* have a positive activator gene that is required for full (normal) expression of *HTS1*, *TOXA*, and HC-toxin production. Interestingly, *TOXC* mRNA expression is normal in 243-7 and therefore *TOXC* does not require this putative activator gene. It is plausible that *TOXA* and *HTS1* are regulated together because their promoters overlap. Co-regulation of *HTS1* and *TOXA* is consistent with the lack of expression of *TOXA/HTS1* when they are transformed into a Tox2$^-$ strain; since these strains lack the putative activator gene, *HTS1* and *TOXA* are not expressed at detectable levels (even if expressed, a Tox2$^-$ strain containing *HTS1* and *TOXA* could not make HC-toxin or show a race 1.5 phenotype because it would still be lacking *TOXC* and any other, unknown biosynthetic genes).

Physical mapping of 243-7 indicates that its 3.5-Mb *TOX2* chromosome has undergone a simple break resulting in the loss of 1.4 Mb. Therefore, the putative regulatory gene is on this 1.4 Mb region. We have refined the position of this putative regulatory gene by analyzing additional strains by Southern blotting for those that have lost one or more of the same genes as 243-7. We have found two additional strains that are missing the same chromosomal region as 243-7 and these have the same race 1.5 phenotype. Another chromosome-break strain that has ca. 70 kb more DNA than 243-7 is still completely Tox2$^+$. This locates the putative regulatory gene to a region of ca. 70 kb in the region of copy 1 of *HTS1* (*HTS1*-1). Various strategies are being pursued to clone this gene.

Table 1. Genomic organization, race, and HC-toxin production in culture of various *C. carbonum* strains.

| Strain | *HTS1* | | *TOXA* | | *TOXC* | | HC-toxin in culture |
|---|---|---|---|---|---|---|---|
| | gene copies | enzyme activity | gene copies | mRNA | gene copies | mRNA | |
| SB111 (race 1) | 2 | + | 2 | + | 3 | + | + |
| SB114 (race 2) | 0 | – | 0 | – | 0 | – | – |
| 243-7 (race 1.5) | 1 | – | 1 | +/– | 2 | + | – |
| 243-7: Δ*HTS1*-2 (race 1.5 → race 2) | 0 | – | 1 | +/– | 2 | + | – |

## Acknowledgments

This paper is dedicated to the memory of Robert P. Scheffer, whose many contributions to our understanding of host-selective toxins included the discovery of HC-toxin. Supported by the U.S. Department of Energy, Division of Energy Biosciences, and the National Institutes of Health.

## Literature Cited

Ahn, J.-H., and Walton, J.D. 1996. Chromosomal organization of *TOX2*, a complex locus controlling host-selective toxin biosynthesis in *Cochliobolus carbonum*. Plant Cell 8:887-897.

Brosch, G., Ransom, R., Lechner, T., Walton, J.D., and Loidl, P. 1995. Inhibition of maize histone deacetylase by HC-toxin, the host-selective toxin of *Cochliobolus carbonum*. Plant Cell 7:1941-1950.

Meeley, R.B., Johal, G.S., Briggs, S.P., and Walton, J.D. 1992. A biochemical phenotype for a disease resistance gene of maize. Plant Cell 4:71-77.

Nikolskaya, A.N., Panaccione, D.G., and Walton, J.D. 1995. Identification of peptide synthetase-encoding genes from filamentous fungi producing host-selective phytotoxins or analogs. Gene 165:207-211.

Panaccione, D.G., Scott-Craig, J.S., Pocard, J.-A., and Walton, J.D. 1992. A cyclic peptide synthetase gene required for pathogenicity of the fungus *Cochliobolus carbonum* on maize. Proc. Natl. Acad. Sci. U.S.A. 89:6590-6594.

Pitkin, J.W., Panaccione, D.G., and Walton, J.D. 1996. A putative cyclic peptide efflux pump encoded by the *TOXA* gene of the plant pathogenic fungus *Cochliobolus carbonum*. Microbiology 142:1557-1565.

Scheffer, R.P., Nelson, R.R., and Ullstrup, A.J. 1967. Inheritance of toxin production and pathogenicity in *Cochliobolus victoriae* and *Cochliobolus carbonum*. Phytopathology 57:1288-1289.

Scott-Craig, J.S., Panaccione, D.G., Pocard, J.-A., and Walton, J.D. 1992. The multifunctional cyclic peptide synthetase catalyzing HC-toxin production in the filamentous fungus *Cochliobolus carbonum* is encoded by a 15.7 kb open reading frame. J. Biol. Chem. 67:26044-26049.

Walton, J.D., and Holden, F.R. 1988. Properties of two enzymes involved in the biosynthesis of the fungal pathogenicity factor HC-toxin. Mol. Plant-Microbe Interact. 1:128-134.

Weiergang, I., Dunkle, L.D., Wood, K.V., and Nicholson, R.L. 1996. Morphogenic regulation of pathotoxin synthesis in *Cochliobolus carbonum*. Fungal Genet. Biol. 20:74-78.

# Structure-function relation studies on AVR9 and AVR4 elicitors of *Cladosporium fulvum*.

P.J.G.M. De Wit[1], M. Kooman-Gersmann[1], R. Vogelsang[2], M.H.A.J. Joosten[1], J.P.M.J. Vossen[1], R.L. Weide[1], R. Laugé[1], G. Honée[1] and J.J.M. Vervoort[3]

[1]  Department of Phytopathology and [3]Biochemistry, Wageningen Agricultural University, Wageningen, The Netherlands.
[2]  RWTH Aachen, Institute of Biology III, Aachen, Germany.

Plant pathogens produce various types of molecules some of which play an essential role in the interaction with their host plant. These include pathogenicity factors and avirulence factors such as race-specific elicitors. So far three fungal race-specific elicitors have been isolated. NIP1 is a race-specific peptide elicitor of the barley pathogen *Rhynchosporium secalis* (Rohe et al., 1995). It is a peptide that specifically induces the accumulation of PR *Hv-1*, a pathogenesis-related protein of barley. AVR4 and AVR9, are race-specific peptide elicitors of the fungal tomato pathogen *Cladosporium fulvum* which induce a hypersensitive response (HR) in tomato genotypes carrying the complementary genes for resistance *Cf-4* (MM-Cf4) or *Cf-9* (MM-Cf9), respectively (De Wit, 1995). Some of the complementary resistance genes of which the products contain leucine-rich repeats which might interact with race-specific elicitors have been cloned (Jones and Jones, 1996).

Not much is known about the secondary and tertiary structure of race-specific elicitors and their receptors in plants, with the exception of the AVR9 elicitor of which the secondary structure has been reported recently (Vervoort et al., 1996). Here we report on structure-function relation studies of the AVR9 and AVR4 elicitor of *C. fulvum*.

## The AVR9 elicitor.

### STRUCTURE OF THE AVR9 ELICITOR.

*C. fulvum* only produces the AVR9 elicitor while growing *in planta* but not *in vitro*. In order to produce sufficient amounts of the AVR9 elicitor *in vitro*, the *Avr9* promoter has been replaced by a constitutive promoter. *Avr9*[+] transgenic strains of *C. fulvum*, however, do not secrete the mature 28 amino acid (aa) AVR9 peptide, which is found in apoplastic fluid of *C. fulvum*-infected tomato, but secrete a mixture of 32, 33 and 34 aa AVR9 peptides. Milligram quantities of the 33 aa AVR9 peptide have been purified. The molecular mass was verified by Electrospray Mass Spectrometry. From the atomic mass it was concluded that 3 disulfide bonds are present in the AVR9 peptide. The structure of the 33 aa AVR9 peptide, as studied by $^1$H NMR, consists of a compact sulfur core, three strands of antiparallel ß-sheet formed by the residues S9-C11, R23-D25 and L29-V32, a large extended loop region T12-Q20 and a region G1-D5 without secondary structure (Figure 1; Vervoort et al., 1996). Residues G1-D5 are not important for necrosis-inducing activity (NIA) on MM-Cf9 plants, as the 28 aa mature peptide is as active as the 33 aa precursor peptide. The secondary structure of AVR9 is related to that of many small cysteine-rich peptides including serine proteinase inhibitors (CMTI-I, EETI-II and CPI), ion channel blockers (X-CgTx GVIA and Kalata B1), and growth factors (Pallaghy et al., 1994; Isaacs, 1995).

```
ω-CqTx GVIA     C-KSZGSSCSZTSYNC-----CRSCNZYT---KRCY
Kalata B        NGLPVC----GETCV-GGT-C-NTPGC-TCSW-----PVCTR
CMTI-I          RVC-PRILMECKK-DSDCL-AE-C-VCLEH----GYCG
EETI-II         GC-PRILMRCKQ-DSDCL-AG-C-VCGPN----GFCG
CPI             EQHADPIC----NKPCKTHD-DCSGAWFCQACW---NSARTCGPYVG
AVR9            GVGLDYC----NSSCTRAF-DCLGQ--CGRCD---FHKLQCVH
```

Figure 1. Various peptide sequences aligned on the basis of half-cystine positions and ß-strand (underlined) hydrogen bonding patterns. Half-cystine residues are in bold (Pallaghy et al., 1994; Isaacs, 1995).

In the three dimensional structure of CPI, EETI-II and X-CgTx GVIA the disulfide bonds are between residues C1-C4, C2-C5 and C3-C6, where C1 is the first cysteine residue at the N-terminus. A common property of the reported peptides and growth factors is the presence of the so-called cystine knot. The cystine knot consists of a ring formed by cystines C1-C4, C2-C5 and the intervening peptide backbone, through which the third disulfide bond (C3-C6) passes. The peptides belonging to the inhibitor cystine knots show large structural similarity over motif residues forming the cystine knot. The aa residues making up this motif are located in a triple-stranded, anti-parallel ß-sheet which consists of a minimum of 10 residues, ($XXC_2$, $XC_5X$, $XXC_6X$; X is any aa residue). The core of the ß-sheets can not be recognized from sequence alone, but the spacing between the 6 cysteine residues is also important. The spacing consensus found in inhibitor peptides including AVR9 is $CX_{3-7}CX_{4-6}CX_{0-5}CX_{1-4}CX_{4-10}C$ (X is any aa residue).

Structurally AVR9 is most related to CPI. AVR9 and CPI can be aligned over a large sequence region (Figure 1). In CPI the C-terminal residues YVG are involved in binding to carboxypeptidase. The loop residues **PRILM** of the trypsin inhibitors CMTI-I and EETI-II are involved in binding to trypsin. We plan to construct a chimeric peptide starting from the AVR9 backbone by adding the **PRILM** and **YVG** aa residues which might have NIA as well as both trypsin and carboxypeptidase inhibiting activities.

Growth factors such as transforming growth factor-ß2, nerve growth factor, and platelet-derived growth factor-BB also contain 3 disulfide bonds and 3 antiparallel ß-sheets, but the motifs are topologically different and cannot be superimposed (Isaacs, 1995). Growth factors are able to form dimers either non-covalent (nerve growth factor), by one intermolecular disulfide bond (transforming growth factor) or by two intermolecular disulfide bonds (platelet-derived growth factor). Dimerization is required before binding to their receptors. Of none of the inhibitor cystine knot peptides described, it is known whether they dimerize before they bind to their receptor. If AVR9 would form dimers through disulfide bonds the C1-C4 disulfide bond might be reduced most easily and could be involved in intermolecular dimerization.

A HIGH AFFINITY BINDING SITE FOR THE AVR9 ELICITOR

Race-specific 28 aa AVR9 elicitor was labelled with [125]Iodine at the N-terminal tyrosine residue and used in binding studies. [125]I-AVR9 showed specific, saturable and reversible binding to plasma membranes isolated from leaves of the tomato cultivar Moneymaker, without *Cf*-resistance genes (MM-Cf0) or from MM-Cf9 containing resistance gene *Cf-9*. The dissociation constant ($K_d$) is 0.07 nM, and the receptor concentration (Rt) was 0.8 pmol per mg of microsomal protein (Kooman-Gersmann et al., 1996). Binding was maximal at pH 6. The ionic strength of the binding buffer and the temperature strongly influenced binding of AVR9 to its receptor, indicating involvement of both electrostatic and hydrophobic interactions. Intact disulfide bridges in AVR9 are required to bind to the receptor as fully reduced AVR9 peptide has very low affinity for the receptor. Binding kinetics and binding capacity were similar for membranes of genotypes MM-Cf0 and MM-Cf9. In all solanaceous plant species tested so far, an AVR9 binding site was present, while in the non-solanaceous species that were analysed no binding site could be identified. Apparently binding of AVR9 is not confined to membranes of MM-Cf9 plants which are the only to respond with HR after treatment with AVR9 elicitor but also occurs to membranes of plants that contain members of the *Cf-9* gene family as was shown by Southern analysis.

SITE-DIRECTED MUTAGENESIS OF AVR9 ELICITOR.

The aa residues within AVR9, crucial for binding and HR inducing activity are not known yet. We have created various mutants in which most aa residues have been exchanged by site-directed mutagenesis (Vogelsang et al., 1996). NIA was assayed with the PVX expression system which has been employed successfully before to express the wild-type *Avr9* gene of *C. fulvum* (Hammond-Kosack et al., 1995). Expression of *Avr9* by PVX in MM-Cf9 plants causes severe necrosis and the plants finally die. Thus, the PVX expression system appears to be very suited to perform structure-function studies on mutants of the AVR9 elicitor peptide. Mutants of the *Avr9* gene were created, expressed via PVX in MM-Cf9 tomato plants and their NIA was determined. Several aa residues with solvent-exposed side chains were exchanged as well as most of the cysteine residues and, residue S05 which is part of the potential N-glycosylation site (NSS). NIAs exhibited by the mutant *Avr9* constructs encoding R08K and R18K were higher than those of the wild-type PVX:*Avr9* construct. D20N and Q25E+V27I mutants showed NIAs comparable to that of the wild-type. NIAs of constructs encoding S05A, D11N, H22L, K23Q and H28L were lower than that of the wild-type. The mutant construct encoding F10S showed hardly any NIA, while the mutant constructs encoding L24S, C02A, C06A en C12A did not show any NIA. The latter phenotypes could not be distinguished from those of control PVX infection (PVX without *Avr9* insert), where only mosaic symptoms were observed. (Kooman-Gersmann et al. unpublished.)

Various mutant AVR9 peptides showing NIA different from the wild-type AVR9 were isolated and purified in microgram quantities from apoplastic fluids of PVX-infected plants. NIA of these peptides after injection into leaves of MM-Cf9 plants correlated with the NIA induced in MM-Cf9 plants by the corresponding PVX:*Avr9* mutant constructs. Molecular mass determination of the purified AVR9 peptides revealed the presence of one molecule of N-acetyl-D-glucosamine attached to the asparagine residue at position 3 of the mutant AVR9 peptides, except for the SO5A 'glycosylation'-mutant. Competition binding assays with microsomal fractions of tomato showed a correlation between affinity to plasma membrane receptor and NIA, indicating that the AVR9 binding site is involved in the process leading to induction of HR.

N-GLYCOSYLATION OF THE AVR9 ELICITOR.

As shown above, the AVR9 elicitor produced via the PVX expression system is glycosylated. However, only one residue of N-acetyl-D-glucosamine was attached to the asparagine residue at the 3-position, the only potential glycosylation site present the *Avr9* sequence. In eukaryotes oligosaccharide moieties are transferred from dolichol precursors into a covalent N-glycosidic linkage to asparagine in nascent peptide chains containing the Asn-X-Ser/Thr consensus sequence as they emerge into the lumen of the endoplasmic reticulum (ER). The oligosaccharide moieties are processed further by trimming or adding monosaccharide residues either in the ER or as the glycoproteins pass through the Golgi apparatus. The precursor molecule initially attached to asparagine usually contains two N-acetyl-D-glucosamines, nine mannose and three glucose residues. The occurrence of only one N-acetyl-D-glucosamine residue in PVX-produced AVR9 is striking. Presumably the precursor glycopeptide is processed by various different glycosidases.

Detailed studies on glycosylation by *C. fulvum* grown *in planta* and *in vitro* revealed that N-glycosylation of the AVR9 peptide does occur in both cases (De Wit et al. unpublished). Two N-acetyl-D-glucosamine residues and up to six, presumably mannose residues, were found to be attached to the asparagine residue of AVR9. However, not all AVR9 peptides that were isolated appeared to be glycosylated. It has to be determined whether a fraction of the AVR9 peptide is never N-glycosylated or whether initially N-glycosylated AVR9 becomes deglycosylated in time. The affinity of AVR9 containing N-acetyl-D-glucosamine to microsomal fractions of MM-Cf9 leaves was twenty fold lower than that of non-glycosylated AVR9 peptide. Preliminar binding studies with AVR9 peptides containing two N-acetyl-D-glucosamines and a variable number of mannose and/or glucose residues showed at least forty fold lower affinity to microsomal fractions than non-glycosylated AVR9.

We plan to purify sufficient amounts of various N-glycosylated AVR9 peptides to be tested for NIA as well as affinity to plasma membrane receptors. $I^{125}$-labeled N-glycosylated AVR9 peptides will be used to prove whether N-glycosylated AVR9 peptides show the same binding characteristics as non-glycosylated AVR9 peptide.

## The AVR4 elicitor

Avirulence gene *Avr4* which conditions avirulence of *C. fulvum* on tomato genotypes carrying resistance gene *Cf-4* (MM-Cf4) is, similar to *Avr9*, specifically induced *in planta*. It encodes an extracellular protein containing eight cysteine residues. Strains virulent on MM-Cf4 tomato genotypes fail to induce HR. The virulent strains contain various single point mutations in the coding region of the *Avr4* gene. Next to substitution of three different cysteine residues by a tyrosine residue and two additional amino acid changes (T into I or T into H) between the fourth and fifth cysteine residue, in one case a frameshift mutation was found (Joosten et al, 1996). Although various *avr4* alleles present in the different strains of *C. fulvum* virulent on tomato genotype MM-Cf4 are also expressed *in planta*, polyclonal antibodies raised against the AVR4 elicitor did not detect AVR4 homologs *in planta*. In order to study whether the AVR4 homologs produced by virulent strains have retained HR inducing activity, the *avr4* alleles were expressed in MM-Cf4 plants using the PVX-based expression system. Inoculation of MM-Cf4 plants with PVX:*Avr4* (wild-type *Avr4* gene) resulted in the development of spreading lesions, eventually leading to plant death, whereas the various PVX:*avr4* derivatives (virulent alleles of the *Avr4* gene) induced symptoms ranging from severe necrosis to no lesions at all. It is concluded that instability of the encoded AVR4 homologs is a crucial factor in circumventing *Cf-4*-mediated resistance (Joosten et al, 1996). Mass determination of the *in planta* secreted AVR4 elicitor revealed that it is processed both at the N- and C-terminus (Joosten et al., 1996).

LITERATURE CITED

De Wit, P.J.G.M. 1995. Fungal avirulence genes and plant resistance genes: unraveling the molecular basis of gene-for-gene interactions Adv. Bot. Res. 21:147-185.

Hammond-Kosack, K.E., Staskawicz, B.J., Jones, J.D.G., and Baulcombe, D.C. 1995. Functional expression of a fungal avirulence gene from a modified potato virus X genome. Mol. Plant-Microbe Interact. 8:181-185.

Isaacs, N.W. 1995. Cystine knots. Cur. Opin. Struct. Biol. 5:391-395.

Jones, D.A., and Jones, J.D.G. 1996. The roles of leucine-rich repeat proteins in plants defences. Adv. Bot. Res. 22: in press.

Joosten, M.H.A.J., Vogelsang, R., Cozijnsen, A.J., Verberne, M.C., and De Wit, P.J.G.M. 1996. The biotrophic fungus *Cladosporium fulvum* circumvents *Cf-4*-mediated resistance by producing instable AVR4 elicitors. Submitted

Kooman-Gersmann, M., Honée, G., Bonnema, G., and De Wit, P.J.G.M. 1996. A high-affinity binding site for the AVR9 peptide elicitor of *Cladosporium fulvum* is present on plasma membranes of tomato and other solanaceous plants. Plant Cell 8:929-938.

Pallaghy, P.K., Nielsen, K.J., Craik, D.J., and Norton, R.S. 1994. A Common structural motif incorporating a cystine knot and a triple-stranded ß-sheet in toxic and inhibitory polypeptides. Protein Sci. 3:1833-1839.

Rohe, M., Gierlich, A., Hermann, H., Hahn, M., Schmidt, B., Rosahl, S. and Knogge W. 1995. The race-specific elicitor, NIP1, from the barley pathogen, *Rhynchosporium secalis*, determines avirulence on host plants of the *Rrs1* resistance genotype. EMBO J. 14:4168-4177.

Vervoort, J.J.M., Vogelsang, R., Joosten, M.H.A.J., and De Wit, P.J.G.M. 1996. Sequence-specific [1]H NMR assignments, secondary structure and global fold of the race-specific elicitor AVR9 of the tomato pathogen *Cladosporium fulvum*. Submitted.

Vogelsang, R., Kooman-Gersmann, M., Vervoort, J.J.M., and De Wit, P.J.G.M. 1996. Molecular, biochemical and functional analysis of wild-type and mutant AVR9 elicitors of *Cladosporium fulvum*. Submitted.

# Cellular and Molecular Mechanisms of Coat Protein and Movement Protein Mediated Resistance Against TMV

Roger N. Beachy, Hal S. Padgett, Ted Kahn, Mohammed Bendahmane, John H. Fitchen, Manfred Heinlein, Yuichiro Watanabe, Bernard L. Epel

Division of Plant Biology, The Department of Cell Biology, The Scripps Research Institute, La Jolla, CA

## Introduction

The use of pathogen derived resistance to develop virus resistant plants was first described in transgenic tobacco plants that exhibited resistance to tobacco mosaic virus (TMV) as a result of expression of the gene encoding the capsid protein (CP) of TMV (Powell-Abel et al. 1986). Since that time a variety of viral sequences from more than 12 different taxons of plant viruses have been expressed in many different types of transgenic plants; in some examples plants exhibit very high levels of resistance while others exhibit low levels or no resistance (reviewed by Fitchen and Beachy 1993). In certain cases there is strong evidence to support the conclusion that certain types of nucleic acid sequences, while not encoding a known protein, can confer resistance or tolerance to infection (against the virus from which the sequences were derived). Although there are a number of examples of pathogen derived resistance (PDR) there is not complete understanding of how pathogen derived resistance is achieved.

We are continuing our studies of the mechanisms of resistance to TMV as conferred by expression of capsid protein (coat protein mediated resistance, CP-MR) and resistance mediated by expression of a defective movement protein (dMP-MR). Here we summarize the status of our understanding of these two types of PDR and suggest additional studies to further elucidate details of resistance mechanisms. It must be emphasized that our knowledge of resistance against TMV may not necessarily be applicable to PDR against other viruses; although some of the principles are likely similar, others clearly are not.

COAT PROTEIN MEDIATED RESISTANCE AGAINST TMV

Transgenic tobacco and tomato plants that accumulate the CP of TMV and exhibit CP-MR are characterized by the following features (see review by Fitchen and Beachy, 1993, for details): (1) Resistance is most effective against TMV, and is less effective against other tobamoviruses; the degree of resistance against other tobamoviruses is related to the degree of identity of amino acid (aa) sequences between the transgene and the coat protein of the challenge virus, and not the RNA genome of the virus (Clark *et al.* 1995b) (2) The degree of resistance is related to the amount of CP that accumulates in transgenic plants; i.e., the more the protein, the greater the resistance. (3) Resistance is largely overcome by infection by viral RNA and by virus that has been swollen by pre-treatment at high pH (Register III and Beachy 1988), a treatment that releases capsid molecules and exposes the viral RNA to initiate the infection process. In experiments carried out in protoplasts it was demonstrated that in transgenic cells that contain CP the association of challenge virus with ribosomes via co-translational disassembly is prevented or delayed when compared with non-transgenic protoplasts (Wu *et al.* 1990). These results implied that resistance was the result of interference with the process of uncoating or disassembly of the virus, the earliest known step in the infection process. Furthermore they showed that once infection is initiated, there is little resistance to local and systemic spread of the infection. It was subsequently shown that spread of TMV between adjacent cells does not require CP (Dawson *et al.* 1988). Because the infection moves as other than virus particles, CP-MR against TMV is somewhat less effective than in cases in which the target virus moves between cells as virions, or in situations where CP interferes with steps in the infection process other than in virus disassembly.

In other studies we investigated CP-MR in plants that accumulate TMV CP in specific cell types. Reimann-Philipp et al ( 1993a) reported that the accumulation of CP in epidermal cells, the site of entry of TMV during mechanical transmission, provided a modicum of resistance; on the other hand, accumulation of CP primarily in leaf mesophyll cells, or in phloem cells provided little or no resistance to TMV.

Based upon the results of these studies we proposed that interactions between the aa sequence of the transgenic CP with the CP of the challenge virus determined the degree of CP-MR. On the basis of the known structure of TMV determined by Namba et al ( 1989), we constructed mutants of the CP that reduced interactions between CP molecules; one such mutant, Thr28Ile, prevented the helical assembly of subunits and thus virus particles. However, the CP aggregated to extended arrays of CP, and in transgenic plants this mutant conferred CP-MR like that of wild type CP molecules (Clark *et al.* 1995a).

These studies led us to conclude that CP-MP against TMV is due to CP:CP interactions, and that if there is high similarity between the trangenic CP and the challenge virus, disassembly is prevented. We suggested that the most likely model is that disassembly, which begins in an ordered manner from the 5' end of the virion, ensues in both CP(+) and CP(-) cells in much the same manner. In CP(-) cells the release of 5-10 capsid protein subunits (exposing 15-30 nucleotides) results in binding of ribosomes and translation of the viral RNA, thereby initiating the infection process. In CP(+) cells the equilibrium between release of CP molecules from the virus is shifted and disassembly is prevented, either by preventing the release of CP molecules, or by re-encapsidating the viral RNA if disassembly begins (Fig. 1).

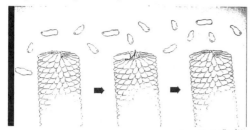

Fig. 1

The studies of Clark *et al.* (1995a) support the hypothesis that the release of 1-4 CP molecules can be blocked by re-encapsidation, whereas more that 5-10 molecules probably results in ribosome binding and infection (Clark *et al.* 1995a). This is in contrast to a second model which proposes that a receptor molecule is blocked or occupied by transgenic CP, thereby limiting access by the challenge virus to the receptor (Register III *et al.* 1989).

## DEFECTIVE MP MEDIATED RESISTANCE AGAINST TMV

The 30 kDa movement protein of TMV and other tobamoviruses is a non-structural protein that is not required for virus replication but is required for local and systemic infection by this group of viruses; local movement of TMV does not require CP while efficient long distance systemic infection requires CP, and the role of MP in such spread is unclear. Other plant viruses also encode proteins that are involved in movement between adjacent cells; in some cases local movement requires multiple non-structural proteins as well as coat protein(s). The known functions of the MP of TMV (recently reviewed by Fenczik *et al.* 1995) include (1) the capacity to accumulate in plasmodesmata in transgenic plants as well as in infected leaves, and to induce changes in size exclusion limits in transgenic plants; (2) the capacity to bind ss nucleic acids *in vitro* (Citovsky *et al.* 1990); (3) the capacity to co-align with microtubules in infected cells (Heinlein et al, 1995) and in the absence of virus infection (McLean *et al.* 1995). In the recent study by Heinlein et al ( 1995) a fusion was constructed between the MP and the green fluorescent protein (GFP) from *Aequorea victoria* and the fusion protein was expressed during TMV infection. A number of different fluorescent structures were observed (by fluorescence microscopy) in protoplasts, and in preliminary studies, in virus infected leaves. On the basis of the microscopic observations and other data it was proposed that the function of the MP during virus infection is to transport viral RNA throughout the cell and to the plasmodesmata between adjacent cells by association with one or more elements of the cytoskeleton; the size exclusion limit of the PD is thereby modified to enable the viral RNA to spread to adjacent cells.

In the early 1990s it was demonstrated that MP that accumulated in transgenic plants could serve as a host for a mutant of TMV that lacked a functional MP. This information, as well as the finding that these transgenic plants had altered plasmodesma(Wolf *et al.* 1989) led us to conclude that some of the functions of the MP could be provided in absence of virus infection. It also led us to propose that accumulation of an appropriate non-functional mutant of the MP may interfere with one or more functions of the MP that is produced during virus infection, and would be another example of pathogen derived resistance (Deom *et al.* 1992). Furthermore, as many viruses encode movement proteins that are expected to possess common functions that are held in common with those of other MPs, it was proposed that certain mutants of MP may provide resistance to a number of different viruses.

Unlike the case of the CP, where the functions of most amino acid sequences and structures are well described, little is known of the relationship between the aa sequence of the MP and the multiple functions of the protein. To gain a greater understanding of the role of selected aa sequences and regions of the MP in protein function a number of different mutants of the MP were constructed, including deletions of aa sequences from the N- and C-termini. Many of the mutant MPs were non-functional when expressed during TMV infection; i.e., virus that contained the mutant MP was unable to cause local or systemic infection in tobacco plants (Gafny *et al.* 1992). Similarly, many of the mutant proteins, when expressed as transgenes in transgenic tobacco plants, did not function to enable MP(-) mutants of TMV to spread locally. However, the accumulation in transgenic plants of a 'non-functional' MP from which aa 3, 4 and 5 were deleted (referred to as dMPΔ3-5 was capable of delaying the local and systemic spread of TMV infection in tobacco plants (Lapidot *et al.* 1993).

To address the question of whether the dMP interfered with a function of the MP that was common amongst tobamoviruses the dMP+ trangenic plants were inoculated with three other tobamoviruses whose MP sequences were different from the sequence of the dMP. The viruses used in this study were tomato mosaic virus (ToMV), Ob tobamovirus, odontoglossum ringspot virus (ORSV) and sunn hemp mosaic virus (SHMV). The results of these studies confirmed that the dMP provided resistance against each of these tobamoviruses that was equal to or greater than the resistance against TMV (Cooper *et al.* 1995; Lapidot *et al.* 1993).

To address the question of whether the dMP(+) transgenic plants interfered with a function(s) of MP that is common amongst different virus groups the resistance studies were extended to viruses that are members of different taxons. In these studies the viruses were also inoculated to transgenic plants that contain the wild type MP and to non-transgenic plants: the table below summarizes the results of these experiments (Data from Cooper *et al.* 1995).

| VIRUS | PHENOTYPE OF INFECTION COMPARED WITH NON-TRANSGENIC PLANTS |
|---|---|
| Alfalfa mosaic ilar virus | Very long delay of symptoms on dMP plants; increased disease ilarvirus severity and virus accumulation on wt MP plants |
| Peanut Chlorotic | Very long delay of symptoms on dMP plants; increased |

| streak caulimovirus | disease severity and virus accumulation on wt MP plants |
| Tobacco rattle tobravirus | Long delay of symptoms on dMP plants; increased disease severity and virus accumulation on wt MP plants |
| Cucumber mosaic cucumovirus | No significant differences in symptoms amongst the plant lines; small delay in virus accumulation in dMP plants |
| Tobacco ringspot nepovirus | Long delay of symptoms and virus accumulation in dMP plants; increase of symptoms and virus in wt MP plants |

The results of these studies are consistent with an hypothesis that the dMP interferes with one or more of the functions of the MP proteins produced by these viruses during infection by these viruses. While the aa sequences of the MPs of these viruses show little or no homology with the TMV MP the resistance studies support the suggestion that MPs of different viruses share common functions. The implications of these observations *vis a vis* the origin of movement proteins are quite significant, and may indicate that MPs have evolved similar functions through different mechanisms. However, until further analyses of the multiple functions of multiple movement proteins are completed and related to protein structure it will not be possible to substantiate these suggestions. Furthermore, greater knowledge of the role of these and other movement proteins will make it possible to create other mutations in the MP that to provide virus resistance in transgenic plants that is both broad (i.e., provide resistance to many different viruses) and durable.

LITERATURE CITED

Citovsky, V., Knorr, D., Schuster, G., and Zambryski, P. 1990. The P30 movement protein of tobacco mosaic virus is a single-strand nucleic acid binding protein. Cell 60:637-647.

Clark, W. G., Fitchen, J., Nejidat, A., and Beachy, R. N. 1995b. Studies of coat protein-mediated resistance to TMV: II. Challenge by a mutant with altered virion surface does not overcome resistance conferred by TMV CP. J. Gen. Virol. 76:2613-2617.

Clark, W. G., Fitchen, J. H., and Beachy, R. N. 1995a. Studies of coat-protein mediated resistance to TMV using mutant CP: I. The PM2 assembly defective mutant. Virology 208:485-491.

Cooper, B., Lapidot, M., Heick, J. A., Dodds, J. A., and Beachy, R. N. 1995. Multi-virus resistance in transgenic tobacco plants expressing a dysfunctional movement protein of tobacco mosaic virus. Virology 206:307-313.

Dawson, W. O., Bubrick, P., and Grantham, G. L. 1988. Modifications of the Tobacco Mosaic Virus Coat Protein Gene Affecting Replication, Movement, and Symptomatology. Molecular Plant Pathology 78:783-789.

Deom, C. M., Lapidot, M., and Beachy, R. N. 1992. Plant virus movement proteins. Cell 69:221-224.

Fenczik, C. A., Epel, B. L., and Beachy, R. N. 1995. *Role of plasmodesmata and virus movement proteins in spread of plant viruses*. In: Plant Gene Research. Ed(s). pp. (in press). Springer-Verlag, Wien - New York.

Fitchen, J. H., and Beachy, R. N. 1993. Genetically engineered protection against viruses in transgenic plants. Annu. Rev. Microbiol. 47:739-763.

Gafny, R., Lapidot, M., Berna, A., Holt, C. A., Deom, C. M., and Beachy, R. N. 1992. Effects of terminal deletion mutations on function of the movement protein of tobacco mosaic virus. Virology 187:499-507.

Heinlein, M., Epel, B. L., Padgett, H. S., and Beachy, R. N. 1995. Interaction of tobamovirus movement proteins with the plant cytoskeleton. Science 270:1983-1985.

Lapidot, M., Gafny, R., Ding, B., Wolf, S., Lucas, W. J., and Beachy, R. N. 1993. A dysfunctional movement protein of tobacco mosaic virus that partially modifies the plasmodesmata and limits virus spread in transgenic plants. Plant J. 2:959-970.

McLean, B. G., Zupan, J., and Zambryski, P. C. 1995. Tobacco mosaic virus movement protein associates with the cytoskeleton in tobacco cells. Plant Cell 7:2101-2114.

Namba, K., Pattanayek, R., and Stubbs, G. 1989. Visualization of protein-nucleic acid interactions in a virus. Refined structure of intact tobacco mosaic virus at 2.9 A resolution by X-ray fiber diffraction. J. Mol. Biol. 208:307-325.

Powell-Abel, P., Nelson, R. S., De, B., Hoffmann, N., Rogers, S. G., Fraley, R., and Beachy, R. N. 1986. Delay of disease development in transgenic plants that express the tobacco mosaic virus coat protein gene. Science 232:738-743.

Register III, J. C., and Beachy, R. N. 1988. Resistance to TMV in transgenic plants results from interference with an early event in infection. Virology 166:524-532.

Register III, J. C., Powell, P. A., and Beachy, R. N. 1989. *Genetically engineered cross protection against TMV interferes with initial infection and long distance spread of the virus*. In: Molecular Biology of Plant-Pathogen Interactions. Ed(s). pp. 269-281. Alan R. Liss, New York.

Reimann-Philipp, U., and Beachy, R. N. 1993a. Coat protein-mediated resistance in trasgenic tobacco expressing the TMV coat protein from tissue specific promoters. Mol. Plant Microbe Inter. 6:323-330.

Wolf, S., Deom, C. M., Beachy, R. N., and Lucas, W. J. 1989. Movement protein of tobacco mosaic virus modifies plasmodesmatal size exclusion limit. Science 246:377-379.

Wu, X., Beachy, R. N., Wilson, T. M. A., and Shaw, J. G. 1990. Inhibition of uncoating of tobacco mosaic virus particles in protoplasts from transgenic tobacco plants that express the viral coat protein gene. Virology 179:893-895.

# ROLE OF PLASMODESMATA AND HOST FACTORS IN CONTROL OF VIRAL INFECTION

William J. Lucas, Section of Plant Biology, University of California, Davis CA 95616, USA

The complexities associated with the establishment of a systemic viral infection, on a susceptible host plant, are well established. If a virus can replicate in an invaded cell, its next challenge is to establish a local infection front, whereby it moves from cell to cell to eventually enter the long-distance transport system of the phloem. Many viruses are known to encode movement proteins (MPs) that are essential in this cell-to-cell trafficking of infectious vRNA/vDNA, which is thought to occur through plasmodesmata, the intercellular organelles that interconnect the cytoplasm of neighboring plant cells. Studies on the molecular interactions between viral MPs and mesophyll plasmodesmata have provided important insights into the events underlying this process. Furthermore, these studies provided the foundation for the subsequent discovery that plasmodesmata constitute an endogenous macromolecular trafficking pathway. Collectively, these studies have provided an expanded conceptual basis for understanding the role of host factors in potentiating and/or restricting viral infection.

## Plasmodesmata Mediate Endogenous & Viral Protein Transport

Pioneering studies on the 30 kDa MP of tobacco mosaic virus (TMV) established that this MP was essential for cell-to-cell movement of the virus in host tissues (Deom et al. 1987). Microinjection experiments performed on transgenic tobacco plants expressing the TMV MP provided the first direct evidence that a viral-encoded protein had the capacity to interact with plasmodesmata to induce a change in their molecular size exclusion limit (SEL) from 0.8 - 1.0 kDa (values in control tissues) to greater than 10 kDa (Wolf et al. 1989). For direct analysis of the interaction between viral MPs and plasmodesmata, E. coli-expressed MPs were fluorescently labeled (using fluorescein isothiocyanate [FITC]) and then injected into mesophyll cells of host plants (Fujiwara et al., 1993; Noueiry et al., 1994). These experiments established the following requirements for efficient viral transport: the MP must (a) bind to the vRNA/vDNA in a non-sequence-specific manner; (b) be targeted to and traffic through plasmodesmata; (c) interact with plasmodesmata to induce an increase in SEL. Dilation within a plasmodesmal microchannel appears necessary to permit the efficient transport of vRNA/vDNA as a viral nucleoprotein complex (reviewed by Lucas and Gilbertson 1995).

As cell-to-cell transport of FITC-MP occurs almost within the instant that it is injected into a cell, and furthermore, as these FITC-MP studies did not involve viral replication, it is clear that the higher plant plasmodesmata (at least as represented by mesophyll plasmodesmata) must always be competent to engage in transport of macromolecules, either of viral or plant origin. This hypothesis has now gained incontrovertible support from recent studies on plant transcription factors (PTF) known to function in vegetative and floral organ development. Microinjection studies performed on FITC-labeled KNOTTED1, a maize

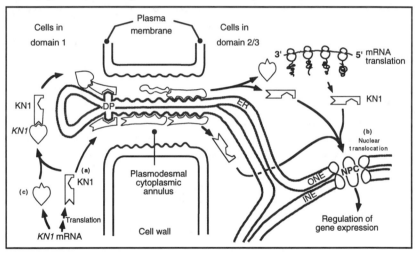

Fig. 1. Plasmodesmata engage in cell-to-cell trafficking of macromolecules. (a) Plant-encoded proteins (e.g. KNOTTED1, a PTF) and MPs contain a motif(s) that allows them to bind to a putative plasmodesmal docking protein (DP). Chaperones may be required to deliver protein to the DP (see also Fig. 3) and to renature it upon its release into the next cell. (b) Presence of a nuclear localization sequence allows the PTF to enter the nucleus via the nuclear pore complex (NPC). Viruses that replicate in the nucleus appear to enter using this same pathway. (c) PTF and viral MPs mediate in RNA/DNA trafficking through the cytoplasmic annulus of the plasmodesma. (From Lucas 1995, with permission)

homeobox transcription factor, revealed that KN1 has the capacity to interact with plasmodesmata to potentiate its rapid movement from cell to cell, and in so doing, KN1 also causes an increase in the plasmodesmal SEL to 40 kDa. Furthermore, KN1 was shown to mediate the cell-to-cell transport of mRNA (Lucas et al., 1995). In contrast to viral MPs, KN1 was found to be selective in this regard, as it appeared to mediate only the transport of its own mRNA. Experimental evidence has now been obtained indicating that higher plants encode for a considerable number of proteins that have the capacity to traffic, via plasmodesmata, within select tissues, organs, or the entire plant (see Mezitt and Lucas 1996). These data support the hypothesis that plasmodesmata establish a supracellular system through which information macromolecules can traffic to exert control over developmental and physiological processes (Lucas 1995). Obviously, it is on this sophisticated network that the plant virus evolved the capacity to 'hitch a ride' (Gilbertson and Lucas 1996).

## Host Factors and Physiological/Developmental Domains

The complex, tissue-specific, controls that orchestrate the transport of such endogenous supracellular control proteins (SCPs; Mezitt and Lucas 1996) may well represent a major barrier (challenge) to MP-mediated transport of a viral nucleoprotein complex. Certainly, it has long been appreciated that, although most plant viruses can gain access to, and replicate in, cells within the shoot (or floral) apex, they are generally excluded from truly meristematic cells. The basis for this exclusion may well reflect the presence of host factors that function to regulate plasmodesmal transport of protein-RNA complexes in order to maintain integrity over a particular developmental domain.

The vascular system of the plant also appears to represent a formidable barrier to an invading virus, commonly restricting it to the inoculated leaf. Similarly, there are other, often insect-borne, virus that appear to be incapable of exiting from the cells of the phloem into the surrounding tissues. Although these viruses have the capacity to replicate in either mesophyll or phloem cells, respectively, they are seemingly incapable of interacting with the plasmodesmata that interconnect the cells located within these physiologically distinct tissues. Thus, viral entry into the phloem appears to involve an interaction with additional host components. Mutant analysis of several viral MPs has indicated that specific motifs are required to permit entry into the companion cell-sieve element complex. The intricate interactions that govern viral entry into the phloem presumably reflect the presence of yet another highly controlled macromolecular trafficking system which operates to control physiological events between the mesophyll and the phloem (see Fig. 2).

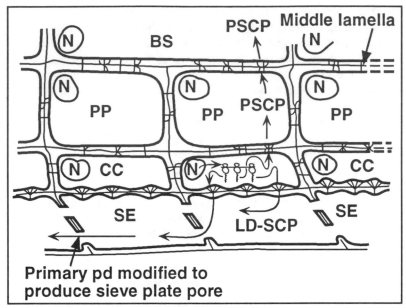

Fig. 2. Role of plasmodesmal trafficking in short- (between CC, PP, BS and MS cells) and long-distance (CC to sink tissues) communication within the plant. Schematic representation of the longitudinal arrangement of the cells within a minor vein in the vicinity of the phloem. Functional sieve elements (SEs) are enucleate and, thus, proteins required for their continued function have to be synthesized within CCs prior to being transported through the special secondary plasmodesmata into the SEs. Other physiological supracellular control proteins (PSCPs) or long-distance SCPs (L-DSCPs; or protein-mRNA complexes) may constitute domain-specific communication networks. Abbreviations are as follows: CC, companion cell; BS, bundle sheath; MS, mesophyll; PP, phloem parenchyma.

Support for the concept that plants use plasmodesmata and PSCPs/L-DSCPs to orchestrate tissue/organ function has been gained from two lines of study. First, we recently performed microinjection experiments with proteins extracted from phloem sap. This study confirmed that the plant encodes a large number of phloem proteins that can increase plasmodesmal SEL and traffic cell to cell. Second, in addition to increasing plasmodesmal SEL, expression of the TMV MP in transgenic tobacco causes pleiotropic effects on the physiology of these plants (see Lucas et al., 1996). During the day, source leaves were found to accumulate sugars and starch, and biomass partitioning into root tissue was reduced, resulting in plants with lowered root-to-shoot ratios. Studies involving grafting, tissue specific promoters, and expression of mutant forms of the MP (Lapidot et al., 1993) identified the mesophyll as the site where the TMV MP likely elicits these changes in plant function. It would appear that the TMV MP competes with host factors, i.e., PSCPs, to alter the endogenous control mechanism, which causes a change in carbon metabolism in the leaf and a shift in biomass partitioning to the root system of these transgenic tobacco plants (Lucas et al., 1996).

Our studies on beet western yellows luteovirus (BWYV) support the hypothesis that viruses can elicit alterations in plant functions by perturbing information exchange between various tissues. Here, although we established that BWYV infection was confined to the CC and PP, the presence of the virus caused major cellular and physiological changes in the mesophyll tissues. Indeed, the virus literature is replete with examples of plant development and physiology being perturbed by infection with 'severe' viral strains. It will be interesting to discover the extent to which such symptoms result from a perturbation to host factors involved in plasmodesmal trafficking of information molecules.

Recent studies using a chimeric MP-green fluorescent protein construct indicated that the plant cytoskeleton may form an integral component of the system that delivers viral (and plant) macromolecules to the plasmodesmal DP (Heinlein et al., 1995; Mclean et al., 1995). Again, if this aspect of MP-host interaction is critical for efficient cell-to-cell movement, minor mutations in a MP could act to enhance, or completely eliminate, the infection process.

Here we have focussed on the manner in which viral MPs have evolved to utilize host constituents ('factors') to potentiate cell-to-cell transport of viral RNA/DNA. There is now an emerging body of evidence that entry into the long-distance pathway of the phloem involves interactions between additional host factors and motifs on the coat protein (Gilbertson and Lucas 1996). Involvement of direct interaction between the coat protein and plasmodesmata is under current investigation.

The above-mentioned studies provide a novel perspective on the complex plasmodesmal 'machinery' in operation within the plant. The number of plant-encoded proteins involved in creating the capacity for cell-to-cell transport of macromolecules, in addition to those associated with regulation, is obviously quite high. Potentially, all of these proteins could function as host factors to either facilitate, or restrict, the spread of viral infectious material. The challenge before us is the identification of the components of this novel supracellular communication system.

Finally, development of an understanding of the complex controls established at the level of plasmodesmal macromolecular trafficking will likely provide new avenues for biotechnology, in terms of the creation of novel resistance mechanisms against the establishment of systemic viral infection.

LITERATURE CITED

Deom, C. M., Oliver, M. J., and Beachy, R. N. 1987. The 30-kilodalton gene product of tobacco mosaic virus potentiates virus movement. Science 237:389-394.

Fujiwara, T., Giesman-Cookmeyer, D., Ding, B., Lommel, S. A., and Lucas, W. J. 1993. Cell-to-cell trafficking of macromolecules through plasmodesmata, potentiated by the red clover necrotic mosaic virus movement protein. Plant Cell 5:1783-94.

Gilbertson, R. L., and Lucas, W. J. 1996. How do viruses traffic on the 'vascular highway'? Trends in Plant Sci. 1 (In Press)

Heinlein, M., Epel, B. L., Padgett, H. S., and Beachy, R. N. 1995. Interaction of tobamovirus movement proteins with the plant cytoskeleton. Science 270:1983-1985.

Lapidot, M., Gafny, R., Ding, B., Wolf, S., Lucas, W. J., and Beachy, R. N. 1993. A dysfunctional movement protein of tobacco mosaic virus that partially modifies the plasmodesmata and limits virus spread in transgenic plants. Plant J. 4:959-970.

Lucas, W. J., and Gilbertson, R. L. 1994. Plasmodesmata in relation to viral movement within leaf tissue. Annu. Rev. Phytopathol. 32:387-411.

Lucas, W.J. 1995. Plasmodesmata: intercellular channels for macromolecular transport in plants. Curr. Opin. Cell Biol. 7:673-680.

Lucas, W. J., Bouché-Pillon, S., Jackson, D. P., Nguyen, L., Baker, L., Ding, B., and Hake, S. 1995. Selective trafficking of KNOTTED-1 homeodomain protein and its mRNA through plasmodesmata. Science 270:1980-1983.

Lucas, W. J., Balachandran, S., Park, J., and Wolf, S. 1996. Plasmodesmal companion cell-mesophyll communication in the control over carbon metabolism and phloem transport: insights gained from viral movement proteins. J Exp Bot (In Press)

McLean, B. G., Zupan, J., and Zambryski, P.C. 1995. Tobacco mosaic virus movement protein associates with the cytoskeleton in tobacco cells. Plant Cell 7:2101-2114.

Mezitt, L. A., and Lucas, W. J. 1996. Plasmodesmal cell-to-cell transport of proteins and nucleic acids. Plant Mol. Biol. (In Press)

Noueiry, A. O., Lucas, W. J., and Gilbertson, R. L. 1994. Two proteins of a plant DNA virus coordinate nuclear and plasmodesmal transport. Cell 76:925-932.

Wolf, S., Deom, C. M., Beachy, R. N., and Lucas, W. J. 1989. Movement protein of tobacco mosaic virus modifies plasmodesmal size exclusion limit. Science 246:377-379.

# Elicitor Functions of Tobamovirus Coat Proteins in *Nicotiana sylvestris*

Zenobia Taraporewala* and James N. Culver*‡

‡Center for Agricultural Biotechnology, University of Maryland Biotechnology Institute, College Park, MD 20742
*Molecular and Cell Biology Program, University of Maryland, College Park, MD 20742

In plants, the hypersensitive response (HR) is a widely occurring active defense mechanism directed against bacteria, fungi, nematodes, and viruses. HR induction involves a complex interplay between plant and pathogen factors and leads to a cascade of biochemical events that cause localized cell death in the area of pathogen invasion. The inducible nature of the HR is largely dependent upon the specific recognition of the invading pathogen. Recognition is typically the result of a gene-for-gene interaction involving a plant resistance (*R*) gene(s) and a pathogen avirulence (*avr*) gene(s) (Flor 1971).

Recognition between plant *R* genes and pathogen elicitors (the direct or indirect products of *avr* genes) has been proposed to occur via a receptor-ligand mechanism (Keen 1992). Support for this model has come from the recent cloning and characterization of several plant *R* genes that encode putative products with conserved leucine rich regions (LRR) and nucleotide-binding domains (Staskawicz et al. 1995), features that suggest a function in protein-protein interaction and signal transduction. In contrast, a variety of pathogen molecules, including proteins, polysaccharides, and low molecular weight compounds, have been shown to act as elicitors (Keen and Dawson 1992). This suggests flexibility within the mechanisms of host recognition. Understanding these mechanisms will require detailed knowledge of the structure of the host and pathogen determinants involved.

Because the three dimensional structure of the tobacco mosaic tobamovirus (TMV) coat protein (CP) has been determined (Bloomer et al. 1978) and it's function as the elicitor of the *N'* gene in *Nicotiana sylvestris* identified (Culver and Dawson 1991), this system is ideal for detailed structure-function analysis of an elicitor molecule. Another unique feature of this system is that the sequences and structures of several tobamovirus CPs are known. Sequence comparisons reveal only 16% identity and 35 % similarity, however, all tobamovirus CPs share a well-preserved three-dimensional fold (Altschuh et al. 1987). This makes the $N'$ gene - TMV CP system ideally suited for studying the relationship/interaction between a plant $R$ gene and related pathogen elicitors. Additionally, the function of the TMV CP is well understood, thus, analysis of it's elicitor function provides an opportunity to investigate the strategies adapted by pathogens to overcome plant disease resistance.

Studies by several groups have conclusively demonstrated the function of the TMV CP as the elicitor of the $N'$ gene HR (Culver and Dawson 1991). The overall focus of this study is to identify and compare tobamovirus CP structural determinants involved in $N'$ gene recognition. The structure of the TMV CP has been resolved to 2.8 Å for a 20S disk aggregate and 2.9 Å for the TMV virion (Bloomer 1978; Namba et al. 1989). The central core of the 17 kDa CP monomer consists of helical bundle composed of four alpha-helices arranged in an anti-parallel fashion. These helices are connected at one end by either a short or long loop that contain the residues involved in RNA binding. A short ß-sheet region braces the opposite end of the helical bundle. The N and C terminus of the CP are on the outer surface of the virus (Fig. 1).

The CP of TMV strain U1 is normally not an elicitor of the $N'$ gene HR. However, specific amino acid substitutions made within the U1 CP can result in HR elicitation (Culver et al. 1994). These substitution mutants exhibited different HR phenotypes ranging from strong, moderate, and weak (necrotic lesions observed at 2-3 days, 4-5 days and 6-7 days post-inoculation, respectively). Within the CP structure these substitutions localize to and would predictably interfere with the interface regions between adjacent CP subunits. This suggests that disruption of normal subunit-to-subunit interactions between U1 CPs results in $N'$ gene recognition.

To investigate the effects of these substitutions on inter-subunit interactions mutant CPs were subjected to *in vitro* aggregation analysis followed by electron microscopy to visualize aggregate forms. Results demonstrated a strong correlation between the strength of the HR and the degree of quaternary structure destabilization. CPs that elicited a strong HR exhibited reduced abilities to form large aggregates. However, not all substitutions designed to destabilize quaternary structure allowed for $N'$ gene recognition. CPs with radical changes in residues important for maintaining tertiary structure could neither assemble *in vitro* into disk or helical aggregates nor elicit the HR. Thus, elicitor conformation in mutants of the U1 CP is the result of interference in quaternary structure but not in the tertiary structure. This implies that folding of the CP into its three dimensional form is critical for host recognition. It was, therefore, hypothesized that aggregate disruption of the U1 CP exposes a structural site "elicitor active site" normally buried within CP quaternary structure.

The next step in the structure-function analysis of the CP was to identify the structural regions that make up the elicitor active site (Taraporewala and Culver 1996). Mapping experiments were performed by adding specific amino acid substitutions to the strong elicitor U1 CP P20L.

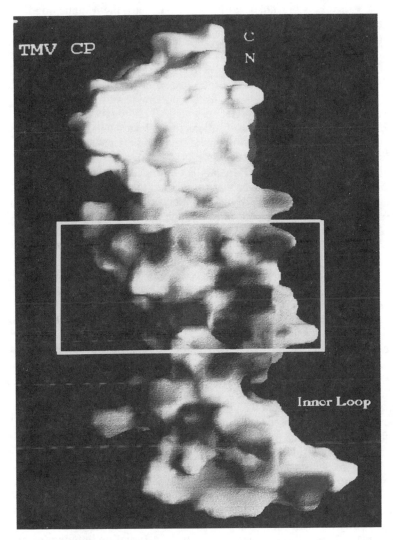

Figure 1. Surface profile of a single coat protein subunit. Box outlines the surface area identified by mapping experiments as important for the elicitation of the $N'$ gene HR.

P20L CP elicits a stable $N'$ gene HR, between 25°C and 35°C, indicating that the molecular features important for HR elicitation are maintained over a broad temperature range. Substitutions were designed to modify various structural features of the P20L CP and were tested for HR elicitation at 25°C and 29°C. Stringency imposed by higher temperatures in this assay helped to discern substitutions that affected the HR greatly from those that had less of an effect or no effect on elicitation.

A total of 32 residues within the N-terminus, C-terminus, long RNA binding loop, ß-sheet, and the four helices were targeted for single-site substitution in the P20L CP. Two CP deletion mutants were also made; 14 residues from the N-terminus were deleted in one mutant (ΔN) and 11 amino acids from the C-terminus were deleted in another mutant (ΔC). Of the 32 substitution mutants, four exhibited total loss of elicitor function (no HR at 25°C), 10 displayed a temperature-sensitive phenotype (no HR at 29°C) whereas 18 had no affect on the HR. ΔN and ΔC CP mutants exhibited total loss of elicitor function. Substitutions that affected the HR were placed into two categories based on the predicted structural effects to the P20L CP. The first category included substitutions that affected surface residues and did not predictably interfere with CP tertiary structure. The second category included substitutions of residues that contributed to structural elements required for the correct folding of the CP. Included in this second category were substitutions of three buried residues, one residue involved in an intra-subunit salt bridge, two residues along the short loop, and the ΔN and ΔC deletion mutants.

The majority of substitutions that interfered with elicitation of the HR were noncontiguous in position and mapped to the right face of the CP's helical bundle (Fig. 1; Taraporewala and Culver 1996). It was therefore proposed that these residues contributed to structural elements important for HR elicitation and together make up an elicitor active site. To examine this and the possibility that residues that map to this site act in concert to facilitate HR induction, two temperature-sensitive substitutions were combined within the same P20L CP. The combination of two 29 °C temperature sensitive substitutions produced an additive effect on HR elicitation, reducing the temperature-sensitive phenotype to 27 °C. This suggests that residues within the elicitor active site contribute independently to elicitor function. Subsequently, studies of the surface and chemical characteristics of this CP region revealed the possible involvement of approximately 25 residues that are 50% non polar, 20% charged and 30% polar in nature, with a total surface area of ≈ 600 Å. This finding is consistent with other known recognition surfaces. Taken together, the data suggests that the elicitor active site possibly delineates a receptor binding site and supports a receptor-ligand model for the *N'* gene - TMV CP interaction.

Within the structure of the assembled virion, the elicitor active site would normally be buried within the subunit interface. Since the wild-type U1 CP does not normally elicit the *N'* gene HR, it is likely that it has evolved a quaternary structure that serves to mask the elicitor active site from host recognition. This explanation is further supported by the observation that substitutions causing U1 CP to elicit the HR, interfere with quaternary structure and lie outside the elicitor active site. Interestingly, we have recently found that most other tobamovirus strains elicit the *N'* gene HR. Our studies show that the CPs of four tobamoviruses (orchid, cucumber, mild-tobacco and ribgrass strains) function as elicitors, comparable in strength and temperature stability to the HR elicited by mutants of the TMV U1 CP. Since tobamovirus CPs exhibit a highly conserved tertiary structure, we have hypothesized that the overall three-dimensional fold of tobamovirus CPs serves as a structural platform for the presentation of the elicitor active site required for *N'* gene recognition.

Based on the premise that identical alterations in structurally similar regions between tobamovirus CPs would produce the same effect on HR induction, mutations were made in the CPs of other tobamovirus. Residues in ORSV (orchid strain), CGMMV (cucumber strain) and U2 (mild-tobacco strain) CPs that map to the elicitor active site and/or are important for maintaining the tertiary fold were targeted for substitution. HR elicitation was tested at 25°C and 29 °C, similar to earlier mapping experiments. These experiments revealed an interesting pattern of effects between the abilities of the different tobamovirus CPs to elicit the *N'* gene HR. In general, amino acid substitutions corresponding to those that interfered with U1 CP elicitation also interfered with elicitation by the other tobamovirus CPs. This indicates that structural features along the right side of the different CP helical bundles are conserved enough to function in *N'* gene recognition, even between CPs that share only 36% sequence identity. However, a one-to-one correspondence between the position of given substitution and its ability to inhibit the induction of the HR was not always observed among the different tobamovirus CPs. These differences in HR phenotypes are possibly due to naturally occurring compensatory changes in tobamoviruses that have evolved in different host systems. Thus, the structure and function of a given residue in one strain is replaced by a different although related residue in another strain. A finding that suggests versatility in the recognition process.

This study indicates that *N'* gene host recognition is targeted towards features common to all tobamovirus CPs. From a functional point of view, this finding implies that host derived molecules involved in pathogen surveillance can exhibit a certain degree of flexibility in elicitor recognition. This allows for the recognition of elicitors that have little similarity in their primary sequences but have conserved structural features. It has been suggested that LRR domains in *R* genes are involved in defining recognitional specificity of the pathogen (Staskawicz et al. 1995). Additionally, the recent structural characterization of an LRR domain has demonstrated it's flexibility in protein-protein interactions (Shapiro et al. 1995). Thus, a single LRR domain may have the capability to interact with a range of structurally similar ligands. The ability of structurally conserved CPs to act as *N'* gene elicitors is consistent with this kind of an interaction. Future studies directed at the molecular identification and characterization of the *N'* gene should provide the insight needed to confirm this possibility.

Literature Cited

Altschuh, D., Lesk, A. M., Bloomer, A. C., and Klug, A. 1987. Correlation of co-ordinated amino acid substitutions with function in virus related to tobacco mosaic virus. J. Mol. Biol. 193: 693-707.

Bloomer, A. C., Champess, J. N., Bricogne, G., Staden, R., and Klug, A. 1978. Protein disk of tobacco mosaic virus at 2.8 Å resolution showing the interactions within and between subunits. Nature 276: 362-368.

Culver, J. N., and Dawson, W. O. 1991. Tobacco mosaic virus elicitor CP genes produce a hypersensitive phenotype in transgenic Nicotiana sylvestris plants. Mol. Plant-Microbe Interact. 4: 458-463.

Culver, J. N., Stubbs, G., and Dawson, W. O. 1994. Structure-function relationship between tobacco mosaic virus coat protein and hypersenstivity in *Nicotiana sylvestris*. J. Mol. Biol. 242: 130-138.

Flor, H. H. 1971. Current status of the gene-for-gene concept. Ann.Rev. Phytopathol. 9: 275-296.

Keen, N. T. 1992. The molecular biology of disease resistance. Plant. Mol. Biol. 19: 109-122.

Keen, N. T., and Dawson, W. O. 1992. Pathogen avirulence genes and elicitors of plant defense. Pages 85-106 in: Plant Gene Research: Genes involved in Plant Defense. T. Boller and F. Meins, ed. Springer-Verlag, New York.

Namba, K., Pattanayek, R., and Stubbs, G. 1989. Visualization of protein-nucleic acid interactions in a virus, refined structure of intact tobacco mosaic virus at 2.9 Å resolution by X-ray fibre diffraction. J. Mol. Biol. 208: 307-325.

Shapiro, R., Riordan, J. F., and Vallee, B. L. 1995. LRRning the RIte of springs. Nature Structural Biology 2 (5): 350-354.

Staskawicz, B. J., Ausubel, F. M., Baker, B. J., Ellis, J. G., and Jones, J.D.G. 1995. Molecular genetics of plant disease resistance. Science 268: 661-667.

Taraporewala, Z. F., and Culver, J. N. 1996. Identification of an elicitor active site within the three-dimensional structure of the tobacco mosaic tobamovirus coat protein. The Plant Cell 8: 169-178.

# Virus-Host Interactions in Southern Bean Mosaic Virus Gene Expression and Assembly

David L. Hacker and Kailayapilla Sivakumaran

Department of Microbiology and Center for Legume Research, University of Tennessee, Knoxville, TN 37996-0845 USA

The cowpea strain of southern bean mosaic virus (SBMV-C) is a positive-sense RNA virus that systemically infects most cultivars of *Vigna unguiculata* (cowpea). The single-component viral RNA is encapsidated in a T=3 icosahedral particle composed of 180 subunits of the 30 kDa coat protein (Abad-Zapatero et al. 1980). The viral RNA is 4,194 nucleotides (nt) in length and is covalently linked at its 5' end to a 12 kDa viral protein (VPg) (Ghosh et al. 1979). The 3' end of the RNA is neither polyadenylated nor folded into a tRNA-like structure (Ghosh et al. 1979; Wu et al. 1987). During infection a viral subgenomic RNA is synthesized that is 953 nt in length and serves as the mRNA for the coat protein. It is coterminal with the 3' end of the viral RNA and is covalently bound at its 5' end to VPg (Ghosh et al. 1981).

The viral RNA contains four open reading frames (ORF) (Wu et al. 1987). To date, little is known about the roles of the viral proteins in the SBMV-C replication cycle or in viral pathogenesis. The function of the 21 kDa protein encoded by ORF1 (nt 49-603) is unknown. ORF2 (nt 570-3437) encodes a polyprotein of 105 kDa that is proteolytically processed into three proteins including VPg, a serine protease, and an RNA-dependent RNA polymerase (Wu et al. 1987). The function of the ORF3 (nt 1895-2380) protein is not known, but mutations which eliminate the expression of this protein abolish the systemic infection of SBMV-C in cowpea (Sivakumaran and Hacker, unpublished). Lastly, ORF4 (nt 3217-4053) encodes the coat protein.

Several aspects of SBMV-C make it an interesting virus for the study of interactions with its host: i) the structure of the virus is known to atomic resolution (Abad-Zapatero et al. 1980), ii) a biologically acitve full-length cDNA clone of the viral genome has been constructed (Sivakumaran and Hacker, unpublished), iii) the virus has a relatively simple genome structure, and iv) several strains of the virus or closely related viruses have been identified which differ in host range and pathogenicity (Hull, 1988). Research in this laboratory has focused on SBMV-C assembly, gene expression, and protein function. Undoubtedly, all of these viral activities will eventually be shown to involve host proteins or subcellular components. Two examples of these interactions are described in this paper.

## SBMV-C Gene Expression

Unlike host cell mRNAs, the SBMV-C RNAs are not capped but are covalently bound to VPg. These viral RNAs, therefore, may have different requirements for translation as compared to host mRNAs. This has proven to be the case for translation of SBMV-C genomic RNA in wheat germ extract. It has been demonstrated that translation of ORF1 and ORF2 is much less sensitive to the presence of methylated cap analogue than is the capped control RNA from brome mosaic virus (BMV) (Sivakumaran and Hacker, unpublished data). This experiment demonstrated a reduced requirement for translation initiation factors by the SBMV-C RNA relative to a capped RNA, but it did not address the mechanism by which the small ribosomal subunit engaged the RNA for the initiation of translation of ORF1 and ORF2. To distinquish between two possible mechanisms for ORF1 translation initiation, 5' end-dependent ribosomal scanning or internal ribosome binding, site-directed mutagenesis of the SBMV-c cDNA clone was used to produce AUG codons at either of two positions within the 5' untranslated region (UTR) of the genomic RNA. Addition of a single AUG codon reduced ORF1 translation greater than 80% but did not affect ORF2 translation (Sivakumaran and Hacker, unpublished data). This result indicated that the small ribosomal subunit enters the genomic RNA near its 5' end and scanns to the ORF1 AUG to initiate translation of this gene. One reason for the reduced requirement for initiation factors for ORF1 translation may be a lack of RNA secondary structure within the SBMV-C 5' UTR which is 68% A+U.

The ORF2 initiation codon is the fourth one from the 5' end of the SBMV-C genomic RNA, making this gene a candidate for expression via internal ribosome binding rather than 5' end-dependent ribosome scanning. In support of this hypothesis, it has been demonstrated in wheat germ extract that ORF2 expression is not affected by the presence of AUG codons within the 5' UTR. Furthermore, addition of an AUG at position 231 within ORF1 did not affect ORF2 translation, and a mutant with two additional initiation codons 5' to the ORF2 AUG produced near-normal levels of ORF2 protein (Sivakumaran and fHacker, unpublished data). These results provided experimental support for the internal ribosome entry model of ORF2 translation. Additional evidence was obtained by *in vitro* translation of an RNA in which the SBMV-C 5'UTR as well as ORFs 1 and 2 were located 3' to the BMV 3a gene. Translation of the BMV 3a ORF and SBMV-C ORF2 but not ORF1 was observed (Sivakumaran and fHacker, unpublished data). These results supported the conclusions that ORF1 was translated by scanning ribosomes and ORF2 by internal ribosomal entry.

What do these translation studies demonstrate about interactions between the virus and the host in the process of viral gene expression? For one thing, there appears to be a reduced requirement for host initiation factors for the translation of ORF1, ORF2, and ORF4. In addition, based on analogy with the animal viruses which are translated by internal ribosomal entry, there may be several host proteins which are required for the internal ribosomal entry of ORF2, and these may be proteins which are not normally involved in the translation of host cell mRNA.

### SBMV-C Assembly

SBMV-C systemically infects *V. unguiculata* but not *Phaseolus vulgaris* (common bean) (Hull, 1988). It has been demonstrated that SBMV-C RNA synthesis does occur in bean protoplasts but that the virus does not move locally in bean. Sunn-hemp mosaic tobamovirus (SHMV) is able to complement the cell-to-cell movement defect of SBMV-C in bean (Fuentes and Hamilton, 1991). In this coinfection of bean, T=1 but not T=3 icosahedral particles of SBMV-C coat protein were observed in the inoculated leaves but not in the systemic leaves (Fuentes and Hamilton, 1993). One interpretation of these results is that the SBMV-C coat protein does not function in the encapsidation of viral RNA in bean and that SBMV-C assembly is required for systemic movement of this virus.

To further define the interactions of SBMV-C with the nonpermissive host, beans were coinfected with SBMV-C and the bean strain of southern bean mosaic virus (SBMV-B). In contrast to SBMV-C, SBMV-B systemically infects bean but not cowpea. Although these are currently classified as strains of a single virus, they are apparently distinct viruses based on the comparative analysis of the SBMV-B and SBMV-C sequences (Othman and Hull, 1995). Following coinfection of bean, virus was recovered from systemic leaves. Only SBMV-B coat protein was present in this virus, but it was able to cause a systemic SBMV-C infection of cowpea (Hacker, unpublished data). This indicated that

SBMV-C systemically infected bean in the presence of SBMV-B and that SBMV-B proteins were able to complement in *trans* the cell-to-cell and long distance movement defects of SBMV-C in bean. Furthermore, SBMV-B coat protein was able to encapsidate SBMV-C RNA in bean. The SBMV-C infection of bean is one of a few examples of a host-specific defect in virus assembly, and it may provide an excellent system for studying host interactions in spherical virus assembly.

## Literature Cited

Abad-Zapatero, C., Abdel-Meguid, S. S., Johnson, J. E., Leslie, A. G. W., Rayment, I., Rossmann, M. G., Suck, D., and Tsukihara, T. 1980. Structure of southern bean mosaic virus at 2.8 Å resolution. Nature 286:33-39.

Fuentes, A. L., and Hamilton, R. I. 1991. Sunn-hemp mosaic virus facilitates cell-to-cell spread of southern bean mosaic virus in a nonpermissive host. Phytopath. 81:1302-1305.

Fuentes, A. L., and Hamilton, R. I. 1993. Failure of long-distance movement of southern bean mosaic virus in a resistant host is correlated with lack of normal virion formation. J. Gen. Virol. 74:1903-1910.

Ghosh, A., Dasgupta, R., Salerno-Rife, T., Rutgers, T., and Kaesberg, P. 1979. Southern bean mosaic virus has a 5' linked protein but lacks a 3' terminal poly(A). Nucleic Acids Res. 7:2137-2146.

Ghosh, A., Rutgers, T., Ke-Qiang, M., and Kaesberg, P. 1981. Characterization of the coat protein mRNA of southern bean mosaic virus and its relationship to the genomic RNA. J. Virol. 39:87-92.

Hull, R. 1988. The sobemovirus group. Pages 113-146 in :The Plant Viruses. Vol. 3. R. Koenig, ed. Plenum Press, New York and London.

Othman, Y., and Hull, R. 1995. Nucleotide sequence of the bean strain of southern bean mosaic virus. Virology 206:287-297.

Wu, S., Rinehart, C., and Kaesberg, P. 1987. Sequence and organization of southern bean mosaic virus genomic RNA. Virology 161:73-80.

# Interactive Roles of Viral Proteins, Viral RNA, and Host Factors in Bromovirus RNA Replication

P. Ahlquist, J. Díez, M. Ishikawa, M. Janda, A. Noueiry, B. D. Price, M. Restrepo-Hartwig, and M. Sullivan.

Institute for Molecular Virology and Department of Plant Pathology
University of Wisconsin, Madison, WI 53706, USA

The majority of viruses infecting plants are positive-strand RNA viruses, i.e., viruses that encapsidate their genomes as single-stranded, messenger-sense RNA and replicate that RNA solely through RNA intermediates, without any natural DNA stage. Achieving a detailed understanding of the mechanisms of RNA replication in these viruses has become a significant goal in virus-host interaction studies for several reasons. Among these are that the nature of RNA replication as a central, infection-specific process in the virus life cycle makes it an attractive target for antiviral strategies, that host specificity determinants for some virus-host combinations map to RNA replication genes (Allison et al., 1988; DeJong and Ahlquist, 1995), and that such RNA viruses and their derivatives can be useful gene expression vectors (French et al., 1986).

Fortunately, despite the large number of different positive-strand RNA viruses and their considerable variation in virion particle morphology and genetic organization, sequence comparisons show that such viruses can be grouped into a small number of "superfamilies," each defined by significant conservation of a set of genes or gene modules involved in RNA replication. One such superfamily is the "alphavirus-like superfamily," which includes a variety of icosahedral, bacilliform and rod-shaped viruses of plants and the animal alphaviruses (Ahlquist et al., 1985; Goldbach, 1991). This superfamily is distinguished by the presence in the genomes of all members of gene segments encoding three protein modules or domains: a polymerase-like module, a helicase-like module, and a module implicated in m7G methyltransferase and possibly guanylyltransferase activities for capping viral RNAs.

# BMV as a Model System for RNA Replication Studies

## BMV-ENCODED RNA REPLICATION FACTORS 1A AND 2A

Brome mosaic virus (BMV) is a representative member of the alphavirus-like superfamily of positive strand RNA viruses and has been used as a model system for fundamental studies of RNA replication (Ahlquist, 1992). Because of the conservation of replication functions within the alphavirus-like superfamily and suggestive evidence that some principles of positive strand RNA virus replication are shared across superfamilies, results on BMV RNA replication should have significance for a wide range of viruses of plants and animals.

BMV encodes two large and apparently multifunctional proteins, 1a and 2a, which are both required and together sufficient to induce full viral RNA replication and transcription of the viral coat protein mRNA in infected cells (French et al., 1986). The N- and C-terminal halves of the 1a protein (109 kDa) contain the putative RNA capping and RNA helicase domains, respectively, conserved among all alphavirus-like viruses, while the conserved polymerase-like domain resides in the central portion of the 2a protein (94 kDa). Site-directed deletion and mutagenesis studies show that all three conserved domains in 1a and 2a are required for RNA replication (Kroner et al., 1990; Traynor et al., 1991). 1a controls at least some aspects of template specificity (Traynor and Ahlquist, 1990) and host specificity (DeJong and Ahlquist, 1995), implying direct or indirect interactions with RNA templates and host factors.

In vitro, the helicase-like domain of 1a interacts with the N-terminal 2a segment preceding the polymerase-like domain, allowing 1a and 2a to form a complex (Kao et al., 1992; Kao and Ahlquist, 1992). Genetic studies show that compatibility between 1a and 2a is essential for RNA replication in vivo, and that synthesis of positive strand genomic RNA and subgenomic mRNA require some aspect(s) of 1a-2a interaction that are distinct from any requirements for synthesis of the negative strand RNA replication intermediate (Dinant et al., 1993). Since positive but not negative strand RNAs are capped, the relevant features of 1a-2a interaction may be related to proper functioning of the putative capping domain of 1a.

1a and/or 2a also interact with host membranes and with host proteins, as the BMV RNA replication complex fractionates from infected cells as a membrane-bound complex of 1a, 2a, and incompletely characterized host proteins (Quadt et al., 1993). Interestingly, it has recently been found that certain BMV RNA sequences have a non-template role in assembling the functional BMV polymerase complex in vivo (Quadt et al., 1995). The non-template role of these RNA sequences conceivably might be to recruit essential host factor(s) into the complex, to somehow activate 1a and 2a, or to directly contribute a non-template function.

CIS-ACTING REPLICATION AND TRANSCRIPTION SIGNALS

The cis-acting signals that direct and regulate BMV RNA replication have been mapped (French and Ahlquist, 1987). Recent work shows that initiation of BMV negative strand synthesis involves interaction of BMV RNA-dependent RNA polymerase with two widely separated cis-acting regions on the positive strand BMV RNA template, one at the 3′-terminal negative strand initiation site and another approximately 1 kb away (Quadt et al., 1995). Since the interior element is upstream of the transcription start site for the subgenomic mRNA for BMV coat protein, this arrangement may allow the virus to distinguish, for replication purposes, between genomic RNA and the subgenomic mRNA. An additional cis-acting region near the 5′ end of the positive strand is required for full RNA replication, including positive strand RNA synthesis. Intriguingly, these replication signals contain multiple tRNA-like sequence motifs, and the 3′ terminal sequences in particular are known to interact with multiple tRNA-specific cellular enzymes, suggesting that the virus may use host factors that normally interact with tRNAs or their genes to facilitate or regulate RNA synthesis (Ahlquist, 1992).

The cis-acting signals that direct initiation of the subgenomic coat protein mRNA have also been mapped and found to include distinct elements specifying the site of initiation and the level of transcription (French and Ahlquist, 1988). Positive, negative and subgenomic RNA synthesis are all differentially regulated and can be altered by cis- or trans-acting mutations. In particular, negative strand RNA accumulation plateaus 6-8 hours post infection, while positive strand RNA synthesis continues unabated beyond 24 hours post infection, yielding an approximately 100-fold excess of positive over negative strand RNA by later stages of infection (Kroner et al., 1990).


## BMV RNA Replication in Yeast

As noted above in part, a variety of findings, including results on the trans-acting BMV RNA-dependent RNA polymerase and the cis-acting BMV RNA replication signals, all imply that host factors play important if not crucial roles in BMV RNA replication. Similar evidence exists for host factor involvement in the replication of some other positive-strand RNA viruses. Unfortunately, until recently little or nothing has been definitively established about such host factors, in large part due to present limitations in the genetic analysis of higher eukaryotes. While promising studies are in progress to use *Arabidopsis thaliana* to identify host factors involved in the replication of some viruses, even more rapid and powerful genetic techniques are available for the yeast *Saccharomyces cerevisiae*.

To utilize the genetic potential of yeast for analysis of virus-host interactions and the nature and roles of host factors in viral RNA replication, we have demonstrated that BMV RNA derivatives can be replicated and transcribed in yeast expressing the BMV 1a and 2a RNA replication factors (Janda and Ahlquist, 1993; Quadt et al., 1995). This replication and transcription of BMV-derived RNAs in yeast appears to closely follow the known features of BMV replication in plant cells. For example, BMV RNA replication in yeast depends on both 1a and 2a in trans, depends on the usual replication and transcripton signals in cis, and yields an over 100-fold excess of positive to negative strand RNA.

BMV RNAs can not only be replicated in yeast but also can be engineered to serve as effective yeast expression vectors for foreign genes. Moreover, such engineered BMV RNA replicons can be maintained in yeast by passage to daughter cells at cell division. In this way, persistent BMV RNA replicons expressing screenable or selectable marker genes can confer heritable phenotypic changes on yeast. BMV-directed expression of the wild type yeast *URA3* uracil biosynthesis gene, e.g., converts *ura3*- yeast to uracil-independent growth (Janda and Ahlquist, 1993). Such BMV-dependent phenotypes can be used as the basis for genetic analysis of the processes of viral RNA replication and mRNA transcription.

These findings allow the facile approaches of yeast genetics to be applied to the identification and characterization of yeast genes required for BMV RNA replication and transcription. Studies pursuing the identification of such yeast genes are in progress and *a priori* considerations and some preliminary results suggest that BMV RNA replication likely is dependent on many host factors that collectively affect many different steps in viral replication. Accordingly, it seems likely that identification of these genes will have dramatic effects on our understanding of viral RNA replication and virus-host interactions.

ACKNOWLEDGMENTS

This research was supported by the National Institutes of Health under Public Health Service grants GM51301 and GM35072.

LITERATURE CITED

Ahlquist, P., Strauss, E., Rice, C., Strauss, J., Haseloff, J. and Zimmern, D. 1985. Sindbis virus proteins nsP1 and nsP2 contain homology to nonstructural proteins from several RNA plant viruses. J. Virol. 53: 536-542.

Ahlquist, P. 1992. Bromovirus RNA replication and transcription. Current Opinion in Genetics and Development 2:71-76.

Allison, R. F., Janda, M. and Ahlquist, P. 1988. Infectious *in vitro* transcripts from cowpea chlorotic mottle virus cDNA clones and exchange of individual RNA components with brome mosaic virus. J. Virol. 62: 3581-3588.

De Jong, W. and Ahlquist, P. 1995. Host-specific alterations in viral RNA accumulation and infection spread in a brome mosaic virus isolate with an expanded host range. J. Virol. 69:1485-1492.

Dinant, M., Janda, M., Kroner, P. and Ahlquist, P. 1993. Bromovirus RNA replication and transcription require compatibility between the polymerase- and helicase-like viral RNA synthesis proteins. J. Virol. 67:7181-7189.

French, R., Janda, M. and Ahlquist, P. 1986. Bacterial gene inserted in an engineered RNA virus: Efficient expression in monocotyledonous plant cells. Science 231: 1294-1297.

French, R. and Ahlquist, P. 1987. Intercistronic as well as terminal sequences are required for efficient amplification of brome mosaic virus RNA3. J. Virol. 61: 1457-1465.

French, R. and Ahlquist, P. 1988. Characterization and engineering of sequences controlling in vivo synthesis of brome mosaic virus subgenomic RNA. J. Virol. 62: 2411-2420.

Goldbach, R., LeGall, O., and Wellink, J. 1991. Alpha-like viruses in plants. Sem. in Virol. 2:19-25.

Janda, M. and Ahlquist, P. 1993. RNA-dependent replication, transcription, and persistence of brome mosaic virus RNA replicons in S. cerevisiae. Cell 72:961-970.

Kao, C., Quadt, R., Hershberger, R. and Ahlquist, P. 1992. Brome mosaic virus RNA replication proteins 1a and 2a form a complex in vitro. J. Virol. 66:6322-6329.

Kao, C. and Ahlquist, P. 1992. Identification of the domains required for direct interaction of the helicase-like and polymerase-like RNA replication proteins of brome mosaic virus. J. Virol. 66:7293-7302.

Kroner, P. A., Young, B. M. and Ahlquist, P. 1990. Analysis of the role of brome mosaic virus 1a protein domains in RNA replication, using linker insertion mutagenesis. J. Virol. 64: 6110-6120.

Quadt, R., Kao, C., Browning, K., Hershberger, R. and Ahlquist, P. 1993. Characterization of a host protein associated with brome mosaic virus RNA polymerase. Proc. Natl. Acad. Sci. USA 90:1498-1502.

Quadt, R., Ishikawa, M., Janda, M. and Ahlquist, P. 1995. Formation of brome mosaic virus RNA-dependent RNA polymerase in yeast requires co-expression of viral proteins and viral RNA. Proc. Natl. Acad. Sci. USA 92:4892-4896.

Traynor, P. and Ahlquist, P. 1990. Use of bromovirus RNA2 hybrids to map cis- and trans-acting functions in a conserved RNA replication gene. J. Virol. 64: 69-77.

Traynor, P., Young, B. M. and Ahlquist, P. 1991. Deletion analysis of brome mosaic virus 2a protein: Effects on RNA replication and systemic spread. J. Virol. 65: 2807-2815.

# Geminivirus Replication

Linda Hanley-Bowdoin*, Patricia A. Eagle*, Beverly M. Orozco*,
Dominique Robertson† and Sharon B. Settlage*

Departments of *Biochemistry and †Botany, North Carolina State
University, Raleigh, NC 27695-7622, USA

Geminiviruses are a family of plant viruses characterized by twin icosahedral particles and single-stranded DNA genomes. They replicate their
small, circular genomes through double-stranded DNA intermediates in
plant nuclei using a rolling circle mechanism and, as such, represent one of
only two known families of plant viruses with true DNA replication cycles.
They encode only a few proteins for their replication and recruit most of
their replication enzymes from their plant hosts. These features make geminiviruses excellent models for studying plant DNA replication and cell cycle
regulation. Our understanding of the molecular and cellular events that
mediate geminivirus replication has increased significantly during recent
years. The goal of this paper is to summarize research progress and to propose a model for initiation of geminivirus replication.

The *Geminiviridae* family consists of three subgroups that differ with
respect to insect vector, host range, and genome structure. There has been
considerable research on the replication of Subgroup III geminiviruses,
which are transmitted by whiteflies and infect dicot plants. Less is known
about replication of Subgroup I and II viruses, which are leafhopper-
transmitted and can have monocot or dicot hosts. We will focus our discussion on replication of Subgroup III geminiviruses, with emphasis on the
well-characterized virus, tomato golden mosaic virus (TGMV).

## Origin of Replication

Geminivirus replication requires two origins – one for (+)-strand synthesis and another for (-)-strand replication. TGMV has a bipartite genome
consisting of two 2.6-kb components designated A and B (Fig. 1). The
TGMV (+)-strand origin is located in the conserved 5' intergenic region or
common region (CR) of both components (Fig. 1; Lazarowitz et al. 1992).
The location of the (-)-strand origin is not known.

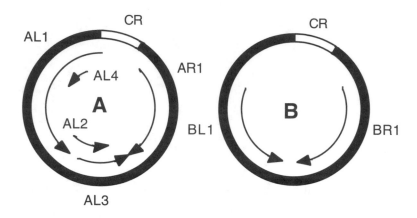

Fig. 1. The TGMV genome. The open boxes indicate the common region. The arrows mark open reading frames and their directions of transcription.

The TGMV (+)-strand origin is delimited by two functional elements (see Fig. 2) – a hairpin structure and a binding site for the viral replication protein AL1 (Fontes et al. 1994a; Orozco and Hanley-Bowdoin 1996). The hairpin contains the cleavage site for initiation of (+)-strand synthesis (Laufs et al. 1995a, Orozco and Hanley-Bowdoin 1996). The AL1 binding site, which consists of two directly repeated motifs, mediates origin recognition (Fontes et al. 1994b). The 34 bps separating the hairpin and AL1 binding site also contribute to origin activity, with both sequence and spacing of this region affecting function (Gladfelter and Hanley-Bowdoin in preparation).

The TGMV promoter for complementary-sense transcription and the (+)-strand origin overlap and share *cis* elements. The AL1 binding site is required for origin recognition and negative regulation of leftward transcription (Eagle et al. 1994). A partial copy of the AL1 binding site upstream of the (+)-strand origin contributes to transcriptional regulation but not to origin function (Orozco et al. in preparation). Two transcription motifs – a TATA box and a G-box – are required for full AL1 gene expression and efficient origin activity (Eagle and Hanley-Bowdoin in preparation).

The (+)-strand origins of other geminiviruses are located in their 5' intergenic regions (see Laufs et al. 1995b and references therein). The hairpin motif is in the 5' intergenic regions of all known geminivirus genomes and contains a sequence in the loop that is evolutionarily conserved (Roberts and Stanley 1994). The initiation site for (+)-strand DNA synthesis has been mapped to the loop sequence for members of all three geminivirus subgroups. Comparison of the 5' intergenic regions of Subgroup II and III geminiviruses also revealed sequences that resemble the TGMV AL1 binding site and transcription elements (Arguello-Astorga et al. 1994; Fontes et al. 1994a,b).

## Viral Replication Proteins

TGMV encodes two replication proteins. AL1, the only viral protein essential for replication (Elmer et al. 1988), also functions as a transcriptional repressor (Eagle et al. 1994) and modifies its host (Nagar et al. 1995). The AL3 protein is not required for replication but greatly increases viral DNA accumulation (Elmer et al. 1988). All geminiviruses encode AL1 homologues but only Subgroup II and III viruses specify AL3-like proteins

Recent biochemical studies have provided insight into AL1 function. AL1 specifically binds double- and single-stranded DNA (Fontes et al. 1994a; Thommes et al. 1993). The double-stranded DNA binding activity mediates virus-specific origin recognition and transcriptional repression. AL1 also hydrolyzes ATP and catalyzes sequence-specific DNA cleavage and ligation (Laufs et al. 1995b). All three enzymatic activities are required for replication, with DNA cleavage and ligation involved in initiation and termination of (+)-strand synthesis.

AL1 and AL3 are involved in multiple protein protein interactions (Settlage et al. 1996). AL1/AL1 interaction is a prerequisite for AL1/DNA binding (Orozco et al. in preparation). AL3 multimerizes with itself and interacts with AL1. Unlike AL1/DNA binding, none of the protein interactions display virus specificity.

## Host Replication Factors

Geminiviruses rely on their plant hosts for most of their replication machinery. TGMV particles and replication proteins have been found in nuclei of differentiated cells (Rushing et al. 1987; Nagar et al. 1995), which normally lack detectable plant DNA replication enzymes. However, TGMV caused the accumulation of the DNA replication processivity factor, proliferating cell nuclear antigen (PCNA; Nagar et al. 1995) in infected cells. AL1 was sufficient to increase PCNA levels in differentiated cells of transgenic plants. Thus, TGMV induces the accumulation of DNA synthesis machinery in quiescent plant cells with AL1 playing a key role in induction.

Geminiviruses resemble some animal viruses in their reliance on host replication machinery and their ability to replicate in differentiated cells. Several mammalian DNA viruses encode proteins that modify cell cycle controls in differentiated cells by interfering with the action of the host tumor suppressor proteins, like retinoblastoma (pRb). Two groups (Collin et al. 1996; Xie et al. 1995) used yeast genetics to show that an AL1 homologue from a Subgroup I geminivirus can interact with human pRb proteins. Interaction was through a LXCXE motif found in the geminivirus protein and in many cell cycle regulatory and animal virus proteins that interact with pRb. Recently, we used biochemical approaches to show that TGMV AL1 interacts with a plant pRb homologue (Ach et al. in preparation). Interestingly, none of the AL1 proteins of Subgroup II and III geminiviruses, including TGMV, contain the LXCXE motif and must interact with pRb through a different amino acid sequence.

## TGMV Replication - A Model

We propose a model for TGMV (+)-strand initiation (Fig. 2) in which [1] AL1 recognizes the (+)-strand origin and binds as a dimer or higher order complex, [2] binding causes a change in origin conformation leading to the extrusion of the cruciform structure, [3] AL3 binds to the cruciform and to AL1, leading to origin bending, [4] the interactions between the viral replication proteins and changes in origin structure are facilitated or stabilized by host transcription factors, [5] this array of interactions brings AL1 bound at its recognition site into close proximity with its nick site to facilitate DNA cleavage and initiation of (+)-strand DNA synthesis.

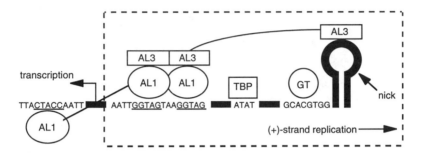

Fig. 2. Model for initiation of (+)-strand TGMV replication. The top strand of the origin is shown with the minimal origin outlined by the dotted lines. The sequences designate protein binding sites whereas the hairpin is diagrammed. The initiation sites and directions of replication and transcription are marked. Proteins known or predicted to interact with the origin are shown (AL1, AL3, TATA binding protein or TBP, G-box transcription factor or GT). AL1 and AL3 protein interactions are indicated, including potential long range interactions (lines) that might change origin structure.

The model explains the essential role of AL1 in TGMV replication and its virus-specific activity (Elmer et al. 1988; Fontes et al. 1994b). The proposed involvement of AL3 in recruiting bound AL1 to the cleavage site is consistent with its enhancement of replication. Separation of the AL1 binding and cleavage sites is likely to reduce the efficiency of (+)-strand replication unless a mechanism exists to direct AL1 to the cruciform. Facilitation of this process by AL3 is consistent with its ability to interact with AL1 and itself and predicts that AL3 also binds DNA. The model is compatible with the virus-nonspecific activity of AL3 (Sunter et al. 1994) because AL1/AL3 protein interactions do not display virus specificity (Settlage et al. 1996). The sequence of the cruciform is highly conserved among Subgroup II and III geminiviruses (Orozco and Hanley-Bowdoin 1996) such that the predicted AL3 interaction with this region is also unlikely to be virus-specific. The model accounts for the lack of an AL3 homologue in Subgroup I viruses, whose C1 binding sites have been proposed to be located in the stems of their hairpins (Arguello-Astorga et al. 1994) and, thus, would not require an AL3-like protein to direct C1 to its cleavage site. Lastly, the model includes plant DNA replication enzymes and provides a mechanism whereby host transcription factors facilitate viral replication.

## Future Experiments

There are many aspects of TGMV replication which are unknown. A major area of interest is the mechanism whereby TGMV modifies its host to render quiescent plant cells competent for viral DNA replication. Does AL1 modify plant cell cycle controls or activate transcription of host DNA synthesis genes directly? Are other viral proteins involved in the induction process? What is the significance of AL1/pRb interaction during geminivirus infection? It is not known which host replication proteins are required for viral DNA synthesis or how they are recruited. Does AL1 or AL3 interact directly with one or more components of the host replication machinery?

There is no information regarding (-)-strand synthesis or the host proteins that mediate the process. How does the initial single- to double-stranded conversion occur in differentiated plant cells lacking DNA replication machinery? Double-stranded DNA is the transcriptionally active form and is necessary for production of the AL1 protein, which induces accumulation of host replication enzymes. Are DNA repair enzymes, which are presumably in all differentiated cells, able to support sufficient (-)-strand synthesis to begin the infection process, with plant DNA replication machinery responsible for amplification of viral DNA to high levels? Alternatively, is geminivirus replication initially confined to phloem-associated cambial cells competent for DNA replication such that fully differentiated cells only support replication later in infection after viral induction? Future studies addressing these questions will contribute to our knowledge of host/ pathogen interactions and basic cellular processes in plants.

## Literature Cited

Arguello-Astorga, G. R., Guevara-Gonzalez, R. G., Herrera-Estrella, L. R. and Rivera-Bustamante, R. F. 1994. Geminivirus replication origins have a group-specific organization of iterative elements: A model for replication. Virology 203:90-100.

Collin, S., Fernandez-Lobato, M., Gooding, P. S., Mullineaux, P. M. and Fenoll, C. 1996. The two nonstructural proteins from wheat dwarf virus involved in viral gene expression and replication are retinoblastoma-binding proteins. Virology 219:324-329.

Eagle, P. A., Orozco, B. M. and Hanley-Bowdoin, L. 1994. A DNA sequence required for geminivirus replication also mediates transcriptional regulation. Plant Cell 6:1157-1170.

Elmer, J. S., Brand, L., Sunter, G., Gardiner, W. E., Bisaro, D. M. and Rogers, S. G. 1988. Genetic analysis of tomato golden mosaic virus II. Requirement for the product of the highly conserved AL1 coding sequence for replication. Nucleic Acids Res. 16:7043-7060.

Fontes, E. P. B., Eagle, P. A., Sipe, P. A., Luckow, V. A. and Hanley-Bowdoin, L. 1994a. Interaction between a geminivirus replication protein and origin DNA is essential for viral replication. J. Biol. Chem. 269:8459-8465.

Fontes, E. P. B., Gladfelter, H. J., Schaffer, R. L., Petty, I. T. D. and Hanley-Bowdoin, L. 1994b. Geminivirus replication origins have a modular organization. Plant Cell 6:405-416.

Laufs, J., Traut, W., Heyraud, F., Matzeit, V., Rogers, S. G., Schell, J. and Gronenborn, B. 1995a. In vitro cleavage and joining at the viral origin of replication by the replication initiator protein of tomato yellow leaf curl virus. Proc. Natl. Acad. Sci. USA 92:3879-3883

Laufs, J., Jupin, I., David, C., Schumacher, S., Heyraud-Nitschke, F. and Gronenborn, B. 1995b. Geminivirus replication: genetic and biochemical characterization of rep protein function, a review. Biochimie 77:765-773.

Lazarowitz, S. G., Wu, L. C., Rogers, S. G., and Elmer, J. S. 1992. Sequence-specific interaction with the viral AL1 protein identifies a geminivirus DNA replication origin. Plant Cell 4:799-809.

Nagar, S., Pedersen, T. J., Carrick, K., Hanley-Bowdoin, L. and Robertson, D. 1995. A geminivirus induces expression of a host DNA replication protein in terminally differentiated plant cells. Plant Cell 7:705-719.

Orozco, B. M. and Hanley-Bowdoin, L. 1996. A DNA structure is required for geminivirus origin function. J. Virol. 270:148-158.

Roberts, S. and Stanley, J. 1994. Lethal mutations within the conserved stem-loop of African cassava mosaic virus DNA are rapidly corrected by genomic recombination. J. Gen. Virol. 75:3203-3209.

Rushing, A. E., Sunter, G., Gardiner, W. E., Dute, R. R. and Bisaro, D. M. 1987. Ultrastructural aspects of tomato golden mosaic virus infection in tobacco. Phytopathology 77:1231-1236.

Settlage, S. B., Miller, A. B., and Hanley-Bowdoin L. 1996. Interactions between geminivirus replication proteins. J. Virol. 270, in press.

Sunter, G., Stenger, D. C., and Bisaro D. M. 1994. Heterologous complementation by geminivirus AL2 and AL3 genes. Virology 203: 203-210.

Thommes, P., Osman, T. A. M., Hayes, R. J., and Buck, K. W. 1993. TGMV replication protein-AL1 preferentially binds to single-stranded DNA from the common region. FEBS Lett. 319:95-99.

Xie, Q., Suarez-Lopez, P. and Gutierrez, C. 1995. Identification and analysis of a retinoblastoma binding motif in the replication protein of a plant DNA virus: Requirement for efficient viral DNA replication. EMBO J. 14:4073-4082.

# Involvement of Rice Dwarf Virus S6 in Symptom Severity and Insect Transmission.

Ichiro Uyeda, Yuko Ando, Yoko Tanji, Hiroki Atarashi, and Ikuo Kimura

Department of Agrobiology and Bioresources, Faculty of Agriculture, Hokkaido University, Sapporo, Japan 060

Rice dwarf *Phytoreovirus* (RDV), a member of Reoviridae, has 12 segmented dsRNAs as a genome and transmitted in a persistent manner by leafhoppers. The infected rice plants are stunted and show white specks on leaves and sheaths. Little attention had been paid for genetic studies of the virus because : 1) there has been no efficient method of isolating genetically homogeneous virus culture, as neither local lesion plant host nor an insect cell culture giving visible cyptopathic effects is available; 2) no distinct phenotypically distinct virus isolates were characterized. We have developed a system for RDV genetics by overcoming above obstacles. Individual RDV isolates can be identified by migration profiles of genomic dsRNAs in a thin and long polyacrylamide gel electrophoresis (PAGE) and homogeneous virus culture is isolated after serial passage using the insect vector ( Murao et al. 1994; Uyeda et al. 1995). It was also shown that reassortants are generated by injecting a mixture of two genomic variants into the insect vector, and they are isolated as genomically homogeneous culture by subsequent inoculation to rice seedlings (Uyeda et al. 1995). Panel of RDV isolates with a range of symptom severity were isolated from fields grown infected plants (Ando et al. 1996). In addition, transmission deficient isolates by a leafhopper *Nephotettix cincticeps* were collected.

The two most biologically interesting features of the virus, stunting of the infected rice plants and transmission by insect vector, are genetically analyzed.

# Severe Stunting of Rice Plants

SYMPTOM VARIANTS

RDV-W-M (Ando et al. 1996) caused unusually severe stunting in rice. Infected plants had short culms and developed progressively shortened leaves with growth of plants. Leaves showed pronounced chlorotic specks and white streaks along the vein, and sometimes curling at the leaf tips. RDV-S (Kimura et al. 1987) caused essentially the same symptoms, but stunting is not as severe as RDV-W-M. RDV-AI and RDV-W-M infected plants showed mild symptoms, and their heights were similar to those of healthy controls. Chlorotic specks on leaves were small in size and number.

Virus accumulation in a newly emerged leaf blade of a main culm 1.5 months after inoculation were measured by ELISA. RDV-S and RDV-W-M infected leaves contained about twice as much as that of RDV-AI and RDV-W-L infected plants.

ANALYSES OF REASSORTANTS

In order to determine genes involved in the severe stunting, artificial reassortants were made from combinations of RDV-AI and RDV-S, and RDV-AI and RDV-W-M. Symptom severity was quantitatively measured by comparing heights of infected rice plants about 1.5 months after inoculation. Parental origin of genomic segments in a particular reassortant was determined by PAGE analyses of dsRNAs directly extracted from the plants. All the genomic segments were differentiated by PAGE between RDV-AI and RDV-S. In the case of combination between RDV-AI and RDV-W-M, parental origin of all but genome segment 2 (S2) could be identified.

Initial screening was made by examining heights of infected plants and parental origin of all the genomic segments present in infected plants after the first inoculation of the mixed viruses. Relative molar ratios of each genomic segments and height of infected plants were scored and plotted. Regression analyses of the plots showed S6 and S11, and S6 is associated with the severe stunting in combinations of RDV-AI and RDV-S, and RDV-AI and RDV-W-M, respectively.

Further isolation of reassortants by serial transfer were made from the combination of RDV-AI and RDV-S. Panel of reassortants showed that severe stunting is associated primarily with S6 (Fig 1).

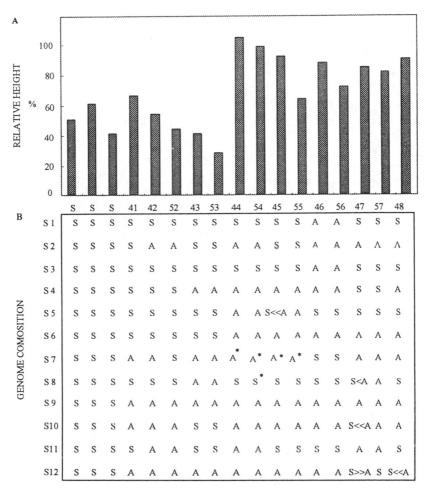

Fig. 1. Association of RDV-S genome segment 6 with severe stunting of rice plants. A: relative height of the infected plants with respect to those with RDV-AI. B: genomic segment composition of reassortants.
* shows genomic segments of abnormal electrophoretic migration.
A and S denote genomic segments from parental RDV-AI and RDV-S, respectively.

Amino acid sequence of the S6-encoded polypeptide of RDV-S and RDV-W-M did not have any common domains nor residues different from that of RDV-AI. RDV-S and RDV-W-M differed in three and five residues from RDV-AI, respectively out of 509 residues.

295

## Insect Transmission

TRANSMISSION DEFICIENT VARIANTS

RDV-CK9401 and RDV-P were not transmitted by *N. cincticeps* when injected, while 60-89% of the injected insect vectors transmitted RDV-N6, RDV-N5 and RDV-AN. ELISA tests showed that RDV-CK9401 was detected in most of the injected insect vectors and only one out of total of 34 injected insect vectors contained RDV-P, suggesting two distinct viral gene functions are involved in insect transmission. They are : 1) ability to multiply in insect vectors; 2) successful transmission to rice plants.

When RDV-P was inoculated to vector cell monolayer, it multiplied as well as the transmission competent isolate. The reason for this is not yet known.

ANALYSES OF REASSORTANTS

Reassortants were made from appropriate combinations of transmission-deficient and -competent isolates as shown in Table 1. When RDV-P was mixed with RDV-N5 or RDV-N6(PN5 and PN6 isolates in Table 1), reassortants recovered from the infected plants always contained S1 and S6 from the transmission competent isolate. S1 and S6 from RDV-N6 could not be replaced with those of RDV-P even when PN6R1-9 was mixed again with an excess of RDV-P in an inoculum and injected (PN6R2-2, PN6R3-9 and PN6R4-2). The reassortants multiply in almost all of the injected insect vectors as revealed by ELISA. Reassortants were made from a combination of RDV-CK9401 and RDV-AN. Genome composition of the selected reassortants that are transmissible contained S6 form RDV-CK9401. These results suggest that S6 from the transmission competent parents required for RDV-P only to multiply in the insect vectors.

Table 1. Transmission of reassortants made from a combination of transmission-deficient and -competent rice dwarf virus isolates

| Reassortant [1] | Genomic Segment Composition [2] | | | | | | | | | | | | Transmission | | ELISA of Insect | |
|---|---|---|---|---|---|---|---|---|---|---|---|---|---|---|---|---|
| | S1 | S2 | S3 | S4 | S5 | S6 | S7 | S8 | S9 | S10 | S11 | S12 | Infected/Tested | % | Positive/Tested | % |
| PN5R1-4 | N5 | N5/P | P | n5/P | n5/P | N5 | P | P | P | N5/P | P | N5/P | 11/22 | 50 | 17/18 | 94 |
| PN5R1-4.6 | N5 | P | P | P | N5/p | N5 | P | P | P | P | P | P | 5/13 | 39 | NT | - |
| PN5R1-4.6.8 | N5 | P | P | P | n5/P | N5 | P | P | P | P | P | P | 5/11 | 46 | 9/10 | 90 |
| PN5R1-4.6.9 | N5 | P | P | P | N5 | N5 | P | P | P | P | P | P | 3/17 | 18 | 16/17 | 94 |
| PN5R2-3 | N5 | P | P | P | n5/P | N5 | P | P | P | P | P | P | 5/10 | 50 | 8/8 | 100 |
| PN5R1-4.6.8.4 | N5 | P | P | P | P | N5 | P | P | P | P | P | P | 3/15 | 20 | 13/14 | 93 |
| PN5R1-4.6.8.6 | N5 | P | P | P | P | N5 | P | P | P | P | P | P | 3/18 | 17 | 13/14 | 93 |
| PN5R1-4.6.10 | N5 | P | P | P | N5 | N5 | P | P | P | P | P | P | 8/14 | 57 | 10/11 | 91 |
| PN6R1-9 | N6 | N6 | P | N6 | N6 | N6 | P | P | P | P | P | P | 8/16 | 50 | 11/12 | 92 |
| PN6R2-2 | N6 | n6/P | P | N6/p | P | N6 | P | P | P | P | P | P | 7/20 | 35 | 18/18 | 100 |
| PN6R3-9 | N6 | P | P | N6/P | P | N6 | P | P | P | P | P | P | 7/14 | 50 | 13/13 | 100 |
| PN6R4-2 | N6 | P | P | P | P | N6 | P | P | P | P | P | P | 1/13 | 8 | 5/5 | 100 |

1) PN5, PN6 : Reassortants made from a combination of RDV-P and RDV-N5, RDV-P and RDV-N6, respectively.

2) N5, N6, P : Parental origin of genomic segments are RDV-N5, RDV-N6, and RDV-P, respectively.
   Small letters indicate that amount of the genomic segment is smaller than that shown by capital letters.

# LITERATURE CITED

Ando, Y., Uyeda, I., Murao, K., and Kimura, I. 1996. Naturally occurring phenotypic variants in symptom severity of rice dwarf Phytoreovirus. Ann. Phytopath. Soc. Japan (In press).

Kimura, I., Minobe, Y., Omura, T. 1987. Changes in nucleic acid and a protein component of rice dwarf virus particles associated with an increase in symptom severity. J. Gen. Virol. 68:3211-3215.

Murao, K., Suda, N., Uyeda, I., Isogai, M., Suga, H., Yamada, N., Kimura, I., and Shikata, E. 1994. Genomic heterogeneity of rice dwarf phytoreovirus field isolates and nucleotide sequences of variants of genome segment 12. J. Gen. Virol. 75:1843-1848.

Uyeda, I., Ando, Y., Murao, K., and Kimura, I. 1995. High resolution genome typing and genomic reassortment events of rice dwarf Phytoreovirus. Virology 212:724-727.

# The molecular basis of host specificity of rhizobia

HP Spaink[1]*, J Bakkers[1], M Bladergroen[1], GV Bloemberg[1], I Dandal[1], CL Diaz[1], L Blok-Tip[2], A Gisel[3], M Harteveld[1], IM Lopez-Lara[1], D Kafetzopoulos[1], E Kamst[1], JW Kijne[1], D Meijer[1], BJJ Lugtenberg[1], AO Ovtsyna[1], I Potrykus[3], NEM Quaedvlieg[1], C Quinto[4], T Ritsema[1], C Sautter[3], HRM Schlaman[1], JE Thomas-Oates[2], JH van Boom[5], K van der Drift[2], GA van der Marel[5], S van Leeuwen[5], A Veldhuis[1], AHM Wijfjes[1]

[1]Leiden University Institute of molecular Plant Sciences, Wassenaarseweg 64, 2333 AL Leiden, The Netherlands, [2]Utrecht University, Department of Mass spectrometry, Sorbonnelaan 16, 3584 CA Utrecht, The Netherlands, [3]E.T.H., Institute Plant Sciences, CH-8092 Zürich, Switzerland, [4] U.N.A.M., Instituto de Biotecnología, Apartado Postal 510-3, Cuernavaca Morelos 62271, Mexico, [5] Leiden University, Leiden Institute of Chemistry, PO Box 9502, 2300 RA Leiden

Factors which determine the host specificity of rhizobia for particular leguminous plants include flavonoids secreted by the host plant roots, the bacterial lipo-chitin oligosaccharides (LCOs) and the host plant lectins. In this paper particular attention will be given to the role of the LCOs. To study the role of structural elements of the LCOs (e.g., a highly unsaturated fatty acyl or a fucosyl moiety) we have made use of the following: (i) our present knowledge of the biosynthesis of LCOs, (ii) synthetically produced chitin oligosaccharide derivatives, (iii) improved methodologies to apply signal molecules to particular parts of the plant root, for instance, using ballistic microtargeting, and (iv) transgenic plants containing reporter gene-fusions or lectin genes. With respect to the latter approach, important new data has been obtained by studying transgenic white clover plants which contain GUS-fusions with the chalcone synthase genes or the auxin responsive promoter GH3. Some of the recent results obtained with this system are described by Mathessius *et al.* (this volume). The approaches which make use of transgenic red clover plants which contain the pea lectin gene are described in this volume by Diaz *et al.*

# Biosynthesis of nod factors

The study of the biosynthesis of the LCOs serves several purposes: (i) It can be used as a model for the study of several fundamental biochemical processes, such as chitin synthesis, chitin modification and specialized fatty acid synthesis and transfer. The system has advantages over the study of these processes in other organisms which are genetically difficult to manipulate or in which these processes are essential for the living cell. (ii) The enzymes involved in LCO synthesis can be used for obtaining radiolabelled derivatives of LCOs or homologs from non-rhizobial origin (for a further discussion of this possibility see Spaink 1996). (iii) Using alternative substrates for the rhizobial enzymes can lead to novel compounds. As substrates signal molecules from other systems which resemble the LCOs or part of the LCO molecule can be used. The resulting derivatives might have altered biological activity. This offers additional opurtunities to study structure-function relationships in several classes of signal molecules. For instance, the enzyme NodZ a fucosyl transferase (as detailed below) can be used to obtain fucosylated derivatives of adhesion factors from vertebrates, such as the compound Lewis-X. (iv) Understanding of the function of the gene products involved in LCO biosynthesis makes it possible to interpret the role of these genes in determining host specificity at a molecular level. The biosynthesis of LCOs has been described in detail in a recent review (Spaink 1996). Therefore, a brief overview only will be presented of our recent advances with regards to the relation with host specificity.

## LCO biosynthesis as determinant of host specificity

### THE NODC PROTEIN

We expressed the cloned nodC genes of *R.meliloti*, *R.l. viciae* and *R.loti* in *E.coli* and analyzed the chitin oligosaccharides produced by these strains. A clear difference in the length of the major chitin oligosaccharides produced by these strains was observed. *R.meliloti* NodC mainly produced chitintetraose, whereas the major product of *R.loti* NodC was chitinpentaose. Expression of the *R.l.* bv viciae nodC led to the production of a mixture of chitintetraose and pentaose as the major products. These differences were found to be independent of the UDP-GlcNAc concentration in vitro,and therefore represent intrinsic properties of the NodC proteins themselfes. Expression of *R.meliloti* *nodC* in a *R.l.viciae nodC* deletion mutant leads to the synthesis of LCOs with oligosaccharide chain lengths identical to *R.meliloti* LCOs, whereas expression of *R.loti nodC* in a *R.l.viciae nodC* deletion mutant leads to an increased synthesis of LCO-V. From these results we conclude that (i) NodC is the main determinant of chitin oligosaccharide chain length in LCOs (ii) The synthesis of mainly LCO-IV in *R.meliloti* is due to NodC. We therefore expect that NodC is also a determinant of host specificity.

## THE NODF AND NODE PROTEINS

In *R. leguminosarum* the multi-unsaturated fatty acyl part of the LCOs is determining host specificity. For biosynthesis of this fatty acid two specialized enzymes NodF (homologous to Acyl Carrier Proteins (ACP)) and NodE (homologous to $\beta$-ketoacyl synthases) are essential. This results in the synthesis of fatty acyl groups of 18 carbon units containing four unsaturations (C18:4) in *R.leguminosarum* biovar viciae. In *R.leguminosarum* biovar trifolii, NodF and NodE are involved in the synthesis of fatty acids which differ from C18:4 in the chain length (C20), the number of trans double bonds or the absence of the *cis* double bond (Spaink *et al* 1995; Bloemberg *et al* 1995;, van der Drift *et al* 1996). Using chimeric proteins of NodF and household ACP, we have shown that the specialized function of NodF is encoded in the second half of the protein. For NodE a central domain was shown to be determining its host specific characteristics (Bloemberg *et al* 1995).

## THE NODA PROTEIN

For transfer of the fatty acid to the chitin backbone NodA is essential. In vitro experiments show that *R.meliloti* and *R.leguminosarum* NodA can use fatty acids that are provided by either ACP or NodF as a donor. In contrast, free fatty acids can not be used as substrates. Also on this level a specificity has been shown. For instance, the NodA protein of *Bradyrhizobium* is not able to transfer a highly unsaturated fatty acyl group (Ritsema *et al.*, 1996).

## THE NODS PROTEIN

The NodS protein has been shown to be a methyl transferase (Geelen *et al* 1995). Furthermore it was shown that NodS of *Azorhizobium* is involved in the addition of a methyl group to the free amino group which results of the enzymatic activity of the NodB protein. We have studied the properties of the *nodS* gene of *R.loti*. This gene is located 300 basepairs downstream from a *nod*-box promoter together with *nodU*. By using a gain of function approach we have analysed the function of the *nodS* gene in the determination of host specificity. In addition to confirming that *nodS* is an important determinant of host specificity, the results show that the function of the *nodL* protein (a transacetylase) (Bloemberg *et al* 1995) was interfering with that of *nodS*. Using an *E.coli* strain which produces high quantities of the NodS protein we have studied its properties in further detail. The results show that NodS of *R.loti* is not able to accept substrates which are *O*-acetylated at the C6 position of the non-reducing saccharide due to the action of NodL.

## THE NODZ PROTEIN

The *nodZ* gene which is present in various rhizobia such as *Bradyrhizobium japonicum, Rhizobium loti* and *R.etli*, is involved in the addition of a fucosyl residue to the reducing *N*-acetylglucosamine residue of lipo-chitin oligosaccharides (LCOs) (Lopez-Lara *et al.* 1996). Using an *E.coli* strain which produces high quantities of the NodZ protein of *B.japonicum* we have purified the NodZ protein to homogeneity. The purified NodZ protein appeared to be active in an *in vitro* transfucosylation assay in which LCOs or chitin oligosaccharides were used as substrates and GDP-fucose as fucosyl donor. Compounds which contain at least one reducing *N*-acetylglucosamine, such as the recognition factor Lewis-x, are also substrates for NodZ, making it possible to obtain novel fucosylated oligosaccharides. The product of the *in vitro* reaction using chitin pentasaccharide as a substrate was studied by mass spectrometry, linkage analysis and composition analysis showing that one fucose residue was added to C6 of the reducing terminal *N*-acetylglucosamine residue. Substrate specificity of NodZ protein was analyzed in further detail using radiolabelled GDP-fucose as a substrate. The results show that COs are much better substrates than LCOs indicating that NodZ in *Rhizobium* is active before the acylation step of chitin oligosaccharides.

## THE NOLL AND NODX PROTEINS

The nucleotide sequence of the *R.loti NolL* gene was recently reported by Scott *et al.* (1996) and was shown to be homologous to the *nodX* gene of *R.leguminosarum*. Our recent results indicate that NolL plays a role in the acetylation of the NodZ-determined fucosyl residue of LCOs. We are cureently testing the presumed transacetylase activity of NolL and NodX proteins by *in vitro* enzymatic analyses. Using a gain of function approach, the *nolL* gene was shown to play an important role in determination of specificity of nodulation of *Lotus* plants. In collaboration with Dr. Scott we are presently analyzing in the host-specific characteristics of the *nolL* gene in more detail.

## THE NODI AND NODJ PROTEINS

Using a new method to distinguish between LCO's inside the bacteria and in the medium, which is based on the external addition of chitinase, we have shown that NodI and NodJ are involved in the secretion of LCOs (Spaink et al. 1995). However the secretion of LCOs in *R.leguminsarum* is not totally abolished in the absence of NodI and NodJ. These results indicate that there are other proteins present in the *Rhizobium* bacteria which can take over the role of NodI and NodJ. NodI and NodJ show homology to other ATP-binding transport proteins, such as DrrA from *Streptomyces peucetius*, BexA from *Haemophilus influenzae*, KpsT from *Escherichia coli* and CtrD from *Neisseria meningitidis* for NodI and DrrB, BexB, KpsM and CtrC for NodJ (Vasquez *et al* 1993). We have used this similarity to obtain DNA fragments by PCR on chromosomal DNA with degenerate primers based on these homologous sequences. These DNA fragments have been sequenced and the results indeed show the presence of genes encoding proteins homologous to NodI. Various amino acids which are typical for the NodI proteins are also present in the deduced protein sequences.

The phenotypic effect of inactivation of the *nodIJ* genes is dependent on the plant species tested (Canter Cremers *et al* 1988). To investigate the molecular basis of this host-specific characterisitcs of *nodIJ* we will test whether chromosomal homologues can influence nodulation and infection efficiency in various plant species.

## Microtargeting of signal molecules

At micromolar concentrations, LCOs are able to induce root nodule primordia when applied externally to leguminous plants provided that they are derived from the corresponding *Rhizobium*. This highly specific process is determined by the decorations of the chitin oligosaccharide and by the composition of the fatty acyl chain. In order to further investigate the role of strain-specific modifications of the LCOs new methods of spot application have been examined (Lopez-Lara *et al* 1995). In future studies, we will apply these methods using various derivatives of LCOs which we have chemically synthesized. To analyze the fate of applied LCOs we can make use of recently developped derivatives which contain fluorescent labels at different positions of the molecule.

To test the hypothesis that the fatty acyl chain is not required for mitogenic activity of LCOs, chitin oligomers were introduced directly into the tissue of *Lotus preslii* and *Vicia sativa* roots using ballistic microtargeting (Sautter *et al.* 1991). In *L. preslii* primordia, visible as a strictly localized red coloration at the surface of the root (Lópes-Lara *et al.* 1995) due to the presence of anthocyanin, were formed with chitin oligomers. The presence of a fucose group on the chitin oligomer appeared not to be a prerequisit for this activity. Frequences were enhanced when uridine, an enhancer of cell proliferation (Smit *et al.* 1995), was co-introduced. Cell divisions were found in the inner root cortex of *V. sativa* as determined by light microscopic examination of roots stained with Schiff reagent, provided that chitin oligomers carried an O-acetylgroup at the C6 position of the non-reducing sugar. In addition, co-targeting of uridine was required for this response. Sectioning revealed that the cell division events, which could comprise only a few cells, were not necessarily found opposite protoxylem poles, as is always the case for lateral root primordia.

## Acknowledgements

H.P.S. was suported by a NWO-PIONIER grant from the Netherlands Organisation of Scientific Research. This work was supported by contracts from the European Union n° BIO2-CT92-5112 (fellowship to I.M.L.-L.) and BIO2-CT93-0400 (DG12 SSMA), the Netherlands Foundation for Chemical Research (SON) (J.T.O., T.R. and G.V.B.). C.Q. was supported by a Marie Curie fellowship from the European Union for a research project at Leiden University.

## Literature cited

Bloemberg, G. V., Kamst, E., Harteveld, M., van der Drift, K. M. G. M., Haverkamp, J., Thomas-Oates, J. E., Lugtenberg, B. J. J., and Spaink, H. P. 1995. A central domain of *Rhizobium* NodE protein mediates host specificity by determining the hydrophobicity of fatty acyl moieties of nodulation factors. *Mol. Microbiol.* **16**: 1123-1136.

Bloemberg, G. V., Lagas, R., van Leeuwen, S., van der Marel, G., van Boom, J. H., Lugtenberg, B. J. J., and Spaink, H. P. 1995. Substrate specificity and kinetic studies of nodulation protein NodL of *Rhizobium leguminosarum. Biochem.* **34**: 12712-12720.

Canter Cremers, H. C. J., Wijffelman, C. A., Pees, E., Rolfe, B. G., Djordjevic, M. A., and Lugtenberg, B. J. J. 1988. Host specific nodulation of plants of the pea cross-inoculation group is influenced by genes in fast growing *Rhizobium* downstream *nodC. J. Plant Physiol.* **132**: 398-404.

López-Lara, I. M., Blok-Tip, L., Quinto, C., Garcia, M. L.,

Stacey, G., Bloemberg, G. V., Lamers, G. E. M., Lugtenberg, B. J. J., Thomas-Oates, J. E., and Spaink, H. P. 1996. NodZ of *Bradyrhizobium* extends the nodulation host range of *Rhizobium* by adding a fucosyl residue to nodulation signals. *Mol. Microbiol.* **in press**

López-Lara, I. M., van den Berg, J. D. J., Thomas-Oates, J. E., Glushka, J., Lugtenberg, B. J. J., and Spaink, H. P. 1995. Structural identification of the lipo-chitin oligosaccharide nodulation signals of *Rhizobium loti*. *Mol. Microbiol.* **15**: 627-638.

Ritsema, T., Wijfjes, A. H. M., Lugtenberg, B. J. J., and Spaink, H. P. 1996. *Rhizobium* nodulation protein NodA is a host-specific determinant of the transfer of fatty acids in Nod factor biosynthesis. *Mol. Gen. Genet.* **251**: 44-51.

Sautter, C., Waldner, H., Neuhaus-Url, G., Galli, A., Neuhaus, G., and Potrykus, I. 1991. Micro-targeting: high efficiency gene transfer using a novel approach for the acceleration of micro-projectiles. *Biotechnology* **9**: 1080-1085.

Scott, D. B., Young, C. A., Collins-Emerson, J. M., Terzaghi, E. A., Rockman, E. S., Lewis, P. E., and Pankhurst, C. E. 1996. Novel and complex chromosomal arrangement of *Rhizobium loti* nodulation genes. *Mol. Plant-Microbe Int.* **9**: 187-197.

Smit, G., de Koster, C. C., Schripsema, J., Spaink, H. P., van Brussel, A. A. N., and Kijne, J. W. 1995. Uridine, a cell division factor in pea roots. *Plant Mol. Biol.* **29**: 869-873.

Spaink, H. P. 1996. Regulation of plant morphogenesis by lipo-chitin oligosaccharides. *Crit. Reviews Plant Sciences* in press

Spaink, H. P., Bloemberg, G. V., van Brussel, A. A. N., Lugtenberg, B. J. J., van der Drift, K. M. G. M., Haverkamp, J., and Thomas-Oates, J. E. 1995. Host specificity of *Rhizobium leguminosarum* is determined by the hydrophobicity of highly unsaturated fatty acyl moieties of the nodulation factors. *Mol. Plant-Microbe Int.* **8**: 155-164.

Spaink, H. P., Wijfjes, A. H. M., and Lugtenberg, B. J. J. 1995. *Rhizobium* NodI and NodJ proteins play a role in the efficiency of secretion of lipochitin oligosaccharides. *J. Bacteriol.* **177**: 6276-6281.

van der Drift, K. M. G. M., Spaink, H. P., Bloemberg, G. V., van Brussel, A. A. N., Lugtenberg, B. J. J., Haverkamp, J., and Thomas-Oates, J. E. 1996. *Rhizobium leguminosarum* bv. *trifolii* produces lipo-chitin oligosaccharides with *nodE*-dependent highly unsaturated fatty acyl moieties: an electrospray ionisation and collision induced dissociation tandem mass spectrometric study. *J. Biol. Chem.* **in press**

Vásquez, M., Santana, O., and Quinto, C. 1993. The NodI and NodJ proteins from *Rhizobium* and *Bradyrhizobium* strains are similar to capsular polysaccharide secretion proteins from gram-negative bacteria. *Mol. Microbiol.* **8:** 369-377.

**For correspondence:**
Dr. H.P. Spaink, TEL: 31715275055, FAX: 3171-5275088,
EMAIL: SPAINK@RULSFB.LEIDENUNIV.NL
INTERNET homepage: http:/wwwbio.leideuniv.nl/~spaink/index.html

# *Nod* Gene Regulation in *Bradyrhizobium japonicum*

John Loh, Minviluz Garcia, Joyce Yuen and Gary Stacey.  Center for Legume
Research, Departments of Microbiology and Ecology and Evolutionary
Biology, University of Tennessee, Knoxville, TN, USA

Bacteria of the genera *Azorhizobium*, *Rhizobium* and *Bradyrhizobium*
infect and establish a root nodule symbiosis with their respective legume host
plants.  This infection process is characterized by a high degree of host
specificity and is controlled by the specific expression of both plant and
bacterial genes.  The bacterial genes required for infection are called the
nodulation genes (*nod, nol,* and *noe*) and are primarily thought to encode
enzymes involved in the synthesis of lipo-chitin oligosaccharide molecules that
function as plant growth regulating components (Carlson et al. 1994).
Addition of these purified compounds to plant roots can induce many of the
early nodulation responses, eg. root hair curling and cortical cell division
found during bacterial infection.  The induction of the *nod* genes in rhizobia
requires the *nodD* gene product and a specific host-produced flavonoid (Peters
and Verma. 1990).  NodD, binds to the conserved *nod* box sequence
preceeding most *nod* operons, and in the presence of the flavonoid inducer,
activates *nod* gene expression (Schlaman et al. 1992).  This generic model of
*nod* gene regulation is, however, only partially applicable to *Bradyrhizobium
japonicum*.  The regulation of *nod* gene expression in *B. japonicum* requires
not only NodD but also a two-component regulatory system, NodVW, as well
as a MerR-type regulator, NolA (Figure 1)

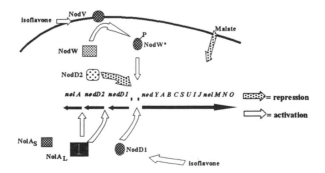

Fig.1 Schematic drawing of the components controlling *nod* gene expression.

*B. japonicum* possesses two *nodD* genes, *nodD₁* and *nodD₂*, (Göttfert et al. 1989) that are arranged in tandem (Figure 1). Of these two genes, *nodD₁* was thought to be the only functional copy as a *nodD₁D₂* deletion mutant could only be phenotypically complemented by *nodD₁* and not by *nodD₂* (Göttfert et al. 1992). Unlike the NodD proteins found in most *Rhizobium* species, *nodD₁* expression in *B. japonicum* requires isoflavones and positively regulates its own expression (Banfalvi et al, 1988). In addition, *nodD₁* can also be induced by glucosylated forms of the isoflavones genistein and daidzein (Smit et al. 1992). These derivatives have no effect on the expression of the common *nod* genes *nodYABC*. This specific induction of *nodD₁* appears to be mediated by a divergent but functional *nod* box sequence 5' of the gene (Wang et al. 1991). Although *B. japonicum nodD₁* mutants are unable to induce *nod* gene expression in response to isoflavone addition, such mutants retain the ability to nodulate host plants (Nieuwkoop et al. 1987). These results were surprising since *nod* gene function is absolutely essential for nodulation ability. The solution to this mystery came when Sanjuan et al (1994) demonstrated that the *nodVW* products are directly involved in the regulation of both *nodD₁* and *nodYABC* induction by isoflavonoids. For example, a *nodW* mutant was found to be incapable of inducing *nod-lacZ* expression in the presence of the isoflavone genistein. This phenotype mimicked that of a mutant defective in *nodD₁*. In the absence of NodD₁, NodV and NodW are essential and sufficient for isoflavone-mediated *nod* gene expression. The importance of *nodVW* in nodulation had previously been demonstrated by Göttfert et al (1990) who showed that mutations in these genes eliminated the ability to nodulate mungbean, cowpea, and siratro, alternative host plants of *B. japonicum*. In contrast, NodV and NodW are not required for the nodulation of soybean. This may be due to the production of isoflavones by soybean that can interact with either NodVW or NodD₁. In support of this idea, *nodD₁* or *nodD₁D₂* mutants also deleted in *nodW* were unable to nodulate soybean (Sanjuan et al. 1994). The requirement for NodV and NodW for nodulation of cowpea, mungbean and siratro suggests that these plants may produce inducers that interact specifically with NodVW and not NodD₁.

Sequence analyses indicate that NodV and NodW fall into a global family of two-component regulators that play an integral role in bacterial signal transduction (Charles et al. 1992). The proteins involved in these systems can be classified into two subclasses: the sensor and the regulator class. On acquiring the appropriate signal, the sensor protein autophosphorylates at a histidine residue. The activated sensor in turn activates the regulator protein by transferring the phosphate to a conserved aspartate residue. Sequence comparisons suggested that NodV is the sensor-kinase, and NodW the regulator protein (Göttfert et al. 1990). To test this notion, *nodV* and *nodW* were cloned and the proteins expressed in *E. coli*. Charcterization of the phosphorylation patterns of these proteins revealed that NodV was phosphorylated in-vitro by $\gamma$-$^{32}$ATP and NodW by the phosphodonor $^{32}$P-acetyl-P. More importantly, NodW could also be phosphorylated in-vitro by its cognate kinase, NodV. The biological significance of phosphorylation on nodulation was further evaluated using a mutant NodW containing an Asp70 to a Asn70 mutation generated by site-directed mutagenesis. By homology to other response regulators, the Asp70 residue was proposed to be the site of NodW phosphorylation. As expected, no detectable phosphorylation of the mutant protein was observed. In addition, when a plasmid harboring either the wild-type or mutant *nodW* was used to complement a *B. japonicum nodW* mutant strain containing a chromosomally encoded *nodC-lacZ* gene, only the transconjugant containing the wild-type NodW showed induced levels of *nod* gene expression in the presence of genistein. In addition, plant infection assays also revealed that transconjugants expressing the mutant NodW were defective in the nodulation of mungbean. These results suggest that phosphorylation is essential and required for both *nod* gene induction and nodulation.

## NOLA AND NODD$_2$

NolA was first identified by its ability to extend the host range of *B. japonicum* serogroup 123 strains to specific soybean genotypes (Sadowsky et al. 1991). Subsequently, Dockendorff et al (1994) reported that *nolA* mediates the repression of *nod* gene expression in *B. japonicum* and that conjugation of a plasmid encoding *nolA* into a *B. japonicum* strain deleted for *nodD$_1$, D$_2$* and *nolA* significantly improved nodulation. Repression of *nod* genes has also been reported in *Rhizobium meliloti* strain AK631, wherein a repressor, NolR, binds to the *nodD$_1$* and *nodABC* promoters and inhibits both *nodD$_1$* and *nodABC* expression (Kondorosi et al. 1991). Mutations in the *nolR* gene lead to increased *nod* gene expression and to a decreased ability of *R. meliloti* to nodulate its host plants. These data for both *nolR* and *nolA* suggest that optimum nodulation requires fine-tuning of *nod* gene expression mediated by both positive and negative regulators.

Recently, we reported the construction of two *nolA* mutations in *B. japonicum* obtained by interposon mutagenesis (Garcia et al. 1996). These *B. japonicum* mutant strains showed grossly defective nitrogen fixation and nodulation phenotypes on cowpea, but were only slightly delayed for nodulation of soybean. Surprisingly, these mutations failed to result in higher expression of either *nodD$_1$* or *nodY* expression. Therefore, contrary to our previous hypothesis, NolA does not appear to act directly to repress *nod* gene expression. However, results indicated that NolA is required for both its expression and that of *nodD$_2$*. This latter observation is interesting in view of reports by Gillete and Elkan (1995) that a protein homologous to NodD$_2$ is able to repress *nod* gene expression in *Bradyrhizobium* sp. (Arachis) NC92. Similar to *B. japonicum, nolA* of *Bradyrhizobium* sp (Arachis) is also required for increased expression of *nodD$_2$*. Given this observation, the possibility that NodD$_2$ can act as a repressor in *B. japonicum* was examined. The expression of NodD$_2$ from a constitutive promoter resulted in a drastic reduction in the expression of a chromosomal *nodC-lacZ* fusion. As NolA is required for NodD$_2$ expression, it is possible that NolA mediates the repression of *nod* genes indirectly, possibly through NodD$_2$. However, as it is obvious that *nod* gene regulation in *B. japonicum* is complex and involves multiple regulatory factors, we are cautious with regard to attributing the repressive effects of NolA simply to an effect on *nodD$_2$*. In a similar fashion, it is unlikely that the nodulation and nitrogen fixation defects observed with the *nolA* mutants are solely due to an effect on Nod signal production, mediated by the *nod* gene products. In all probability, NolA regulates the expression of yet unidentified bacterial determinants which are required for the infection of host plants.

NolA belongs to the MerR family of transcriptional regulators (Sadowsky et al. 1991). Members of this family are functionally similar in that they are all involved in regulating the expression of genes that provide resistance to toxic compounds (eg. O'Halloran et al. 1989). These regulatory proteins have been shown to bind to a site between the -10 and -35 consensus sequence within their target promoters. These promoters have the unique feature in that the -10 and -35 regions are separated by 19-bp, rather than the usual 16/17-bp. An inverted repeat sequence is contained within the 19-bp. Examination of the sequences upstream of both *nolA* and *nodD$_2$* revealed putative NolA binding sites sharing similar characterisitcs to those recognized by the MerR regulators. Recent data have shown that *nolA* and *nodD$_2$* are transcribed immediately downstream of these putative NolA binding sites. These results, though not conclusive, do provide evidence suggesting that NolA regulates the expression of these genes by binding to these promoter sequences.

Further examination of the *nolA* coding sequence reveals that, in addition to the ATG start codon proposed by Sadowsky et al (1991), a second ATG start codon preceded by a conserved ribosome binding site can be identified within the *nolA* coding region. Translation from the first ATG would give a full-length protein (NolA$_L$) possessing an N-terminal helix-turn-helix DNA binding motif. Translation from the second ATG would result in a N-terminally truncated protein (NolA$_S$) devoid of the DNA binding domain. Both proteins are identical in the C-terminal region since the same reading frame is utilized. Consistent with these observations, western blots of *B. japonicum* cell extracts using antibody made against NolA revealed two proteins of the expected size for NolA$_L$ and NolA$_S$. Moreover, primer extension experiments indicated that *nolA* is transcribed from two promoters. It now appears that the expression of NolA$_L$ and NolA$_S$ is controlled transcriptionally by formation of two mRNAs that are translated at alternative ATG start codons. The longer transcript, presumably leading to NolA$_L$, is inducible by isoflavones. The shorter transcript, giving rise to NolA$_S$ requires NolA (ie. NolA$_L$) for expression.

DICARBOXYLIC ACIDS REPRESS *NOD* GENE EXPRESSION

Recent observations in our laboratory have shown that specific dicarboxylic acids can inhibit *nod* gene expression in *B. japonicum* (Yuen et al. 1996).

Effective nodulation seems to require the repression of *nod* genes in planta (Knight et al. 1986). Indeed, constitutive expression of *nod* genes in *R. leguminosarum* bv. *vicae* resulted in the formation of Fix⁻ nodules. The mechanism for *nod* gene expression in planta is currently unknown. However, as dicarboxylic acids are the primary carbon source for rhizobia in planta (Streeter 1991), the possibility exists that dicarboxylic acids may play a critical role in the repression of *nod* genes in bacteriods. In this connection, it has been reported that *Rhizobium* strains defective in TCA cycle enzymes were found to form Fix⁻ nodules. More recently, analysis by Mavridou et al (1995) of a *R. leguminosarum* bv. *vicae dctB* mutant that demonstrated increased dicarboxylic acid transport showed a significant reduction in *nodD* and *nodABC* gene expression. These results suggest that the inhibition of *nod* genes by dicarboxylic acids may be important for effective nodulation.

## FINAL COMMENTS

It is now clear that *nod* gene regulation in *B. japonicum* is more complex than expected, and distinctly different from regulatory systems elucidated in *Rhizobium* species. Indeed, our present model (Figure 1) shows that the control of *B. japonicum nod* gene expression involves members of three global regulatory families; namely, the Lys-R family (ie. $NodD_1$), two-component regulatory family (ie. NodVW) and the MerR-type family (ie. NolA). Further complexity is added by the fact that specific regulators can exist in two forms; for example, phosphorylated and unphosphorylated NodW and $NolA_L$ and $NolA_S$.

The use of these three different regulators raises the question as to the purpose of such complexity. For example, given that NodVW is alone sufficient for *nod* gene expression, why then does $NodD_1$ appear to provide a similar function? Does this redundancy reflect some other primary function for $NodD_1$, or does the combined use of both positive regulators provide a "competitive"-edge for this slow-growing rhizobial strain? Also, NolA is involved in the regulation of *nod* genes whose products are involved in the synthesis of lipochitin-oligosaccharides. These nod signal molecules are involved in the initiation of the early nodulation steps. Yet, NolA also appears to be involved in the later stages of nodule development, suggesting that it regulates a set of genes that have yet to be identified. Our recent work, though emphasizing much progress in our understanding of the regulation of *nod* genes in *B. japonicum*, does underscore the reality that many gaps still exist in our understanding of this complex biological system.

## LITERATURE CITED

Banfalvi, Z., Nieuwkoop, A. J., Schell, M. G., Besl, L., and Stacey, G. 1988. Regulation of *nod* gene expression in *Bradyrhizobium japonicum*. Mol. Gen. Genet. 214:420-424.
Carlson R. W., Price, N. P. J., and Stacey, G. 1994. The biosynthesis of rhizobial lipo-oligosaccharide nodulation molecules. Mol. Plant-Microbe Int. 7:684-695.

Charles, T. C., Jin, S., and Nester, E. W. 1992. Two component sensory transduction systems in phytobacteria. Ann. Rev. Phytopathol. 30:463-484.

Dockendorff, T. C., Sanjuan, J., Grob., P., and Stacey, G. 1994. NolA represses nod gene expression in *Bradyrhizobium japonicum*. Mol. Plant-Microbe Int. 7:596-602.

Garcia, M. L., Dunlap, J., Loh, J., and Stacey, G. 1996. Phenotypic characterization and regulation of the nolA gene of *Bradyrhizobium japonicum*. Mol. Plant-Microbe Int. (in press)

Gillette, W. K., and Elkan, G. H. 1995. *Bradyrhizobium* sp. (Arachis) NC 92 contains two *nodD* genes involved in the repression of *nodA* and *nolA* genes required for the efficient nodulation of host plants. J. Bacteriol. 178:2757-2766.

Göttfert, M., Lamb, J. W., Gasser, R., Semenza, J., and Hennecke, H. 1989. Mutational analysis of the *Bradyrhizobium japonicum* common *nod* genes and further *nod*-box-linked genomic DNA regions. Mol. Gen. Genet. 215:407-4145.

Göttfert, M., Grob, P., and Hennecke, H. 1990. Proposed regulatory pathway encoded by the *nodV* and *nodW* genes, determinants of host specificity in *Bradyrhizobium japonicum*. Proc. Natl. Acad. Sci. (USA) 87:2680-2684.

Göttfert, M., Holzduser, D., and Hennecke, H. 1992. Structural and functional anlysis of two different *nodD* genes in *Bradyrhizobium japonicum* USDA 110. Mol. Plant-Microbe Int. 5:257-265.

Kondorosi, E., Pierre, M., Cren, M., Haumann, U., Buire, M., Hoffman, B., Schell, J., and Kondorosi, A. 1991. Identification of NolR, a negative transacting factor controlling the nod regulon in *Rhizobium meliloti*. J. Mol. Biol. 222:885-896.

Knight, C. D., Rossen, L., Robertson, J. G., Wells, B., and Downie, J. A. 1986. Nodulation inhibition by *Rhizobium leguminosarum* multicopy *nodABC* genes and analysis of early stages of plant infection. J Bacteriol. 166:552-558.

Mavridou, A., Barny, M-A., Poole, Pl, Plaskitt, K., Davies, A. E., Johnston, A. W. B., and Downie, J. A. 1995. *Rhizobium leguminosarum* nodulation gene (*nod*) expression is lowered by an allele-specific mutation in the dicarboxylate transport gene *dctB*. Microbiology (U.K.) 141:103-111.

Nieuwkoop, A. J., Banfalvi, Z., Deshmane, N., Gerhold, D., Schell, M. G., Sirotkin, K. M., and Stacey, G. 1987. A locus encoding host range is linked to the common *nod* genes of *Bradyrhizobium japonicum*. J. Bacteriol. 169:2631-2638.

O'Halloran, T. V., Frantz, B., Shin, M. K., Ralston, O. M., and Wright, J. G. 1989. The MerR heavy metal receptor mediates positive activation in a topologically novel transcription complex. Cell 56:119-129.

Peters, N. K. and Verma, D. P. S. 1990. Phenolic compounds as regulators of gene expression in plant-microbe interactions. Mol. Plant-Microbe Int. 3:4-8

Sadowsky, M. J., Cregan, P. B., Gottfert, M., Sharma., Gerhold, D., Rodriguez-Quinones, F., Keyser, H. H., Hennecke, H., and Stacey, G. 1991. The *Bradyrhizobium japonicum nolA* gene and its involvement in the genotyp-specific nodulation of soybeans. Proc. Natl. Acad. Sci. (USA) 88:637-641

Sanjuan, J., Grob, P., Göttfert, M., Hennecke, H., and Stacey, G. 1994. NodW is essential for full expression of the common nodulation genes in *Bradyrhizobium japonicum*. Mol. Plant-Microbe Int. 7:364-369.

Smit, G., Puvanesarajah, V., Carlson, R. W., Barbour, W. M., and Stacey, G. 1992. *Bradyrhizobium japonicum nodD₁* can be specifically induced by soybean flavonoids that do not induce the *nodYABCSUIJ* operon. J. Biol. Chem. 267:310-318.

Streeter, J. G. 1991. Transport and metabolism of carbon and nitrogen in legume nodules. Adv. Bot. Res. 18:129-187.

Wang, S-P., and Stacey, G. 1991. Studies of the *Bradyrhizobium japonicum nodD₁* promoter: a repeated structure for the *nod* box. J. Bacteriol. 173:3356-3365.

Yuen, J., and Stacey, G. 1996. Inhibition of *nod* gene expression in *Bradyrhizobium japonicum* by organic acids. Mol. Plant-Microbe Int. 9:424-428.

# Signal Peptidases of *Bradyrhizobium japonicum* as new symbiosis-specific proteins

Peter Müller, Andrea Bairl, Anja Klaucke, Christian Sens, Till Winzer

Philipps Universität Marburg, Marburg, Germany

The interaction between various *Rhizobium* strains and their homologous legume host plants has been analyzed on a molecular level mostly by bacterial mutants which are blocked at different steps of nodule formation and symbiotic Nitrogen fixation (Jacobi et al. 1995). While the early events of nodule intitation as well as the regulation and expression of $N_2$ fixation have been intensely studied, there are only a few examples of rhizobial mutants defective in nodule organogenesis and bacteroid differentiation. The use of Tn*phoA* for random mutagenesis of *B. japonicum* 110*spc*4 allowed the direct screening on Agar-plates for mutations within genes for secreted proteins. Two mutant strains, 132 and 184, were analyzed in more detail. They fixed $N_2$ at a markedly reduced level, but their *ex planta* nitrogenase activities were maintained, indicating that essential *nif* and *fix* genes were still intact (Müller et al. 1995a).

## MICROSYMBIONT GENES AFFECT THE SYMBIOSOME MEMBRANE

The symbiotic phenotypes of mutants 132 and 184 were analyzed by nodule sections both on the light microscopic and electron microscopic level. Mutant 184 is characterized by an abnormal pattern of nodule colonization. The mutants infect the plant root via infection threads, but the bacteria are released only in a few cells of the central tissue of the emerging nodule. As a consequence the majority of cells of the central nodule tissue remain uninfected. Nodule organogenesis procedes with some delay and the infected cells of the central nodule tissue contain fewer bacteroids compared to wild-type induced nodules. This phenotype might be explained by a longer generation time of *B. japonicum* mutant 184.

Single mutant 184 bacteroids are enveloped by symbiosome membranes, but they are irregularly shaped and undergo bacterial lysis. Despite the instability of the symbiotic compartments it is possible to isolate symbiosome membranes from mutant 184 infected nodules 4 - 6 weeks after inoculation. As reported earlier (Werner et al. 1995) 2-D SDS-PAGE in combination with a Western analysis using a symbiosome membrane nodulin-specific antibody revealed that a number of proteins was detectable

313

at a lower level, whereas other proteins were present in higher amounts. The overall expression pattern of symbiosome membrane proteins was similar to results obtained after induced senescence by cutting the plant shoots.

When one of the protein spots which was expressed at a markedly lower level in mutant 184-induced soybean nodules was analyzed in more detail, it was found that three distinguishable proteins of about 53 kDa could be identified. All were detected by a symbiosome membrane nodulin specific antiserum. The most prominent protein was enriched and purified and finally subjected to amino acid sequence analysis. Since the entire protein was N-terminally blocked, Lys-C protease cleavage products were used for Edman degradation. The resulting partial amino acid sequences did not have any significant homologies to other proteins of the NIH database. The protein function remains to be determined.

Compared to *B. japonicum* 184 the symbiotic phenotype of strain 132 is more severe, since the symbiotic nitrogenase activity is only about 10% of the wildtype level. Nodule structure analysis revealed that mutant 132-infected cells were extensively vacuolized. Electron microscopic analyses show that cells of the invading mutant 132 are stably maintained as long as they are located within infection threads, but soon after bacterial release they commence to lyse. Finally the majority of the "symbiosomes" are devoid of bacteroids but contain cell debris and these "garbage packages" fuse and form large central vacuoles. Immunohistochemical studies showed that the concentration of PEP-carboxylase, a nodulin providing carbon skeletons for ammonium assimilation, was clearly reduced. Furthermore, the polyclonal antibody directed against symbiosome membrane-specific nodulins resulted in weaker signals in thin sections of mutant induced nodules, whereas chitinase was detected at higher amounts in infected nodule tissue. In addition, callose formation was observed in some central nodule cells which were in contact with infection threads, indicating plant defense reactions. This phenomenon was most clearly visible 24 days after inoculation.

## THE ALTERATIONS IN THE SYMBIOSOME MEMBRANE ARE CAUSED BY MUTATIONS IN TWO DIFFERENT SIGNAL PEPTIDASE GENES OF *B. japonicum*

The mutated DNA fragments of *B. japonicum* strains 132 and 184 were cloned and used for vector integration mutagenesis of the wild-type strain, and it was verified that the original Tn*phoA* insertions were responsible for the phenotypic characteristics of the mutant stains. DNA sequence determination and subsequent comparative analysis with data in sequence databanks revealed that in both instances Tn*phoA* was inserted in genes homologous to prokaryotic type I signal peptidases (Müller et al. 1995b). The genes were therefore designated *sipS* (132) and *sipF* (184). The deduced amino acid sequences of *sipS* and *sipF* share marked homologies, but no cross hybridization signals were detected by Southern analyses. A comparison of both genes at the nucleotide level revealed a less pronounced similarity. Since the *phoA* reporter gene is fused in frame with *sipS*, mutant 132 forms a translational fusion product which is expressed in aerobically grown free-living cells as indicated by intensely blue stained colonies on peptone salt yeast extract agar plates containing X-Phosphate as the chromogenic substrate. This result suggested that *sipS* is expressed constitutively and does not require plant derived inducers. In contrast, Tn*phoA* was found to be inserted opposite to the orientation of *sipF*, and therefore mutant 184 formed white colonies on this medium.

The genetic analyis in the vicinity of the Tn*phoA* insertion site in strain 132 was further extended with a 5.9 kb *Hin*dIII/*Bam*HI subclone. On this wild-type DNA-fragment three complete ORFs are encoded and further upstream a truncated ORF is located which all are oriented in the same direction (Fig. 1). ORF1 does not show significant similarity to other sequences of the databank, but its hydrophobicity profile indicates that it has a potential transmembrane helix close to the C-terminus. ORF2 comprises 480 amino acids which form a cytoplasmatic protein that has 43% identity to an *E. coli* ORF of 481 aa. The identity of the second half of the protein is remarkably higher (60% identity), indicating that this domain is particularly important for protein function. ORF3 is equivalent to *sipS*. The stop codon of ORF2 overlaps with the start codon of ORF2, indicating a close translational coupling. ORF4 encodes a 376 amino acid protein and overlaps with the *sipS* gene, suggesting that at least these four ORFs form one common operon. The operon apparently ends with ORF4, since another truncated ORF lies downstream and is divergently oriented. ORF5 is highly homologous to γ-glutamyl cysteine synthetase of chloroplasts of *Arabidopsis thaliana* (May et al. 1994), the primary enzyme in the glutathione biosynthesis pathway. Fragments of the putative ORF4 gene product resemble the *secD* gene product, known to be involved in the general secretion pathway of Gram negative bacteria, and the cyclosporin A- binding site of a yeast peptidyl-prolyl-*cis/trans*-isomerase, also known as rotamase. Therefore, the ORF4 protein, which also has two potential transmembrane regions, could interact with the signal peptidase. Since mutations in this gene do not result in a detectable phenotype, the mutant phenotype of strain 132 is due to a truncated *sipS* gene product and not to a polar effect on the expression of downstream genes.

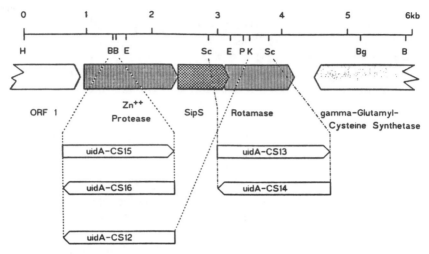

Fig. 1 Physical and genetic map of *B. japonicum sipS* DNA-region. Putative ORFs and homologous proteins are indicated below the thick arrows. Replacements of deletions by transcriptional *uidA* reporter gene fusions are represented by thin arrows. Abbreviations: **B** *Bam*HI, **Bg** *Bg*lII, **E** *Eco*RI, **H** *Hin*dIII, **K** *Kpn*I, **P** *Pst*I, **Sc** *Sac*I.

The putative biochemical function of the *sipS* gene product was substantiated by a heterologous complementation analysis of an *E. coli* temperature sensitive leader peptidase mutant (*E. coli* IT41, *lep*[ts], Inada et al., 1989). Hybrid plasmids carrying *B. japonicum* DNA fragments with an intact *sipS* gene and providing a vector-borne promoter in the proper orientation complemented the mutation and allowed groth at 42°C. This experiment proves that *B. japonicum sipS* indeed encodes a signal peptidase which is expressed and functionally active in an heterologous background. These results indicate that the substrate specificity of SipS is low, at least in *E. coli*. Furthermore it can be concluded that the indigenous promoter is located upstream of the 5'-end of the cloned DNA fragment.

RHIZOBIAL SIGNAL PEPTIDASES MIGHT BE A NOVEL REGULATION MECHANISM FOR THE CONTROL OF SYMBIOSIS-SPECIFIC PROTEINS

Since genetic evidence indicated that the indigenous promoter of the putative operon was located further upstream, a gene bank of *B. japonicum* 110*spc*4 was screened for a cosmid covering a larger DNA fragment of this genetic region. Cosmid C-27 was isolated and the *sipS* gene was mapped on the 1.6 kb *Eco*RI fragment at the 3'-end. DNA sequence analysis of various shot gun subclones of the entire C-27 insert revealed some interesting sequence homologies to other proteins in the databank. Total sequence identity to *B. japonicum coxA* (Bott et al. 1990) allowed the exact location of the insert within the *B. japonicum* genome, based on the map of Kündig et al. 1992. Furthermore homologies to cytochrome oxidase subunits II and III were detected, as well as to mitochondrial processing peptidases of different origins (e.g. Hawlitschek et al. 1988). Most exciting was the existence of a second copy of ORF2, which is homologous to a $Zn^{2+}$ protease. Both allelic forms, which have highly conserved but not identical C-terminal amino acid sequences, could function as a heterodimer, and they could be involved in the processing of several different preproteins. Whether SipS is directly involved in protein processing of cytochrome subunits or any auxillary proteins remains to be determined. More likely appears the connection to other genes, which encode a novel response regulator (Anthamatten et al. 1991) and a metabolite transporter (Seol et al. 1991) although the exact location on the C-27 insert is not yet clear.

In order to study the expression of the *sipS* operon during the symbiotic interaction several deletions were replaced by a promoterless *uidA* gene encoding the β-glucuronidase. The resulting transcriptional fusions (Fig. 1) were introduced into *B. japonicum* wild type by single cross-over. Depending on the site of recombination, white or blue colonies of the transformant strains were obtained on selective agar plates containing X-gluc as the chromogenic indicator substrate. In plant assays the inoculant strains behaved similarly within nodules. In addition, the inoculant strains were grown as microspots on agar plates in parallel aerobically (air) and microaerobically ($1\% \ O_2$). After 8 to 10 days the microaerobically grown colonies turned blue whereas the areobically grown colonies were still white. This result suggests that the *sipS* operon is expressed preferably under microaerobic (i. e. symbiotic) conditions, and the response regulator close by might be involved in the regulation of this operon.

ACKNOWLEDGMENTS

We would like to thank D.P.S. Verma, Ohio State University, for providing the SM-specific nodulin antiserum, Drs. Linder, Justus-Liebig-University, Gießen for amino acid sequence determination of the 53 kDa protein, and K. Wilson, CAMBIA for sending *uidA* reporter gene constructs.

LITERATURE CITED

Anthamatten, D., Hennecke, H. 1991. The regulatory status of the *fixL*- and *fixJ*-like genes in *Bradyrhizobium japonicum* may be different from that in *Rhizobium meliloti*. Mol. Gen. Genet. 225:38-48

Bott, M., Bolliger, M., and Hennecke, H. 1990. Genetic analysis of the cytochrome c-aa₃ branch of the *Bradyrhizobium japonicum* respiratory chain. Molecular Microbiology 4:2147-2157.

Inada, T., Court, D.L., Ito, K., and Nakamura, Y. 1989. Conditionally lethal amber mutations in the leader peptidase gene of *Escherichia coli*. J. Bacteriol. 171:585-587.

Hawlitschek, G., Schneider, H., Schmidt, B., Tropschug, M., Hartl, F.-U., Neupert, W. 1988. Mitochondrial protein import: Identification of processing peptidase and of PEP, a processing enhancing protein. Cell 53:795-806.

Jacobi, A., Werner, D., Müller, P. 1995. Molekulare Mechanismen der Symbioseentwicklung von *Rhizobium/Bradyrhizobium* und Leguminosen. BIO*spektrum* 2:21-28.

Kündig, Ch., Hennecke, H. and Göttfert, M. 1992. Correlated physical and genetic map of the *Bradyrhizobium japonicum* 110 genome. J. Bacteriol. 175:613-622.

May, M.J., and Leaver, C.J., 1994. Arabidopsis thaliana γ-glutamylcysteine synthetase is structurally unrelated to mammalian, yeast, and *Escherichia coli* homologs. Proc. Natl. Acad. Sci. U.S.A. 91:10059-10063.

Müller, P., Klaucke, A., and Wegel, E. 1995a. Tn*phoA*-induced symbiotic mutants of *Bradyrhizobium japonicum* that impair cell and tissue differentiation of *Glycine max* nodules. Planta 197:163-175.

Müller, P., Ahrens, K., Keller, T. and Klaucke, A. 1995b. A Tn*phoA* insertion within the *Bradyrhizobium japonicum sipS* gene, homologous to prokaryotic signal peptidases, results in extensive changes in the expression of PBM-specific nodulins of infected soybean (*Glycine max*) cells. Molecular Microbiology 18:831-840.

Seol, W., Shatkin, A.J. 1991. *Escherichia coli kgtP* encodes an α-ketoglutarate transporter. Proc. Nat. Acad. Sci. U.S.A. 88:3802-3806.

Werner, D., Jacobi, A., Müller, P., and Winzer, T. 1995. Symbiosomes under control of plant and microsymbiont genes. Pages 449-453 in: Nitrogen Fixation: Fundamentals and Applications. I. A. Tikhonovich, N. A. Provorov, V. I. Romanov, and W. E. Newton, eds. Kluwer Academic Publishers Dordrecht/Boston/London.

# Rhizobium species NGR234 Host-specificity of Nodulation Locus III contains *nod-* and *fix-*genes

S. Jabbouri[1], M. Hanin[1], R. Fellay[1], D. Quesada-Vincens[1], B. Reuhs[2], R.W. Carlson[2], X. Perret[1], C. Freiberg[3], A. Rosenthal[3], D. Leclerc[4], W.J. Broughton[1], and B. Relic'[1,4]

[1]LBMPS, Université de Genève, Geneva, Switzerland; [2]Complex Carbohydrate Research Centre, University of Georgia, Athens, GA, USA; [3]MB Institut für Molekular Biotechnologie, Jena, Deutschland; [4]Friedrich Miescher Institut, Basel, Switzerland.

**Abstract** Most nodulation genes of the broad host-range *Rhizobium* sp. NGR234 are located on the 536 kb symbiotic plasmid. Complete sequencing of the replicon revealed over 400 open-reading frames (ORF), many of which correspond to known genes. Combined physical/genetic analysis of the host-specificity of nodulation (*hsn*) locus III revealed many nodulation genes, some of which are essential for Nod-factor production (*nodABC*) and others which modify the Nod-factor structure in specific ways (*noeE*). When mutated, two other genes, ORF3 and *fixF* gave Fix⁻ phenotypes on *Vigna unguiculata*. *fixF* has a nuclear localisation sequence (NLS) which is functional in nuclear targeting in tobacco protoplasts. Mutations in *fixF* abolish production of a novel, rhamnose-rich lipopolysaccharide that is produced under conditions of flavonoid induction.

*Rhizobium* sp. NGR234 nodulates more than 110 genera of legumes as well as the non-legume *Parasponia andersonii* (S.G. Pueppke and W.J. Broughton unpublished). Most of this capacity to nodulate a broad spectrum of plants may be attributed to a family (> 80) lipo-oligosaccharide Nod-factors which are excreted into the legume rhizosphere (Price et al.

319

1992; S. Jabbouri, unpublished). Nod-factors provoke curling of the root-hairs, cortical cell-division, and entry of rhizobia into the root (Lerouge et al. 1990; Truchet et al. 1991; Relic´ et al. 1994). The genome of NGR234 is partitioned between a 5,700 kb chromosome and a 536 kb plasmid on which most of the symbiotic genes are located. Complementation experiments identified three plasmid-borne *hsn* loci, one of which (HsnIII) contains the *nodABC* genes (Broughton et al. 1986; Lewin et al. 1987). As the plasmid has now been completely sequenced (C. Freiberg, R. Fellay, A. Bairoch, W.J. Broughton, A. Rosenthal, and X. Perret, unpublished) all the ORF's in the *hsn*-loci have been identified. Here we present a molecular-genetic and biochemical analysis of *hsn*III.

MATERIALS AND METHODS Standard chemical, genetic, and molecular techniques were used (see Carlson et al. 1978; Fellay et al. 1995; Jabbouri et al. 1995; Negrutiu et al. 1987).

RESULTS AND DISCUSSION There are thirteen ORF's in the *hsn*III locus, arranged in what appears to be five operons (Fig. 1).

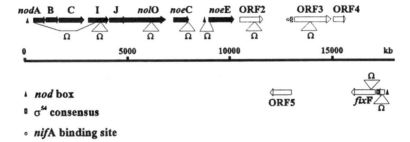

Fig. 1 Genetic and physical map of the *hsn*III locus of *Rhizobium* sp. NGR234. *nod*, *nol*, and *noe* stand for nodulation- and *fix* for nitrogen fixation genes. Ω is the Omega antibiotic resistance cassette. Arrows show the direction of transcription.

Mutations in *nodABC* are Nod⁻ on all plants tested, while those in *nodI* or *nolO* (part of the same operon) affect Nod-factor structure (Fig. 2) but not the phenotype on *Vigna unguiculata*. Unfortunately, the Omega insertion in *noeC* is in the last codon of the ORF and thus does not give a distinct phenotype. Introduction of a plasmid containing *nolO* or *noeC* confers on *R. fredii* USDA257 the capacity to nodulate *Calopogonium caeruleum* however.

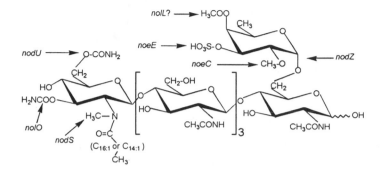

Fig. 2. Structures of Nod-factors produced by *Rhizobium* sp. NGR234 and its mutants.

NoeC seems to be a 2-*O*-methyltransferase. Although the function of ORF2 is unknown, it has homology to ChaA, a calcium/proton antiporter of *Escherichia coli*. NoeE appears to be the enzyme which transfers sulphate to the *O*-methylfucose moiety of Nod-factors, and its synthesis is under the control of a flavonoid-inducible *nod*-box. As mutations in ORF4 and ORF5 have yet to be made, little can be said about their function. ORF3 and *fixF* however, are interesting in that they may be regulated by the $\sigma^{54}$ and that mutations in either gene are Fix⁻ on *Vigna*. ORF3 has homology to ABC transporter genes. In contrast, *fixF* shares homology with the *kpsS* gene of the *E. coli* antigen cluster and contains the nuclear location sequence KRSDYEFLTTGNVLRKKIRLP (requirements for nuclear localisation are underlined).

NLS activity was tested by fusing the first 300 bp of *fixF* which contains the bi-partite NLS in frame to the gene encoding the green fluorescent protein (GFP)(Haseloff and Amos, 1995). *Nicotiana plumbaginifolia* protoplasts were transformed using the polyethylene glycol method, and fluorescence excited using blue light. Green fluorescence was only observed in the nucleus of protoplasts transformed with the intact NLS while the fluorescence was evenly dispersed amongst the cytoplasm of those protoplasts transformed with GFP alone (not shown). These data suggest that the FixF protein may be either transported out of the

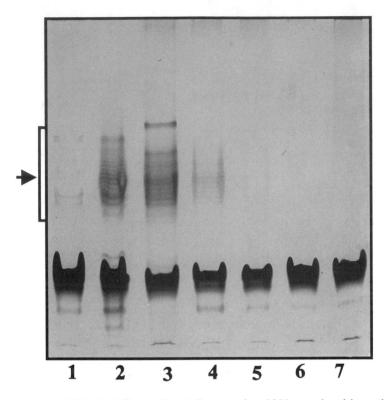

Fig. 3. DOC-PAGE profiles (silver-stain, 18% acrylamide gels, deoxycholic acid as the detergent) of the phenol-water extracts (water layer) of polysaccharides of *Rhizobium* sp. NGR234 and its mutants. Arrow indicates migration of the inducible rhamnose-rich cell wall compound. Lanes: **1** - NGR234 uninduced; **2** - NGR234 induced by $10^{-6}$ M apigenin; **3** bacteria isolated from nodules of *V. unguiculata* inoculated with wild-type NGR234; **4** - NGRΔ*syrM*; **5** - NGRΔ*nodD2*; **6** - NGRΔ*nodD1*, and **7** - NGRΔ*fixF*.

bacterium, or that the protein itself is a transporter of other substances. Its role was further investigated by examining the polysaccharides produced by *fixF⁻* and various regulatory mutants. Bacterial polysaccharides were extracted with hot phenol-water from apigenin induced ($10^{-6}$ M) or non-induced cultures. After separation by poly-acrylamide gel electrophoresis, the different bands were visualised by silver staining. Under induced conditions, NGR234 produced large quantities of a polymeric substance which is not found in the regulatory mutant, *nodD1⁻*. Interestingly, smaller quantities were still found in cultures of $syrM^-$, and $nodD2^-$ mutants. What is important however is the finding that deletion of the $fixF$-gene leads to complete abolition of polysaccharide production (Fig. 2). Preliminary chemical analysis has shown that the polysaccharide is rich in rhamnose.

In summary, these data suggest that two genes, ORF3 and $fixF$, the activity of which is possibly controlled by $\sigma^{54}$, are essential for nitrogen fixation. And, although this remains to be proven, it is possible that FixF is involved in transport of a substance from Rhizobium to the nucleus of the host-plant.

## ACKNOWLEDGEMENTS

We wish to thank J. Kim and M. Müller for their help with this work. J. Haseloff kindly provided pBIN35SmGFP4. Financial assistance was provided by the Fonds National Suisse (Grant # 31-36454.92), the Université de Genève, the Fonds Marc Birkigt, and the Sociéte Academique de Genève.

## REFERENCES

Broughton W.J., Wong, C.-H., Lewin, A., Samrey, U., Myint, H., Meyer z.A., H., Dowling, D.N., and Simon, R. 1986. Identification of Rhizobium plasmid sequences involved in recognition of Psophocarpus, Vigna, and other legumes. J. cell Biol. 102:1173-1182.

Carlson, R.W., Sanders, R.E., Napoli, C., and Albersheim, P. 1978. Host-symbiont interactions. III. Purification and characterisation of Rhizobium lipopolysaccharides. Plant Physiol. 63:912-917.

Fellay, R., Perret, X., Viprey, V., Broughton, W.J., and Brenner, S. 1995. Organisation of host-inducible transcripts on the symbiotic plasmid of Rhizobium sp. NGR234. Mol. Microbiol. 16:657-667.

Haseloff, J., and Amos, B. 1995. GFP in plants. Trends in Genet. 11:328-329

Jabbouri, S., Fellay, R., Talmont, F., Kamalaprija, P., Burger, U., Relic´, B., Promé, J.-C., and Broughton, W.J. (1995)

Involvement of *nodS* in *N*-methylation and *nodU* in 6-*O*-carbamoylation of *Rhizobium* sp. NGR234 Nod-factors. J. biol. Chem. 270:22968-22973.

Lewin, A., Rosenberg, C., Meyer z.A., Wong, C.-H., Nelson, L., Manen, J.-F., Dowling, D.N., Dénarié, J., and Broughton, W.J. 1987. Multiple host-specificity loci of the broad host-range *Rhizobium* sp. NGR234 selected using the widely compatible legume *Vigna unguiculata*. Plant Mol. Biol. 8:447-459.

Negrutiu, I., Shillito, R.D., Potrykus, I., Biasini, G., and Sala, F. 1987. Hybrid genes in the analysis of transformation conditions I. Setting up a simple method for direct gene transfer in plant protoplasts. Plant Mol. Biol. 8:363-373.

# Exopolysaccharides and Their Role in Nodule Invasion

Gregory M. York, Juan E. González and Graham C. Walker

Department of Biology, Massachusetts Institute of Technology, Cambridge, MA 01239

## Introduction

Bacterial exopolysaccharide (EPS) is required for establishment of the nitrogen fixing symbiosis between *Rhizobium meliloti* and its host plant *Medicago sativa* (alfalfa). The symbiosis is initiated by the exchange of small signaling molecules between the plant and the bacterium. Flavonoids secreted by plant roots into soil are detected by bacteria and trigger the expression of bacterial *nod* genes (Fisher et al., 1992). The *nod* genes are involved in synthesis and modification of lipochitooligosaccharides (LCOS) (Denarie et al., 1992). LCOS secreted by the bacteria are detected by the plant, triggering root hair curling and the initiation of nodule primordia. The bacteria become trapped in curled root hairs and proceed to invade the developing nodule through infection threads. Bacterial EPS is crucial for this step; mutants which fail to produce EPS can not invade plant root nodules (Leigh et al., 1994).

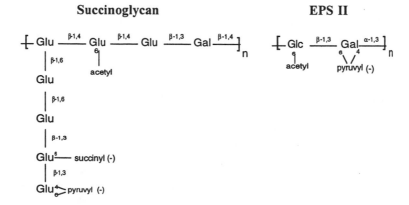

Fig. 1- *Rhizobium meliloti* exopolysaccharides

Although it is now well established that flavonoids and LCOS play signaling roles early in establishment of the symbiosis, the precise role of EPS is not yet clear. EPS may play a structural role, enabling the bacterium to attach to nodule tissue or providing protection against plant defense responses. Alternately, EPS may function as a signaling molecule, triggering a developmental response in the plant or regulating host defense responses. We are analyzing the synthesis, processing, and activity of two types of EPS produced by R. meliloti, succinoglycan and EPS II, with the aim of distinguishing between these roles for EPS in the symbiosis. A growing body of evidence suggests that various forms of EPS may act analogously to flavonoids and LCOS as small signaling molecules in establishment of the symbiosis.

## Results and Discussion

### Succinoglycan Biosynthesis

The wild type strain Rm1021 produces an EPS termed succinoglycan, which is crucial for nodule invasion by this strain. Succinoglycan is a polymer of an octasaccharide subunit, consisting of a backbone of one galactose and three glucose residues, a side chain of four glucose residues, and succinyl, acetyl, and 1-carboxyethylidene (pyruvyl) modifications in a ratio of approximately 1:1:1 (see Fig. 1) (Reuber et al., 1993a). Mutants defective in the production of succinoglycan fail to invade alfalfa root nodules (Leigh et al., 1994). Use of the laundry whitener Calcofluor, which fluoresces brightly upon binding succinoglycan, permitted the isolation of a large series of Tn5 mutants (exo) of Rm1021 which fail to fluoresce under ultraviolet light on growth medium containing Calcofluor and thus fail to produce succinoglycan. Plants inoculated with exo mutants form non-nitrogen fixing nodules which are devoid of differentiated, nitrogen fixing bacteroids (Leigh et al., 1994). Characterization of the nodules elicited by an exo mutant show that root hair curling is significantly delayed and that infection threads can form, but abort at a very early stage, so the bacterium never reaches the interior of the nodule. Furthermore, these exo mutants are able to elicit only two of 17 of the nodule-specific plant proteins, termed nodulins, which are elicited by Rm1021 (Norris et al., 1988).

Through a combination of genetic and biochemical approaches, we have developed a model for succinoglycan biosynthesis. The octasaccharide subunits are assembled on a lipid carrier, and in vitro labeling studies showed that the sequence of assembly is galactose, β-1,3-glucose, then the rest of the glucose residues and other substituents (Tolmasky et al., 1982). Our lab has been able to assign biochemical functions to many of the exo gene products by isolating lipid-linked biosynthetic intermediates from exo mutant strains, hydrolyzing the oligosaccharides from the lipid carriers, and characterizing the oligosaccharide intermediates to determine the biochemical block of each mutant (Reuber et al., 1993b). Based on this work, we have proposed that the products of the *exoY* and *exoF* genes function in the addition of the first sugar, galactose, to the lipid carrier, and the products of the *exoA, exoL, exoM, exoO, exoU* and *exoW* genes function in subsequent sugar additions (Reuber et al., 1993b). The products of the *exoH, exoZ,* and *exoV* genes are needed for the addition of succinate, acetate and pyruvate respectively (Reuber et al., 1993a; Reuber et al., 1993b). *exoP, exoQ,* and *exoT* mutants appear to make complete succinoglycan subunits, but these mutants are defective in nodule invasion, and therefore, the products of these genes are postulated to affect polymerization of the octasaccharide subunits or transport of the completed polymer (Reuber et al., 1993b). We and others have determined the DNA sequence of the 25 Kb *exo* gene cluster (Becker et al., 1993a; Glucksmann et al., 1993b).

LOW MOLECULAR WEIGHT SUCCINOGLYCAN

*R. meliloti* succinoglycan can be separated into fractions of high molecular weight (HMW) polymer and low molecular weight (LMW) monomers, trimers, and tetramers of octasaccharide. Two lines of evidence suggest that LMW succinoglycan is the form which is active in mediating nodule invasion. First, a mutant termed *exoH* produces HMW but not LMW succinoglycan and fails to invade nodules (Leigh et al., 1987). However, the basis of the invasion defect is not certain because *exoH* mutants also fail to add the succinyl modification to the HMW polysaccharide. Secondly, the addition to plant roots of LMW but not HMW succinoglycan is sufficient to restore invasion by *exoA, exoB, exoF,* and *exoH* strains (Battisti et al., 1992; Urzainqui et al., 1992). The active fraction of LMW succinoglycan was reported to be tetramers of a highly charged form of the octasaccharide (Battisti et al., 1992).

We are investigating how *R. meliloti* produces LMW succinoglycan. One possibility is that HMW succinoglycan is depolymerized to LMW forms by bacterial glycanases or lyases. A second possibility is that LMW succinoglycan is produced *de novo* by limited polymerization as in the case of the *Escherichia coli* and *Salmonella enterica* O-antigen component of lipopolysaccharide (Bastin et al., 1993). The two possibilities are not mutually exclusive, and indeed we have obtained evidence supporting both.

We have determined that *R. meliloti* produces at least two succinoglycan depolymerizing activities. The *R. meliloti* gene *exoK*, which is located within a cluster of genes involved in succinoglycan biosynthesis, encodes a protein homologous to bacterial endo-1,3-1,4-$\beta$-glycanases (Becker et al., 1993a; Glucksmann et al., 1993b) which would be predicted to depolymerize succinoglycan. On growth medium containing Calcofluor, *exoK* mutant colonies produce a delayed fluorescent halo relative to the wild type strain, implying reduced or delayed production of diffusible, partially depolymerized succinoglycan. On the basis of the Calcofluor halo phenotype, we have conducted a genetic screen and identified three additional genes which are involved in an *exoK* independent pathway for depolymerization of HMW succinoglycan. Partial nucleotide sequence implies that two of the genes are homologous to genes involved in Type I protein secretion; the third gene may encode the depolymerizing activity. *R. meliloti* mutants which are defective in both the *exoK* dependent and *exoK* independent succinoglycan depolymerizing pathways produce little or no detectable LMW succinoglycan in minimal medium cultures and exhibit a proportional increase in production of HMW succinoglycan. However, these mutants are still capable of invading alfalfa nodules, which seems to indicate that the strains can use an alternate means to produce sufficient LMW succinoglycan for invasion.

We have obtained evidence suggesting that *R. meliloti* can produce LMW succinoglycan by direct synthesis. To explore this possibility, we have analyzed the products of *in vitro* polymerization from various mutants thought to be involved in the export and/or polymerization of succinoglycan. Employing improved techniques for analyzing the biosynthesis of succinoglycan that involve electroporating radiolabelled sugar nucleotides in cells, (Semino et al., 1993) we have determined that an *exoP* mutant only releases octasaccharide monomers in solution, an *exoQ* mutant releases octasaccharide monomers, trimers and tetramers but no high molecular weight material, and an *exoT* mutant releases higher molecular weight material but no octasaccharide trimers and tetramers (González *et al.*, 1996). These results suggest that different genes are involved in the production of LMW and HMW polymerization products, and that *R. meliloti* can make the LMW fraction by direct synthesis.

LOW MOLECULAR WEIGHT EPS II RESCUES THE NODULE INVASION DEFICIENT MUTANT

The symbiotic defects of *exo* mutants can be suppressed by the presence of a mutation, *expR101* (Glazebrook et al., 1989), that derepresses synthesis of a second exopolysaccharide, EPS II. EPS II consists of a repeating disaccharide subunit of glucose and galactose, modified with an acetyl and a pyruvyl group (see Fig. 1) (Her et al., 1990). Genetic analyses have demonstrated that the products of a cluster of at least six *exp* genes located on the second symbiotic megaplasmid, as well as the product of the *exoB* gene, are required for EPS II synthesis. The presence of the *expR101* mutation causes increased transcription of the *exp* genes, resulting in production of large quantities of EPS II. We have constructed strains which produce succinoglycan or EPS II exclusively, or neither EPS. *M. sativa* plants inoculated with various *expR101 exo* strains, which produce only EPS II, form nitrogen-fixing nodules (Glazebrook et al., 1989).

EPS II production has also been reported for strains containing a mutation in the *mucR* locus (Keller et al., 1990; Zhan et al., 1989) and in strains with the recombinant cosmid pMuc (Zhan et al., 1989) The *mucR* mutation maps to a different location on the chromosome than the *expR101* mutation and each class of mutation seems to increase production of EPS II independently of the other (Glazebrook et al., 1989; Keller et al., 1990). The EPS II isolated from a strain carrying either an *expR101* or a *mucR* mutation is identical in terms of linkages and modifications (Her et al.,

1990; Levery et al., 1991). Nevertheless, only *expR101 exo* mutants (Glazebrook et al., 1989) but not *mucR exo* mutants (Pühler et al., 1993), are invasion proficient. We found that the difference between these two classes of EPS II-producing strains is due to specific production of a low molecular weight form of EPS II by *expR101* strains. A low molecular weight EPS II fraction of 15-20 disaccharide subunits efficiently allows nodule invasion by non EPS-producing strains when present in amounts as low as 7 picomoles per plant, suggesting that low molecular EPS II may act as symbiotic signal during infection (González et al., 1996).

## Conclusions

Experimental results increasingly support a specific signaling role for *R. meliloti* EPS. Both genetic and biochemical evidence indicates that LMW but not HMW forms of succinoglycan and EPS II are crucial for nodule invasion. EPS II in particular is active at very low concentrations. These results are consistent with the hypothesis that LMW EPS are small molecule signals which mediate nodule invasion, or more generally modulate plant defense or developmental responses.

## References

Bastin, D. A., Stevenson, G., Brown, P. K., Haase, A., and Reeves, P. R. 1993. Repeat unit polysaccharides of bacteria: a model for polymerization resembling that of ribosomes and fatty acid synthetase, with a novel mechanism for determining chain length. Mol. Microbiol. 7(5): 725-734.

Battisti, L., Lara, J. C., and Leigh, J. A. 1992. Specific oligosaccharide form of the *Rhizobium meliloti* exopolysaccharide promotes nodule invasion in alfalfa. Proc. Natl. Acad. Sci. USA 89: 5625-5629.

Becker, A., Kleickmann, A., Arnold, W., and Pühler, A. 1993a. Analysis of the *Rhizobium meliloti exoH, exoK, exoL* fragment: ExoK shows homology to excreted endo-β-1,3-1,4 glucanases and ExoH resembles membrane proteins. Molec. Gen. Genet. 238: 145-154.

Denarie, J., Debelle, F., and Rosenberg, C. 1992. Signalling and host range variation in nodulation. Ann. Rev. Microbiol. 46: 497-531.

Fisher, R. F., and Long, S. R. 1992. *Rhizobium*-plant signal exchange. Nature 357: 655-660.

Glazebrook, J., and Walker, G. C. 1989. A novel exopolysaccharide can function in place of the Calcofluor-binding exopolysaccharide in nodulation of alfalfa by Rhizobium meliloti. Cell 56: 661-672.

Glucksmann, M. A., Reuber, T. L., and Walker, G. C. 1993b. Genes needed for the modification, polymerization, export, and processing of succinoglycan by *Rhizobium meliloti*: A model for succinoglycan biosynthesis. J. Bacteriol. 175: 7045-7055.

González, J. E., Reuhs, B. L., and Walker, G. C. 1996. Low molecular weight EPS II of *Rhizobium meliloti* promotes nodule invasion in *Medicago sativa*. Proc. Natl. Acad. Sci. USA, in press.

Her, G.-R., Glazebrook, J., Walker, G. C., and Reinhold, V. N. 1990. Structural studies of a novel exopolysaccharide produced by a mutant of *Rhizobium meliloti* strain Rm1021. Carbohydr. Res. 198: 305-312.

Keller, M., Arnold, W., Kapp, D., Müller, P., Niehaus, K., Shmidt, M., Quandt, J., Weng, W. M., and Pühler, A. 1990. *Rhizobium meliloti* genes involved in exopolysaccharide production and infection of alfalfa nodules. 91-97 In *Pseudomonas*: Biotransformations, Pathogenesis, and Evolving Biotechnology,S. Silver, A. M. Chakrabarty, B. Iglewski and S. Kaplan, eds. American Society of Microbiology, Washington D.C.

Leigh, J. A., Reed, J. W., Hanks, J. F., Hirsch, A. M., and Walker, G. C. 1987. Rhizobium meliloti mutants that fail to succinylate their Calcofluor-binding exopolysaccharide are defective in nodule invasion. Cell 51: 579-587.

Leigh, J. A., and Walker, G. C. 1994. Exopolysaccharides of *Rhizobium*: synthesis, regulation and symbiotic function. Trends Genet. 10: 63-67.

Levery, S. B., Zhan, H., Lee, C. C., Leigh, J. A., and Hakomori, S. 1991. Structural analysis of a second acidic exopolysaccharide of *Rhizobium meliloti* that can function in alfalfa root nodule invasion. Carbohydr. Res. 210: 339-347.

Norris, J. H., Macol, L. A., and Hirsch, A. M. 1988. Nodulin gene expression in effective alfalfa nodules and in nodules arrested at three different stages of development. Plant Physiol. 88: 321-328.

Pühler, A. M., Arnold, W., Becker, A., Roxlau, A., Keller, M., Kapp, D., Lagares, A., Lorenzen, J., and Niehaus, K. 1993. The role of *Rhizobium meliloti* surface polysaccharides in nodule development. 207-212 In New Horizons in Nitrogen Fixation, R. Palacios, J. Mora and W. E. Newton, eds. Kluwer Academic Publishers, Netherlands.

Reuber, T. L., and Walker, G. C. 1993a. The acetyl substituent of succinoglycan is not necessary for alfalfa nodule invasion by *Rhizobium meliloti* Rm1021. J. Bacteriol. 175: 3653-3655.

Reuber, T. L., and Walker, G. C. 1993b. Biosynthesis of succinoglycan, a symbiotically important exopolysaccharide of Rhizobium meliloti. Cell 74: 269-280.

Semino, C. E., and Dankert, M. A. 1993. *In vitro* biosynthesis of acetan using electroporated *Acetobacter xylinum* cells as enzyme preparations. J. Gen. Microbiol. 139: 2745-2756.

Tolmasky, M. E., Staneloni, R. J., and Leloir, L. F. 1982. Lipid-bound saccharides in *Rhizobium meliloti*. J. Biol. Chem. 257: 6751-6757.

Urzainqui, A., and Walker, G. C. 1992. Exogenous suppression of the symbiotic deficiencies of *Rhizobium meliloti exo* mutants. J. Bacteriol. 174: 3403-3406.

Zhan, H., Levery, S. B., Lee, C. C., and Leigh, J. A. 1989. A second exopolysaccharide of *Rhizobium meliloti* strain SU47 that can function in root nodule invasion. Proc. Natl. Acad. Sci. USA 86: 3055-3059.

# Acidic capsular polysaccharides (K antigens) of *Rhizobium*.

Bradley L. Reuhs

Complex Carbohydrate Research Center and Dept. of Biochemistry, University of Georgia, Athens, Georgia, USA.

The cell-associated and extracellular polysaccharides of rhizobia and gram-negative plant pathogens are routinely studied as potential factors in symbiosis. These include the acidic capsular polysaccharides, or K antigens ("Kapselantigene"), lipopolysaccharides (LPS) and extracellular polysaccharides (EPS), which are freely excreted by the cells. In contrast to EPS, the K antigens are tightly associated with the bacterial cells, and do not impart a mucoid colony morphology to plated cells. K antigens are produced by many well studied bacteria, most notably *E. coli* and *Neisseria* spp., and in some cases the capsules have been shown to be important in the virulence of specific pathogenic strains, by protecting the bacteria from host defense responses (Jann and Jann, 1990).

The K antigens of *E. coli* are divided into two subclasses, group I and group II, based on several criteria, including structure (Jann and Jann, 1990): Group I K antigens are characterized by a temperature insensitive expression, the lack of a phospholipid anchor, the predominance of large polysaccharide repeating units (7-10 residues), which contain uronic acids and pyruvate as the acidic components, and a high molecular weight (HMW) range (>100,000 da).

The group II K antigens include those that contain 3-deoxy-D-*manno*-2-octulosonic acid (Kdo), uronic acids, sialic acid, amino acids, or phosphate as the acidic components. In contrast to the group I K antigens, the group II polysaccharides are relatively low molecular weight (LMW), and usually comprise relatively small repeating units of only two or three sugars. In addition, the expression of group II capsules is temperature regulated, with capsular expression occurring at $\geq 20°C$ or higher. The group II polysaccharides often possess a phospholipid anchor linked to the reducing end via Kdo. The production of the group II K antigens is under the control of the *kps* genes, which are tightly clustered in three distinct regions (Pazzani et al. 1994).

# The K antigens of *Rhizobium* spp.

BIOCHEMISTRY.

The production of K antigens by rhizobia was first demonstrated in *R. fredii* USDA205 (Reuhs et al. 1993). A novel polysaccharide was isolated from that strain and shown to consist of sequential repeating units of Kdo and galactose (Gal): $[\rightarrow)$-$\alpha$-D-Gal$p$ $(1\rightarrow5)$ $\beta$-D-Kdo$p$ $(2\rightarrow]_n$, which is structurally analogous to the group II K antigens of *E. coli*. Strain USDA205 also produces a second polysaccharide, which consists of 2-*O*-methyl-Man and Kdo. The extreme lability of these polysaccharides was responsible for difficulties in their analysis, and is probably the reason that they had not been identified earlier.

*R. fredii* USDA257 has been the subject of numerous biological and genetic studies. Structural studies of the K antigens showed that strain USDA257, like strain USDA205, produces two distinct polysaccharides (Forsberg and Reuhs, 1996); the primary polysaccharide consists of Kdo and mannose (Man): $[\rightarrow)3$-$\beta$-D-Man$p$-$(1\rightarrow5)$-$\beta$-D-Kdo$p$-$(2\rightarrow]$, and the secondary polysaccharide of Kdo and 2-*O*-methylMan: $[\rightarrow)3$-$\beta$-D-2-*O*-methylMan$p$-$(1\rightarrow5)$-$\beta$-D-Kdo$p$-$(2\rightarrow]$.

A similar polysaccharide was isolated from *R. meliloti* AK631, and present data indicate that the primary K antigen consists of disaccharide repeats of 4-deoxy-4-aminuronic acid and a variant of Kdo, 2-keto-3,5,7,9-tetradeoxy-5,7-diaminononulosonic acid (Reuhs et al. 1993); Reuhs et al., in preparation). Each subunit carries acetyl and $\beta$-hydroxybutyrate substitutions.

A polyacrylamide gel electrophoresis (PAGE) technique, which employs a differential staining technique, can be used to distinguish between LPS and the K antigens (Reuhs et al. 1993). PAGE gels are prestained with Alcian blue, a cationic dye, and the acidic K antigens appear as an Alcian blue-specific ladder pattern. This allows for the analysis of crude extracts from 2-5 ml cultures, without the degradation of the K polysaccharides that typically occurs during purification processes. This has facilitated the study of more than twenty strains of *R. fredii*, *R. meliloti*, and *Rhizobium* sp. NGR234. Each strain was found to produce K antigens and subsequent analyses have shown that the polysaccharides from this group of closely related species consistently contain Kdo or Kdo variants in the repeating units (Reuhs et al. 1996).

Despite the conserved motif for K antigen production in the *R. fredii/R. meliloti* group of rhizobia, these capsular polysaccharides are highly variable in structure. Each polysaccharide examined thus far varies in composition, linkage, substitutions, or size range. Thus, in contrast to EPS, which is conserved within species, the K antigens are strain-specific antigens. It may be that there are some differences in the structural motif between species, as a K antigen isolated from *R. leguminosarum* contains phosphate as the acidic unit (Reuhs et al., unpublished).

K ANTIGEN EXPRESSION.

Genetic investigations indicate that capsule expression in *R. meliloti* AK631 involves at least five separate capsule-specific gene regions, comprising a minimum of fifteen genes, but most likely many more. Three of these gene regions are presently under investigation (Kereszt et al., unpublished), and two have been characterized. First, it was shown that the plasmid-borne *rkpZ* (formerly *lpsZ*) gene product promotes the export of LMW polysaccharide. The activity of RkpZ is not dependent on K antigen structure, as the introduction of *rkpZ* into *R. fredii* USDA257, which does not carry *rkpZ* and produces a structurally distinct K antigen, resulted in a reduction in the size range of that polysaccharide (Reuhs et al. 1995). The predicted amino acid sequence of RkpZ is homologous to KpsC, which is involved in the polymerization and export of the group II K antigens of *E. coli* (Pazzani et al. 1994).

The second capsule-specific locus is the chromosomal fix-23 gene region. The first complementation unit of fix-23, which comprises six ORFs (termed *rkpABCDEF*), appears to encode a lipid carrier involved in the biosynthesis and export of the capsule (Petrovics et al. 1993). The remaining four genes (*rkpGH, I*, and *J*) in the last three complementation units encode products that appear to modify the lipid carrier or the polymerization and export processes. The predicted amino acid sequence of RkpJ shows significant homology to KpsS of *E. coli*, another protein involved in polymerization and export of the group II K antigens (Kiss et al. 1996; Pazzani et al. 1994). Mutations in the fix-23 region result in the intracellular accumulation of incompletely polymerized K polysaccharide. Importantly, Southern analysis of the total DNA extracts from several species of rhizobia, including *A. tumefaciens*, using *rkpABCDEF* as a probe has shown that this K antigen-specific gene region is ubiquitous in these soil bacteria (Reuhs et al., unpublished).

A gene region that controls host range in *R. fredii*-soybean interactions has been shown to affect the expression of the capsule (Kim and Reuhs, 1996; Balatti and Pueppke, 1990). Mutations in the *nolWXBTUV* genes of *R. fredii* USDA257 have been shown to extend the host range of this strain to improved cultivars of soybeans, which are not normally infected; these extended host-range mutants are significantly altered in their surface chemistry, with clear changes in both the LPS and K antigens. The exact mechanism by which the *nol* gene products effect these changes is unclear at this time, but the process appears to be secondary to normal capsule expression, as the genes are specific to *R. fredii* and are not present in *R. meliloti*.

A model for K antigen expression in rhizobia is suggested by the data gathered thus far from studies of *R. meliloti* and *R. fredii*: (1) The disaccharide repeating units are produced by gene products from the, as yet, uncharacterized gene regions; and an initial polymerization process, which takes place on a common lipid carrier, yields polysaccharide sub-units of ~5000-7000 daltons (8-15 repeating units). (2) Exportable polysaccharides are then assembled on a capsule (fix-23)-specific lipid carrier, resulting in HMW polymers of the subunits. (3) The polymerization process is modified by RkpZ, which promotes the export of the smaller polysaccharides. RkpZ may function in the modification or inhibition of a *cld* (chain length determinant) system. (4) Upon termination of the polymerization process, the K antigens are exported to the cell surface, and the lipid carrier is recycled. It is probable that the lipid carrier also functions in the transport of the polysaccharide across the CM. This model is complicated by the effects of various external factors on capsule expression; e.g., host-derived compounds induce changes in K antigen production in *R. meliloti* and *R. fredii* (Reuhs et al. 1995; Reuhs et al. 1994).

BIOLOGY.

Mutations that disrupt the bacterial production of specific polysaccharides often result in a lack of infectivity in *Rhizobium*-legume interactions. For example, EPS production by *R. meliloti* is normally required for the infection of alfalfa, an indeterminate host (Noel, 1992). The addition of exogenous EPS, at levels as low as $7 \times 10^{-12}$ M, to the inoculum of an EPS$^-$ mutant of *R. meliloti* promotes the infection of alfalfa (Gonzalez et al. 1996), indicating an active function for the EPS in the infection process. *R. meliloti* AK631 is an EPS$^-$ strain that is able to infect normally; mutations in the *rkp* genes, however, result in a Fix$^-$ phenotype, demonstrating that the K antigen of that strain is able to substitute for the EPS in the infection process (Reuhs et al. 1995; Petrovics et al. 1993). Given the structural differences between the EPS and the K antigens of *R. meliloti* AK631, it is probable that the two polysaccharides act to promote infection by different pathways, and some evidence has been found to support this premise.

Whole cells of *R. meliloti* Rm41, strain AK631, and three capsule mutants (*rkpA*, *rkpH*, and *rkpJ*), as well as purified K polysaccharides from strain AK631 were employed in a study of plant genes involved in the biosynthesis of flavonoids and plant defense compounds (Becquart-de Kozak et al. 1996). The infusion of the whole cells of AK631 or the purified K polysaccharide into the leaves of alfalfa seedlings resulted in a significant accumulation of chalcone synthase (CHS) mRNA, which is in the flavonoid pathway, suggesting a signal-based response of the host-plant to this bacterial product. In contrast, there was no response to whole cells of the *rkp* mutants, which lack the capsule and cannot infect the host plant, or to purified EPS. Importantly, the kinetics of CHS induction by *R. meliloti* or the K antigens were unique from the pattern of elicitation resulting from the infusion of plant pathogens, indicating that the response to the K antigen is not a typical defense response. Studies employing MAbs have shown that production of the *R. meliloti* capsule may ultimately be modified or shut down in the bacteroid (Olsen et al. 1992); Reuhs and Kim, in preparation). It is possible that the host-plant, in response to the K polysaccharide, produces another signal molecule, which is required for a further modification of the bacterial cell surface, as the cell undergoes morphogenesis to the bacteroid form.

SUMMARY

The K antigens are common products of rhizobia, and probably many other plant-microsymbionts. Studies of *R. fredii* and *R. meliloti* have shown that these capsular polysaccharides are highly variable strain-specific, surface antigens. Capsule expression involves at least fifteen genes in five distinct regions, and is regulated by host-derived compounds. Finally, the K antigens appear to perform infection related functions in certain host plants.

## Bibliography

Balatti P. A. and Pueppke S. G. 1990. Nodulation of soybean by a transposon-mutant of *Rhizobium fredii* USDA257 is subject to competitive nodulation blocking by other rhizobia. Plant Physiol. 94:1276-1281.

Becquart-de Kozak I., Reuhs B. L., Buffard D., Breda C., Kim J. S., Esnault R. and Kondorosi A. 1996. Role of the K-antigen subgroup of capsular polysaccharides in the early recognition process between *Rhizobium meliloti* and alfalfa leaves. MPMI. (Submitted).

Forsberg L. S. and Reuhs B. L. 1996. Characterization of the capsular antigens (K polysaccharides) from *Rhizobium fredii* USDA257 shows a conserved motif for capsule structure in *Rhizobium*. J. Biol. Chem. (Submitted).

Gonzalez J. E., Reuhs B. L. and Walker G. C. 1996. Low molecular weight EPS II of *Rhizobium meliloti* allows nodule invasion in *Medicago sativa*. Proc. Natl. Acad. Sci. USA. (In press).

Jann B. and Jann K. 1990. Structure and biosynthesis of the capsular antigens of *Escherichia coli*. Curr. Top. Microbiol. Immunol. 150:19-42.

Kim J. S. and Reuhs B. L. 1996. Extended host range mutants of *Rhizobium fredii* USDA257 show modified expression of the K antigens and lipopolysaccharides. 8th International Congress on Molecular Plant-Microbe Interactions (Abstract)

Kiss E., Reuhs B. L., Kim J. S., Kereszt A., Petrovics G., Putnoky P., Dusha I., Carlson R. W. and Kondorosi A. 1996. The fix-23 gene region encoding for *rkp* genes involved in expression of capsular polysaccharides in *Rhizobium meliloti*. Mol. Microbiol. (Submitted).

Noel K. D. 1992. Rhizobial polysaccharides required in symbioses with legumes. Pages 341-357 in: Molecular signals in plant-microbe communications. D. P. S. Verma, ed. CRC Press, Boca Raton, FL.

Olsen P., Collins M. and Rice W. 1992. Surface antigens present on vegetative *Rhizobium meliloti* cells may be diminished or absent when cells are in the bacteroid form. Can. J. Microbiol. 38:506-509.

Pazzani C., Rosenow C., Boulnois G. J., Bronner D., Jann K. and Roberts I. S. 1994. Molecular analysis of Region 1 of the *Escherichia coli* K5 antigen gene cluster: a region encoding proteins involved in cell surface expression of capsular polysaccharide. J. Bacteriol. 175:5978-5983.

Petrovics G., Putnoky P., Reuhs B., Kim J., Thorp T. A., Noel K. D., Carlson R. W. and Kondorosi A. 1993. A novel type of surface polysaccharide involved in symbiotic nodule development requires the expression of a new fatty acid synthase-like gene cluster in *Rhizobium meliloti*. Mol. Microbiol. 8:1083-1094.

Reuhs B. L., Carlson R. W. and Kim J. S. 1993. *Rhizobium fredii* and *Rhizobium meliloti* produce 3-deoxy-D-*manno*-2-octulosonic acid-containing polysaccharides that are structurally analogous to group K antigens (capsular polysaccharides) found in *Escherichia coli*. J. Bacteriol. 175:3570-3580.

Reuhs B. L., Badgett A., Kim J. S. and Carlson R. W. 1994. Production of the cell-associated polysaccharides of *Rhizobium fredii* USDA205 is modulated by apigenin and host root extract. MPMI 7:240-247.

Reuhs B. L., Williams M. N. V., Kim J. S., Carlson R. W. and Cote F. 1995. Suppression of the Fix⁻ phenotype of *Rhizobium meliloti exoB⁻* by *lpsZ* is correlated to a modified expression of the K polysaccharide. J. Bacteriol. 177:4289-4296.

Reuhs B. L., Geller D. G., Kim J. S., Fox J. F. and Pueppke S. G. 1996. Structural conservation in the cell-associated polysaccharides of *Rhizobium* sp. NGR234, R. *fredii*, and R. *meliloti*. Appl. Environ. Microbiol. (Submitted).

# *Rhizobium etli* Lipopolysaccharide Alterations Triggered by Host Exudate Compounds

K. Dale Noel, Dominik M. Duelli, and Valerie J. Neumann

Department of Biology, Marquette University, Milwaukee, WI, USA

The O-polysaccharide (OPS) portion of the lipopolysaccharide (LPS) is the outermost portion of the Gram-negative bacterial outer membrane and, as such, is a likely interface for interactions with other organisms. Over the years there has been a great deal of interest in the possible roles of LPS and other surface polysaccharides of *Rhizobium* species in the process of infection of legume hosts during the development of nitrogen-fixing root nodules. Indeed, mutants from several rhizobial species that lack the OPS are defective in nodule infection (Noel, 1992). However, it remains unclear exactly why.

Previous work in various laboratories has established that in the nodule the LPS structure differs from that of the bacteria grown in normal laboratory media. Very persuasive evidence of this has been marshalled by Brewin and colleagues, using monoclonal antibodies (Brewin 1991). In *Rhizobium etli* CE3 the differences in bacteroid LPS are subtle and most easily detected by monoclonal antibodies generated against this strain by Brewin (Tao et al. 1992). Certain conditions of growth ex planta also result in LPS alterations that can be detected with these antibodies. In particular, exudate from host *Phaseolus vulgaris* seeds and roots induces changes in the LPS that may be similar to those found in the nodule (Noel et al 1996). During the past year particular exudate compounds that induce this effect have been identified. In addition, a method for isolating mutants unable to undergo this response has been developed, and several such mutants have been isolated.

## Inducing Compounds in Seed Exudate

The inducing activity released from germinating seeds was highest during the first 12 h of imbibition of water. The crude exudate collected during the first 24 h of germination was resolved by reverse-phase high performance liquid chromatography (HPLC) (Fig. 1). Several fractions that eluted consecutively, those with absorbance maxima between 530 and 560 nm, had activity. These combined fractions were hydrolyzed with acid, extracted into amyl alcohol, and resolved by HPLC conditions optimized for anthocyanidin separation. Fractions with activity were the ones having the spectral properties of anthocyanidins, with the most abundant one having the properties of delphinidin (Fig. 1).

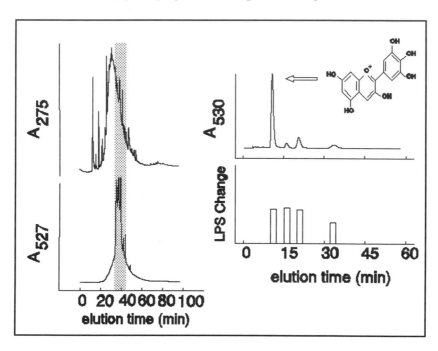

**Fig. 1. Anthocyanins as the inducers of LPS modification in seed exudate.**
Left half: Seed exudate resolved by $C_{18}$ HPLC and monitored by absorbance at 275 nm, detecting many types of aromatics (top), and 527 nm, mainly due to anthocyanins (bottom). The hatched box indicates fractions that triggered modification of LPS. Right half: aglycones released by hydrolysis of the anthocyanin fraction of the exudate were resolved by HPLC monitored at 530 nm (top). Delphinidin, whose structure is shown, accounted for more than half of the total $A_{530}$. The other anthocyanidins were identified as (left to right) cyanidin, petunidin, and malvidin. Each purified compound was tested at 100 $\mu$M for induction of LPS modification (bottom). The conversion of LPS I to the modified form ranged from 58% by malvidin to greater than 95% by the others.

To test whether the activity might be due to a co-eluting contaminant, delphinidin was obtained from a commercial source and, in addition, purified from eggplant peels. Both sources gave specific activities that were similar to that of the material purified from bean exudate.

The anthocyanidins were found in the crude exudates as glycolysated derivatives (anthocyanins), as determined by using extraction methods and a purification scheme designed specifically for anthocyanins. The anthocyanidin in each purified anthocyanin was identified by comparison of HPLC elution with anthocyanidin standards, thin-layer chromatographic behavior, UV/visible spectra, spectral shifts in AlCl₃, and gas chromatography. The sugars were identified by derivatization and gas chromatography. The most abundant anthocyanin (57 % of the total) was delphinidin-3-O-glucoside. In descending order of abundance the other major anthocyanins were petunidin-3-O-glucoside, a cyanidin diglycoside, malvidin-3-O-glucoside, and a delphinidin diglycoside. Each was active in inducing the LPS antigenic change, although the malvin had lower specific activity. The anthocyanidins exhibited half-maximal activity at 10 to 50 $\mu$M, depending on the compound.

Root exudates also triggered the LPS antigenic change, but *nod*-inducers naringenin and genistein were very poor inducers of this effect (Noel et al 1996). Anthocyanins are not known to be present in roots, and HPLC analysis indicated that the active root compounds eluted from the column before seed anthocyanins would and much earlier than narigenin or genistein. A white variety of *Phaseolus vulgaris*, which does not produce anthocyanins, does not have activity in its seed exudate, but its root exudate is active.

## Isolation of Mutants Defective in this Response

In previous work it has been determined that mutants lacking the Sym plasmid of *R. etli* CE3 carry out this response in a fashion that appears to be identical with the wild type. Therefore, *nodD* and any other genes found only on this plasmid apparently are not involved. This effect joins the few known examples of responses to legume exudates that do not involve *nodD*.

In order to identify the protein components required and determine what role this phenomenon plays in the physiology of *R. etli*, particularly in its symbiosis with *Phaseolus vulgaris*, modification-deficient mutants are being isolated. A screening procedure based on transposon mutagenesis and immuno-staining of colony lifts has been devised. Two types of mutants are sought. One referred to as Lpm⁻ (Lipopolysaccharide modification) still binds strongly to JIM28 antibodies after growth in the presence of exudate or purified anthocyanins, whereas the wild type no longer binds to the antibody after this treatment (Fig. 2). The other (Lpe⁻, Lipopolysaccharide epitope) lacks the antibody epitope under all conditions.

The reason for isolating this latter type of mutant is that its mutations may target the modifying enzyme itself or master regulators of the decorations that seem to be responsible for the epitopes recognized by the antibodies and for the modifications that prevent antibody recognition after growth in conditions such as exposure to exudate compounds. The two mutants with this phenotype are symbiotically proficient, as might be predicted, since they have the antigenic type observed in bacteroids.

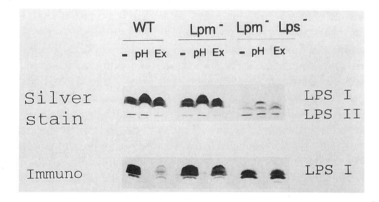

**Fig. 2. Lpm⁻ mutant CE396 and Lpm⁻ Lps⁻ mutant CE397.** Each mutant and wild type CE3 were grown under standard conditions (-), at pH 5.0 (**pH**), or in the presence of seed exudate (**Ex**). After SDS-PAGE of SDS extracts of the bacterial cells, the nitrocellulose blot ("immuno") was reacted with JIM28 antibodies and the residual material in the gel was stained by the silver-periodate procedure.

Two mutants of the first type are shown in Fig. 2. Strain CE397 is an example of the more common Lpm⁻ phenotype among the mutants thus far isolated and analyzed. This type is referred to as Lpm⁻ Lps⁻ because it exhibits obvious abnormalities in LPS banding on SDS polyacrylamide gel electrophoresis, regardless of the conditions of growth. The two mutants of this type that have tested for symbiotic properties (strains CE394 and CE395) were severely defective in infection. However, either the alteration in structure or the deficiency in LPS I (the O-polysaccharide-containing form of LPS) could have been responsible; i.e, it is not possible to attribute its symbiotic properties merely to the lack of response to exudate. Moreover, it seems likely that the alteration in basal structure is responsible for the absence of the antigenic conversion seen in the wild type. For instance, in these mutants the altered LPS may not be recognized by modifying enzymes, even though these enzymes may still be activated or induced by exposure to exudate compounds.

Strain CE396 (Fig. 2), however, appears to be the type of mutant sought; its LPS I gel mobility and abundance appear normal, yet the binding to antibody JIM28 is not lost at the concentration of seed exudate used in this experiment. Tests of its symbiotic properties are underway.

Reactivity of the LPS with antibody JIM28 is lost in the wild type also after growth at low pH, low oxygen, limiting phosphate, or high temperature (Tao et al. 1992). As shown in Fig. 2, although Lpm⁻ mutants respond abnormally to exudate (no loss of JIM28 binding), each responds to growth at low pH in normal fashion. Therefore, the pathways by which LPS is modified in response to these two conditions are at least partially independent.

The insertions of Lpm⁻ Lps⁻ mutants CE394 and CE395 map in the long chromosomal *lps* region α (Cava et al. 1990, Noel 1992) near the *lps* mutation found in strain CE374. This latter mutant has an LPS I that is only very subtly different structurally from the wild type (as compared with the obvious alterations in CE394 and CE395), does not respond normally to exudate, and is partially deficient in symbiosis (Noel et al. 1996). The insertion of Lpe⁻ mutant CE367 does not map in *lps* α region or any of the other three *lps* regions that have been cloned from *R. etli* CE3 (Noel 1992). The existence of such a region for decorating the LPS basal structure so that it becomes antigenically complete was predicted previously from studies in which *lps* α was transferred into Lps mutants of *R. leguminosarum* (Brink et al. 1990).

## Conclusions

As with *nod* induction (Hungria et al. 1991), major effectors of this effect on LPS structure from *Phaseolus vulgaris* seeds are anthocyanins. However, higher anthocyanin concentrations are required for triggering LPS modification. Since *nodD* apparently is not required for this effect, it may provide a means of uncovering a second signal transduction pathway for responses in rhizobia to exudate compounds. The modification of LPS triggered by growth at low pH is at least partially independent of the mechanism induced by exudates. The physiological role(s) of these LPS alterations have still not been established, but the existence of mutants deficient in this response provides a basis for beginning to understand why the bacteria respond in this way.

# Literature Cited

Brewin, N.J. 1991. Development of the legume root nodule. Ann. Rev. Cell Biol. 7:191-226.

Brink, B.A., Miller, J., Carlson, R.W., and Noel, K.D. 1990. Expression of *Rhizobium leguminosarum* CFN42 genes for lipopolysaccharide in strains derived from different *Rhizobium leguminosarum* soil isolates. J. Bacteriol. 172:548-555.

Cava, J.R., Tao, H., and Noel, K.D. 1990. Mapping of complementation groups within a *Rhizobium leguminosarum* CFN42 chromosomal region required for lipopolysaccharide synthesis. Mol. Gen. Genet. 221:125-128.

Hungria, M., Joseph, C.M., and Phillips, D.A. 1991. Anthocyanidins and flavonols, major *nod* gene inducers from seeds of a black-seeded common bean (*Phaseolus vulgaris* L.). Plant Physiol. 97:751-758.

Noel, K.D. 1992. Rhizobial polysaccharides required in symbiosis with legumes. Pages 341-357 in:Molecular Signals in Plant-Microbe Communications. D.P.S. Verma, ed. CRC Press, Boca Raton, FL.

Noel, K.D., Duelli, D.M., Tao, H., and Brewin, N.J. 1996. Antigenic change in the lipopolysaccharide of *Rhizobium etli* CFN42 induced by exudates of *Phaseolus vulgaris*. Mol. Plant-Microbe Interact. 4:332-340.

Tao, H., Brewin, N.J., and Noel, K.D. 1992. *Rhizobium leguminosarum* CFN42 lipopolysaccharide antigenic changes induced by environmental conditions. J. Bacteriol. 174:2222-2229.

# Analysis of the Secretion of Symbiosis-Related Proteins by *Rhizobium leguminosarum* biovar *viciae*

C. Finnie, G. Dean, J. M. Sutton, S. Gehlani, and
J. A. Downie

John Innes Centre, Colney, Norwich, NR4 7UH, UK

Although there is evidence describing a role in pathogenesis for proteins secreted by bacterial plant pathogens (see Salmond 1994), there are relatively few cases describing a role for secreted proteins in rhizobial-legume symbiotic interactions. During the course of our work on nodulation signalling by *Rhizobium leguminosarum* biovar *viciae* we have identified a role for a nodulation protein secreted via a haemolysin-type secretion system.

Lipo-oligosaccharide nodulation (Nod) factors are the primary determinants of signalling in rhizobial-legume interactions (Dénarié and Debellé 1996). *R. l. viciae* normally makes four Nod factors which consist of oligomers of four or five N-acetyl glucosamine residues carrying an N-linked $C_{18:4}$ or $C_{18:1}$ acyl group on the terminal non-reducing glucosamine (Spaink et al., 1991). Mutation of *nodE* abolishes the formation of the $C_{18:4}$-containing Nod factors and reduces (but does not block) nodulation of peas or vetch (Downie et al., 1985). Although mutation of *nodO* alone had little effect on nodulation (Economou et al., 1990), in a *nodE* mutant background, mutation of *nodO* strongly affected nodulation (Downie and Surin 1990; Economou et al., 1994). NodO is a secreted $Ca^{2+}$-binding protein that forms ion-selective pores in membranes (Sutton et al., 1994). It shows homology to the $Ca^{2+}$-binding domain of the so called RTX proteins that include the *Escherichia coli* haemolysin A. (Economou et al., 1990). Purified NodO protein must be secreted by a mechanism analogous to that involved in secretion of haemolysin A because *E. coli* strains carrying the *hlyBD, tolC* genes (involved in haemolysin A secretion) acquire the ability to secrete NodO (Scheu et al., 1992). In general the genes encoding such secretion systems are closely linked to the genes encoding the secreted proteins (Salmond 1994). However, this is not the case for NodO because the *nodO* gene is on the symbiotic plasmid but the genes encoding its secretion are not (Scheu et al., 1992). In this work we have investigated the domains of NodO required for its secretion and identified the genes involved in its secretion. This has led us to identify other proteins secreted by this system.

## Results

NodO SECRETION SIGNAL

Proteins secreted via the haemolysin-type mechanism normally have a C-terminal secretion signal. We constructed a deletion derivative of *nodO* encoding a protein lacking the C-terminal 24 residues. This was achieved by cloning an oligonucleotide into a unique *Sph*1 site in *nodO*. This derivative of *nodO* was introduced into *E. coli* carrying the haemolysin secretion genes *hlyBD*, *tolC* and into *R. l. viciae* lacking the wild-type *nodO* gene. The culture supernatants of both strains were analysed using antibodies to detect the deleted NodO. In each case the amount of protein detected was less than 2% of that seen with isogenic strains carrying wild-type *nodO*. Analysis of *R. l. viciae* cell extracts revealed that the deleted form of NodO was present. These results indicate that the C-terminal 24 residues of NodO are essential for secretion.

We constructed a series of deletion derivatives of *nodO* encoding proteins that retained the C-terminal 24 residues of NodO but lacked up to 132 amino acids (residues 128-259). These derivatives of NodO were found to be secreted at between 60% and 90% of wild-type levels. This suggested that the secretion signal is contained within the C-terminal 24 residues of NodO.

To determine if passenger proteins could be secreted using NodO as a carrier, the *phoA* gene from *E. coli*, (lacking the DNA encoding the N-terminal transit peptide) was cloned in frame into *nodO* at a position equivalent to amino acids 221-235 of NodO (Fig. 1). The NodO-PhoA-NodO fusion protein was secreted by *R. l. viciae*; the growth-medium supernatant contained an appropriate sized fusion protein that reacted to NodO and to PhoA antiserum and alkaline phosphatase activity could be detected in the growth medium supernatant. A deleted derivative was then constructed near the C-terminal domain resulting in a fusion protein that retained the 24 C-terminal residues of NodO (Fig. 1); this fusion was also secreted. This confirms that these C-terminal residues are crucial for secretion. However, when a *phoA*-*nodO* fusion was constructed, in which the only parts of *nodO* were the promoter, translational start and the region encoding the C-terminal 24

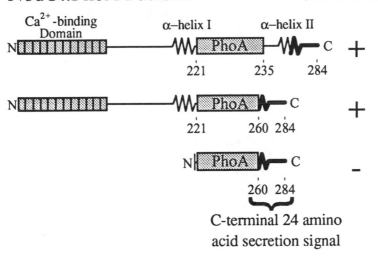

Fig. 1. Secretion of NodO-PhoA fusions. The NodO-PhoA fusions are drawn schematically showing the $Ca^{2+}$-binding domain and two α-helical regions of NodO. The numbers represent amino-acid residues of NodO and the heavy line indicates the secretion signal.

amino acids (Fig. 1), the PhoA-NodO fusion was not detectable in the growth-medium supernatant. This indicates that the N-terminal $Ca^{2+}$-binding domain of NodO may be important for the secretion of passenger proteins.

IDENTIFICATION OF A NodO SECRETION MUTANT

An immunological screen was set up to discriminate between colonies of *R. l. viciae* that could or could not secrete NodO using as controls, strains carrying the plasmids as described above encoding secreted and non-secreted NodO proteins. A Tn5-induced mutant was identified that could not secrete wild-type NodO. The mutation was transduced into different backgrounds and the NodO secretion defect cotransduced with kanamycin resistance (Tn5). The gene affected was cloned and sequenced and its product is clearly homologous to the ATP-binding component of proteins involved in secretion of proteins such as haemolysin.

Based on previous work (Downie and Surin 1990; Economou *et al.*,

1994) it was predicted that a strong effect on pea or vetch nodulation might only be seen in a NodO secretion mutant if the *nodE* gene was also mutated. Therefore, we constructed an isogenic series of mutants carrying the mutation blocking NodO secretion in strains retaining or lacking *nodE*. Nodulation tests revealed that, compared with a *nodE* mutant, nodulation of pea and vetch was greatly reduced in double mutants lacking both NodO secretion and *nodE*. However, what was not anticipated was that nodulation by the secretion mutant carrying a normal copy of *nodE* was increased (by about 30%) compared with the wild-type control. Analysis of the nodules revealed that this was probably due an absence of nitrogen fixation in nodules induced by the NodO secretion mutant. (It is common to see increased nodulation in Fix⁻ mutants). Since *nodO* structural gene mutants are Fix⁺ it seems likely that the defect in $N_2$-fixation might be due to the inability of the secretion mutant to secrete protein(s) (other than NodO) that are essential for establishing a $N_2$-fixing symbiosis.

## CHARACTERISATION OF THE NodO SECRETION MUTANT

Proteins were concentrated from the growth-medium supernatants from control strains and mutants defective in NodO secretion. This was done using strains both carrying and lacking the symbiotic plasmid pRL1JI and in the presence or absence of a flavanoid to induce *nod* gene expression. The only flavonoid-inducible secreted protein that was absent from the secretion mutant was NodO. However, comparison of proteins secreted by a Sym plasmid-deleted control strain with those of an isogenic secretion mutant revealed that at least three proteins were absent from the mutant. We are interested in understanding the roles of these proteins.

One unusual characteristic of the protein-secretion mutant is that the colonies are sticky. Furthermore, growth-medium supernatants tended to be more viscous. This suggested that surface polysaccharides of the mutant might be altered. We compared the overall lengths of the exopolysaccharide (eps) produced in culture by the mutant and control strains by measuring in each the ratio of reducing (end) sugars to the total sugar content. Although the total amounts of eps produced by the mutant and control strains were similar, it was clear that the mutant eps had significantly lower levels of reducing sugars. This suggests that the eps might normally be cleaved into shorter lengths by an extracellular glucanase. Therefore one of the proteins secreted by the same pathway as NodO may have a role in processing extracellular eps.

On the basis of the results described above it seemed likely that an eps hydrolysing endoglucanase is normally secreted by *R. l. viciae*. We devised a screen to identify mutants defective in such activity. This involved isolating eps from the NodO secretion mutant and incorporating the eps into agar plates. Wild-type colonies grown on such plates hydrolysed the eps and this could be detected by staining the plates with congo red. In contrast the NodO (and endoglucanase) secretion mutant could not hydrolyse the eps.

A Tn*5* mutagenised population of *R. l. viciae* was plated on the eps-agar plates and mutants defective in eps hydrolysis were identified. Two classes of mutants were found; one corresponded to mutants defective in NodO (and endoglucanase) secretion and $N_2$ fixation. A second class of mutants was identified as being deficient in cleavage of the eps in the agar plates but normal with respect to NodO secretion and $N_2$ fixation. These mutants might be affected in the structural gene for the secreted endoglucanase.

An endoglucanase gene *(egl)* encoding a protein that hydrolyses carboxymethyl cellulose (CMC) was recently described in *Azorhizobium caulindans* (Geelen et al., 1995). In *A. caulinodans* this gene normally causes hydrolysis of CMC in CMC-agar plates. Using CMC-agar plates we found that *R. l. viciae* has a similar CMC endoglucanase activity, and that neither the protein secretion mutants nor the eps-hydrolysing endoglucanase mutant could hydrolyse CMC. When the *A. caulinodans egl* endoglucanase gene was transferred into the *R. l. viciae* endoglucanase mutant, the CMC hydrolysis activity was restored. No such restoration of CMC hydrolysis was seen with the secretion mutant carrying the cloned *egl* gene.

### Conclusions

The results presented here show that in *R. l. viciae* a non-sym-plasmid located gene encodes a protein required for NodO secretion. Mutation of this gene blocks NodO secretion and also blocks secretion of other proteins. One protein normally secreted is an endoglucanase that appears to hydrolyse the bacterial eps and this activity is absent from secretion-defective mutants. A mutant that retains NodO secretion but is defective in secreted endoglucanase activity can be restored for endoglucanase activity using a cloned endoglucanase gene from *A. caulinodans* whereas a secretion mutant cannot. This indicates that like the *R. l. viciae* endoglucanase the *A. caulinodans* endoglucanase is secreted via a haemolysin-type system. The

observation that the secretion mutant of *R. l. viciae* is Fix⁻ whereas the
endoglucanase-deficient mutant is Fix⁺ indicates that an additional as
yet uncharacterised secreted protein may be essential for a late stage in
development of the symbiosis.

## References

Dénarié, J., Debellé, F. 1996. *Rhizobium* lipo-chitooligosaccharide
nodulation factors: signalling molecules mediating recognition and
morphogenesis. Ann. Rev. Biochem. 65:503-535.

Downie, J. A., Knight, C. D., Johnston, A. W. B. 1985. Identification
of genes and gene products involved in nodulation of peas by
*Rhizobium leguminosarum*. Mol. Gen. Genet. 198:255-262.

Downie, J. A., Surin, B. P. 1990. Either of two *nod* gene loci can
complement the nodulation defect of a *nod* deletion mutant of
*Rhizobium leguminosarum* bv. *viciae*. Mol. Gen. Genet. 222:81-86.

Economou, A., Davies, A. E., Johnson, A. W. B., Downie, J. A. 1994.
The *Rhizobium leguminosarum* biovar *viciae nodO* gene can enable
a *nodE* mutant of *Rhizobium leguminosarum* biovar *trifolii* to
nodulate vetch. Microbiol. 140:2341-2347.

Economou, A., Hamilton, W. D. O., Johnston, A. W. B., Downie, J. A.
1990. The *Rhizobium* nodulation gene *nodO* encodes a $Ca^{2+}$-binding
protein that is exported without N-terminal cleavage and is
homologous to haemolysin and related proteins. EMBO J. 9:349-354.

Geelen, D., Van Montagu, M., Holsters, M. 1995. Cloning of an
*Azorhizobium caulinodans* endoglucanase gene and analysis of its
role in symbiosis. Appl. Environ. Microbiol. 61:3304-3310.

Salmond, G. P. C. 1994. Secretion of extracellular virulence factors by
plant pathogenic bacteria. Ann. Rev. Phytopathol. 32:181-200.

Scheu, A. K., Economou, A., Hong, G-F., Ghelani, S., Johnston, A.
W.B., Downie, J. A. 1992. Secretion of the *Rhizobium
leguminosarum* nodulation protein NodO by haemolysin-type
systems. Mol. Microbiol. 6:231-238.

Spaink, H. P., Sheeley, D. M., Van Brussel, A. A. N., Glushka, J. N.,
York, W. S., Tak, T., Geiger, O., Kennedy, E. P., Reinhold, V. N.,
Lugtenberg, B. J. J. 1991. A novel, highly unsaturated, fatty acid
moeity of lipo-oligosaccharide signals determines host specificity of
*Rhizobium leguminosarum*. Nature 354:125-130.

Sutton, J. M., Lea, E. J. A., Downie, J. A. 1994. The supernatants
nodulation-signaling protein NodO from *Rhizobium leguminosarum*
biovar *viciae* forms ion channels in membranes. Proc. Natl. Acad.
Sci. USA 91:9990-9994.

# Symbiotic suppression of the *Medicago sativa* defense system - the key of *Rhizobium meliloti* to enter the host plant?

Karsten Niehaus, Ruth Baier, Anke Becker, and Alfred Pühler

Universität Bielefeld, Bielefeld, Germany

The symbiotic interaction between *Rhizobium meliloti* and alfalfa results in the formation of root nodules. Within this specialized plant organ nitrogen fixation is carried out by bacteria differentiated into bacteroids. Rhizobial surface polysaccharides play an important role in the invasion of alfalfa root nodules.

*R. meliloti* Rm2011 produces two structurally different exopolysaccharides. Succinoglycan (EPS I) constitutes a polymer of octasaccharide repeating units composed of one galactose and seven glucose residues (Reinhold et al. 1994). It is decorated by acetyl, succinyl and pyruvyl groups. A high molecular weight (HMW) and a low molecular weight (LMW) fraction is produced. HMW EPS I represents a polymer of $10^6$ to $10^7$ D, whereas LMW EPS I contains monomers, trimers and tetramers of the repeating unit (Battisti et al. 1992). Galactoglucan (EPS II) is composed of alternating galactose and glucose residues and is acetylated and pyruvylated (Her et al. 1990). The regulatory gene *mucR* is responsible for the activation of EPS I and the repression of EPS II biosynthesis (Keller et al. 1995). *R. meliloti* mutants unable to produce EPS I induce white uninfected alfalfa pseudonodules exhibiting pronounced symptoms of plant defense (Niehaus et al. 1993).

In this paper we describe the production of LMW EPS I by a specific *R. meliloti exoP\** mutant and demonstrate that this EPS can act as a suppressor of plant defense in alfalfa cell suspension cultures.

## *R. meliloti* LMW EPS I is exclusively produced by a specific *exoP* mutant

The membrane topology of the ExoP protein involved in EPS I biosynthesis was determined using translational fusions of *phoA* and *lacZ* reporter genes to the *exoP* gene (Becker et al. 1995). The ExoP protein contains an N-terminal domain mainly located in the periplasm and a C-terminal cytoplasmic domain (Fig. 1). The N-terminal domain resembles CLD proteins involved in the determination of O-antigen chain length in several enterobacteria. A conserved amino acid sequence motif (ExoP motif) was identified in ExoP and CLD proteins (Fig. 1). The C-terminal domain of ExoP is characterized by a nucleotide binding motif (Fig. 1). This domain was not found in any enterobacterial CLD protein, but was identified in proteins homologous to ExoP from several bacteria, e.g. *Erwinia amylovora*, *Klebsiella pneumoniae*, *Bradyrhizobium japonicum* and *Streptococcus thermophilus*. As is the case for ExoP, the homologous proteins from *E. amylovora* and *K. pneumoniae* are encoded by a single gene. In contrast, two consecutive genes encoding two proteins homologous to the N-terminal and the C-terminal ExoP domain were identified in *B. japonicum* and *S. thermophilus*.

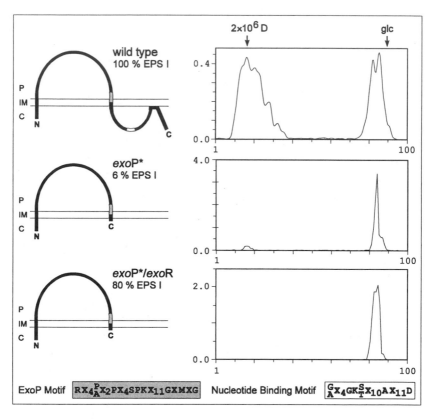

**Fig. 1.** Membrane topology of the *R. meliloti* ExoP protein and EPS I fractions produced by the wild type Rm2011, the *exoP\** mutant and the *exoP\*/exoR* double mutant.

A model of the membrane topology of ExoP and the structure proposed for the mutated ExoP protein encoded by the truncated *exoP\** gene is shown. Conserved amino acid sequence motifs are indicated. HMW and LMW EPS I was separated by gel chromatography on Bio-Gel A5m columns and was detected by the HCl/cysteine method after total hydrolysis.

Abbreviations: C, cytoplasm; glc, glucose; IM, inner membrane; P, periplasm.

*R. meliloti* Rm2011 mutants carrying truncated *exoP** genes exclusively encoding the N-terminal ExoP domain produced a reduced amount of EPS I (Becker et al. 1995). This reduction was suppressed by a mutation of the gene *exoR* encoding a negative regulator of EPS I biosynthesis. The ratio of HMW EPS I to LMW EPS I was significantly decreased in *exoP** mutants, whereas LMW EPS I was exclusively detected in culture supernatants of *exoP**/*exoR* double mutants (Fig. 1). It is tempting to speculate that ExoP might be implicated in processes determining the ratio of HMW EPS I to LMW EPS I.

**LMW EPS I of *R. meliloti* acts as a suppressor of yeast elicitor induced alkalinization of *M. sativa* cell suspension cultures**

Mutants of the symbiotic soil bacterium *R. meliloti* that fail to synthesize EPS I were unable to induce effective root nodules on 19 different *Medicago* species and cultivars (Niehaus et al. 1994), indicating a general importance of this surface carbohydrate for the establishment of the symbiosis. A detailed microscopical and biochemical analysis of noninfected pseudonodules, induced by an EPS I defective *R. meliloti* mutant on *M. sativa,* revealed strong evidence for the induction of a plant defense response by the mutated microsymbiont (Niehaus et al. 1993). Battisti et al. (1992) reported that the defect in invasion of the *R. meliloti* EPS I nonproducing mutants could be restored by the exogenous addition of purified LMW EPS I. From these observations we propose that LMW EPS I acts as a suppressor of the plant defence system, enabling the symbiont *R. meliloti* to infect the host plant. In order to test this hypothesis we established elicitor responsive cell cultures of the host plant *M. sativa* (alfalfa) and as a control the non host plants *Nicotiana tabacum* (tobacco) and *Lycopersicum esculentum* (tomato). Apart from other defense related reactions, all cell cultures reacted to the addition of small amounts of the non specific yeast-clicitor with a strong transient alkalinization of their culture medium. Using this assay system purified homologous and heterologous EPS were analysed for possible suppressor functions in the three plant suspension cultures. In alfalfa cell cultures the elicitor induced alkalinization could be suppressed by the simultaneous application of LMW EPS I (Fig. 2). Neither HMW EPS I, HMW EPS II nor the heterologous EPS xanthan from *Xanthomonas campestris* provoked a reduction of the elicitor response. None of the carbohydrate preparations were able to suppress the elicitor induced alkalinization in the cell cultures of the non host plants tobacco and tomato. These data provide strong evidence for a specific recognition of LMW EPS I by the host plant as a suppressor of the plant defense system enabling the microsymbiont to infect the plant.

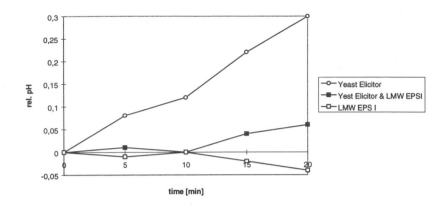

**Fig. 2.** Alkalinization of the alfalfa cell culture medium in response to the application of yeast elicitor (O), isolated LMW EPS I (□), and the simultaneous application of yeast elicitor and LMW EPS I (■).

## Literature cited

Battisti, L., Lara, J.C., and Leigh, J.A. 1992. Specific oligosaccharide form of the *Rhizobium meliloti* exopolysaccharide promotes nodule invasion in alfalfa. Proc. Natl. Acad. Sci. 89:5625-5629.

Becker, A., Niehaus, K., and Pühler, A. 1995. Low molecular weight succinoglycan is predominantly produced by *Rhizobium meliloti* strains carrying a mutated ExoP protein characterized by a periplasmic N-terminal and a missing C-terminal domain. Mol. Microbiol. 16:191-203.

Her, G.-R., Glazebrook, J., Walker, G.C., and Reinhold, V.N. 1990. Structual studies of a novel exopolysaccharide produced by a mutant of *Rhizobium meliloti* Rm1021. Carbohydr. Res. 198:305-312.

Keller, M., A. Roxlau, W.M. Weng, M. Schmidt, J. Quandt, K. Niehaus, D. Jording, W. Arnold, and A. Pühler. 1995. Molecular analysis of the *Rhizobium meliloti mucR* gene regulating the biosynthesis of the exopolysaccharide succinoglycan and galactoglucan. Mol. Plant-Microbe Interact. 2:267-277.

Niehaus, K., Kapp, D., and Pühler, A. 1993. Plant defence and delayed infection of alfalfa pseudonodules induced by an exopolysaccharide (EPS I)-deficient *Rhizobium meliloti* mutant. Planta 190:415-425.

Niehaus, K., Kapp, D., Lorenzen, J., Meyer-Gattermann, P., Sieben, S., and Pühler, A. 1994. Plant defence in alfalfa pseudonodules induced by an exopolysaccharide (EPS I) deficient symbiont. Acta Horticulturae 381:258-264.

Reinhold, B.B., Chan, S.Y., Reuber, T.L., Marra, A., Walker, G.C., and Reinhold, V.N. 1994. Detailed structural characterization of succinoglycan, the major exopolysaccharide of *Rhizobium meliloti* Rm1021. J. Bacteriol. 176:1997-2002.

# New Tools for Investigating Nodule Initiation and Ontogeny: Spot Inoculation and Microtargeting of Transgenic White Clover Roots Shows Auxin Involvement and Suggests a Role for Flavonoids

U. Mathesius, H.R.M. Schlaman[1], D. Meijer[1], B.J.J. Lugtenberg[1], H.P. Spaink[1], J.J. Weinman, L.F. Roddam, C. Sautter[2] and B.G. Rolfe and M. A. Djordjevic.

PMI Group, Research School of Biological Science, Australian National University, Canberra, Australia, 2601. [1]Institute of Molecular Plant Sciences, Leiden University, Wassenaarseweg 64, 2333 AL Leiden, The Netherlands. [2]ETH Zürich, Dept of Plant Science, Zürich, Switzerland.

## SUMMARY

We have developed new approaches and tools to investigate the roles of the phytohormone auxin and plant flavonoids in root nodulation. Whereas auxin plays a central role in plant growth, cell division and development, a role in root nodulation has not been demonstrated. In addition, specific flavonoids have roles (a) as microbial signals (including the Ini response) (Recourt et al, 1991, Lawson et al, 1996) and (b) as auxin transport inhibitors (Jacobs and Rubrery, 1988), but a role in nodule organogenesis has not been demonstrated. We show evidence for roles of both auxin and flavonoids during root nodule morphogenesis. We have assessed the effect of the precise application of R. l. bv. trifolii strains, auxin, NPA (N-(1-naphthyl)phthalamic acid), flavonoids and Nod factors upon the expression of reporter gene constructs in transgenic white clovers. The transgenic plants contain the GUS gene fused to either the auxin responsive promoter GH3 (Guilfoyle et al 1993) or to chalcone synthase (CHS) promoters. Our results show that the first recognisable effect of the addition to roots of Rhizobium or Nod factors is an induction of CHS1 at the application site and an induction of CHS3 in the inner cortex and endodermis both within 5 h. Non-nodulating rhizobia do not induce this CHS expression. These location-specific CHS inductions appear to precede the detectable changes in GH3:GUS expression following Rhizobium or Nod factor addition which occur 17 h after their application. The addition of Rhizobium, Nod factors, flavonoids or NPA cause localised up regulation of GH3:GUS expression but a striking down regulation of in developmentally younger parts of the root between the induced site of activity and the root tip. We interpret these results to suggest that flavonoids resulting from CHS expression coordinate nodule formation by modulating auxin distribution and that inhibition of auxin transport is an integral part of nodule formation.

# BACKGROUND

Transgenic clovers (*T. repens* cv. *Haifa*) were used which contained the *GUS* reporter gene fused to either the soybean *GH3* (auxin responsive) promoter (Larkin et al, 1996) or to clover *CHS* (chalcone synthase) promoters (Arioli et al 1993; Howles et al 1994). Recently we showed that there are three features of *GH3:GUS* expression which indicate that expression from this construct is a rapid and reliable measure of auxin mediated responses in the transgenic plants (Larkin et al 1996). *GH3:GUS* expression in untreated transgenic white clover is consistently induced in the outer cortex of the root effectively marking the site of an initiating lateral root beneath. A developing lateral root invariably grows towards, and through, this area of *GH3:GUS* expression. In addition, *GH3:GUS* is expressed within 60-90 min on the non-elongating side of gravistimulated roots concurrently with root curvature, in accordance with the role of auxin in gravistimulation. Finally expression of the *GH3:GUS* construct is significantly enhanced over endogenous levels by the addition of auxin but not other phytohormones (Larkin et al 1996).

## NODULE VS LATERAL ROOT FORMATION USING TRANSGENIC PLANTS CONTAINING *GH3:GUS*

We compared nodule initiation by *Rhizobium* to lateral root initiation as a control because lateral root formation is a process where the role of auxin is firmly established (Hirsch, 1992). We used root systems which develop spontaneously from the petiole of leaves from transgenic plants (called "rooted leaves"). Rooted leaves support the nodulation of *R. l.* bv. trifolii and are a useful bioassay system for studying lateral root and nodule initiation (Rolfe and McIver, 1996). Spot inoculation (Bhuvaneswari et al, 1981) with a 20 nL suspension containing about 5000 *R. l.* bv. trifolii cells (to prevent over-inoculation) was done using strains ANU843 (wild-type) or the Nod⁻ strains mutant 277 (*nodC*::Tn5) or ANU845 (pSym⁻). GUS activity was visualised as described after tissue fixation (Larkin et al, 1996).

The GUS activity pattern seen at different times during nodule initiation is consistent with the requirement of auxin to initiate the very early (but not subsequent) cell divisions and also in the formation of vascular traces (Fig 1 and 2). In roots inoculated with ANU843, staining occurred in the initial dividing cells in the inner cortex seen at 24 and 50 h but not in the cells that subsequently divide to form the primordium (Fig 2). By 70 h, staining of *GH3:GUS* occurs in a row of cells in the outer layers of the nodule in cells which ultimately differentiate into vascular traces (Fig 2). As the nodule emerges from the root (7 days post inoculation) staining occurs in the cells of the vascular traces and also in the nodule meristem. Nod⁻ mutants ANU845 or 277 did not induce a detectable response on these plants. Finally, whereas *Rhizobium* inoculation led to localised effects on *GH3:GUS* expression, it also led to a clear down regulation of *GH3:GUS* in the area between the site of induced *GH3:GUS* activity and the root tip (Fig 1 and 2) detected between 17 and 50 h post inoculation.

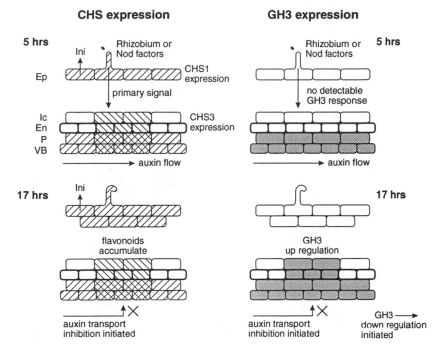

**CHS expression**      **GH3 expression**

Fig 1. Expression of *CHS1:GUS*, *CHS3:GUS* and *GH3:GUS* after exposure to *Rhizobium* and Nod Factors. *CHS1:GUS* ▨ shows rapid induction in the epidermal (Ep) and outer cortical tissues and constitutive expression in the vacsular tissue (P = pericycle; VB = Vascular bundle). *CHS3:GUS* ▨ shows rapid induction in the inner cortex (Ic), endodermis and vascular tissue. Cross hatched areas indicate that both *CHS1:GUS* and *CHS3:GUS* are expressed. *GH3:GUS* (shaded areas) shows an up regulation in the inner cortex and endodermis at 17 h and is constitutively expressed in the vascular tissue. The rapid induction of *CHS3:GUS* may lead to the production of flavonoids which cause auxin transport inhibition, up regulation of *GH3:GUS* and a focused concentration of auxin needed to mediate nodule formation.

Parallels in the staining pattern of *GH3:GUS* occur during lateral root and nodule formation. GUS activity occurs first in pericycle cells of initiating lateral roots before and during the first cell divisions but, like nodules, larger primordia show no expression. As the lateral root begins to penetrate the inner cortex, staining occurs in the cells which form the defined lateral root boundary and simultaneously in the outer cortex forming the "island" of cells through which the lateral root ultimately penetrates, consistent with previous observations (Larkin et al 1996). Finally, the cells which differentiate into the vascular traces are stained before and during lateral root emergence, as in nodule formation.

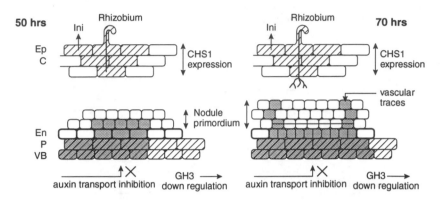

Fig 2. *CHS1:GUS* and *GH3:GUS* during early nodule formation initiated by *Rhizobium*. *CHS1:GUS* ▨ is induced in deeper layers of the outer cortex and remains constitutively expressed in the vascular tissues. The nodule primordium continue to grow towards the area of *CHS1:GUS* expression. *GH3:GUS* (shaded) is up regulated in cells which participate in early cell divisions and later in cells which give rise to the nodule vascular tissues. *GH3:GUS* is strikingly down regulated between the site of up regulation and the root tip. NPA, Nod factors and specific flavonoids induce similar responses in *GH3:GUS* and *CHS1:GUS* expression.

## NODULE INITIATION USING TRANSGENIC PLANTS CONTAINING *CHS1:GUS* or *CHS3:GUS*

Transgenic plants containing *CHS1:GUS* or *CHS3:GUS* were used to assess the involvement of the flavonoid pathway in nodule formation. During nodule initiation, a clear dichotomy between *CHS1* and *CHS3* expression is seen after 5 h incubation with ANU843. Firstly, *CHS1:GUS* expression is clearly induced in the epidermis and outer cortex, whereas *CHS3:GUS* is expressed in the inner cortex beneath the site of inoculation of ANU843. These results suggest that *CHS:GUS* expression is induced preceding root hair penetration and inner cortical cell division at two distinct sites and is dependent upon the presence of nodulating rhizobia (Fig 1). Neither ANU845 or mutant 277 induce any of the *CHS* promoters. *CHS1:GUS* expression becomes localised in deeper layers of the outer cortex at 20, 50 and 70 h. The nodule primordia induced do not show *CHS1:GUS* expression and grow towards the outer cortical layers showing *CHS1:GUS* expression.

## SPOT INOCULATION AND MICROTARGETING OF NPA, AUXIN, NOD FACTORS AND FLAVONOIDS ONTO AND INTO CLOVER ROOTS.

Having examined the effect of *Rhizobium* addition upon transgene expression in early nodule organogenesis, we used microtargeting (Sautter et al, 1991) and spot inoculation techniques to apply the following compounds to the transgenic roots: auxin, flavonoids, NPA (*N*-(1-naphthyl)phthalamic acid) or *R.l.* bv. trifolii Nod factors (Spaink et al 1995).

Using *GH3:GUS* plants the microtargeting of auxin (1 x $10^{-6}$M NAA in a 60 nl sample load) to the roots resulted in increased expression in the vascular tissue (after about 20 h) both at the site of inoculation and in the vascular tissue between this site and the root tip. The effect was much weaker or absent with spot inoculation of 20 nl of NAA at the same or higher concentrations although the addition of agar blocks (containing 1 x $10^{-6}$ M NAA) or bathing of roots in NAA at the same or lower concentrations, led to an increased expression of *GH3:GUS* in the vascular tissue. This result was determined by comparing quantitative *in vitro* measurements of GUS activity with the location of enhanced GUS activity *in situ* using identically treated plant material. In contrast, the application of NPA, flavonoids or Nod factors resulted in similar alterations of *GH3:GUS* expression in the vascular tissue of roots. The microtargeting or spot inoculation of Nod factors (1 x $10^{-8}$ M in 60 nl for microtargeting or 20 nl for spot inoculation), NPA (1 x $10^{-6}$ M) or the flavonoids apiginen, kaempferol or quercetin (all at 3x $10^{-5}$M, or individually at 10 µM using spot inoculation only), all led to a localised increase in *GH3:GUS* expression at or just above the site of application and to a down regulation of *GH3:GUS* expression in the areas between the site of induced activity and the root tip. Preliminary results indicate that a more rapid induction in *GH3:GUS* expression occurs with NPA and flavonoids (at all tested concentrations) than with Nod factors and *Rhizobium*. After 50 - 70 h, the down-regulation of GH3 activity was no longer detected in the developmentally younger parts of the root just behind the root tip. Collectively, these results suggest that these compounds block auxin transport transiently and result in a focusing of auxin at the site of nodule initiation (Hirsch, 1992).

As with the application of *Rhizobium*, rapid (< 5 h), localised and identical increases in *CHS* expression occur after the spot inoculation of Nod factors (approx. 1 x $10^{-8}$ M) to the roots of plants containing either *CHS1:GUS* or *CHS3:GUS* at 5 and 17 h post inoculation. As with *Rhizobium* inoculation, *CHS1:GUS* expression occurred in the epidermis and cortex and *CHS3:GUS* expression occurred in the inner cortex. These results indicate that a precise and differential induction of *CHS* expression is an early result of *Rhizobium* inoculation mediated by Nod factors. Since *CHS* induction precedes changes in *GH3* expression it is possible that endogenous flavonoids made as a result of *CHS* induction are naturally occurring auxin transport inhibitors involved in mediating nodulation and that Nod factors exert their effect, in part, through the action of these flavonoids (Figs 1 and 2). We are currently examining this hypothesis.

CONCLUSIONS

It is clear that the application of precise amounts of biologically active material to transgenic plants carrying reporter gene constructs at defined and marked sites has many advantages in the analysis of nodule initiation and ontogeny. The results we have obtained show the first experimental evidence consistent with the hypothesis that flavonoids can be produced in the inner cortical cell region due to a site specific induction of the flavonoid pathway. This site specific induction of the flavonoid pathway results from events initiated early during the infection of *Rhizobium*. Flavonoids that accumulate may then act as auxin transport inhibitors, as proposed by Hirsch (1992) (Fig 1 and 2), resulting in a localisation of active auxin and induction of the *GH3* gene. Auxin, as a secondary signal, may then promote cell division to generate the nodule primordium. The expression of the *GH3* gene in cells destined to become the vascular elements of the nodule may result from a specific flow of auxin from the stele in a manner consistent with the canalisation effect of auxin proposed by Sachs (1981). The *GH3* gene may be a useful marker to follow nodule ontogeny.

REFERENCES

Arioli T., Howles P.A., Weinman J.J. and Rolfe B.G. (1994) Gene **138**, 79-86.
Bhuvaneswari T.V., Bhagwat A.A. and Bauer W.D. (1981) Plant Physiol. **68**:1144-49.
Guilfoyle T.J., Hagen G., Li Y., Ulmasov T., Liu Z. and Gee M. (1993). Aust. J. Plant Physiol. **20**: 489-502.
Hirsch AM. (1992) New Phytol. **122**: 211-237.
Howles P.A., Arioli T. and Weinman J.J. (1995). Plant Physiol. **107**, 1035-36.
Jacobs M. and Rubrery P.H. (1988) Science **241**: 346-49.
Larkin P.J. Gibson J.M., Mathesius U., Weinman J.J., Gartner E., Hall E., Tanner G.J., Rolfe B.G. and Djordjevic M.A. (1996) Transg. Res.. **5**: 1-11
Lawson C.G.R. Rolfe B.G. and Djordjevic, M.A.(1996) Aust. J. Plant Physiol. **23**: 93-101.
Recourt K., Schripsema J,, Kijne J.W., van Brussel A.A.N. and Lugtenberg B.J.J. (1991) Plant Molec. Biology. **16**: 841-852.
Rolfe B.G. and McIver J. (1996) J Aust. J. Plant Physiol **23**: 271-83.
Sachs T. (1981).Advances in Botanical Research **9**: 151-262
Sautter C., Waldner K., Neuhaus-Url G., Galli A., Neuhaus G. and Potrykus I. (1991). Biotechnology **9**: 1080-1085.
Spaink H.P., Blomberg B.V., van Brussel A.A.N., Lugtenberg B.J.J., van der Drift K.M.G.M., Haverkamp J. and Thomas-Oates J.E. (1995). Molec. Plant Microbe Interact. **8**: 155-164.

# SEPARATION AND CHARACTERISATION OF *RHIZOBIUM* AND *TRIFOLIUM* PROTEINS USING PROTEOME ANALYSIS TO STUDY GLOBAL CHANGES IN GENE EXPRESSION.

N. Guerreiro, J.J. Weinman, S. Natera, A.C. Morris, J. W. Redmond, M. A. Djordjevic and B.G. Rolfe.

PMI Group, Research School of Biological Sciences, Australian National University, PO Box 475, Canberra, ACT Australia, 2601.

## INTRODUCTION

We have used proteome analysis (Wilkins 1996) to monitor global changes in gene expression in *Rhizobium* and *Trifolium subterraneum* in response to defined stimuli. Proteome analysis uses 2-D gel electrophoresis for protein separation followed by several post separation techniques, such as N-terminal sequencing and amino acid composition analysis (Fig 1). We have applied these techniques to the analysis of *Rhizobium* proteins induced in response to flavonoid addition and to the analysis of proteins synthesised in a mite-resistant variety of clover in response to mite exposure. Using immobilised pH gradient (IPG) gels for the first dimension separation, we have generated reproducible protein separation patterns of over 1000 *Rhizobium* proteins, giving the first detailed 2-D map of gene products of *R. l.* bv. trifolii. Three flavonoid-induced proteins were observed in the pH 4–7 range and the identity of one, NodE, was determined. The addition of flavonoids did not result in global changes in gene expression, and several constitutive proteins were identified in order to establish internal mobility standards. Several of these proteins revealed post-translational protein processing. A comparison of acidic proteins from mite-resistant varieties of *Trifolium* before and after mite feeding revealed the induction or up-regulation of up to fifteen proteins. N-terminal microsequencing strongly suggests that one of these is superoxide dismutase.

## METHODS FOR PROTEOME ANALYSIS

Strain ANU843 was grown to late lag phase before 7,4'dihydroxy flavone (DHF) was added to 2 μM and incubated for 6 h before harvesting. The cells were sonicated and protein samples prepared and separated by 2-D gel electrophoresis (Gorg 1995) using IPG strips. Analytical gels loaded with 75 μg of protein were silver stained and the images analysed using

# Proteome analysis

## Two-dimensional electrophoresis
## and associated post separation techniques

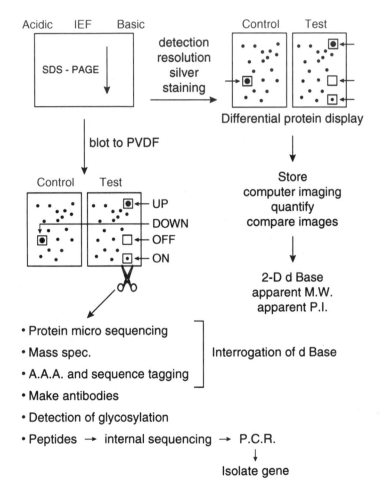

Fig. 1: Proteins separated by 2-DE for proteome analysis. Protein spots presented in two ways; analytical silver stained spot for image analysis or as a preparative spot blotted onto PVDF membrane for post separation analysis.

MELANIE II software (Bio Rad). Preparative gels loaded with 500 μg of protein, were transferred to PVDF membranes and stained with Coomassie Brilliant Blue R250. For N-terminal sequencing, selected spots were excised from the PVDF membranes and analysed using a PROCISE 494-01 sequencer. Amino acid composition analysis of excised spots and database matching (Wilkins 1996) was also employed.

ANALYSIS OF FLAVONOID INDUCED AND CONSTITUTIVE PROTEINS IN STRAIN ANU843.

Over 1000 well resolved silver-stained proteins were reproducibly detected in control (–DHF) and treated (+DHF) protein samples. Little variation in the pattern of proteins detected was evident, although there was some variation in spot intensity. Only 3 induced proteins of apparent molecular weights of 50, 29.8 and 24 kD were detected in the pH 4–7 range. These and 18 other constitutive proteins were subjected to N-terminal sequence analysis using the FASTA and TFASTA programs for the screening of protein and nucleic acid databases, respectively. Two of the three induced proteins were present in insufficient quantities to give sequence data, whereas the 50 kd protein was identified as NodE. The constitutive proteins were analysed to provide internal standards and 8/18 exhibited high sequence homology to previously identified proteins. One was N-terminally blocked and the remainder yielded unique sequences. Reliable sequence data can be achieved with 200 fmol of protein (although sequence has been generated from as little as 60 fmol) and the proteins are remarkably pure. The constitutive proteins included aspartate amino transferase (predicted mol wt, 48 kD), two malate dehydrogenases (PMW 41.8 kD and 38.5 kD respectively), ChvE (PMW 37.5 kD), a homologue of the orf 240 protein of *R. leguminosarum* (PMW 11.8 kD, GroES A (PMW 13.9 kD) and the 50s ribosomal protein (PMW 15.6 kD).

Amino acid composition analysis combined with estimates of the predicted molecular weight and isoelectric point were used to analyse 88 constitutive proteins and provide data on protein identity. The data were compared to the theoretical amino acid composition of proteins in the Swiss-Prot database using the AACompID program. Because the number of *Rhizobium* proteins is under-represented in the currently available databases, the single species and cross-species matching was undertaken. Thus far this analysis has not offered a reliable means of establishing protein identities with only two proteins (aspartate aminotransferase of *R. meliloti* and a procaryotic 50s ribosomal protein) being identified with confidence. Similar results were reported previously (Wasinger 1995). Further analysis is required to establish whether this technique is of value for matching proteins across species.

## ANALYSIS OF *TRIFOLIUM* PROTEINS PRODUCED IN RESPONSE TO MITE PREDATION

Redlegged Earth Mite (RLEM) is a major cause of clover pasture failure in temperate regions of Australia. In order to undertake molecular approaches to combatting these mites, we have begun an analysis of proteins induced by RLEM feeding in RLEM resistant subterranean clovers from the Australasian Subterranean Clover and Alternative Legume Improvement Program. Seed from the lines DGI007, S3615H and EP145 SubD was kindly supplied for testing by Drs Gillespie and Collins (Agriculture Western Australia). Freshly collected and sorted RLEM were added to ten day-old seedlings at an infection density greater than 10,000/m$^2$. After four days the boxes were removed and tapped to remove any RLEM climbing on the seedlings. These were cut at ground level, snap-frozen in liquid nitrogen and then stored at $-80^{\circ}$C until required. Following purification, extracted acidic proteins were separated using an 11 cm, pH 3–10 IPG in the first dimension and a 12–14% acrylamide gradient in the second dimension. Following silver staining, these gels revealed significant differences in the presence or levels of a number of proteins after 4 days of RLEM infestation.

Current research has now scaled up the 2-D separations for 500 µg samples, with 18 cm pH 3–10 IPG in the first dimension (focused for 200,000 V/h), and 12–14% SDS PAGE for the second. Proteins were electro-blotted to PVDF membrane. Protein micro-sequencing has characterized the N-terminal 10 amino acids of a number of these proteins. One of the induced proteins from cultivar DGI007 has the N-terminal sequence AAKKAVAVLK, and yields a perfect match to amino acids 49-58 of the Cu-Zn superoxide dismutase from *Populus tremuloides* in a database comparison. It is possible that this protein plays a role in protecting the plant during the oxidative burst associated with plant defense.

REFERENCES

Wilkins M R, et al (1996) BIO/TECHNOLOGY 14: 61-65
Gorg A, et al (1995) Electrophoresis 16: 1079-1086
Wasinger V C, et el (1995) Electrophoresis 16: 1090-1094

ACKNOWLEDGEMENTS

We acknowledge support from the Australian Meat Research Corporation and thank the ANU Biomolecular Resource Facillity for protein micro-sequencing.

# Control of nodule organogenesis in *Medicago*

Crespi M.[1], Charon C.[1], Johansson C.[1], Frugier F.[1], Coba T.[1], Bauer P.[1], Fehér, A.[1], Lodeiro, A.[1], Poirier S.[1], Brown S.[1], Ratet P.[1], Staehelin, C.[1], Trinh, T.H.[1] Schultze M.[1], Kondorosi E.[1], Felle, H.H.[3] and Kondorosi A.[1,2]

[1]Institut des Sciences Végétales, CNRS, Gif-sur-Yvette, France, [2]Biological Research Center, Szeged, Hungary and [3]Botanisches Institut I, Universität Giessen, Germany

To study the molecular mechanisms implicated in nodule organogenesis in the *Rhizobium meliloti-Medicago* symbiosis we have studied root hair responses and susceptibility of *Medicago* cells to the mitotic activity of the Nod factors. Moreover, we identified three early nodulin genes *Msenod12A*, *Mscal* and *Msenod40*, that appear to be molecular markers of nodule initiation among which *Mtenod40* might be one of the critical elements involved in the control of root cortical cell division.

## Nod signal transduction in alfalfa root hairs

It is at the tip of growing root hairs where Nod signals most likely are sensed for the first time by the host plant. Subsequently, Nod factors or secondary signals may be transported to the inner cortical cell layers where they induce cell divisions. In order to unravel the first events in Nod signal transduction, we have started to analyze intracellular ion fluxes in root hairs of *M. sativa* (alfalfa) using ion-selective microelectrodes. In a first series of experiments, we have recorded changes in intracellular pH, in parallel with measuring changes in the plasma membrane potential of root hair cells. Concomitant with a transient plasma membrane depolarization, root hairs react with an intracellular alkalinization of more than 0.2 pH units (Felle et al., 1996). Whereas both responses are rapid with a delay of only 15 sec, their kinetics is different, the pH change persisting in the presence of Nod factors. Moreover, pH change and depolarization can be uncoupled, indicating that the two responses are not causally linked. In general, the ability of different Nod factors to induce both depolarization and alkalinization correlated well with their symbiotic activity (root hair deformation, formation of nodule primordia). However, *O*-acetylation did not contribute significantly to Nod signal activity (Felle et al., 1995). Moreover, intracellular alkalinization, but not membrane depolarization, was also observed in response to elevated concentrations of non-sulfated Nod factors. In addition, whereas we observed the occurrence of a refractory state for pairs of different types of sulfated Nod factors, non-sulfated Nod factors did not affect the responsiveness of sulfated ones, and vice versa. These data indicate that alfalfa root hairs possess two independent perception systems for Nod factors. The one responding to non-sulfated molecules may represent an ancestral perception system that is still partially functional and might be related to those of other legumes that do not require Nod factor sulfation. A role of intracellular alkalinization in root hair curling is suggested by the finding that intracellular pH changes in the range of 0.2 units can influence root hair tip growth (Herrmann and Felle, 1995).

### Susceptibility of *Medicago* cells to the mitogenic activity of the *R. meliloti* Nod factors

Competence for cell division in response to the Nod factor is controlled by the level of combined nitrogen in the soil and by the position of the cells in the roots (Hirsch, 1992). Only cells in the inner cortex, located opposite to the protoxylem poles in the zone of emerging root hairs, can undergo cell division upon addition of Nod factors (Yang et al., 1994). On the other hand, *Medicago* cell cultures are also capable to respond to Nod factor by differential expression of genes as well as by re-entering the cell division cycle (Savouré et al., 1994). Moreover, synthetic Nod factors, were shown to stimulate growth of tobacco protoplasts and substitute for auxin and cytokinin required for growth (Röhrig et al., 1995).

In rapidly growing *Medicago* cell cultures the Nod factor has no effect. Therefore, we investigated whether protoplasts, isolated from *Medicago* cell cultures arrested in growth by hormone starvation, respond to the Nod factors or to a combination of Nod factors with NAA or kinetin. While no increase in cell number was detected when the protoplasts were cultivated in the absence or limiting concentration of NAA, growth of the protoplasts in the latter case was stimulated by the addition of either the Nod factor or kinetin. This result indicates that in this experimental system Nod factor can act as substituent for kinetin. The active Nod factor structure exhibited growth stimulation at 5 orders of magnitude lower concentrations than the non-sulfated molecules or chitotetraose indicating that host specific recognition of Nod factors is maintained in *Medicago* protoplasts.

### Hormonal and metabolic controls of nodule initiation by the host plant

In transgenic alfalfa plants carrying a promoter-*gus* fusion, expression of *Msenod12A* , coding for a putative proline-rich protein, was induced in the dividing cortical cells by *R. meliloti* or Nod factor. These transgenic plants served as a useful tool to follow nodule initiation and the hormonal and metabolic control exerted by the plant at the onset of nodulation. By treating the roots with various growth regulators, cytokinins (BAP or kinetin at 1 μM) were found to induce similar pattern of cortical cell division, as Nod factors, and under the same control by the carbon/nitrogen metabolism of the root (Bauer et al., 1996) indicating that nitrogen limitation  sensitizes the root cortex to the action of both growth regulators.

In addition to *Msenod12A* expression and cortical cell division, Nod factor treatment results in amyloplast deposition in the inner cortex indicating the accumulation of carbon translocated from the leaves (Ardourel et al., 1994). This was further supported by the observation that in the absence of rhizobia certain alfalfa cultivars develop nodules spontaneously accumulating large amounts of amyloplasts in the central region (Truchet et al., 1989). The spontaneous nodules may serve as carbon storage organs elicited during nitrogen starvation and be the ancestors of the *Rhizobium*-induced nodules. Then, the latter organs may initially function as a carbon "sink" under nitrogen control by the plant host. We have recently identified a carbonic anhydrase gene, *Msca1,* expressed in the nodule primordium and in spontaneous nodules that might be related to this function. *Msca1* as well as amyloplast deposition was induced by cytokinin treatment of roots. Thus, exogenous application of purified Nod factors and cytokinins had similar effects with respect to the induction of cortical cell division, early nodulin expression and amyloplast deposition. In contrast, the responses of the epidermal root hairs to these compounds were totally different. These results suggest that lipochitooligosaccharides and cytokinins may share certain signalling elements in the activation of the root cortical cells, further strengthening the role of phytohormonal imbalances in the elicitation of nodule initiation (Hirsch, 1992; Cooper and Long, 1994).

### *Msenod40* induces cortical cell division in alfalfa roots

The spatial and temporal expression pattern of *Msenod40* revealed that it was induced before the division of cortical cells and hence it may have a function acting very early in nodulation. The *enod40* genes code for 700 nucleotide long RNAs without any long ORF. Comparison of several *enod40* genes showed that only a small ORF (12 or 13 aminoacids) is common among them (Vijn et al., 1995), despite a strong conservation of the nucleotide sequence. Moreover, the *enod40* sequences form particularly stable secondary structures, a property characteristic of biologically active RNAs. We proposed that these genes might act as "riboregulators"(Crespi et al., 1994), a class of RNAs involved in the control of cell division and differentiation.

To analyse the action of this gene, plants ectopically expressing *enod40* from the constitutive 35S promoter were constructed. Initial experiments with *M. sativa* showed that overexpression of *enod40* affected the regeneration of somatic embryos, rendering them hypersensitive to cytokinins. Recently, we succeeded in regenerating transgenic *M. truncatula* plants, a diploid autogamous species, overexpressing *enod40*. These plants expressed the gene at various levels and individual analysis of each plant and its progeny was carried out. Plants strongly overexpressing *enod40* exhibited major alterations in growth and development and could not be maintained. Fortunately, we were able to recover several fertile transgenic plants with relatively elevated levels of transgene expression that were stably inherited. These F2 transgenic plants did not show major alterations in growth by comparison to plants transformed with a control vector. Since the *Msenod40* gene was identified by its strong expression in spontaneous nodules, we tested the behaviour of these plants during nitrogen limitation, a condition required for nodule development. The roots of the transgenic plants showed a significant increase in the number of dividing cells after 15 days of nitrogen starvation. Detailed microscopical analysis and screening revealed two types of cell division: a) divisions localised close to the vascular tissue without significant amyloplast accumulation ("putative" lateral root primordia) and b) divisions clearly located in the cortex with net amyloplast accumulation. The latter type of cell division was significantly enhanced (5-8 fold) in the transgenic plants. These results suggest that *enod40* might be involved in the induction of cortical cell division in legume roots. Moreover, cortical cell division was not detected if plants were grown in the presence of combined nitrogen, reinforcing the relation of this phenotype with nodulation.

We tested whether *enod40* may act upstream of *Msenod12A* in the signal transduction pathway leading to nodule development. To this end, we developed a transient assay in alfalfa roots using particle gun bombardment of transgenic alfalfa plants carrying the *Msenod12A* promoter-*uidA* fusion. Bombardment of *Msenod40* DNA induced cortical cell division and *Msenod12A* expression in roots. These effects required nitrogen starvation and intact plants. Taken together, these results indicate that in alfalfa roots under nitrogen-limiting conditions, the overexpression of *enod40*, either stably or transiently, lead to cortical cell division. Therefore, *enod40* mimics the phytohormonal imbalances in the inner root cortex leading to cell division, and may be an element involved in nodule initiation.

It has recently been shown that both the *enod40* gene and the small encoded peptide render tobacco protoplasts tolerant to auxin (Van de Sande et al., 1996). However, deletion of the sequence corresponding to the small peptide did not affect auxin tolerance. It was proposed that the shortened sequence was capable of inducing *in trans* the production of a homologous endogenous peptide in tobacco. In our transient assay on legume roots, we tested different regions of the *enod40* RNA. After two days, both the region spanning the small peptide as well as the 3'UTR of *enod40* elicited cortical cell division and *Msenod12A* expression. Bombardment of control DNAs did not provoke any of these two responses. These results indicate that whereas the small peptide might be the *enod40* gene product, the 3'UTR region may play an important regulatory role.

To gain further insght into *enod40* function, the *Mtenod40* gene was overexpressed in *Arabidopsis thaliana*. The homozygous transgenic plants showed a conditional "short root" phenotype depending on the medium, correlating to the level of transgene expression. Thus, the work of Van de Sande et al. (1996) and ours suggest that *enod40* may have a more general action in plant growth. The gene function might imply a tight control of the production of the small peptide, possibly mediated by the long 3'UTR as has been reported for certain differentiation processes occurring in animal cells (Rastinejad et al, 1993).

## Does Nod signal inactivation contribute to the coordination of cell division in nodule organogenesis?

Responsiveness of cells to the highly mitogenic Nod signals must be tightly regulated in order to avoid uncontrolled tissue proliferation (callus formation). Two ways in which this is achieved are the regulation of cell division competence and desensitization of perception systems as described above. Additional regulation can be envisioned at the level of signal inactivation. We have recently identified a Nod signal degrading activity that is stimulated in Nod factor treated alfalfa roots (Staehelin et al., 1995). New data indicate that this activity is due to a novel root and nodule-specific glycosyl hydrolase that lacks chitinase activity and prefers Nod factors as substrate. Our results suggest that this enzyme may play a role in limiting the mitogenic activity of Nod factors.

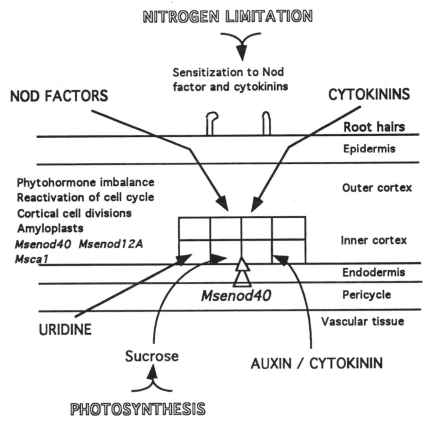

Fig. 1. Working model for the control of nodule initiation.

### Conclusion

Rapid changes in intracellular pH of root hair cells may constitute one element in a Nod factor signal transduction cascade. Whether ion fluxes would also occur in division-competent cells in the inner root cortex remains to be shown. It is possible that Nod factors do not reach the inner cortex in intact form, due to their rapid inactivation by specific cell wall bound hydrolases. Thus secondary signals may be generated from the activated epidermal cells establishing a gradient of morphogens that overlaps with corresponding gradients in the concentration of a variety of other signals necessary for triggering cell division at a well defined position.

The model presented in Fig. 1 highlights the various components proposed to be important in nodule initiation. Nitrogen limitation leads to sensitization of root cells towards Nod factors and cytokinins. Moreover, the endogenous morphogen gradients of auxin, cytokinin and uridine determine the division competence of the inner cortical cells. In these cells, the Nod factors and cytokinin induce similar responses, like cell division, amyloplast accumulation or activation of the early nodulin expression, suggesting convergence of the two signaling pathways or sharing certain signalling components. These components may influence or interact with the phytohormones or uridine, whose concentration gradients might be important in nodule initiation, or with the *enod40* gene product to activate the initial cells of the organ. Analysis of possible interactions between these components is needed in order to understanding of the mechanisms implicated in the control of nodule organogenesis.

## References

Ardourel, M., Demont, N., Debellé, F., Maillet, F., de Billy, F., Promé, J.C., Dénarié, J. and Truchet, G. 1994. Plant Cell 6: 1357-1374.

Bauer, P., Ratet, P., Crespi, M.D., Schultze M. and Kondorosi, A. 1996. Plant J., in press .

Cooper, J.B., and Long, S.R. 1994. Plant Cell 6: 215-225.

Crespi, M., Jurkevitch, E., Poiret, M., D'Aubenton-Carafa, Y., Petrovics, G., Kondorosi, E., and Kondorosi, A. 1994. EMBO J. 13: 5099-5112.

Felle, H.H., Kondorosi, É., Kondorosi, Á. and Schultze, M. 1995. Plant J. 7: 939-947.

Felle, H.H., Kondorosi, É., Kondorosi, Á. and Schultze, M. 1996. Plant J., in press.

Herrmann, A. and Felle, H.H. 1995. New Phytol. 129: 523-533.

Hirsch, A.M. 1992. New Phytol. 122: 211-237.

Rastinejad, F., Conboy, M.J., Rando, T.A., and Blau H.M. 1993. Cell 75: 1107-1117.

Röhrig, H., Schmidt, J., Walden, R., Czaja, I., Miklasevics, E., Wieneke, U., Schell, U. and John, M. 1995. Science 269: 841-843.

Savouré, A., Magyar, Z., Pierre, M., Brown, S., Schultze, M., Dudits, D., Kondorosi, A. and Kondorosi, E. 1994. EMBO J. 13: 1093-1102.

Staehelin, C., Schultze, M., Kondorosi, É. and Kondorosi, Á. 1995. Plant Physiol. 108: 1607-1614.

Truchet, G., Barker, D.G., Camut, S., Billy, F.D., Vasse, J., and Huguet, T. 1989. Mol. Gen. Genet. 219: 65-68.

Van de Sande, K., Pawlowski, K., Czaja, I., Wieneke, U., Schell, J., Schmidt, J., Walden, R., Matvienko, M., Wellink, J., Van Kammen, A., Franssen, H. and Bisseling, T. 1996. Science in press.

Vijn I., Yang W., Pallisgard N., Ostergaard E., van Kamen A. and Bisseling T. 1995. Plant Mol. Biol. 28: 1111-1119.

Yang, W.C., De Blanck, C., Meskiene, I., Hirt, H., Bakker, J., Van Kammen, A., Franssen, H., and Bisseling, T. 1994. Plant Cell 6: 1415-1426.

# Gene discovery in early plant nodulation responses and systemic regulation of nodulation

Peter M. Gresshoff, Gustavo Caetano-Anollés, Roel P. Funke, Farshid Ghassemi, Jaime Padilla, Gabrielle Crüger, Sanjeev Pillai, Jiri Stiller, Ruju Chian, Anatoli Filatov, Raymond McDonnell, Sunil Tuppale, Qunyi Jiang, Lisa Calfee-Richardson and Debbie Landau-Ellis.

Plant Molecular Genetics, Center for Legume Research and Institute of Agriculture, The University of Tennessee, Knoxville TN 37901-1071, USA

That nodulation of legumes is controlled by plant genes was confirmed by genetic and biochemical means. For example, many gene products, called nodulins, are preferentially or exclusively expressed in nodules (Legocki and Verma, 1979). Mutants altered in symbiotic steps such as infection and nodule initiation (Nod⁻; Mathews *et al.*, 1990), symbiotic establishment and senescence (Fix⁻; Sagan *et al.*, 1995) as well as nodulation number control (super- or hypernodulation, Nts⁻; Carroll *et al.*, 1985a,b) were isolated in several legumes (see Caetano-Anollés and Gresshoff, 1991; Gresshoff, 1993). The plant controls the colonization of the rhizosphere, the potential of root hair infection, the positional control of cell division initiation in the cortex and along the root, phytohormone and sucrose supply to the meristem, defense responses, membrane growth, nodule architecture, nodule oxygen, nitrogen and carbon metabolism and finally senescence. The microsymbiont supplies the genetic ability to convert nitrogen gas to ammonia and to synthesize a new plant growth regulator group called lipo-oligosaccharides (or nod-factors). The latter may represent a biochemical mimicry or evolutionary "pirating" of endogenous plant functions suggested by the fact that some genotypes of alfalfa and white clover develop nodules spontaneously (Truchet *et al.*, 1989; Joshi *et al.*, 1991; Blauenfeldt *et al.*, 1994), and that some plants possess chitin-related molecules and chitin-processing biochemistry (see Spaink *et al.*, this volume).

Genetics and molecular biology define two ends of the structure-function continuum. The plant mutants define single genetic changes in gene sequences which are essential for the symbiosis; yet for none the causative mutation event is understood, nor is the gene isolated. Thus it is impossible to speculate on biochemical function. In contrast, cloned plant genes for nodulins define precise stretches of DNA involved, but not necessarily essential, for the symbiosis. Indeed, many nodulins are only defined by their DNA sequence, their possible site of action as demonstrated by *in situ* hybridization (detecting mRNA not protein!), and presumptive function as determined by database searches and domain/motif matching. Through more refined detection methods such as differential display (Goormachtig *et al.*, 1995), more "orphan" sequences have been isolated, increasing the need to determine whether the gene product is a consequence of, or necessity for, nodulation.

The question of essential function for nodulins was tested in alfalfa, where Csanádi *et al.* (1993) were able to breed a plant line which lacked entirely the signal for nodulin Enod12; yet the plant was normal, suggesting biochemical redundancy or non-essentiality of Enod12.

## Nodule-enhanced gene expression and molecular physiology

The recent past has seen considerable activity trying to overcome these shortcomings. Nodulin cDNAs were transformed into plants as antisense constructs to interfere with endogenous gene function (Cheon *et al.*, 1993). Preliminary results showed negative conclusions; for example, antisense transgenics of enod2 did not eliminate nodulation nor nitrogen fixation (Hunt *et al.*, 1995). Alternatively nodulin promoters were fused to reporter genes to detect cell-specific gene expression after differential environmental stimulation (e.g., phytohormone or nod-factor exposure). This form of molecular physiology is powerful (Jensen *et al.*, 1986) as it helps define promoter elements involved in gene regulation and organ-specificity (Stougaard *et al.*, 1987), but sofar it tends to confirm only what was known already from physiology, Northern analysis, classical biochemistry, or RT-PCR expression analysis. Plants transgenic for developmental promoter-reporter gene constructs, however, may become tools to study the interaction of genes and their products in two-component systems or multi-step biochemical or regulatory pathways. Re-transformation of such transgenics (either stably or transiently) can provide insight into new genetic components.

But the paradigm of nodule specificity is being challenged. Early nodulin research compared nodules with roots and not other tissues. Even there, the biologically appropriate control is difficult to achieve. A 4 day old uninoculated seedling root should not be compared to a 14 day old nodule. As resolution increased and other tissues were surveyed, nodulins were found to exist in other tissues or pre-existing in roots at lower than nodule levels. Uricase is found in soybean embryos, enod40 is expressed at low levels in the vasculature of uninfected roots (phloem associated; Kouchi and Hata, 1993), enod5 is found in flowers, and even leghemoglobin appears to have a non-symbiotic form with a yet unknown function (perhaps oxygen transport to sub-endodermal lateral root primordia or even oxygen level sensing?). Clearly nodulin genes did not arise *de novo*, but are "pirated" sequences reflecting evolutionary analogs of perhaps related function.

The question remains: where did the genes needed for nodule initiation and function come from? Many genes surely have parallel functions in nodule and root growth. Some will be members of multigene families which have diverged from their ubiquitous analog (such as glutamine synthase, or chalcone synthase), while others may be the chance product of unequal crossovers, exon shuffling, deletion and duplications (perhaps enod2). This set of metabolic and structural genes requires coordinate expression to form an organ. Where did the genes controlling the nodule manufacturing genes arise? At first one must clarify, which genes are controlling, and not being controlled by, nodulation? The classical query of cause versus response needs to be clarified. Does nodulation control mirror plant disease responses (the controlled disease paradigm), or is it reflecting altered lateral root developmental control? It is likely that many of the control circuits governing nodule initiation exist in non-legumes (for example note the recent discovery of autoregulation of nodulation in actinorhizal plants, and the shoot control of tuber numbers in potato), and that in legumes, regulatory patterns involving lateral root and nodule development are superimposed (c.f., Caetano-Anollés *et al.*, 1991b).

Nodulation in the absence of *Rhizobium* (NAR) provides a clue towards the evolutionary origins of nodule ontogeny (Caetano-Anollés *et al.*, 1991a). First, one needs to distinguish nodulation and nitrogen fixation, being regulated by different parameters, although pleiotropy frequently confuses causality. For example, many Fix⁻ plant mutants may actually be altered in a pre-fixation-related step (e.g., membrane biogenesis), which leads to a premature break-down of the symbiotic balance, leading to symbiosome decay and a fixation-deficient phenotype.

Plants autoregulate nodule number independent of nitrogen fixation (as demonstrated by nodules induced with *nifH⁻* or *nifA⁻* bacterial mutants). NAR nodules on alfalfa possess extremely interesting nodule properties, some of which were long considered to be part of the bacterial contribution to the functional symbiosis. For example, beside being subject to autoregulation, NAR nodules are able to elicit it (Caetano-Anollés *et al.*, 1991a). NAR nodules are subject to nitrate inhibition. They develop different cell types; central cells are filled with starch grains, illustrating that the NAR nodule functions as a major carbon sink (previously attributed to the respiring and N-fixing bacterium). Even different cell types are found in the central zone (large starch-filled) and small (no starch), reminiscent of infected and uninfected cells. Did the development of starch grains in amyloplasts stimulate cell enlargement through increased auxin sensitivity, biosynthesis, or transport? *Rhizobium*-induced nodules were long known to possess transfer cell on their xylem vessels. These function through "vili" with an associated increased density of mitochondria, possible increasing surface area and permease biochemistry and energetics for high solute export (amides, ureides) into the xylem. NAR nodules also possess transfer cells associated with the phloem, consistent with the high carbon transport needed for amyloplast loading (Joshi et al., 1993). Thus nodulation is possible in the absence of *Rhizobium* in some legume genotypes, giving rise functional organs, but without nitrogen fixation activity. Can we understand nodulation better by trying to understand carbon metabolism and its role in development, than nitrogen metabolism? Does a plant monitor sucrose levels (or hexoses) to induce nodulation? If nitrate levels are high, the excess carbon will be used for growth; it needs to be low, when carbon supply is high, that nodules are induced. Many simple experiments may have gone astray because researchers tried to study early nodulation related steps in plants which were so nitrogen starved, that they did not have the ability to develop sufficient photosynthetic ability to trigger nodule initiation in cortical tissues. *Rhizobium* seemingly has learnt to mimic this internal regulatory system using lipo-oligosaccharides as signals.

## Co-Mapping

About 15 symbiotic mutations have been mapped sofar on maps of pea, soybean, sweet clover, *Lotus japonicus* and *Medicago truncatula*, together with about 10-15 cDNAs, that are conceptually or functionally connected to nodulation and N fixation. For example, one of the two ENOD2 genes is on linkage group A of soybean, close to the loci controlling seed coat color (chalcone synthase), seed hardness and a QTL for pod filling period. A major gene controlling soybean cyst nematode (*Rhg4*) maps in the same region (Ghassemi *et al.,* 1997). In pea many symbiotic genes cluster on chromosome 1; this was not observed in soybean where *rj₁ , Rj₂, rj₆, nts-1,* leghemoglobin (*lbc3*), *enod2b, cdc2M* and *cdc2N* protein kinase appear to map on separate linkage groups. *Rj₂*, however, controlling strain specific nodulation interestingly maps in a cluster of other resistance genes including resistance to bacterial, nematode and fungal pathogens (Polzin *et al.*, 1994).

The approach is to find absolute co-segregation of a mapped phenotype with a known gene sequence, producing a candidate gene. As yet this approach has not led to success for symbiotic genes, although increased activity using more mutants and more clones (such as resistance gene analogues, RGAs) may eventually define a symbiotic (like *Rj₂*) or resistance gene.

## Map-based cloning

Since the chances of finding co-segregation of a molecular marker (either cDNA, random genomic RFLP, arbitrary primer marker, or microsatellite) with a symbiotic gene locus is larger than detecting a direct coupling, efforts have been extended into this direction. The goal is to use the anchored molecular marker as a starting point to isolate a large fragment of DNA, cloned either into YAC or BAC vectors, then to screen the candidate clone for either coding abilities or complementation activity using retrofitted YACs or BACs, or subclones thereof. The risk of this procedure lies in the need to have sufficiently close markers to assure that the isolate YAC/BAC harbors the gene of interest. Alternatively neighboring fragments need to be isolated and searched. The existence of repeated interspersed DNA complicates this matter, as in soybean about half of the endclones of YACs were highly repeated. Many technical hurdles for this work have been nearly overcome with several BAC libraries available in soybean, *Lotus japonicus* and *Medicago truncatula*. YAC cloning was achieved for soybean (Funke *et al.,* 1994), and YACs were used in molecular cytology, valuable for chromosome painting, analysis of repeated DNA units, and tests for chimerism (Zhu *et al.,* 1996).

The need for close markers has increased the pressure for faster, more accurate and frequent molecular marker detection. The last decade has seen a shift from isozymes and RFLPs to RAPD and DAF markers and recently to SSRs (microsatellites; Rongwen *et al.*, 1995) and AFLP (selected restriction fragment amplification; Vos *et al.*, 1995). These are combined with large and specially designed breeding populations such as phenotypic or genotypic bulks (usually of $F_2$ or $F_3$ material) in bulked segregant analysis, near-isogenic lines (NILs), or recombinant inbred lines (RILs). RILs offer the advantage of having fixed gene frequencies (local homozygocity) for recombination events in the $F_2$ (useful for marker technology, like DAF, RAPD and AFLP, unable to distinguish heterozygotes), Screening large populations with marker technologies that detect many markers increases the chances of finding close linkage.

However, genetic distance is a function of recombination, and eukaryotic genomes are characterized by local variation in cross-over frequencies. For soybean the conversion of genetic to physical distance in one location close to the supernodulation locus (Funke *et al.*, 1993) gave 500 kb equivalent to no more than 1 cM. Many maps such as in Frenchbean, alfalfa, *Medicago truncatula*, Melilotus and pea were produced by distant relatives. This leads to suppression of recombination in the map, giving low total map sizes, but high polymorphism frequencies. Choosing more related parents for a mapping population achieves the opposite; for example, the *Lotus japonicus* map was made from line Gifu and Funakura, which show about 5% polymorphic banding pattern for AFLP and DAF analysis alike (Jiang and Gresshoff, 1997). Similarly the RIL map of Lark *et al.* (1993) gives different distances than the more widely used *G. soja* x *G. max* maps.

### Current mapping status of supernodulation in soybean

The *nts-1* locus of soybean was mapped on linkage group H (Iowa-USDA map) close to the pUTG-132a RFLP marker ($0.7\pm 0.5$ cM) and pA-381 ($4.8\pm1.5$ cM). By extrapolation this region of the genome is equivalent to linkage group U23 of the Utah RIL map. Synteny exists between *Phaseolus vulgaris* and soybean, and two linked RFLPs were mapped distal to pA-381. BSA coupled with DAF failed to detect closely linked markers, although the approach was verified by the detection of at least two proximal as well as distal markers (Kolchinsky *et al.*, 1997; Caetano-Anollés and Gresshoff, 1996). The genomic region appears to be sparsely heterogeneous among soybean varieties and even *Glycine soja*. Recently isolated YACs and BACs will now need to be searched for homology to the pUTG-132a RFLP.

### Gene tagging

This approach utilizes high efficiency gene transfer techniques that are absent in soybean and pea. Significant emphasis is given to model legumes such as *Lotus japonicus* and *Medicago truncatula*, characterized by fast and high frequency gene transfer. Approaches involve both transposable elements (Ac/Ds) and T-DNA either from *Agrobacterium tumefaciens* or *A. rhizogenes*. Both gene trapping and enhancer trapping strategies are now possible, especially since primary transgenic roots induced by *A. rhizogenes* nodulate in *Lotus japonicus* and can be screened for reporter gene activation (Stiller *et al.*, in preparation). Stougaard *et al.* (this volume) as well as J. Webb (pers. comm.) used T-DNA to isolate symbiotic mutants, suggesting that the approach is feasible. Whether these are caused by the insertion into an essential gene requires further tests, because in Arabidopsis, many morphological mutations isolated after T-DNA mutagenesis are not tagged by the insert DNA.

### New genes from differential display, fast neutrons and EMS

While there is a large number of interesting mutations and cloned nodulin genes, the search for more continues. Comparisons of mRNA populations by arbitrary primer technology (called differential display) has revealed new nodule enhanced genes in both Sesbania and *Lotus japonicus*. Similarly several new mutants were described in pea, *Lotus japonicus, Medicago truncatula* as well as sweet clover (*Melilotus alba*) following chemical mutagenesis. Many of these are tested for differential responses to phytohormones. We are presently screening $M_2$ seeds of *L. japonicus* irradiated with fast neutrons (H. Brunner, IAEA, Vienna) in an attempt to find symbiotic mutants caused by chromosomal deletions, which could serve as tools for differential screening and mapping.

All the above strategies have the potential to find new genes involved in signal reception, signal transmission as well as whole plant, organ, and cellular responses. Most importantly they may help find genes, for which symbiotic functions are known, and find functions for the genes only known by their sequence. The near future hopefully will bring this important connection to the field of plant control of nodulation and N fixation.

### References

**Bauer, P., Coba de la Pena, T., Frugier, F., Poirier, S., McKhann, H.I., Ratet, P, Brown, S., Crespi, M and Kondorosi, Á.** (1995) Role of plant hormones and carbon/nitrogen metabolism in controlling nodule initiation on alfalfa roots. In: Nitrogen Fixation: Fundamentals and Applications. eds. I.A. Tikhanovich, N. Provorov, V. I. Romanov and W.E. Newton., Kluwer Academic Publishers, Dordrecht, The Netherlands, pp 443-448.
**Blauenfeldt, J. Joshi, P.A., Gresshoff, P.M. and Caetano-Anollés, G.** (1994) Nodulation of white clover (*Trifolium repens*) in the absence of *Rhizobium*. Protoplasma **179**, 106-110.
**Caetano-Anollés, G. and Gresshoff, P. M.** (1991) Plant genetic control of nodulation. Annu. Review of Microbiology **45**, 345-382.

**Caetano-Anollés, G. and Gresshoff, P.M.** (1996) Generation of sequence signatures from DNA amplification fingerprints with mini-hairpin and microsatellite primers. Bio/Techniques **20**, 1044-1056.

**Caetano-Anollés, G., Joshi, P.A., and Gresshoff, P.M.** (1991a) Spontaneous nodules induce feedback suppression of nodulation in alfalfa. Planta **183**, 77-82.

**Caetano-Anollés, G., Paparozzi, E.T. and Gresshoff, P.M.** (1991b) Mature nodules and root tips control nodulation in soybean. J. Plant Physiol. **137**, 389-396.

**Carroll, B.J., McNeil, D.L. and Gresshoff, P.M.** (1985a) Isolation and properties of soybean mutants which nodulate in the presence of high nitrate concentrations. Proc. Natl. Acad. Sciences (USA) **82**, 4162-4166.

**Carroll, B.J., McNeil, D.L. and Gresshoff, P.M.** (1985b). A supernodulation and nitrate tolerant symbiotic (nts) soybean mutant. Plant Physiol. **78**, 34-40.

**Cheon, C-I., Lee, N-G., Siddique, A-B-M., Bal, A.K. and Verma, D.P.S.** (1993) Roles of plant homologs of *Rap1p* and *Rap7p* in the biogenesis of the peribacteroid membrane, a subcellular compartment formed (*de novo*) during root nodule symbiosis. EMBO J. **12**, 4125-4135.

**Csanádi, G., Szécsi, J., Kaló, P., Endre, G. Kondorosi, Á., Kondorosi, É. and Kiss, G.B.** (1993) ENOD12, an early nodulin gene is not required for nodule formation and efficient nitrogen fixation in alfalfa. Plant Cell **6**, 201-213.

**Funke, R.P., Kolchinsky, A.M., and Gresshoff, P.M.** (1993) Physical mapping of a region in the soybean (*Glycine max*) genome containing duplicated sequences. Plant Mol. Biol. **22**, 437-446.

**Funke, R.P., Kolchinsky, A.M., and Gresshoff, P.M.** (1994).High EDTA concentrations cause entrapment of small DNA molecules in the compression zone of pulsed field gels, resulting in smaller than expected insert sizes in YACs prepared from size selected DNA. Nucl. Acids Res. **22**, 2708-2709.

**Ghassemi, F. and Gresshoff, P.M.** (1997) The early nodulin gene *enod2* clusters in a region close to soybean cyst nematode resistance, seed hardness, and chalcone synthase, but segregates independently from leghemoglobin and a gene controlling nodule number. (in preparation).

**Goormachtig, S., Valerio-Lepiniec, M., Szczyglowski, K., van Montagu, M., Holsters, M., and de Bruijn, F. J.** (1995) Use of differential display to identify novel *Sesbania rostrata* genes enhanced by *Azorhizobium caulinodans* infection. Mol. Plant Microbe Interactions **8**, 816-824.

**Gresshoff, P.M.** (1993). Molecular genetic analysis of nodulation genes in soybean. Plant Breeding Reviews **11**, 275-318.

**Gresshoff, P.M.** (1995) Moving closer to the positional cloning of legume nodulation genes. *In*: Nitrogen Fixation: Horizons and Application. eds. I. Tikhonovich, N. Provorov, V. Romanov, and W.E. Newton, Kluwer Academic Publishers, Dortrecht, Netherlands, pp. 431-436.

**Jensen, J.S., Marcker, K.A., Otten, L. and Schell, J.** (1986) Nodule-specific expression of a chimaeric soybean leghaemoglobin gene in transgenic *Lotus corniculatus*. Nature **321**, 669-674.

**Jiang, Q. and Gresshoff, P.M.** (1997) Classical and molecular genetics of the model legume *Lotus japonicus*. Molec. Plant Microbe Interactions (in press).

**Joshi, P.A., Caetano-Anollés, G., Graham, E.T. and Gresshoff. P. M.** (1991) Ontogeny and ultrastructure of spontaneous nodules in alfalfa (*Medicago sativum*). Protoplasma **162**, 1-11.

**Joshi, P.A., Caetano-Anollés, G. Graham, E.T. and Gresshoff. P.M.** (1993) Ultrastructure of transfer cells in spontaneous nodules in alfalfa (*Medicago sativum*). Protoplasma **172**, 64-76.

**Kolchinsky, A., Landau-Ellis, D. and Gresshoff, P.M.** (1997) Map order and linkage distances of molecular markers close to the supernodulation (*nts-1*) locus of soybean. Mol. Gen. Genet. (in review).

**Kouchi, H. and Hata, S.** (1993) Isolation and characterization of novel cDNAs representing genes expressed at early stages of nodule development. Mol. Gen. Genet. **238**, 106-119 .

**Landau-Ellis, D., Angermüller, S., Shoemaker, R.C. and Gresshoff, P.M.** (1991) The genetic locus controlling supernodulation in soybean (*Glycine max* L) co-segregates tightly with a cloned molecular marker. Mol. Gen. Genet. **228**, 221-226.

**Lark, K.G., Weisemann, J.M., Mathews, B.F., Palmer, R., Chase, K., and Macalma, T.** (1993) A genetic map of soybean (*Glycine max* L.) using an interspecific cross of two cultivars 'Minsoy' and 'Noir 1'.Theor. Appl. Genet. **86**, 901-906.

**Legocki, R. and Verma, D.P.S.** (1979) A nodule-specific plant protein (Nodulin-36) from soybean. Science **205**, 190-193.

**Mathews, A., Carroll, B.J. and Gresshoff, P.M.** (1990) The genetic interaction between non-nodulation and supernodulation in soybean: an example of developmental epistasis. Theor. Appl. Genet. **79**, 125-130.

**Polzin, K.W., Lohnes, D.G, Nickell, C.D. and Shoemaker, R.C.** (1994) Integration of *Rps2*, *Rmd*, and *Rj2* into linkage group J of the soybean molecular map. Journal of Heredity **85**, 300-303.

**Rongwen, J., Akkaya, M.S., Bhagwat, A.A., Lavi, U. and Cregan, P.B.** (1995) The use of microsatellite DNA markers for soybean genotype identification. Theor. Appl. Genetics **90**, 43-48.

**Sagan, M., Morandi, D., Tarengui, E., and Duc, G.** (1995) Selection of nodulation and mycorrhizal mutants of the model legume *Medicago truncatula*(Gaertn.) after gamma ray mutagenesis. Plant Science **111**, 63-71.

**Stougaard, J., Sandal, N.N., Gron, A., Kuhle, A. and Marcker, K.A.** (1987) 5' Analysis of the soybean leghaemoglobin Lbc3 gene: regulatory elements required for promoter activity and organ specificity. EMBO J. **6**, 3565-3569

**Truchet, G., Barker, D.G., Camut, S., deBilly, F., Vasse, J. and Huguet, T.** (1989) Alfalfa nodulation in the absence of *Rhizobium*. Mol. Gen. Gen. **219**, 65-68.

**Vos, P., Hogers, R., Bleeker, M., Reijans, M., van de Lee, T., Hornes, M., Frijters, A., Pot, J., Peleman, J., Kuiper, M. and Zabeau, M.** (1995) AFLP: a new technique for DNA fingerprinting. Nucl. Acid Res. **23**, 4407-4414.

**Zhu, T., Funke, R.P., Shi, L. Gresshoff, P.M. and Keim, P.** (1996) Cytogenetic localization of soybean YACs by degenerate oligonucleotide-primed PCR and fluorescence *in situ* hybridization. Mol. Gen. Gen. (in press).

# Plant genes controlling the fate of bacteria inside the root.

Tikhonovich I.A., Borisov A.Y., Lebsky V.K., Morzhina E.V., Tsyganov V.E.

All-Russia Research Institute for Agricultural Microbiology, Saint-Petersburg-Pushkin, 189620, Russia. E-mail: biotec@riam.spb.su

Introduction.

The legume-Rhizobium symbiosis provides a model for studying the intimate mechanisms of plant-microbe interactions. The initial stages of symbiotic interaction are being studied intensively while little is known about the plant genetic control of the bacteria fate inside the root. Those genes can be identified using symbiotic mutants of macrosymbionts. About 200 independent plant symbiotic mutants have been isolated in different legume species (1) and their number is increasing permanently.

As a result of our work several mutations have been obtained in the pea (*Pisum sativum* L.) genes which control the process of infection thread (**IT**) and symbiosome development and hence they control the transformation of bacteria into specialized endosymbiotic form, termed bacteroid (**B**).

Development of symbiotic structures in nodules of different pea lines.

*Symbiosome formation in wild-type pea nodules.* Several morphological characters of the infection thread and symbiosome formation in wild-type nodules are important for understanding the nature of mutations induced (Fig. 1a). **IT** processing provides delivery of bacteria (**b**) into the nodule, formation of infection droplets (**ID**) and causes the individual endocytosis of bacteria into plant cell cytoplasm where bacteria are transformed into **B**s being individually enclosed by membrane of plant-origin (2). Large pleomorphic **B**s are clearly distinguished from bacterial cells inside the **IT**s by shape, size and electron density of matrix. Newly formed symbiosomes are capable of maintaining the process of nitrogen fixation during some period without senescence and degradation.

*Mutations affecting the* **IT** *functioning.* Two novel mutants have been obtained using the laboratory line SGE. The mutation in line SGEFix⁻-2 (Fig. 1b) determines the formation of two types of nodules on the root. The

Figure 1. Comparative analysis of nodule ultrastructure in mutant lines
SGEFix⁻-1 (b), SGEFix⁻-2 (c) and initial line SGE (a).

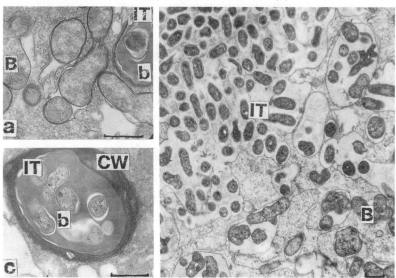

Figure 2. The scheme representing the succession of mutant genes
functioning in different mutant lines studied.

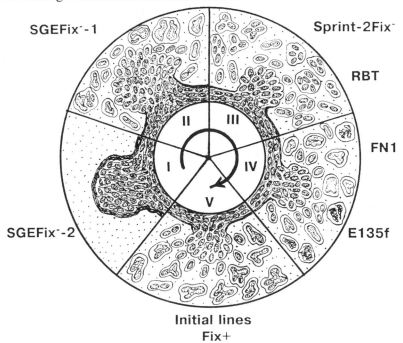

first one consists of nodules in which the bacteria delivered into the central part of the nodule remain locked in the **IT**s with abnormally thick cell walls (**CW**). Such thickness of plant **CW**s is thought to be the reason for the absence of bacterial release and can be caused by plant defense reactions. The second nodule type includes those nodules in which there is endocytosis of bacteria into the host-plant cell cytoplasm. Those **B**s become the target of plant defense reaction which is similar to Hypersensitive Response (**HR**).

The mutation in line SGEFix⁻-1 (Fig. 1c) determines formation of the nodules with deviations in differentiation of infection droplets (**ID**). In such nodules **ID**s occupy sufficient part of the infected cell. Those **ID**s are not stable and mass-endocytosis looks like "an explosion". In addition to that defect the symbiosomes with partly differentiated **B**s are subjected to premature degradation. This process is also similar to **HR**.

*Mutation affecting symbiosome formation.* Line Sprint-2Fix⁻ described earlier (**3**) and carrying mutation in gene *sym-31* represents the mutant phenotype with the absence of **B** differentiation following the endocytosis and with an abnormal symbiosome structure: several bacterial cells per one symbiosome unit. Such symbiosome structure is thought to be caused by bacterial multiplication inside the symbiosome membrane or symbiosome fusion. Premature degradation of symbiosomes as well as other processes which are similar to plant defense reactions were not revealed in this mutant.

*Mutations which cause active plant defense reactions and premature degradation of symbiotic structures.* Two mutant lines E135f (mutation in gene *sym-13*) and FN1 well characterized earlier (**4,5**) represent the group of "early senescent" mutants. At present some more mutants can be also referred to this group (**6,7**). In the nodules of mutants E135f and FN1 the process of infection thread and symbiosome differentiation looks morphologically like in wild-type nodules but the plant defense reaction similar to **HR** destroys symbiotic structures about the moment when nitrogen fixation is switched on.

Action of plant symbiotic genes during symbiosis development.

All five mutants mentioned above are not allelic to each other. The genes identified represent at least a part of the gene system controlling the fate of bacteria inside the root (nodule). The existence of about 30 Fix⁻ mutants of pea (*Pisum sativum* L.) raises a question about the succession of plant gene actions during symbiosis development. For two such genes the "double" mutant line RBT has been constructed (**8**). Morphological study of RBT line nodules revealed that mutation in locus *sym-31* suppresses phenotypic manifestation of *sym-13* mutation. Therefore the gene *sym-31* functions at an earlier stage of symbiosis development than the gene *sym-13*. As for other symbiotic genes identified the order of their functioning is based on morphological data only without combining those data together with genetic ones. Using the principle of the **earliest morphological abnormality observed** we can arrange mutant genes in the order of their functioning (Fig. 2). The mutant gene of line SGEFix⁻-2 (Locked **ITs**) can be placed first, the second place can be taken by the mutant gene of line SGEFix⁻-1 (Hypertrophied **ID**s), the third - by gene *sym-31* (Abnormal symbiosome structure) and the fourth - by the group of "early senescent" mutants (E135f and FN1). The proposed sequential functioning of plant symbiotic genes reflects existence of the complicated gene system responsible for the bacteria transformation into **B**s. A more precise analysis of molecular characteristics of the mutations in comparison with the initial lines (**9**) can allow to reveal the molecular mechanisms affecting the fate of microsymbionts inside the host plant root (nodule).

*Acknowledgments.* This work was supported by the grants of Russian Foundation of Fundamental Research (95-04-12573) and INTAS (94-1058).

References:

1. Phillips D.A. and Teuber L.R, 1992. In: Biological Nitrogen Fixation (Stacey G., Burris R. and Evans H.J., eds.). New York, Chapman and Hall: 200-235.

2. Brewin N.J., 1991. Ann. Rev. Cell Biol. 7: 191-226.

3. Borisov et al., 1992. Symbiosis 14: 297-313.

4. Kneen B.E, LaRue T.A., 1990. Plant Physiol. 94: 899-905.

5. Postma J.G. et al., 1990. Plant Sci. 68: 151-161.

6. Sagan M. et al., 1993. Plant Sci. 95: 55-66.

7. Novak K. et al., 1995. Annals of Botany 76: 303-313.

8. Borisov et al., 1995. In: Nitrogen Fixation: Fundamentals and Applications. Proceedings of X-th International Congress on Nitrogen Fixation (Tikhonovich I.A., Provorov N.A., Romanov V.I., Newton W.E., eds.) Kluwer Academic Publishers, Netherlands, 488.

9. Tikhonovich et al., ibid., 461-466.

*Note for figures.* Bar represents 1 μm.

# Transposon tagging in *Lotus japonicus* using the maize elements *Ac* and *Ds*.

Eloisa Pajuelo, Leif Schauser, Thomas Thykjær, Knud Larsen and Jens Stougaard.

Laboratory of Gene Expression, Department of Molecular and Structural Biology, University of Aarhus, Denmark.

Molecular genetic analysis in cultivated plants is often hampered by the complexity and the size of the plant genomes involved. The small genome and a number of other model plant characteristics has consequently qualified the crucifer *Arabidopsis thaliana* as the best choice for an international effort directed at comprehensive genetic mapping and genome sequencing. The aim for the large conglomerate of research groups participating is to provide a backbone for improving the molecular, biochemical and physiological understanding of plants and to provide background knowledge for plant breeding. This concerted effort has already contributed substantially to the progress of plant molecular biology and the scale of the advance may best be illustrated by the ongoing sequencing of the total genome estimated to be finished shortly into the next century.

Some important plant processes can however not be studied in *Arabidopsis*. One of these processes is the organogenic process leading to the development of nitrogen fixing root nodules on legumes. After considering several candidates for a model legume, a close relative of Birdsfoot trefoil called *Lotus japonicus* was suggested as model plant for legumes (Handberg and Stougaard 1992).

## Insertion mutagenesis in *Lotus japonicus*

### The model plant

The ideal model plant fulfilling all the requirements for simple and fast genetic, biochemical and physiological studies does probably not exist in nature. To be recommendable for the scientific community, a model plant should however offer substantial advantages compared to the alternative cultivated plants. *Lotus japonicus* is diploid, the genome size is small, it is selffertile, seed production is ample, large flowers make hand pollination possible, the seed to seed generation time is 3-4 months, *Agrobacterium tumefaciens* transformation followed by hygromycin or geneticin selection of transgenes is effective and transgenic plants can be regenerated with a simple procedure giving a high frequency of shoot formation (Handberg and Stougaard 1992, Handberg et al. 1994). These characteristics suggest that studies on *Lotus japonicus* could accelerate progress in understanding the plant contribution to symbiotic nitrogen fixation. In order to develop the genetics of *L. japonicus* we are at present attempting to tag symbiotic genes using transposable elements from maize.

Strategies for transposon mutagenesis

Classical DNA transposable elements have not been identified in many plant species including *L. japonicus*. Elements isolated form maize have therefore been introduced by transformation to establish transposon mutagenesis in some of these plant species. Both the *Ac* transposon family and the *Spm* transposon family have been employed for this purpose (Bancroft et al. 1992, Whitham et al. 1994, Aarts et al. 1995, and references therein). Several T-DNA segments for launching transposons from both families have been constructed. The approaches taken for introducing elements can be divided into a few general strategies. An outline of some of these are shown in Figure 1. Most vehicles have the element located within a marker that allows excision activity to be monitored. The element is generally located in the untranslated leader between a constitutive promoter and a reporter gene. Excision results in expression of the reporter which can then be monitored by its particular activity, for example Spectinomycin resistance or β-glucuronidase activity. The simplest tagging approach introduces an autonomous element. Excision events passing through the germline is then scored and mutant screening follows in the next generation. After excision the activity of autonomous elements can only be followed using time consuming DNA methodology or mutant/revertant phenotypes. To allow reinsertions to be selected, to follow the element genetically and to obtain stable insertions, inactive elements are often used in a two component strategy. A selectable marker is loaded onto the autonomous element inactivating the transposase gene(s). Transposase is then supplied from a transposase source gene located in *trans* on a separate T-DNA or in *cis* from the same T-DNA launching construct. In the *trans* configuration the inactive element is activated after crossing to a plant carrying the transposase source. In both configurations the marked inactivated element remains active until the transposase source is segregated away. The two component approach can be arranged with a negative marker located on the launching T-DNA segments allowing elements in unlinked positions to be selected directly by a combined positive selection for reinsertion and negative selection against the T-DNA segments (Sundaresen et al. 1995). The transposase source may also be counterselected alone using negative markers. The inactive element in the two component systems can also be modified to carry a reporter gene close to one of the termini. Examples of promoter trap and gene trap modified *Ds* elements carrying a GUS reporter used in *Arabidopsis* was presented by Sundaresen et al 1995.

Behaviour of *Ac* in *Lotus japonicus*

The autonomous maize element *Ac* or the derived *Ac*ΔNaeI was transferred into *L. japonicus* via *Agrobacterium tumefaciens* mediated transformation. Using the Spectinomycin resistance excision marker (Thykjær et al. 1995) the behaviour of *Ac* has until now been scored in progeny from 123 independent transformants. The results are summarised in Table 1. Excision activity was detected in progeny from 58 % of the lines indicating that *Ac* is quite active in *L. japonicus* compared to for example *Arabidopsis*. In 62 % of these lines a segregation approximating the expected 3:1 ratio for a dominant marker gene was detected. This again supports the notion of quite high transposition activity and suggests that the penetrance of the Spectinomycin

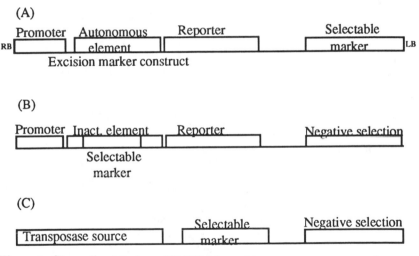

Figure 1. Generalised design of T-DNA launching segments representing different strategies for transposon tagging in plant species without known endogenous elements. (A) Construct carrying an autonomous element for example *Ac* or *Spm* located within an excision marker. (B) A genetically marked inactivated element (for example *Ds* or *dSpm*) located within an excision marker and linked to a negative selection marker. A positive negative selection scheme will allow selection of reinsertions unlinked to the T-DNA after crossing to a plant carrying the transposase source. (C) A transposase source construct. The transposase gene(s) can be expressed from a promoter of choice. The presence of a negative selection marker on the T-DNA allows counterselection of the transposase source resulting in stabilisation of the inactive element.

resistance in *L. japonicus* is sufficient to make the Spectinomycin selection robust. A number of lines demonstrate segregation different from 3:1. These lines represent either plants carrying T-DNA in more than one position or lines with less activity than expected from the predicted segregation ratios. In a smaller material of 38 transgenes a germinal excision frequency of 12 % was measured as the frequency of fully green progeny from variegated plants (Thykjær et al. 1995). This germinal excision activity of *Ac* is comparable to the germinal excision frequency in maize. From these experiments it is concluded that the overall excision activity of *Ac* in *L. japonicus* occurs with a frequency making transposon tagging feasible. Thykjær et al. 1995 estimated the reinsertion frequency to be 42 % indicating that transposon tagged mutants could be identified using this approach. There is some variation in *Ac* activity between transgenes. One of the variations we have observed is the inability of certain lines producing plenty of variegated plants to produce fully green offspring. If there is no fully green plants in the progeny from the primary transgene, fully green plants are in most cases also absent in the progeny from the variegated plants of these lines. The transposon is apparently not active in the germline cells of these lines.
The two component approach has also been started in *L. japonicus*. Several hundred transgenic lines carrying a *Ds* element marked with a Hygromycin resistance gene (Bancroft et al. 1992) were generated. Assuming random insertion of the T-DNA launching segments this number of individual transformants should allow subsequent transpositions into most of, if not the entire genome. For activation of the *Ds* element two different transposase sources (Bancroft et al. 1992) will be provided in *trans*. One transposase source is a stabilised *Ac* element where one of the termini required for transposition has bee removed. In the other transposase source the transposase gene is expressed from the constitutive 35S promoter. Preliminary data indicate that the *Ds* element can be mobilised in *L. japonicus* using the two component system.

Screening for symbiotic mutants in *L. japonicus*

In order to identify tagged mutations with phenotypes expected for symbiotic mutants *L. japonicus* tagging populations are screened after growth on a medium without reduced nitrogen. The clearly visible nitrogen deficiency symptoms of *L. japonicus* (small plant, light green or white leaves and completely bright red stem) make the identification of symbiotic mutants possible by visual inspection of shoot appearance. Growth in inert material (Leca), easily removed from the roots, makes the subsequent inspection of the root system manageable. Under greenhouse conditions the screening can be done after six to eight weeks of growth without nitrogen source. True symbiotic mutants are then rescued on nitrogen containing medium where pleiotrophic mutants impaired in other processes giving a secondary phenotype on nitrogen fixation can be separated from more strict symbiotic mutants. A large scale screening of progeny from approximately 5000 fully green plants with *Ac* excision events going through the germline is currently on the way to asses the efficiency of the maize *Ac* element in *L. japonicus*.

The preferential insertion of *Ac* at linked positions in maize and other plants is assumed to be an intrinsic feature of the transposition mechanism and Table 1. Summary of the excision activity detected in the progeny from 98 primary transformants containing *Ac* and progeny from 25 transformants containing *Ac*ΔNael. The *Ac*ΔNael element is derived from *Ac* by removal of 537 bp from the untranslated leader of the transposase gene. Activity was scored as presence of variegated or fully green plants in the progeny. The inheritance was calculated from the number of variegated and fully green plants compared to the number of completely bleached Spectinomycin sensitive plants.

| | *Ac*ΔNael | *Ac* |
|---|---|---|
| Total number of lines analysed | 25 | 98 |
| Number or lines showing activity | 11 | 60 |
| Number of lines without activity | 14 | 38 |
| Number of lines with Mendelian segregation (3: 1) | 6 | 38 |
| Number or lines with segregation 3:1 with non-Mendelian ratio | 5 | 22 |
| Number of lines without fully green offspring | 2 | 18 |

To improve the chances of finding tagged symbiotic mutants the 5000 fully green plants were chosen to represent 65 independent transgenics carrying the T-DNA launching construct at different positions in the genome. The effective transformation regeneration procedure available for *L. japonicus* made it possible to make this number of transgenics. Preliminary screening and mutant characterization emphasised another important biological parameter for a model plant - the seed production. The ability of the model plant to produce large amounts of seeds within a reasonable time period is very important and this parameter should not be underestimated when choosing a plant species for this type of genetic research. Once the genetic and phenotypic characterization of a mutant is initiated a lot of seeds is needed for the various experiments required. Profuse and continuous flowering of the perennial *L. japonicus* is one of the advantages of this model system. Preliminary screening for mutants tagged *by Ac* in a small population have identified one or two putatively tagged symbiotic mutants that are now carefully checked for cosegregation of the *Ac* insert and the mutant phenotype in the progeny of a backcross.

Conclusions and future perspectives

Several plant pathogen resistance genes and other types of genes have already been isolated from various plant species with tagging approaches using *Ac/Ds* or *Spm/dSpm* (Whitham et al 1995 and references therein). With the activity of *Ac* demonstrated in *L. japonicus* (Thykjær et al. 1995) it can therefore also be expected that tagging of the many genes presumed to participate in the organogenic process forming the root nodule should be possible. The number of interesting target genes should be relatively high. One of the conclusions from the preliminary experience with tagging in *L. japonicus* is that efficiency in selection of the valuable material is important. Identification of putatively tagged mutants requires handling of the large numbers of seedlings and plants in intermediate generations and screening populations. The transposon launching constructs should therefore be designed to diminish the amount of labour involved. The selection scheme should be designed to eliminate uninteresting plant material as early as possible in the procedure and the selection markers used should be chosen so that minimal handling of the plants is involved. The preliminary experience with *Ac* in *Lotus japonicus* appears to indicate that selection for reinsertion may be an advantage in this species. The main emphasis is therefore right now on the optimisation of the two component systems.

Aarts, M.G.M.,Keijzer, C.J., Stiekema, W.J. and Pereira A. 1995. Molecular characterisation of the *CER1* gene of *Arabidopsis* involved in epicuticular wax biosynthesis and pollen fertility. The Plant Cell 7:2115-2127.
Bancroft, I., Bhatt, A.M., Sjodin, C., Scofield, S., Jones, J.G.J., Dean, C. 1992. Development of an efficient two-element transposon tagging system in *Arabidopsis thaliana*. Mol. Gen. Genet. 233:449-461.
Handberg, K.and Stougaard, J. 1992. *Lotus japonicus*, an autogamous, diploid legume species for classical and molecular genetics. Plant J 2:487-96.
Handberg, K., Stiller, J., Thykjær, T. and Stougaard, J. 1994. Transgenic plants: *Agrobacterium* mediated transformation of the diploid legume *Lotus japonicus*. in Cell Biology: A Laboratory Handbook. Celis ed. Acad. Press.
Sundaresen, V., Springer, P.,Volpe, T., Haward, S., Jones, J.D.G., Dean, C., Ma, H. and Martienssen, R. 1995. Patterns of gene action in plant development revealed by enhancer trap and gene trap transposable elements. Genes & Development 9:1797-1810.
Thykjær, T., Stiller, J., Handberg, K., Jones, J. and Stougaard, J. 1995. The maize transposable element Ac is mobile in the legume *Lotus japonicus*. Plant Molec. Biol. 27:981-993.
Whitham, S., Dinesh-Kumar, S.P., Choi, D., Hehl, R., Corr, C., Baker B. 1994. The product of the Tobacco Mosaic Virus resistance gene N: Similarity to Toll and the Interleukin-1 receptor. Cell: 78:1101-1115.

# *Nicotiana tabacum* SR1 contains two *ENOD40* homologs

Martha Matvienko, Karin van de Sande, Katharina Pawlowski, Ab van Kammen, Henk Franssen and Ton Bisseling
Agricultural University, Wageningen, the Netherlands

## Isolation of tobacco *ENOD40* clones

*ENOD40* clones have been isolated from several legumes, eg. *Glycine max* (Kouchi and Hata 1993, Yang *et al.* 1993), *Phaseolus vulgaris* (Papadopoulou *et al.* 1996), (forming determinate nodules) and *Vicia sativa* (Vijn *et al.* 1995), *Pisum sativum* (Matvienko *et al.* 1994), *Medicago truncatula* and *M. sativa* (Asad *et al.* 1994, Crespi *et al.* 1994) (forming indeterminate nodules). When *ENOD40* clones from soybean (*GmENOD40-1* and *GmENOD40-2*) and alfalfa (*MsENOD40*, Matvienko unpublished results) where transiently expressed in SR1 protoplasts, this caused division of these protoplasts in the presence of high, normally inhibitory levels of NAA (Van de Sande *et al.* 1996). The activity of the heterologous legume genes in tobacco is indicative of the presence of *ENOD40* homolog(s) in tobacco. Therefore we attempted to isolate tobacco *ENOD40* clones.

The largest homology between the different legume *ENOD40* sequences is found in two small stretches, one located at the 5' end (region 1), the other more centrally (region 2) in the *ENOD40* clones. Region 1 contains a small ORF encoding a peptide of 12 (*Glycine, Phaseolus*) or 13 (*vicia, Medicago, Pisum*) amino acids (Table 1), while region 2 does not contain a conserved ORF (Fig. 1). Interestingly, the sequence conservation in region 2 is even higher than in region 1 (Fig. 1; van de Sande *et al.* 1996)).

Degenerate primers were designed against region 1 and the 3' half of region 2 of the legume *ENOD40* clones, with the sequences 5'-GGC(A/T) (C/A)(A/G)(C/A)A(A/T)C(C/A)ATCCATGGTTCTT-3' and 5'-GGA(G/A)T CCATTGCCTTTT-3', respectively. This way, the 5' part of region 2 could confirm the identity of the isolated tobacco clones. cDNA clones were obtained via RT-PCR, using poly A RNA from *Nicotiana tabacum* cv. Petit Havana SR1 flowers, and *ENOD40*-like clones were selected by sequencing. Two different cDNA clones were isolated containing the 5' part of region 2. To obtain full length clones named pNtENOD40-1 and pNtENOD40-2, 5' and 3' racing was used. For the 5' race two specific primers, 5'-GCTTTT GCCAACATCCTTTC-3' and 5'-CTATTAGTGTGATTATCAATC-3' and two universal primers 5'-CTCGAGGATCCGCGGCCGCTTTTTTTTTTT TTTTTTTT-3' and 5'-GCTCGAGGATCCGCGGC-3' were used. For the 3' race primers 5'-CAAGTTTGTTCATACTTTGCC-3' and 5'-GCTAGAAT TCCAGAAAATGC-3' were used. The nucleotide sequence of the resulting cDNA clones (Fig. 1) was confirmed by designing primers against their 3' and 5' end and perform PCR on genomic DNA from tobacco. The 5' primer used had the sequence 5'-GACTAGCTTGTCTCAAGAAC-3'. The insert of the cDNA clone pNtENOD40-2 was shorter than that of pNtENOD40-1. So two different 3' primers were designed, 5'-ATGACAATCTTAACAACTCT-3' and 5'-TATTCGGTAATAATTGGTGTG-3', for pNtENOD40-1 and pNtENOD40-2, respectively. The nucleotide sequence of the genomic clones and the cDNA clones was identical, confirming the nucleotide sequence of the cDNA clones and showing that like in legumes, the tobacco *ENOD40* genes do not contain introns.

Table 1: Amino acid sequence comparison of the *ENOD40* encoded peptides of different leguminous plants and from tobacco. iep = iso-electric point.

| peptide from | sequence | iep |
|---|---|---|
| *GmENOD40-1* | M.ELCWQTSIHGS | 5.41 |
| *GmENOD40-2* | M.ELCWLTTIHGS | 5.41 |
| *PvENOD40* | MKF.CWQASIHGS | 8.44 |
| *PsENOD40* | MKFLCWQKSIHGS | 9.62 |
| *MsENOD40* | MKLLCWQKSIHGS | 9.62 |
| *MtENOD40* | MKLLCWEKSIHGS | 8.43 |
| *VsENOD40* | MKLLCWQKSIHGS | 9.62 |
| *NtENOD40* | M...QWDEAIHGS | 4.23 |

## Comparison of tobacco and legume *ENOD40* clones

The overall nucleotide homology between the legume *ENOD40* clones and the tobacco clones is low whereas *NtENOD40-1* and *NtENOD40-2* are 96% identical (Fig. 1). The most striking feature of both tobacco clones, is the presence of two sequence stretches that are highly homologous to the legume region 1 and 2. These are the only regions with significant homology between legume and tobacco *ENOD40* clones. Just like in the legume *ENOD40* clones, the tobacco *ENOD40* region 1 sequences encode an oligo peptide. Region 1 from the tobacco clones is 100% identical at the nucleotide level and so the peptide encoded by *NtENOD40-1* and *NtENOD40-2* is identical. Sequence comparison of the legume and tobacco peptides shows that they are highly homologous (Table 1). The tobacco peptide is ten amino acids in lenght, a bit smaller than the legume peptides. A centrally located tryptophan and the four C-terminal amino acids (ile, his, gly and ser) are conserved in all ENOD40 peptides. However, in iso-electric point (iep) are differences. The peptides from pea, alfalfa and *Vicia* have an iep of 9.62, while those from soybean and tobacco have an iep of 5.41 and 4.23,

```
                                                                  *                                    *    *** *    * * ******** ***
pGmENOD40-2                                                   TGAA.TCTGGTGA.................GCAAATATGGAGCTTTGTTGGCAAACATCCATCCATGGTTCT
pPsENOD40                      CTTTAACCTCTAAACCAA...TCCA........TTATCAA.AACTTTGAATCTTTGTTATT....AGCAATGAAGTTTCTTTGTTGGCAAAAATCAATCCATGGTTCT
pMsENOD40                            CCTCTAAACCA....TCCA...........TCAA.GACTT.GAATCTT.GTTTGTAATAAGG.ATGAAGCTTCTTTGTTGGCAAAAATCAATCCATGGTTCT
pNtENOD40-1     GACTAGCTTGTCTCAAGAACACACTAAC...TATCAGTTTCAAACTCTGA.TAGAT................TATTAAGTGAGAAAAAATGCAGTGGGATGAAGCAATCCATGGGTCT
pNtENOD40-2   ACCACTAGCTTGTCTGAAGAGCACACTAACTATTATCAATTTCAAACTCTGAGTAGAT................TATTAAGTGAGAAAAGATGCAGTGGGATGAAGCAATCCATGGGTCT
                                                                                                          ----------region 1-----------

              *      **     **         * ***                                                      *
pGmENOD40-2   TGAA..GAAGCATGGAGATAAAGGGGTGTGAGAGG...................GTCCTCACCACTCACACTCCTCCACTT.....................AAAACAGTTTGTT
pPsENOD40     TAAAAACAAATATGGAG....TGAAGTGTGAGAGGGCTAATTAGTTGCTA.........ATTACTACTAATTACATCATCCTTTACACTCTCCCTCCATTTTTACAAAAACAGTTTGCT
pMsENOD40     TAAAA.CAAACATGGAG....AGAAGTGTGAGAGGGT..ATTAAACAAAAAC..........................CCTACACACTCTCCCTCCATTTT..CCTAAACAGTTTGCT
pNtENOD40-1   TAGGA.AAAGAGAGGCAATTAGTTGGAGTGAATTAAGGACTT..TTAAATGCCCTTCCTCCAAATACAATTT.CAAGTTTGTTCATACT...TTGCCAATTT..CTCTGCTAGAATTCC
pNtENOD40-2   TAGGA.AAAGATAGGCAATTAGTTGGAGTG....AAGGACTT..TTAAATGGCCTTCCTCCAAATACAATTT.TAAGTTTGTTCATACT...CTGCCAATTT................
              -

                     ***                  *  **** *   ** *      *                    *                                ** *   *   *   ****
pGmENOD40-2   TTG..GCTTAGCT.TTGGCTTCTCTAATCAACAAGGGATG.TGTTCTAACATTCTCTCTTGAGTGGCAGAAGCAGATATGCATTCTCCAAAGGAGGAGAGGCTTTGGCTACAGCCTGGC
pPsENOD40     TTGTGCTTTAGCTATTGGCTTCTCATATCAAAAAGGGATG.TGCTTTTTTCTGAGTAGCAGAAGCAAATAATTA.AGCATTTTTCT.TCAAAGATCAGAAGCTTTGGTTATTGTATGGC
pMsENOD40     TTGTGCTTTAGCTTTTGGCTTCTCATATCACAAAGGGATTATGCTTTTTTCTGAGTAGCAGAAGCAAATAATTA.AGTATTTTTCTCCAAAGGATCAGAAGCTTTGTTATAGCATGGC
pNtENOD40-1   AGAAAATGCAGC................AAAAAAGGAAAG.TGTTTATGGATTGATAATCACACTAATAGCAAAGGAT.........................GTTGGCAAAAGCATGGC
pNtENOD40-2   ..CAAATGCAGC................AAAAAAGGGAAG.TGTTTATGGATTGATAATCACACGAATAGCAAAGGAT.........................GTTGGCAAAAGCATGGC
                                                                                                                                    ----

              ************** *** ********* ***    **** ****   * *   *
pGmENOD40-2   AAACCGGCAAGTCACGAAAAGGCAATGGACTCC..ATTGGGGTCTCTATGGCTATCTATTGCTCATCTATGTAGTTCTTCTTGCTGTAGAATGTAATA..AAACAAAGTTGGCCTTCC.
pPsENOD40     AAACCGGCAAGTCACAAAAAGGCAATGGATTCCTTTTTGGAGTCTTAATGGCTATGTATCAATCACTCTATCTGTTATTATCA.....................TCTATGCAGACACTT
pMsENOD40     AAACCGGCAAGTCACAAAAAGGCAATGGATTCCTTTTTGGAGTCTTAATGGCTATGTATCAATCACTCTATCTAT............................GTAGCACTGACACT
pNtENOD40-1   GAACCGGCAAGTCACGAAACGGCAATGGACTCCGTTTTGGAGTCTTTCTTGGCCT.................TTTACTTATTGTTGTACTTTATATTGTTATATCCAGTCTTCTATAAT
pNtENOD40-2   GAACCGGCAAGTCACAAAACGGCAATGGACTCCGTTTTGGAGTCTTTCTTGGCTT.................TTTACTTATTGTTGTAC.........TTATATCCAGTCTTCTGTAAT
              ---------------region 2--------------------

                                                                                                          *         *
pGmENOD40-2   .TTTGAGAAGTTACCAACTTTTGCTGTCCAA......ATTACTCAATTTGCAGCTGACTAGAATTCCTTTCTCTCTGATCAGTTTCTGCAGATGAG..TAGGTAGGTAATTTGTGATCA
pPsENOD40     CAGATTGAAG.GTTGTCTGGTGTCTGACACGTGTGTTAGTTT..ACGGCGCAGTCGTCTGGTGTCTGTGCTTCGTAG....ATTATTATAGTTCTT..CTTGTAG.TAGAATGTAATAA
pMsENOD40     TGAGATTGTAGACTTGTCCGGTGTCTG.............................TGTCAGTGCTTATAG....ATTGTTATAGTTATTTTCTTGCAG.TAGAATGTAATAA
pNtENOD40-1   TGTGTTTTATTTGTAAGTCACATGCAGTGTG...........................................TAGTTAATATGGCTTATCAAAAAT.TTATAA
pNtENOD40-2   TGTGTTTTATTTGTAAATCACATGCATTGTG...........................................TGGTTAATATGGCTTATCA

pGmENOD40-2   CT.............CCCTTCCCTTCATGTCTGTGTTCCCTTTTCCATGCTTGTTTGTGTTGTT.AGTTATAACCTTATGAGG.AATAAAAGAATAGTACAATTCTAGTCCCTCAAAGT
pPsENOD40     TAAACATAAAGATGGTGTTGTCTTCCTTTGAGAAGTTTCCA.ACTTTGTGATGTACTTCAAGTTCACTCAATTTGCAGCTGATCCTAGAGTCTGTTTCTTGTTTCAGTTTCTGCA....
pMsENOD40     TAAACATAAAGATGGTGTTGTATTCCTTTGAGAAATTGCCAACTTT.ATGATGTACTTCAA.TTCACTCAATTTGCAGCTGA..CTAGAGTCTGTT.CTTGTTTCAGTTTCTGCAGATG
pNtENOD40-1   TAGAGTTGTTAAGATTGTCAT

pGmENOD40-2   TTAGGATTGTATT.CTATTGAACTTTAATAG
pPsENOD40     TGTAAGGTAGGTAACTGTTATCATTAATTCCTGTTTCTTTTCTTC
pMsENOD40     AGTAAGGTAGGTCACTCTTATCATTAATTCATGTTCCTTTTCTTCTG
```

Figure 1: Nucleotide sequence comparison of the *ENOD40* clones from soybean (*GmENOD40-2*), pea (*PsENOD40*), alfalfa (*MsENOD40*), and tobacco (*NtENOD40-1* and *NtENOD40-2*). * indicates nucleotides conserved between all *ENOD40* clones used.

respectively.

Region 2 is highly conserved in the legume and tobacco *ENOD40* clones. Between the tobacco clones there is only one basepair substitution, centrally in region 2. Like in the legume clones, region 2 from tobacco *ENOD40* does not contain a conserved ORF (Fig. 1). The already mentioned difference in size between both tobacco *ENOD40* clones is caused by four deletions in *NtENOD40-2*, making it 56 nucleotides shorter. From these deletions, two are located between the conserved regions, and the largest of 32 bp is at the 3' end.

## Characterization of tobacco *ENOD40* clones

The expression of *NtENOD40* in tobacco could not be detected with Northern blot hybridization. Using RT-PCR (data not shown) *NtENOD40-1* and *NtENOD40-2* expression was detectable in stems, roots and flowers, and a markedly lower level of expression was detectable in leaves. In the tissues tested, *NtENOD40-1* and *NtENOD40-2* were expressed at similar levels. In legumes *ENOD40* is also expressed in non-symbiotic organs albeit at a markedly lower level than in nodules (Crespi *et al*. 1994, Matvienko *et al*. 1994, Papadopoulou *et al*. 1996, Yang *et al*. 1993)

The activity of *ENOD40* genes in non symbiotic tissues indicates a function of ENOD40 in common plant development. *ENOD40* genes are expressed in some non-symbiotic organs, for instance in flowers and at a low level in stems (Asad *et al*. 1994, Crespi *et al*. 1994, Kouchi and Hata 1993, Matvienko *et al*. 1994, Papadopoulou *et al*. 1996, Yang *et al*. 1993. The very big difference in *ENOD40* expression levels, between symbiosis associated expression, or common expression is puzzling, but not unique for *ENOD40*. For example *ENOD12* is also expressed in nodule, stem and flower ans in the non-symbiotic organs it is also expressed at a markedly reduced level.

Several lines of evidence suggest that signals and genes involved in nodule formation are recruited from common plant development. Several genes that initially where thought to be nodule specific turned out to be expressed at low levels in non-symbiotic organs (e.g. *ENOD12* and *ENOD40*) and are even found in non-legumes (e.g. *ENOD40*). Even the Rhizobial signal molecules, Nod factors, could have their counterparts in common development. Röhrig *et al*. (1995) demonstrated that *in vitro* synthesised lipo chito-oligosaccharides (LCOs), resembling rhizobial Nod factors, could induce cell divisions in tobacco protoplasts in the absence of either auxin or cytokinin in the growth medium. LCOs are capable of. Thus, rhizobial Nod factors might be related to endogenous plant signal molecules and in the *Rhizobium*-legume interaction induce the expression of *ENOD40* from which the gene products appears to be a new kind of plant peptide hormone, influencing the respons to the plant growth regulator auxin (Van de Sande *et al*. 1996).

# References

Asad, S. Fang Y., Wycoff, K.L., and Hirsch, A. 1994. Isolation and characterization of cDNA and genomic clones of *MsENOD40*; transcripts are detected in meristematic cells of alfalfa. Protoplasma 183:10-23.

Crespi, M.D., Jurkevitch, E., Poiret, M., d'Aubenton-Carafa, Y., Petrovics, G., Kondorosi, E., and Kondorosi, A. 1994. *ENOD40*, a gene expressed during nodule organogenesis, codes for a non-translatable RNA involved in plant growth. EMBO J. 13:5099-5112.

Kouchi, H., and Hata, S. 1993. Isolation and characterization of novel nodulin cDNAs representing genes expressed at early stages of soybean nodule development. Mol. Gen. Genet. 238:106-119.

Matvienko, M., van de Sande, K., Yang, W.-C., van Kammen, A., Bisseling, T., and Franssen, H. 1994. Comparison of soybean and pea *ENOD40* cDNA clones representing genes expressed during both early and late stages of nodule development. Plant Mol. Biol. 26:487-493.

Papadopoulou, K., Roussis, A., and Katinakis, P. 1996. *Phaseolus ENOD40* is involved in symbiotic and non-symbiotic organogenetic processes: expression during nodule and lateral root development. Plant Mol. Biol. 30:403-417.

Röhrig, H., Schmidt, J., Walden, R., Czaja, I., Miklasevics, E., Wieneke, U., Schell, J., and John, M. 1995. Growth of tobacco protoplasts stimulated by synthetic lipo-oligosaccharides. Science 269:841-843.

Van de Sande, K., Pawlowski, K., Czaja, I., Wieneke, U., Schell, J., Schmidt, J., Walden, R., Matvienko, M., Wellink J., van Kammen, A., Franssen, H., and Bisseling, T. 1996. Modification of phytohormone response by a peptide encoded by *ENOD40* of legumes and a non-legume. Science, in press.

Vijn, I., Das Neves, L., Van Kammen, A., Franssen, H., and Bisseling, T. 1993. Nod factors and nodulation in plants. Science 260:1764-1765.

Vijn, I., Yang W.-C., Pallisgård, N., Østergaard Jensen, E., van Kammen, A., and Bisseling, T. 1995. *VsENOD5*, *VsENOD12* and *VsENOD40* expression during *Rhizobium*-induced nodule formation on *Vicia sativa* roots. Plant Mol. Biol. 28:1111-1119.

Yang, W.C., Katinakis, P., Hendriks, P., Smolders, A., De Vries, F., Spee, J., Van Kammen, A., Bisseling, T., and Franssen, H. 1993. Characterization of *GmENOD40*, a gene showing novel patterns of cell-specific expression during soybean nodule development. Plant J. 3:573-585.

# Calcium-dependent Phosphorylation of the Nodulin 26 Channel by a Symbiosome Membrane Protein Kinase

Daniel M. Roberts, C. David Weaver, and Jung Weon Lee

Department of Biochemistry, Cellular and Molecular Biology, Center for Legume Research, University of Tennessee, Knoxville, TN 37996.

Nodulin 26 is a major symbiosome membrane-specific protein found in soybean root nodules infected by *Bradyrhizobium japonicum* (Fortin et al. 1987). Analysis of the nodulin 26 sequence reveals that it belongs to an ancient membrane channel family known as the "major intrinsic protein" (MIP) family or more recently as the aquaporins (reviewed in Agre et al. 1995). These proteins facilitate the flux of water or solutes (e.g., glycerol) across membranes (Agre et al. 1995). Nodulin 26 has the structural hallmarks of this family including six putative transmembrane α-helical domains, as well as characteristic "NPA" motifs located at symmetrical positions within the loop regions (Fig 1.).

The targeting of a nodule-specific channel protein to this unique membrane compartment suggests that it plays a role in membrane transport during the symbiosis. With a variety of biophysical techniques, we have investigated the channel properties of nodulin 26, as well as the regulatory significance of calcium-dependent phosphorylation.

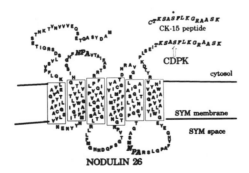

Fig. 1. Sequence and proposed topology of soybean nodulin 26. Highly conserved NPA motifs and the unique CDPK phosphorylation site, ser 262 are noted. The CK-15 synthetic peptide sequence is also shown.

## Nodulin 26: A Major Symbiosome Membrane Phosphoprotein

Analysis of isolated symbiosome membranes (SM) revealed that: **1.** nodulin 26 is a major component, constituting about 10% of the total SM protein; and **2.** it is the major phosphorylated protein detected in this membrane (Weaver et al. 1991). Phosphorylation is calcium-dependent, and is catalyzed by a SM-associated protein kinase (Weaver et al. 1991). Sequence analysis of nodulin 26 showed that serine 262 within the carboxyl terminal domain (Fig. 1), is the only residue phosphorylated (Weaver et al. 1992). The site recognized by the protein kinase resides within the 14 COOH-terminal residues of nodulin 26 (Fig. 1). A synthetic peptide (CK-15) containing this sequence is phosphorylated by a $Ca^{2+}$-dependent protein kinase activity in nodule extracts. This activity is found both in the soluble, membrane-free fraction as well as in isolated symbiosome membrane fractions (Weaver et al. 1991). We have purified the protein kinase from the soluble fraction. The kinase shows complete $Ca^{2+}$ dependence without a requirement for calmodulin or lipids (Weaver et al. 1991). SDS-PAGE of the purified kinase shows a major 60,000 MW band, which exhibits $Ca^{2+}$-dependent kinase activity upon *in situ* renaturation in the gel (Fig. 2). This band cross reacts (Fig. 2) with an antibody against the calmodulin-like domain protein kinase (CDPK) (Putnam-Evans et al. 1990) suggesting that the nodule kinase belongs to this family.

CDPKs are a unique class of calcium-regulated protein kinases that are widely distributed in plant tissues (Roberts and Harmon, 1992). CDPKs have a protein kinase domain that is homologous to the catalytic domain

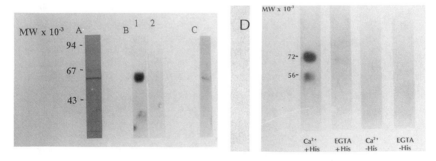

Fig. 2. Characterization of nodule CDPKs. SDS-PAGE was done on 14% polyacrylamide gels. **A.** Silver stain of soluble kinase; **B.** Renatured soluble kinase activity assayed with histone H1 in 1 mM $CaCl_2$ (lane 1) or EGTA (lane 2); **C.** Western blot of soluble kinase with anti-CDPK antibodies. **D.** Renatured symbiosome membrane protein kinase activity in the presence or absence of histone H1, and 1 mM $CaCl_2$ or EGTA.

of calmodulin-dependent protein kinases. However, in contrast to calmodulin-dependent kinases, CDPKs are not bound and regulated by calmodulin, but instead have a fused calcium binding regulatory domain that is homologous to calmodulin. Thus, CDPK binds directly to $Ca^{2+}$ resulting in activation.

Soybean has multiple CDPK genes encoding structurally, and possibly functionally, distinct kinase isoforms (Harmon et al. 1996). Several observations suggest that the SM and soluble CDPKs of soybean nodules may indeed be different isoforms. Unlike the soluble activity, the SM-CDPK is tightly associated with the membrane and requires detergents for solubilization. This suggests that the SM-CDPK possesses a membrane association domain, or perhaps a lipid tag that enables insertion into the bilayer. *In situ* renaturation studies show two SM-CDPK activity bands that differ in molecular weight from the soluble CDPK (Fig. 2). Whether the SM CDPK represents a unique CDPK gene product which is selectively targeted to the SM with the purpose of phosphorylating nodulin 26 requires further investigation.

## Planar Lipid Bilayer Studies of Nodulin 26

The reconstitution of purified membrane proteins into planar lipid bilayers has facilitated studies of their transport and channel activities under defined conditions with purified components. Since nodulin 26 is a major SM component, and considering the importance of SM in the transport of metabolites and ions (e.g., dicarboxylates) that are essential for the symbiosis (Whitehead et al. 1995; Streeter 1995), we investigated the properties of purified soybean nodulin 26 in planar lipid bilayers (Weaver et al. 1994).

Upon reconstitution into bilayers, purified nodulin 26 forms a high conductance ion channel (single channel conductance of 3.1 nanosiemens in 1 M KCl) that transports $K^+$ as well as $Cl^-$ across the bilayer, but shows a slight preference for the anion. Malate appears to also to be transported, but larger ions such as HEPES or Tris salts are poorly transported (Weaver et al. 1994). At low potentials (e.g., 30 mV), the channel is pore-like, remaining in a completely open state allowing the symmetrical flux of ions, However, at high voltage potentials (e.g., 70-100 mV), the channel exhibits gating behavior, fluctuating between several lower subconductance, partially "closed" states that have restricted ion conductance (Weaver et al. 1994). The potential significance of this observation is discussed below.

The effect of phosphorylation on the nodulin 26 channel became apparent when recombinant nodulin 26 was investigated (Lee et al. 1995).

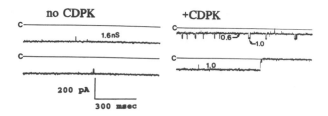

Fig. 3. CDPK phosphorylation of nodulin 26 affects channel conductance at +70 mV. Shown are representative one second recordings of a single recombinant nodulin 26 channel in 20 mM MOPS-NaOH, pH 7.4, 0.2 M KCl, 0.1 mM ATP before and after CDPK addition. Single channel conductance values for the various states are shown.

Expression of nodulin 26 in *E. coli* results in a preparation that is unphosphorylated at ser 262 (Lee et al. 1995). Reconstitution of this protein into planar lipid bilayers yields a channel activity that is identical to the native soybean nodulin 26 with one exception: voltage-sensitive gating was not observed (Fig. 3). However, upon *in situ* phosphorylation of nodulin 26 by CDPK, voltage-sensitive gating was restored and partial or complete channel closure occurred (Fig. 3). Since the SM is electrogenic and establishes a potential via a $H^+$-proton pumping ATPase (Udvardi and Day 1989), it is conceivable that voltage potential can influence nodulin 26 activity and that calcium-dependent phosphorylation of ser 262 on a carboxyl terminal "gate" domain on nodulin 26 could confer voltage-sensitive behavior. Thus, membrane potential and phosphorylation could control flux through the nodulin 26 channel.

Planar lipid bilayer experiments provide important information regarding channel behavior and regulatory phenomena with purified biochemical components. However, various studies (Whitehead et al., 1995; Streeter, 1995) show that the SM is a selectively permeable barrier and it is not likely that nodulin 26 exists in the pore-like state observed in planar lipid bilayers. MIP (a nodulin 26 homolog from bovine lens fibers) also forms a high conductance ion channel pore in planar lipid bilayers, however this activity can be can be attenuated by the types of lipids present in the membrane and by extrinsic proteins (Shen et al., 1991; Ehring et al. 1990). Thus, factors on the SM may contribute to the control of this activity. Investigation of the channel behavior *in situ* by patch clamp of isolated symbiosomes (Whitehead et al. 1995) will aid in determining whether a large conductance channel with selectivity and voltage-dependent subconductance states similar to isolated nodulin 26 is present.

## Nodulin 26 Aquaporin Activity

Many members of the aquaporin family are reported to form water selective channels or channels for uncharged solutes such as glycerol (Agre et al. 1995). Recently, it was found that MIP forms a low activity water channel upon expression in *Xenopus* (Agre et al. 1995). We also have found that nodulin 26 expressed in *Xenopus* oocytes forms a aquaporin-like channel that fluxes water in hypoosmotic conditions (Chandy et al. 1995). This activity shows a lowered activation energy and is inhibited by $Hg^{2+}$, similar to several aquaporin activities (Agre et al. 1995). Preliminary swelling experiments of SM vesicles show a similar $Hg^{2+}$-sensitive aquaporin activity, suggesting that nodulin 26 has this function *in situ* on SM (Rivers, R.L., Dean, R.M., Roberts, D.M., and Zeidel, M.L., unpublished observations). Interestingly, SM also show $Hg^{2+}$-sensitive transport of glycerol and other uncharged solutes, suggesting that the nodulin 26 aquaporin activity may be similar to aquaporin 3 of kidney which also shows broad specificity (Agre et al., 1995).

## Summary

Since its initial discovery as a major SM protein, and speculation about a membrane channel function, our understanding of nodulin 26 activity has progressed. Although we can now conclude that nodulin 26 is a channel based on several criteria, it is clear that the activities associated with the protein are complex and many questions remain regarding the nature of the molecules that are transported *in vivo*. While a role in metabolite or ion transport is possible based on measurements of the purified reconstituted protein, it is also possible that nodulin 26 acts as an aquaporin involved in osmotic/turgor control in symbiosomes. From a regulatory perspective, nodulin 26 is phosphorylated, and likely regulated, by CDPK. It is essential that the physiological and environmental signals controlling $Ca^{2+}$ fluxes and CDPK activation be defined. These elements are presently poorly understood in the soybean nodule.

Work was supported by USDA grants 92-37304-7874 and 94-37305-0619.

## LITERATURE CITED

Agre, P., Brown, D., and Nielsen, S. 1995. Aquaporin Channels: unanswered questions and unresolved controversies. Curr. Op. Cell Biol. 7:472-483.

Chandy, G., Roberts, D.M., and Hall, J.E. 1996. Water channel activity of wild type and mutant nodulin 26 expressed in *Xenopus laevis* oocytes. Biophys. J. 70:A196.

Ehring, G.R., Zampighi, G., Horwitz, J., Bok, D., and Hall, J.E. 1990. Properties of channels reconstituted from major intrinsic protein of lens fiber membranes. J. Gen. Physiol. 96:631-664

Fortin, M.G., Morrison, N.A., and Verma, D.P.S. 1987. Nodulin 26, a peribacteroid membrane nodulin is expressed independently of the development of the peribacteroid compartment. Nucleic Acids Res. 15:813-824

Harmon, A.C., Yoo, B.-C., Lee, J.-Y., Zhang, Y., and Roberts, D.M. 1996. Molecular and biochemical properties of calmodulin-like domain protein kinases. pp. 267-277. in: Protein Phosphorylation in Plants, Shewry, P.R., Halford, N.G., Hooley, R., eds. Claredon Press, Oxford.

Lee, J.W., Zhang, Y., Weaver, C.D., Shomer, N.H., Louis, C.F., and Roberts, D.M. 1995. Phosphorylation of nodulin 26 on serine 262 affects its voltage-sensitive channel activity in planar lipid bilayers. J. Biol. Chem. 270:27051-27057.

Putnam-Evans, C.L., Harmon, A.C., and Cormier, M.J. 1990. Purification and characterization of a novel calcium-dependent protein kinase from soybean. Biochemistry 29:2488-2495.

Roberts, D.M. and Harmon, A.C. 1992. Calcium modulated proteins. Annu. Rev. Plant Physiol. Plant Mol. Biol. 43:375-414.

Shen, L., Shrager, P., Girsch, S.J., Donaldson, P.J., and Peracchia, C. 1991. Channel reconstitution in liposomes and planar bilayers with purified MIP of bovine lens. J. Membrane Biol. 124:21-32.

Streeter, J.G. 1995. Recent developments in carbon transport and metabolism in symbiotic systems. Symbiosis 19:175-196.

Udvardi, M.K. and Day, D.A. 1989. Electrogenic ATPase activity on the peribacteroid membrane of soybean (*Glycine max* L.) root nodules. Plant Physiol. 90:982-987.

Weaver, C.D., Crombie, B., Stacey, G., and Roberts, D.M. 1991. Calcium-dependent phosphorylation of symbiosome membrane proteins from nitrogen-fixing soybean nodules. Plant Physiol. 95:222-227.

Weaver, C.D. and Roberts, D.M. 1992. Determination of the site of phosphorylation of nodulin 26 by the calcium-dependent protein kinase from soybean nodules. Biochemistry 31:8954-8959.

Weaver, C.D., Shomer, N.H., Louis, C.F., and Roberts, D.M. 1994. Nodulin 26, a nodule-specific symbiosome membrane protein from soybean, is an ion channel. J. Biol. Chem. 269:17858-17862.

Whitehead, L.F., Tyerman, S.D., Salom, C.L., and Day, D.A. 1995. Transport of fixed nitrogen across symbiosome membranes of legume nodules. Symbiosis 19:141-154.

# Sugar signals and legume lectins

Clara L. Díaz, Herman P. Spaink and Jan W. Kijne

Institute of Molecular Plant Sciences, Leiden University, Leiden, The Netherlands

Several signals for plant cells contain essential sugar residues. Examples are oligogalacturonide elicitors from the extracellular pectic matrix of plant cells (Farmer et al., this Volume), glucan elicitors such as the hexaglucosyl glucitol elicitor from *Phytophtora megasperma* f.sp. *glycinea* (Ebel et al., this Volume) and, to be discussed in this chapter, lipochitin oligosaccharides (LCOs) produced by *Rhizobium* bacteria (Spaink et al., this Volume). Sugar-binding proteins are likely to play a role in recognition of these signal molecules.

Legume plants produce a characteristic group of sugar-binding proteins, the legume lectins. Lectins are proteins with only one property in common: presence of at least one noncatalytic domain that can bind reversibly to a specific saccharide (Peumans and Van Damme 1995). Consequently, lectins occur with an enormous variation in structure, size and sugar-binding specificity. Within this large collection, the legume lectins form a homogeneous group of similar proteins, the structure of which is dominated by ß-sheets and connecting loops (a ß-barrel structure). Obviously, legumes produce other types of lectin as well (Quinn and Etzler 1987, Etzler et al., this Volume). In this short chapter, awaiting a more specific nomenclature, we will refer to legume lectins as the classical group including, for example, Concanavalin A, phytohemagglutinin and, to be highlighted, PSL, the homodimeric lectin from pea.

PSL is encoded by a single gene, which results in the presence of identical proteins in seed and root (Hoedemaeker et al. 1994). Apart from being secreted into the rhizosphere, PSL is located on the pea root surface, more specifically on the outer surface of the plasmamembrane in the tip growth area of young root hairs and root hair-forming epidermal cells. At these sites, PSL is present with at least one of its two sugar-binding domains available for binding of an appropriate ligand (Diaz et al. 1995b). The pattern of PSL location on pea roots entirely corresponds with the susceptibility of root epidermal cells to infection by *R. leguminosarum* biovar viciae, the pea symbiont (Diaz *et al.* 1986, Diaz 1989). Transformation of white clover roots with the *psl* gene results in the presence of PSL at root surface sites similar to those on pea roots (Diaz *et al.* 1995b). Taken together, these results demonstrate that PSL is ready to play a role as a binding factor for sugar signals, in particular signals produced by rhizobia.

White clover hairy roots transformed with the *psl* gene have acquired the ability to be nodulated by the pea symbiont *R. leguminosarum* biovar viciae (*Rl viciae*)(Diaz *et al.* 1995a), in contrast to hairy roots transformed with a PSL mutant containing defective sugar-binding sites (Van Eijsden *et al.* 1995). These results suggest that the sugar-binding domain of PSL is involved in the nodulation process, and provide evidence for a role of PSL in the legume-*Rhizobium* interaction. Signals of *Rl viciae* essential for their ability to nodulate pea and *psl*-transformed white clover hairy roots are LCOs carrying a specific polyunsaturated acyl group and a certain O-acetyl group (Diaz *et al.* 1995a, Spaink *et al.*, this Volume). In a recent set of experiments, we have tested whether heterologous nodulation of *psl*-transformed clover roots results from the acquisition of susceptibility to LCOs produced by pea rhizobia. As a test plant, we selected red clover, because red clover does not visibly respond to LCOs from *Rl viciae* whereas white clover does so at a background level.

Twelve days after application of LCOs from *Rl viciae* to the zone of emerging root hairs, loci of cortical cell divisions resembling root nodule primordia were observed in 60% of the red clover plants transformed with the *psl* gene. Loci were only observed at the site of LCO application. This response was not observed in control plants carrying hairy roots induced without the lectin gene. Rather surprisingly, application of LCOs from *R. meliloti* or *R. loti* also resulted in induction of cell division loci in *psl*-transformed roots, in contrast to the situation with control roots. These LCOs differ from those of *Rl viciae* by carrying different modifications at the reducing and nonreducing ends of the LCOs and by differences in length and degree of saturation of the acyl moiety. Apparently, *psl*-transformed red clover hairy roots can respond to LCOs regardless of the presence of specific substituents.

In view of the location of PSL, one may hypothesize that this sugar-binding protein functions as a binding factor for LCOs and that overproduction of the lectin in transgenic roots enables a response to heterologous LCOs. However, the sugar-binding domain of PSL is well-characterized and can not accommodate any of the LCOs tested in the way established for a specific ligand such as mannose (discussed by Kijne *et al.* 1994 for LCOs of *Rl viciae*). As an alternative, PSL may bind LCOs in a nonspecific way, either inside or outside the sugar-binding domain. Amar *et al.*

(1995) have shown that sulfates and phosphates can enter the sugar-binding site of legume lectins, resulting in nonspecific binding. Whether the sugar-binding domain of legume lectins can accommodate certain LCO substituents remains to be studied into detail. Such nonspecific binding would also explain the interaction of PSL with acidic extracellular polysaccharides produced by *Rl viciae* and *R. meliloti* (Kamberger *et al.* 1979). Furthermore, legume lectins contain a conserved hydrophobic pocket, which may interact with the acyl moiety of LCOs. If so, inactivity of a non-sugar-binding mutant of PSL may be explained by the involvement of the sugar-binding domain in passing the signal to a subsequent link in the signal transduction chain.

PSL is readily soluble in aqueous solutions of low osmolarity. However, the protein is able to insert into a phospholipid monolayer (Booij *et al.* 1996). This property is likely to be relevant for a putative binding factor of LCOs. Recently, Hervé *et al.*(1996) characterized an *Arabidopsis thaliana* receptor-like serine/threonine kinase gene, the predicted polypeptide sequence of which contains a legume lectin-like extracellular domain. As judged from this sequence, the protein does not contain a functional sugar-binding site of the legume lectin type. Referring to our earlier results, the authors speculate that the extracellular domain may form a heterodimer with a soluble lectin and suggest that "oligomerization between a pre-existing clover lectin-receptor kinase and a pea soluble lectin could modify the perception of lipo-oligosaccharide nodulation factors". At the present state of research, various other explanations can be proposed.

Induction of cortical cell divisions by LCOs in hairy roots of red clover transformed with the pea lectin gene may be related to the hairy root phenotype. The hormonal status of hairy roots is different from that of normal roots, which may enhance a response to heterologous LCOs. It should be mentioned however, that the nodulation process of the homologous symbiont *Rl trifolii* in red clover hairy roots is normal in every respect studied so far. Hirsch *et al.*(1995) transformed lotus with the soybean seed lectin gene, by using *Agrobacterium rhizogenes*. Transgenic lotus roots responded to the presence of the soybean symbiont *Bradyrhizobium japonicum* by induction of cortical cell divisions. The specificity of this response awaits further study. Our experiments indicate that under certain conditions red clover roots can respond to heterologous LCOs, and that legume lectin can play a role in signal transduction. In normal legume roots, lectin may be also involved in induction of cortical cell divisions by LCOs, provided that the signal is able to induce a proper response..

Amar,M.,Dumeirain,F., Barre,A., Chatelain,C.,and Rougé,P.1995.Interaction of legume lectins with sulfated and phosphorylated monosaccharides Book of Abstracts Interlec 16, June 26-30, Toulouse.

Booij,P., Demel,R.A., De Pater,B.S.,and Kijne,J.W.1996.Insertion of pea lectin into a phospholipid monolayer. Plant Mol. Biol. (in press).

Diaz,C.L., Van Spronsen,P.C., Bakhuizen,R., Logman,G.J.J., Lugtenberg,E.J.J.,and Kijne,J.W.1986. Correlation between infection by Rhizobium leguminosarum and lectin on the surface of Pisum sativum L. roots. Planta 168:350-359.

Diaz,C.L.1989.Root lectin as a determinant of host-plant specificity in the Rhizobium-legume symbiosis.PhD Thesis, Leiden University, The Netherlands.

Diaz,C.L., Spaink,H.P., Wijffelman,C.A.,and Kijne,J.W.1995. Genomic requirements of Rhizobium for nodulation of white clover hairy roots transformed with the pea lectin gene. Mol. Plant-Microbe Interact. 8:348-356.

Diaz,C.L., Logman,G.J.J., Stam,J.C.,and Kijne,J.W.1995. Sugar-binding activity of pea lectin expressed in white clover hairy roots. Plant Physiol. 109:1167-1177.

Hervé,C., Dabos,P., Galaud,J.-P.,Rougé,P.,and Lescure,B.1996. Characteri zation of an *Arabidopsis thaliana* gene that defines a new class of putative plant receptor kinases with an extracellular lectin-like domain. J.Mol.Biol 258:778-788.

Hirsch,A.M., Brill,L.M., Lim,P.O., Scambray,J.,and Van Rhijn, P.1995 Steps toward defining the role of lectins in nodule development in legumes.Symbiosis 19:155-173.

Hoedemaeker,Ph.J., Richardson,M., Diaz,C.L., De Pater,B.S.,and Kijne,J.W.1994.Pea *(Pisum sativum* L.) seed isolectins 1 and 2 and pea root lectin result from carboxypeptidase-like processing of a single gene product. Plant Mol. Biol. 24:75-81.

Kamberger,W.1979.An Ouchterlony double diffusion study on the interaction between legume lectins and rhizobial cell surface anti-gens.Arch.Microbiol.121:83-90.

Kijne,J.W., Diaz,C.L., Van Eijsden,R.R., Booij,P., Demel,R., Van Wor-kum,W.A.T., Wijffelman,C.A., Spaink,H.P., Lugtenberg,B.J.J.,and De Pater,B.S.1994.Lectin and Nod factors in Rhizobium-legume symbiosis. Pages 106-110, in: Proceedings of the Ist European Nitrogen Fixation Conference. G.B.Kiss, G.Ende, eds. Officina Press, Szeged.

Peumans,W.J.,and Van Damme,E.J.M.1995.Lectins as plant defense proteins. Plant Physiol. 109:347-352.

Quinn,J.E. and Etzler,M.E.1987.Isolation and characterization of a lectin from the roots of *Dolichos biflorus*.Arch.Biochem.Biophys.258:535-544.

Van Eijsden,R.R., Diaz,C.L., De Pater,B.S.,and Kijne,J.W.1995. Sugar-binding activity of pea *(Pisum sativum)* lectin is essential for hetero-logous infection of transgenic white clover hairy roots by Rhizobium leguminosarum biovar viciae. Plant Mol. Biol. 29:431-439.

# Molecular Communication in Cyanobacterial-Plant Symbioses

Bergman, B., Matveyev, A., Rasmussen, U. & Viterbo-Fainzilber, A.

Department of Botany, Stockholm University, S-106 91 Stockholm, Sweden

Cyanobacteria are prokaryotic phototrophs of great antiquity which readily form symbioses with eukaryotic organisms. In addition to oxygen-evolving photosynthesis, some cyanobacteria fix atmospheric nitrogen under conditions of combined nitrogen deficiency. The fixed nitrogen is a most innocuous source of nitrogen for plants forming symbiosis nitrogen-fixing cyanobacteria.

The only natural interaction in which cyanobacteria establish an intra-cellular symbiosis with a higher plant is the *Gunnera-Nostoc* symbiosis, recently investigated by our group (Bergman et al. 1992a; 1996, Rasmussen et al. 1994, 1996). Since free-living cyanobacteria fix nitrogen only to meet their own immediate needs, the normal regulatory mechanisms governing nitrogen fixation must be subverted during cyano-bacterial symbioses so that sufficient fixed nitrogen will escape the cyanobacterium to feed the host plant. Cell differentiation in compatible cyanobacteria is usually stimulated on contact with the host plant. The normally photoautotrophic cyanobacteria also move towards the darker interior of the plant and initiate a heterotrophic mode of nutrition when in symbiosis. These events are probably directed by the eukaryotic host and there is now evidence that hosts such as *Gunnera* control cyanobacterial gene expression during the establishment of successful nitrogen-fixing symbiosis (Bergman et al. 1996).

## Establishment of the *Nostoc-Gunnera* symbiosis

Representatives of the heterocystous cyanobacterium genus *Nostoc* infect stem glands of the host plant *Gunnera*. These red stem glands, formed at the base of each new leaf stalk, secrete a viscous acidic mucilage (Berg-man et al. 1992a). The mucilage carries signal molecules that specifically induce differentiation of motile hormogonia, *de facto* performers of the infection. In addition the mucilage induces symbiosis specific proteins and stimulates growth of compatible cyanobacteria (Rasmussen et al. 1994), indicating that there are several host compounds in the mucilage which target compatible *Nostoc* strains. Cell divisions are also obvious in the host plant tissue near the invading compatible cyanobacterium, indicating the existence of cyanobacterial signals capable of acting on the eukaryotic host (Bergman et al. 1996).

## Identification of symbiosis related genes

Accumulated data suggest that cyanobacteria in *Gunnera* symbioses receive chemical signals from the host plant which may induce gene expression. Our aim is to identify these genes, their gene products, and to characterize regulatory mechanisms and underlying functions. Currently three categories of genes are being investigated.

A. GENES OF KNOWN FUNCTION IN FREE-LIVING CYANOBACTERIA
DIFFERENTIALLY EXPRESSED IN SYMBIOSIS

In this context genes involved in nitrogen fixation and heterocyst differentiation are obvious. Heterocysts are specialized cells differentiated by a number of strains in order to overcome the effects of oxygen, deleterious for the nitrogen-fixing enzyme nitrogenase. Due to the formation of multilayered envelopes around the heterocysts, to the elimination of a functional oxygen-producing photosystem II, and to additional changes in their physiology, an anaerobic environment is provided inside the heterocysts. In return for fixed nitrogen the photosynthetic vegetative cells provide heterocysts with carbohydrates (Fay 1992).

As in other symbioses involving plants and cyanobacteria, the heterocyst frequency increases drastically in *Nostoc* after the *Gunnera* symbiosis is established. In this symbiosis hetetrocysts constitute up to 65-80% of the total cell number (Bergman et al. 1992b), compared to 5-10% in most free-living filaments. At least in young symbiotic tissues, elevated heterocyst frequencies correlate positively with nitrogenase activities indicating an apparent input of the multi-heterocyst phenotype into feeding the host with nitrogen. Among the genes involved in heterocyst differentiation, *hetR* appears to play an essential role (Buikema and Haselkorn 1991). During symbiosis, it is therefore likely that the host plant influences the expression of *hetR* in the invading *Nostoc*.

Consequently we have cloned and characterized the *hetR* gene and adjacent regions from *Nostoc* PCC 9229, a symbiotic isolate from *Gunnera*.

For the purpose of characterizing the entire *hetR* gene and adjacent DNA sequences, a genomic library of *Nostoc* PCC 9229 was constructed, by using approximately 15-22 kb fractions of genomic DNA, partially digested with *Sau*3A, and phage λ-EMBL3 arms cut with *Bam*HI (Matveyev et al., unpublished). When compared to the *hetR* gene from the non-symbiotic strain *Anabaena* PCC 7120 (Buikema and Haselkorn 1991, 1994), significant differences were observed in the sequence organization adjacent to the gene (Lotti et al. 1995). Also, Southern analysis of the genomic DNA and the *hetR*-containing fragments from the λ genomic library reveal the presence of an additional copy of the *hetR* gene in *Nostoc* upstream from the original (Matveyev et al., unpublished).

Unlike in *Anabaena* PCC 7120, only one of the multiple *hetR* transcripts was expressed as a result of nitrogen stepdown in *Nostoc*, the other were expressed irrespective of the nitrogen source. Accordingly, the expression of *hetR* in response to the environment experienced by the cyanobacterium during symbiosis is now being analysed.

B. GENES HOMOLOGOUS TO THOSE OF OTHER PLANT-INTERACTIVE
BACTERIA

Homologies to infection-related rhizobial and agrobacterial genes have been detected in the symbiotic cyanobacterium *Anabaena azollae* (Plazinski et al. 1991) and in the actinomycete *Frankia* (Chen et al. 1991). Common and host specific *nod* genes have been detected also in the bacterium *Azospirillum* (Fogher et al. 1985, Vieille and Elmerich 1990). In *Nostoc* PCC 9229, DNA sequences homologous to the rhizobial *nodEF*, *nodMN* and *exoY* genes, along with the *nod* box, were identified by heterologous hybridization, while no homologies were found to *nodABC*, *nodD1* or *nodD2* (Rasmussen et al. 1996). Probes containing the agrobacterial *chvA*, *chvB* and *picA* genes also hybridized to *Nostoc* DNA, while *virA* and *virG* genes did not (Rasmussen et al. 1996).

The *nodM* gene is characterized in *Rhizobium melitoti* as a second copy of the *glmS* (glucosamine-fructose-6-phosphate aminotransferase) gene. The *nodM* gene product is involved in the synthesis of glucosamine a constituent of the host interactive Nod factor released by *Rhizobium* (van Rhijn and Vanderleyden 1995). Degenerate primers were designed after alignment of the rhizobial *nodM* and *E. coli glmS* sequences. An approximately 800 bp product was isolated from DNA of *Nostoc* PCC 9229 after PCR amplification, cloned and 611 bp were sequenced. Comparison of the predicted protein sequence verified homology to the NodM protein of *R. leguminosarum* and *R. melitoti* (64-65% similarity in a 170 amino acid overlap). The PCR product was then used as a probe for Southern analysis of *Nostoc* DNA. The data obtained, together with restriction and sequencing analyses, show that the *nodM* gene is present as a single copy in symbiotic competent *Nostoc*. A 3-kb clone was isolated as a subclone from the genomic *Nostoc* library after two screenings with the *nodM* probe and at present, the fragment is under analysis for characterization of flanking regions.

*Gunnera* mucilage is capable of inducing β-galactosidase activity in rhizobia with *nod-lacZ* fusions, albeit at fairly low levels (Rasmussen et al. 1996). However, northern hybrydization to RNA extracted from mucilage treated *Nostoc* PCC 9229 did not reveal any upregulation in *nodM* expression. Flavonoids including the *nod* gene inducers kaempferol and quercetin have been isolated from *Gunnera* leaves (Patricia et al. 1989). Preliminary attempts to induce the expression of the *Nostoc nodM* homologue by such substances were negative. The absence of *nodD1* and *nodD2* homologues may indicate that other compounds regulate *nodM* expression in *Nostoc* symbiosis.

## C.  GENES  UNIQUE FOR CYANOBACTERIA IN PLANT SYMBIOSIS

As an alternative approach, we are searching for cyanobacterial genes specifically induced by the acidic mucilage of *Gunnera*. We expect to isolate genes responsible for hormogonia formation in general, for early communication and for determining the symbiotic competence of the cyanobacterium. Mini-cDNA libraries have been constructed using a subtractive technique. Total RNA from mucilage-induced and non-induced *Nostoc* PCC 9229 was reverse transcribed into cDNAs. The population of unique cDNAs obtained after subtraction were cloned and are currently used as probes for transcription analyses, larger-scale cloning  from the genomic *Nostoc* library and for sequencing in order to identify any particular gene.

## Remarks and Perspectives

Besides giving us valuable insight into regulatory events involving nitrogen fixation and differentiation in cyanobacteria, an understanding of the molecular basis for symbiosis may enable us to extend the host range and to create artificial nitrogen-fixing symbioses through genetic engeneering of plants and cyanobacteria.

## Acknowledgements

Financial support was provided by the Swedish Natural Sciences Research Council, the Wenner-Green Foundation, and the C. Trygger's Foundation.

## References

Bergman, B., Johansson, C. and Söderbäck, E. 1992a. The *Nostoc-Gunnera* symbiosis. New Phytol. 122:379-400.

Bergman, B., Rai, A.N., Johansson, C. and Söderbäck, E. 1992b. Cyanobacterial-plant symbioses. Symbiosis 14:61-81.

Bergman, B., Matveyev, A. and Rasmussen, U. 1996. Chemical signalling in cyanobacterial-plant symbioses. Trends in Plant Science 1:191-197.

Buikema, W. and Haselkorn, R. 1991. Characterization of a gene controlling heterocyst differentiation in the cyanobacterium *Anabaena* 7120. Genes & Dev. 5:321-330.

Buikema, W. and Haselkorn, R. 1994. Controlled expression of *hetR* in *Anabaena* 7120 and adjacent genome organization. Page 156 in: International Symposium on Phototrophic Prokaryotes. G. Tedioli, S. Ventura and D. Zannoni, eds. Abstracts, VIII, CNR, Urbino.

Chen, L., Cui, Y., Wang, Y., Bay, X. and Ma, Q. 1991. Identification of a *nodD*-like gene in *Frankia* by direct complementation of a *Rhizobium nodD* mutant. Mol. Gen. Genet. 233:311-314.

Fay, P. 1992. Oxygen relations of nitrogen fixation in cyanobacteria. Microbiol. Rev. 56:340-373.

Fogher, C., Dusha, H., Barbot, P. and Elmerich C. 1985. Heterologous hybridization of *Azospirillum* DNA to *Rhizobium nod* and *fix* genes. FEMS Microbiol. Lett. 30:245-249.

Lotti, F., Matveyev, A.V. and Bergman, B. 1995. The *hetR* gene from *Nostoc* sp. PCC 9229, a symbiotic cyanobacterium: cloning, sequencing and expression. Page 73 in: Abstracts, Fifth Cyanobacterial Molecular Biology Workshop, Asilomar.

Patricia, P., Crawford, D.J., Stuessy, T.F. and Silva, M. 1989. Flavonoids of *Gunnera* subgenera *Misandra, Panke* and *Perpensum* (Gunneraceae). Am. J. Bot. 76(6 suppl.):264.

Plazinski, J., Croft, L., Taylor, R., Zheng, Q., Rolfe, B.G. and Gunning, B.E.S. 1991. Indigenous plasmids in *A. azollae*: their taxonomic distribuition and existence of regions of homology with symbiotic genes in *Rhizobium*. Canad. J. Microbiol. 37:171-181.

Rasmussen, U., Johansson, C. and Bergman, B. 1994. Early communication in the *Gunnera-Nostoc* symbiosis: plant-induced cell differentiation and protein synthesis in the cyanobacterium. Mol. Plant-Microbe. Interact. 4: 563-570.

Rasmussen, U., Johansson, C., Renglin, A., Petersson, C. and Bergman, B. 1996. A molecular characterization of the *Gunnera-Nostoc* symbiosis: comparison with *Rhizobium-* and *Agrobacterium-*plant interactions. New Phytol. 133 (in press).

Van Rhijn, P. and Vanderleyden, J. 1995. The *Rhizobium-*Plant symbiosis. Microbiol. Rev. 59:124-142.

Vieille, C. and Elmerich, C. 1990. Characterization of two *Azospirillum brasilense* Sp7 plasmid genes homologous to *R. melitoti nodPQ.* Mol. Plant Microbe Interact. 3:389-400.

# Specific Flavonoids Stimulate Intercellular Colonization of Non-legumes by *Azorhizobium caulinodans*

Clare Gough[1], Gordon Webster[1], Jacques Vasse[1], Christine Galera[1], Caroline Batchelor[2], Kenneth O'Callaghan[2], Michael Davey[2], Shanker Kothari[2], Jean Dénarié[1] and Edward Cocking[2].
[1]Laboratoire de Biologie Moléculaire des Relations Plantes-Microorganismes, INRA-CNRS, BP27, 31326 Castanet-Tolosan, France. [2]Plant Genetic Manipulation Group, Department of Life Science, University of Nottingham, Nottingham NG7 2RD, U.K.

## Introduction

A major limitation of associative nitrogen fixation is that bacteria only colonize the surface of roots and remain vulnerable to competition from other micro-organisms of the rhizosphere. In contrast, a great advantage of symbiotic systems is that the nitrogen-fixing bacteria internally colonize the plant and become endophytic, thus being protected from competition with rhizospheric micro-organisms and having the possibility of more intimate metabolic exchange with host plants.

With the aim of trying to extend biological nitrogen fixation to non-legume crops, an important area of research is to look for natural endophytic diazotrophs of these plants and to optimise such endophytic nitrogen fixation. This approach is currently being exploited with sugar cane (Boddey *et al.* 1991) and other crops such as rice (Ladha and Reddy 1995). An alternative line of research is to try to see whether any rhizobial strains can internally colonize non-legumes. Much interest has recently been shown in this field, particularly with cereals. Rhizobia have been reported to enter rice, maize, wheat and oilseed rape plants at low frequency at the points of emergence of lateral roots (Cocking *et al.* 1992,1994). Among the rhizobia used in this latter work, *Azorhizobium caulinodans*, which forms root and stem nodules on the tropical legume *Sesbania rostrata*, is especially interesting. *A. caulinodans* (i) forms nodules on its host plant following crack entry infection *ie* intercellularly, between adjacent cells, (ii) has the unusual ability among rhizobia to fix nitrogen in free-living conditions *ie* does not need to be differentiated into bacteroids and (iii) can tolerate up to 12 μM dissolved oxygen while fixing nitrogen in culture (de Bruijn 1989).

In this present work, we have adopted an interdisciplinary approach, combining bacterial genetics and cytology to study interactions between *A. caulinodans* ORS571 and the monocot wheat (a major cereal) and the dicot *Arabidopsis thaliana* (a model plant). Strain ORS571 and derivatives, containing a constitutive *lacZ* reporter gene fusion, were used to localise bacteria.

## Results and Discussion

INTERCELLULAR COLONIZATION AT LATERAL ROOT CRACKS OF *Arabidopsis thaliana* AND WHEAT BY *Azorhizobium caulinodans*

Plants (wheat variety Canon; *A. thaliana* ecotype Columbia), were grown in tubes with soft agar Fahreus medium (without nitrate for wheat; 0.25 mM nitrate for *A. thaliana*), and inoculated after 2 or 3 days with ORS571(pXLGD4), the strain containing a constitutive *lacZ* reporter gene fusion (Leong *et al.* 1985). ß-galactosidase activity associated with roots of inoculated plants was visualised after 1 (*A. thaliana*) and 2 (wheat) weeks, by light microscopy of the dark blue precipitate resulting from the degradation of X-Gal (Boivin *et al.* 1990). Bacteria were observed on the surface of root tips and inside lateral root cracks (LRCs), which result from lateral root emergence and which are found at the bases of lateral roots. LRCs of *A. thaliana* were colonized at a very early stage in lateral root formation, and as the laterals developed bacteria were seen to enter deeper into the cracks. At the points of emergence of older lateral roots, bacteria had progessed further into the plant, mainly being seen as blue pockets or files inside the main root, adjacent to points of lateral root emergence. Sections of such blue-staining LRCs of *A. thaliana* showed bacteria in the elongated intercellular space between the cortex and the endodermis, in the main root. Sections of blue-staining LRCs of wheat showed intercellular pockets of bacteria in the cortex of the main root, as in the first stages of root nodulation of *S. rostrata* (Ndoye *et al.* 1994) and as in wheat roots inoculated with *Azospirillum brasilense* (Levanony *et al.* 1989). ß-galactosidase activity was still detected at colonization sites several weeks after inoculation, suggesting the presence of viable bacteria. In addition, no obvious signs of plant defense reactions were seen and slight expression of a *nif* promoter-*lacZ* fusion (Pawlowski *et al.* 1987), was detected in both plants.

A high proportion of plants had colonized LRCs (including both crack colonization and intercellular colonization); 75-100% of wheat and 63-100% of *A. thaliana*. This high level of colonization enabled us to develop a scoring system, by counting the number of colonized LRCs per plant and calculating the percentage of LRCs which were colonized; 9.2% for wheat and 10.8% for *A. thaliana* (means of four independent experiments). The use of a reporter gene, in this present study, has facilitated detection of bacteria at points of lateral root emergence and suggests that previous reports of interactions between wheat and *A. caulinodans* (Cocking *et al.* 1992,1994), underestimated the proportion of plants colonized.

A certain specificity was found in the colonization ability of *A. caulinodans* when we looked at the colonization of LRCs of wheat and *A. thaliana* roots by various other rhizobia and plant-associated bacteria. Among these, *Rhizobium meliloti* was virtually unable to and *Azospirillum brasilense* was as able to colonize LRCs of wheat. For *A. thaliana*, both *R. meliloti* and *A. brasilense* were able to colonize LRCs, but the frequency and level of colonization were lower than with *A. caulinodans*.

SPECIFIC FLAVONOIDS STIMULATE INTERCELLULAR COLONIZATION

Phenolic compounds secreted in the rhizosphere by plants, induce expression of a variety of genes in plant-associated bacteria. Certain bacteria have evolved specific catabolic pathways enabling them to use these compounds as energy sources and phenolic compounds can also act as chemo-attractants (Peters and Verma 1990). Flavonoids, a class of plant phenolics, induce rhizobial nodulation (*nod*) genes that are required for symbiotic associations with legumes. Flavonoids also enhance the growth rate of certain rhizobia (Hartwig *et al.* 1991).

Four families of flavonoids (flavones, flavonols, flavanones and isoflavones), have been shown to have *nod* gene inducing activity (Dénarié *et al.* 1992). One of each family was tested for their effect on LRC colonization of *A. thaliana* by ORS571(pXLGD4). Acetosyringone, a plant phenolic which induces expression of virulence genes in *Agrobacterium tumefaciens*, and succinate, one of the best carbon sources for ORS571, were also tested. Figure 1 shows that the flavanone naringenin and the isoflavone daidzein (both at $5 \times 10^{-5}$ M) significantly stimulated (at the P=0.01 level), colonization of LRCs of *A. thaliana*. Succinate, at the same concentration, had no significant effect on colonization, indicating that these flavonoids do not act as simple carbon sources. Naringenin (at $10^{-4}$ M) also significantly stimulated (at the P=0.01 level), the colonization of LRCs of wheat by

ORS571(pXLGD4), approximately to the same degree as for *A. thaliana*.

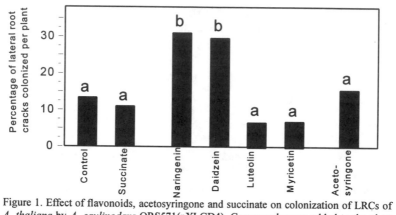

Figure 1. Effect of flavonoids, acetosyringone and succinate on colonization of LRCs of *A. thaliana* by *A. caulinodans* ORS571(pXLGD4). Compounds were added to the plant growth medium at $5 \times 10^{-5}$ M. Statistical analysis: treatments with different letters differ significantly at the P=0.01 level. Analysis of variance with Fisher's test (Snedecor and Cochran 1980). 16 plants were scored per treatment.

*nod* GENES ARE NOT INVOLVED IN INTERCELLULAR COLONIZATION OF WHEAT AND *A. thaliana*

Naringenin is one of the most efficient inducers of the expression of *nod* genes in ORS571 and *nod* genes are involved in the intercellular crack entry infection of *S. rostrata* (Goethals *et al.* 1989). It was therefore important to determine whether naringenin stimulates colonization of these non-legumes via *nod* genes. In the presence of specific flavonoids, NodD proteins activate the transcription of *nod* operons, which in turn control the synthesis of specific lipo-chitooligosaccharides, the so-called Nod factors.

We first assessed the possible role of Nod factors in colonization of LRCs of wheat and *A. thaliana* by ORS571. For this, a *nodC* mutant of ORS571 (Goethals *et al.* 1989), which does not produce Nod factors, was tested. No difference was found between the levels of colonization by the *nodC* mutant and the wild type strain. Colonization by the *nodC* mutant was still stimulated by naringenin.

As it is possible that NodD could activate genes other than *nod* genes in the presence of naringenin, an ORS571 *nodD* mutant was tested (Goethals *et al.* 1990). Again, this mutant was able to colonize

LRCs of wheat and *A. thaliana* as well as the wild type strain. Colonization by the *nodD* mutant was also stimulated by naringenin, indicating that the mechanism by which naringenin stimulates LRC colonization of non-legumes is not mediated by the NodD protein. The fact that daidzein also stimulates LRC colonization of *A. thaliana*, but does not induce *nod* gene expression in ORS571 (Goethals *et al.* 1989), also supports this conclusion.

## General Conclusions

This work has shown that *A. caulinodans* can colonize the cracks at the points of emergence of lateral roots of wheat and *A. thaliana* and can subsequently intercellularly colonize adjacent cortical tissue. This colonization is *nod* gene-independent, unlike infection by crack entry in *S. rostrata*. The demonstration of this in the monocot wheat, and the dicot *A. thaliana,* implies that it is likely to be a general phenomenon of non-leguminous plants. This is supported by studies with rice inoculated with ORS571, which also show large intercellular pockets of bacteria at emerging lateral roots.

The flavonoids naringenin and daidzein were found to stimulate colonization of LRCs and intercellular colonization adjacent to LRCs. Since stimulation does not appear to be due to a carbon source effect and as naringenin and daidzein are active at low concentrations, these flavonoids probably act as signals and not as substrates. Phillips and Streit (1996) have proposed that plant signals, including flavonoids, are involved in the selective stimulation of specific microbial communities in the rhizosphere, explaining why plant roots are only colonized by certain types of soil bacteria. Our results support this hypothesis. The precise colonization site that we have described for *A. caulinodans*, at LRCs, is one of the principal sites of entry into roots of soil-borne fungal and bacterial pathogens. There are therefore important potential applications of the discovery of plant signals that stimulate colonization of this site, including applications in the areas of root colonization by plant growth-promoting rhizobacteria and by biocontrol bacteria against root pathogens, in addition to the presently reported interactions with rhizobia.

The induction by flavonoids of rhizobial genes which are not *nod* genes, has already been reported (Sadowsky *et al* 1988). These genes of *Rhizobium fredii* are induced by flavonoids present only in root exudates of plants nodulated by *R. fredii*. One of the active flavonoids is daidzein, which also has *nod* gene-inducing activity in *R. fredii*. This control by host specific flavonoids of rhizobial genes other than *nod* genes, may be important for early steps in symbiosis. It will therefore be interesting to determine whether naringenin and daidzein induce a new class of genes in *A. caulinodans*, which could be involved both in the symbiotic interaction between *A. caulinodans* and *S. rostrata* and in interactions between *A. caulinodans* and non-legumes.

The ultimate aim of establishing stable endophytic interactions between diazotrophs and non-legumes is that the diazotrophs should fix nitrogen and transfer this fixed nitrogen to the plant. We were able to show that conditions are appropriate for nitrogen fixation at colonization sites in wheat and *A. thaliana*, by detecting expression of a *nif* promoter-*lacZ* fusion. However, we could only detect nitrogenase activity (by the acetylene reduction assay) when succinate was added to the plant growth medium, indicating that, in our experimental conditions, carbon is limiting for nitrogen fixation. Now that we have defined conditions which give reproducible LRC colonization at high frequency, it should be possible to clearly define the various genetic and physiological factors which are limiting, both for rhizobial colonization and for endophytic nitrogen fixation in non-legumes.

ACKNOWLEDGMENTS

G.W. was supported by the grant "Assessing opportunities for biological nitrogen fixation in rice" from the Danish International Development Agency, C.G. by a fellowship from the Institut National de la Recherche Agronomique, C.B. by the O.D.A. (Plant Sciences Programme), K.O'C. by M.A.F.F. and E.C. by the Leverhulme Trust and the Rockefeller Foundation. S.K. is a Rockefeller Foundation Biotechnology Career Fellow.

LITERATURE CITED

Boddey, R.M., Urquiaga, S., Reis, V. and Döbereiner, J. 1991. Biological nitrogen fixation associated with sugar cane. Plant Soil. 137:111-117.

Boivin, C., Camut, S., Malpica, C.A., Truchet, G. and Rosenberg, C. 1990. *Rhizobium meliloti* genes encoding catabolism of trigonelline are induced under symbiotic conditions. Plant Cell 2:1157-1170.

Cocking, E.C., Davey, M.R., Kothari, S.L., Srivastava, J.S., Jing, Y., Ridge, R.W. and Rolfe, B.G. 1992. Altering the specificity control of the interaction between rhizobia and plants. Symbiosis 14:123-130.

Cocking, E.C., Webster, G., Batchelor, C.A. and Davey, M.R. 1994. Nodulation of non-legume crops. A new look. Agro-Food-Industry Hi-Tech. January/Febuary 21-24.

de Bruijn, F. 1989. The unusual symbiosis between the diazotrophic stem-nodulating bacterium *of Azorhizobium caulinodans* ORS571 and its host, the tropical legume *Sesbania rostrata*. Pages 457-504 in: Plant-Microbe Interactions: Molecular and Genetic Perspectives. Vol. 3. T. Kosuge and E.W. Nester, ed. McGraw-Hill Publishing Company.

Dénarié, J., Debellé, F. and Rosenberg, C. 1992. Signaling and host range variation in nodulation. Annu. Rev. Microbiol. 46:497-531.

Goethals, K., Gao, M., Tomekpe, K., Van Montagu, M. and Holsters, M. 1989. Common *nodABC* genes in *Nod* locus 1 of *Azorhizobium caulinodans*: Nucleotide sequence and plant-inducible expression. Mol. Gen. Genet. 219:289-298.

Goethals, K., Van Den Eede, G., Van Montagu, M. and Holsters, M. 1990. Identification and characterization of a functional *nodD* gene of *Azorhizobium caulinodans* ORS571. J. Bacteriol. 172:2658-2666.

Hartwig, U.A., Joseph, C.M. and Phillips, D.A. 1991. Flavonoids released naturally from alfalfa seeds enhance growth rate of *Rhizobium meliloti*. Plant Physiol. 95:797-803.

Ladha, J.K. and Reddy, P.M. 1995. Extension of nitrogen fixation to rice - necessity and possibilities. GeoJournal 35.3:363-372.

Leong, S.A. Williams, P.H. and Ditta, G.S. 1985. Analysis of the 5' regulatory region of the gene for δ-aminolevulinic acid synthetase *of Rhizobium meliloti*. Nucl. Acids Res. 13:5965-5976.

Levanony, H., Bashan, Y., Romano, B. and Klein, E. 1989. Ultrastructural localisation and identification of *Azospirillum brasilense* Cd on and within wheat root by immuno-gold labeling. Plant Soil 117:207-218.

Ndoye, I., De Billy, F., Vasse, J., Dreyfus, B. and Truchet, G. 1994. Root nodulation of *Sesbania rostrata*. J. Bacteriol. 176:1060-1068.

Pawlowski, K., Ratet, P., Schell, J. and de Bruijn, F.J. 1987. Cloning and characterization of *nifA* and *ntrC* genes of the stem nodulating bacterium ORS571, the nitrogen fixing symbiont of *Sesbania rostrata*: Regulation of nitrogen fixation (*nif*) genes in the free living versus symbiotic state. Mol. Gen. Genet. 206:207-219.

Peters, N.K. and Verma, D.P.S. 1990. Phenolic compounds as regulators of gene expression in plant-microbe interactions. Mol. Plant-Microbe Interact. 3:4-8.

Phillips, D.A. and Streit, W. 1996 Legume Signals to rhizobial symbionts: a new approach for defining rhizosphere colonization. Pages 236-271 in Plant-Microbe Interactions. G. Stacey and N.T. Keen, ed. Chapman and Hall.

Sadowsky, M.J., Olson, E.R., Foster, V.E., Kosslak, R.M. and Verma, D.P.S. 1988. Two host-inducible genes of *Rhizobium fredii* and characterization of the inducing compound. J. Bacteriol. 170:171-178.

Snedecor, G. and Cochran, W. 1980. Statistical Methods. (Ames, IA, Iowa State University Press).

# Actinorhizal nodules from different plant families

K. Pawlowski, A. Ribeiro, C. Guan, A. van Kammen, A.M. Berry[1] and T. Bisseling

Department of Molecular Biology, Agricultural University, 6703 HA Wageningen, The Netherlands
[1]Department of Environmental Horticulture, University of California, Davis CA 95616, U.S.A.

In nitrogen-fixing plant-bacterial symbioses, upon interaction with the microsymbiont the plants form special organs, the root nodules. Within these nodules, the bacteria become hosted inside plant cells, where they form the $O_2$-sensitive enzyme nitrogenase to fix nitrogen, exporting ammonia to the plant cytoplasm (Mylona et al. 1995).

The nitrogen-fixing actinomycete *Frankia* can induce the formation of root nodules on a diverse group of dicotyledonous plants, collectively called actinorhizal plants (Benson and Silvester 1993). *Frankia* sp. are gram-positive soil bacteria which normally grow in hyphal form. Under nitrogen-limiting conditions and atmospheric oxygen tension, they can form spherical vesicles at the ends of hyphae or short side branches. In these vesicles, nitrogenase is formed, protected from $O_2$ by the vesicle envelope. Actinorhizal nodules consist of multiple lobes, each of which represents a modified lateral root without root cap, with a superficial periderm and with infected cells in the expanded cortex. Due to the activity of the meristem at the tip of a nodule lobe, the cortical cells are arranged in a developmental gradient. Four zones can be distinguished (Ribeiro et al. 1995). The meristematic zone (zone 1) is followed by the infection zone (zone 2), were some of the cortical cells are invaded and progressively filled by branching *Frankia* hyphae. The nitrogen fixation zone (zone 3) is characterized by the expression of *Frankia nif* (nitrogen fixation) genes. In this zone, vesicles have been formed and *Frankia* is fixing nitrogen. In the senescence zone (zone 4), plant cytoplasm and *Frankia* material are degraded.

While a single origin has been suggested for the predisposition of plants to enter symbioses with nitrogen-fixing bacteria (Soltis et al. 1995), several independent direct origins of the symbiosis have been proposed among legumes (Doyle 1994) and among actinorhizal plants (Swensen 1996). These independent origins are reflected by a considerable degree of structural variability between nodules of actinorhizal plants from different families. Due to this diversity, the system seems ideal to study the principles of symbiotic nitrogen-fixation. Previously, studies on legume nodules had led to the impression that the problems imposed by bacterial nitrogen fixation, have to be solved by the plant in one particular way. Now, the comparison of nodules formed by different plant families can show divers options.

For instance, to cope with the necessity for the protection of $O_2$-sensitive bacterial nitrogenase while providing enough $O_2$ for respiration, legume nodules contain a peripheral $O_2$ diffusion barrier, while the $O_2$ transport protein leghemoglobin in the infected cells facilitates $O_2$ diffusion to the sites of respiration. Amongst actinorhizal systems, where also bacterial symbiont can contribute to $O_2$ protection, different strategies have been taken to solve the $O_2$ dilemma. *Alnus* nodules (Fig. 1A) are well aerated via lenticells, and *Frankia* forms spherical vesicles at the periphery of the infected cells (Fig. 2A; Silvester et al. 1990). Maybe the plant is facilitating oxygen diffusion, but although the concentration of hemoglobin in nodules is higher than in roots, it is not comparable to that found in *Casuarina* nodules (Suharjo and Tjepkema 1995). *Casuarina* nodules are aerated via agraviotropical nodule roots which grow out of the tips of nodule lobes (Silvester et al. 1990). Here, the $O_2$ dilemma seems to be solved by the plant similarly to the situation in legume nodules. The walls of infected cells are lignified to provide an $O_2$ diffusion barrier (Fig. 2B; Berg and McDowell 1987), and high amounts of hemoglobin in the infected cells (Fleming et al. 1987; Gherbi et al. 1996), facilitate oxygen diffusion to the sites of respiration. In *Casuarina* nodules, *Frankia* does not form vesicles (Fig. 2B). In nodules of both *Casuarina* (Fig. 2B) and *Alnus* (Fig. 1A), the infected cells are distributed over the cortex, interspersed with uninfected cells. In nodules of *Datisca* (Fig. 1B), the infected cells occupy a region on one side of the acentric stele, not interspersed with uninfected cells (Newcomb and Pankhurst 1982). *Datisca* nodules are well aerated via nodules roots and lenticells and do not contain significant amounts of hemoglobin (Silvester et al. 1990). Here, *Frankia* forms lanceolate vesicles which point toward the central vacuole of the infected cell (Fig. 2C; Hafeez et al. 1984). The fact that the shape and location of vesicles formed in infected cells depends on the host plant indicates that this symbiosis-specific differentiation depends on plant signalling.

Meanwhile, some actinorhizal symbioses have been examined using the tools of molecular biology, enabling us to extend the comparison between the different nodule types. Plant genes showing elevated expression levels in nodules compared to roots have been cloned, mostly via differential screening of cDNA libraries with nodule and root cDNA, respectively, from *A. glutinosa* (Goetting-Minesky and Mullin 1994; Pawlowski et al. 1994; Ribeiro et al. 1995) and *C. glauca* (Jacobsen-Lyon et al. 1995; Gherbi et al. 1996), and *D. glomerata* (K. Pawlowski and A.M. Berry, unpublished). These genes will be termed "actinorhizal nodulin genes" in this article. Based on their deduced amino acid sequences and their expression patterns, the genes identified thus far can be implicated in several processes crucial to

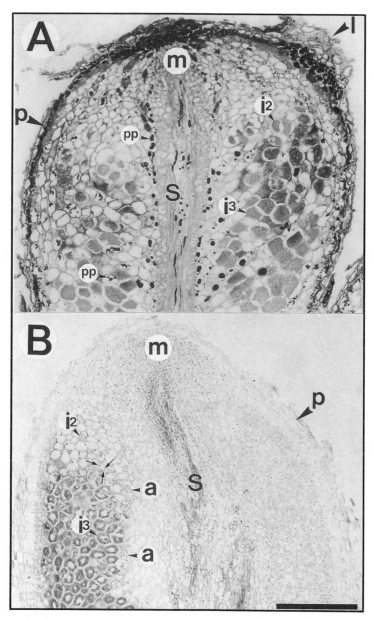

**Figure 1.** Longitudinal sections of actinorhizal nodule lobes. **(A)** *Alnus glutinosa*; **(B)** *Datisca glomerata*. Both lobes contain a central stele (s) and a meristem at the tip (m) and are surrounded by a periderm (p), which in (A) is interrupted by a lenticel (l). In *Alnus* nodules, cells of the pericycle, endodermis and some cortical cells contain polyphenols (pp). The infected cells (i2: from infection zone; i3: from nitrogen fixation zone) in *Alnus* are dispersed over the cortex, while in *Datisca*, they form a patch at one side of the stele, surrounded by uninfected cells containing big amyloplasts (a). In *Datisca*, cells become multinucleate upon infection by *Frankia* (arrows; Hafeez et al. 1984). Bar = 250 μm.

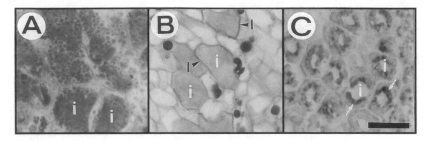

**Figure 2.** Nitrogen-fixing infected cells (I) of **(A)** *A. glutinosa,* **(B)** *Casuarina glauca,* and **(C)** *D. glomerata. Frankia* forms spherial vesicles in *Alnus* (A), while no vesicles are formed in *Casuarina* (B). The lignified cell walls of *Casuarina* infected cells are indicated (l). The *Datisca* section has been hybridized with *Frankia nifH* antisense RNA to visualize *nifH* expression (arrows pointing at dark stain), which is restricted to vesicles, in order to show the localization of vesicles around the central vacuole. Bar = 500 μm.

the symbiotic interaction. Several genes have been found to encode enzymes involved in the metabolic specialization of the nodules, e.g. ammonium assimilation, the biosynthesis of nitrogen transport forms, and glycolytic processes (Pawlowski et al. 1994). Other genes were implicated, based on their expression patterns, in the process of infection of cortical cells, or in the senescence of infected cortical cells.

### The metabolic specialization of actinorhizal nodules

Expression of plant glutamine synthetase (GS), the enzyme catalyzing the first step in ammonium assimilation, was compared in nodules of *A. glutinosa, C. glauca,* and *D. glomerata.* While GS expression in nodules of *Alnus* and *Casuarina* was confined to the nitrogen-fixing infected cells and the pericycle of the nodule vascular system, in *Datisca* high levels of expression were found in the uninfected cortical cells. While *A. glutinosa* is a citrulline exporter, *C. glauca* and *D. glomerata* are both amide exporters (Schubert 1986). Therefore, the different tissue specialization cannot be explained by the requirements for the synthesis of different nitrogen transport forms as it is the case in determinate versus indeterminate legume nodules (Mylona et al. 1995). Thus, in actinorhizal nodules from different plant genera, tissue specialization can be strikingly different, indicating that plants have more than one option to handle the high amounts of ammonium exported by their bacterial endosymbionts.

### Genes encoding products involved in the interaction with *Frankia*

Several actinorhizal nodulin genes were found to be expressed in infected cortical cells in specific stages of development. For instance, a gene encoding a putatively extracellular serine protease (*ag12*) is expressed at high levels in *A. glutinosa* nodules in infected cells of zone 2 (Ribeiro et al. 1995). *ag12* is also expressed in young infected cells of ineffective, i.e. not nitrogen-fixing *Alnus* nodules (Guan et al. 1996). A homolog of *ag12* was isolated from *C. glauca* by RT-PCR using degenerate oligonucleotides derived from the amino acid sequences conserved between Ag12 and its Arabidopsis homolog Ara12 (Ribeiro et al. 1995). The *C. glauca* homolog, termed *cg12*, shows a similar expression pattern as *ag12*, i.e. nodule-specific expression with high amounts of mRNA present infected cells of zone 2. For Ag12, a role in processing of a proteinaceous component of the cell wall-like matrix surrounding *Frankia* in infected cells has been proposed. The data on Ag12/Cg12 suggest that the protein components of the matrix surrounding *Frankia* share some similarity in both plants.

However, there is evidence that the proteinaceous compounds of the matrix surrounding intracellular *Frankia* can differ in nodules from different actinorhizal plant genera. cDNAs representing an actinorhizal nodulin gene family expressed in infected cells of zone 2 have been isolated from *A. glutinosa* and shown to encode small putative extracellular metal-binding proteins. Non-homologous putative metal-binding proteins have previously been identified as components of the peribacteroid space in several *Rhizobium*-legume symbioses, i.e. the nodulin-A family in soybean (Jacobs et al. 1987), supporting the idea that such proteins are an essential part of the matrix surrounding endosymbiotic bacteria. A cDNA homologous to those of *A. glutinosa* was isolated from *D. glomerata*. This homolog was not expressed during infection of cortical cells, but mainly in the periderm of nodules as well as roots.

Thus, the results of molecular analysis of actinorhizal nodules support the hypothesis that there is more than one direct phylogenetic origin of the symbiosis. The analysis of regarding cell types involved in metabolic specialization and putative components of the matrix surrounding intracellular Frankia, *Datisca* seems strikingly different from *Alnus* and *Casuarina*. The latter two, however, while different in nodule $O_2$ protection/transport systems and endosymbiont differentiation, seem to share a lot of similarity in molecular aspects, indicating a common origin. Further research is needed to integrate the molecular features with the data on phylogeny of actinorhizal symbioses.

### References

Benson, D.R., Silvester, W.B. 1993. Biology of *Frankia* strains, actinomycete symbionts of actinorhizal plants. Microbiol. Rev. 57:293-319.

Berg, R.H., McDowell, L. 1987. Cytochemistry of the wall of infected cells in *Casuarina* actinorhizae. Can. J. Bot. 66:2038-2047.

Doyle, J.J. 1994. Phylogeny of the legume family: An approach to understanding the origins of nodulation. Annu. Rev. Ecol. Syst. 25:325-349

Fleming, A.I., Wittenberg, J.B., Wittenberg, B.A., Dudman, W.F., Appleby, C.A. 1987. The purification, characterization and ligand-binding kinetics of hemoglobins from root nodules of the non-leguminous *Casuarina glauca-Frankia* symbiosis. Biochim. Biophys. Acta 911:209-220.

Gherbi, H., Duhoux, E., Franche, C., Pawlowski, K., Berry, A., Bogusz, D. 1996. Cloning of a full-length symbiotic hemlglobin cDNA and in situ localization of the corresonding mRNA in *Casuarina glauca* root nodule. Physiol. Plant., in press.

Goetting-Minesky, M.P., Mullin, B.C. 1994. Differential gene expression in an actinorhizal symbiosis: Evidence for a nodule-specific cysteine proteinase. Proc. Natl. Acad. Sci. USA 91:9891-9895.

Guan, C., Wolters, D.J., van Dijk, C., Akkermans, A.D.L., van Kammen, A., Bisseling, T., Pawlowski, K. 1996. Acta Bot. Gall., in press.

Hafeez, F., Akkermans, A.D.L., Chaudhary, A.H. 1984. Observations on the ultrastructure of *Frankia* sp. in root nodules of *Datisca cannabina* L. Plant Soil 79:383-402.

Jacobs, F.A., Zhang, M., Fortin, M.G., Verma, D.P. 1987. Several nodulins of soybean share structural domains but differ in their subcellular locations. Nucl. Acids Res 15:1271-1280.

Jacobsen-Lyon, K., Jensen, E.Ø., Jørgensen, J.-E., Marcker, K.A., Peacock, J., Dennis, E. 1995. Symbiotic and nonsymbiotic hemoglobin genes of *Casuarina glauca*. Plant Cell 7:213-223.

Mylona, P., Pawlowski, K., Bisseling, T. 1995. Symbiotic nitrogen fixation. Plant Cell 7:869-885.

Newcomb, W., Pankhurst, C.E. 1982. Fine structure of actinorhizal nodules of *Coriaria arborea* (Coriariaceae). New Zealand J. Bot. 20:93-103.

Pawlowski, K., Guan, C., Ribeiro, A., van Kammen, A., Akkermans, A.D.L., Bisseling, T. 1994. Genes involved in *Alnus glutinosa* nodule development. Pages 220-224 in Proceedings of the 1st European Nitrogen Fixation Conference. G.B. Kiss. G. Endre, eds. Officina Press, Szeged.

Ribeiro, A., Akkermans, A.D.L., van Kammen, A., Bisseling, T., Pawlowski, K. 1995. A nodule-specific gene encoding a subtilisin-like protease is expressed in early stages of actinorhizal nodule development. Plant Cell 7:785-794.

Schubert, K.R. 1986. Products of biological nitrogen fixation in higher plants: Synthesis, transport, and metabolism. Annu. Rev. Plant Physiol. 37:539-574.

Silvester, W.B., Harris, S.L., Tjepkema, J.D. 1990. Oxygen regulation and hemoglobin. Pages 157-176 in The Biology of *Frankia* and Actinorhizal plants. C.R. Schwintzer, J.D. Tjepkema, eds. Academic Press, New York.

Soltis, D.E., Soltis, P.S., Morgan, D.R., Swensen, S.M., Mullin, B.C., Dowd, J.M., Martin, P.G. 1995. Chloroplast gene sequence data suggest a single origin of the predisposition for symbiotic nitrogen fixation in angiosperms. Proc. Natl. Acad. Sci. USA 92:2647-2651.

Suharjo, U.K.J., Tjepkema, J.D. 1995. Occurrence of hemoglobin in the nitrogen-fixing root nodules of *Alnus glutinosa*. Physiol. Plant. 95:247-252.

Swensen, S. 1996. The evolution of actinorhizal symbioses: Evidence for multiple origins of the symbiotic association. Am. J. Bot., in press.

# *In vitro* expression of actinorhizal nodulin AgNOD-GHRP and demonstration of its toxicity to *Escherichia coli*

Svetlana V. Dobritsa[1] and Beth C. Mullin

The Department of Botany and the Center for Legume Research, The University of Tennessee, Knoxville, TN 37996 USA

In the nitrogen-fixing actinorhizal symbiotic system the presence of a compatible *Frankia* strain alters the developmental pathway of lateral roots in such a way that they become nitrogen-fixing root nodules capable of supplying the plant with reduced nitrogen. It has been shown that nodule development in *Alnus* is accompanied by the expression of genes not expressed in the roots or leaves of host plants (Mullin et al., 1993; Twigg, 1993; Goetting-Minesky and Mullin, 1994; Ribeiro et al., 1995). One nodule-specific cDNA, pAgNt84, has the capacity to code for a 99 amino acid-residue glycine/histidine-rich protein, AgNOD-GHRP (Fig. 1), with a putative hydrophobic signal peptide of 31 amino acid residues which is predicted to direct the protein to a membrane system (Twigg, 1993).

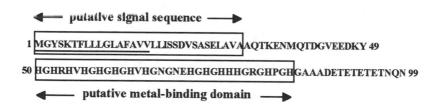

Figure 1. Derived amino acid sequence of AgNOD-GHRP showing the putative signal sequence and the putative metal-binding domain. The underlined sequence represents the portion of the signal sequence missing in protein expressed from the truncated clone III-41.

[1] Present address: Institute of Biochemistry and Physiology of Microorganisms, Russian Academy of Sciences, Pushchino, Russia

AgNOD-GHRP as a whole does not share high sequence identity with proteins in either GenBank or EMBL databases. It does however share a glycine and histidine-rich domain with two known metal-binding proteins, COT1 from yeast (Conklin et al., 1992) and WHP from *E. coli* (Wülfing et al., 1987). A second *Alnus* homolog of this cDNA has been isolated by K. Pawlowski and C.H. Guan (Agricultural University, Wageningen) and a *Datisca* homolog has been isolated by A. Berry (UC, Davis) and K. Pawlowski. This paper describes experiences encountered during the course of *in vitro* expression of AgNOD-GHRP in *E. coli* and documents its toxicity in this expression system.

*Construction of the vector expressing the MBP-AgNOD-GHRP fusion protein.* The protein-coding sequence of pAgNt84 was subcloned into the pMAL-c2 expression vector to obtain a fusion protein in which the protein of interest was linked to the C-terminal region of MBP (Maltose-Binding Protein). The fragment to be subcloned was amplified from pAgNt84 using primers designed to amplify a 410-bp PCR fragment from pAgNt84 which codes for the 99 amino acids of AgNOD-GHRP and has a unique *Hind*III site at its 3' end, downstream of the termination codon. The recombinant pMAL-c2 plasmid was transformed into DH5α competent cells (GibcoBRL), and transformants were selected on carbenicillin/X-gal (5-bromo-4-chloro-3-indolyl-ß-D-galactoside) plates.

*Analysis of transformants.* Analysis of DH5α-derived clones transformed with ligation mixtures, which were selected on carbenicillin plates, showed that about 35% of the clones examined produced white colonies on carbenicillin/X-gal/IPTG plates and thus had ß-galactosidase α-fragment activity of the vector inactivated by inserts. At the same time, about 50% of the clones did not grow or grew very poorly on carbenicillin/X-gal/IPTG plates so that their LacZ phenotype could not be determined. Since IPTG induces protein expression from the $P_{tac}$ promoter of the pMAL-c2 vector, this suggested to us that expression of the MBP-AgNOD-GHRP fusion protein might be toxic to these latter clones.

Sequence analyses of 5 of the "white" clones, whose plasmids, based on restriction endonuclease digestions, contained inserts of the correct size showed that all of them had deletions at the start of the pAgNt84 coding sequence, with the first 1, 2, 4 or 13 nucleotides missing. No other nucleotide changes were detected either in the entire insert sequence or in those vector sequences that were determined 5' of the *Xmn*I and 3' of the *Hind*III cloning sites. All of the deletions found resulted in the pAgNt84 coding sequence being out of the *malE* gene reading frame. This selection of clones with frame shift mutations further supported the suggestion that expression of AgNOD-GHRP is lethal to the *E. coli* cells.

Due to the occurrence of only mutant clones among "white" transformants and the specific targeting in those of the 5' end of the cloned fragment as shown above, we looked for an easier tool than sequencing for rapid screening of the IPTG-sensitive recombinant clones. We found by computer search among 406 restriction enzymes available in the restriction enzyme database REBASE (version 502) that one enzyme, *Fok*I, had a recognition site (5' GGATG 3') which turned out to be useful in screening clones. This restriction endonuclease has no cleavage sites within the cloned fragment of pAgNt84 but does have a cleavage site at the fusion juncture. Any mutation involving the ATG start codon of the cloned fragment would eliminate this *Fok*I cleavage site in the recombinant plasmids, which would allow us to distinguish between mutant clones and those that have the correct AgNOD-GHRP reading frame.

Using this approach, we examined 22 IPTG-sensitive clones and were able to show that 11 of them might be expected to have the correct AgNOD-GHRP reading frame. Sequence analysis of two of these clones showed that one clone, designated 1-13, did have the correct insert sequence. The other clone, designated 1-1, had two nucleotide substitutions within the insert sequence (both A for G in the positions 256 and 343 of pAgNt84), i.e. outside of the 5' end of the cloned fragment. The substitutions are both in the 3rd positions of the codons, resulting in GGG instead of GGA (both coding for Gly) and ACG instead of ACA (both coding for Thr), respectively, and thus do not potentially change the amino acid sequence of the protein encoded. To test whether the toxic effect of protein expression is due to the membrane-active signal peptide of AgNOD-GHRP and if deleting the signal sequence may result in production of a non-toxic protein, we looked for a recombinant clone which had a larger deletion at the start of the insert sequence than those described above. A clone, designated III-41, was shown by sequence analysis to have the first 48 base pairs missing from the cloned pAgNt84 coding sequence. Because this deletion does not change the reading frame of the sequence, we could expect to obtain a fusion protein with the first 16 amino acid residues of the AgNOD-GHRP signal peptide missing.

*Expression and analysis of the fusion proteins.* Proteins of the expected sizes for chimeric proteins were observed in cultures of clones 1-1 and 1-13 (about 53.7 KDa) as well as clone III-41 (about 51.9 KDa), after IPTG induction. A second expressed protein whose mobility corresponded to MBP (about 42.7 KDa) was also observed in clones 1-1 and 1-13, while only the fusion protein was expressed in clone III-41. Levels of expression of the fusion proteins in cultures of clones 1-1 and 1-13 were markedly lower as compared to that in the culture of clone III-41 at the same cell density. Immunoblotting using anti-MBP serum demonstrated strong reaction with both induced proteins expressed in clones 1-1 and 1-13, as well as with the only induced protein expressed in clone III-41. The high levels of expression of the 51.9-KDa fusion protein in clone III-41 resulted in protein yields of at least 100 mg of fusion protein purified from the cell soluble fraction on an amylose resin column per liter of culture. Cleavage of the fusion protein with the protease factor Xa to generate the truncated AgNOD-GHRP protein with no MBP-derived amino acids involved production of an intermediate peptide (about 15.1 KDa) which was further cut with the enzyme to give a protein band of the expected size (about 9.2 KDa). The identity of the 9.2-KDa peptide with truncated AgNOD-GHRP was confirmed by protein microsequencing.

*Toxicity of AgNOD-GHRP to E. coli.* The toxicity of the full-length fusion protein was demonstrated using two approaches, one involved a determination of plating efficiencies and the other involved measuring the effect of protein expression on bacterial growth. To determine plating efficiencies, isolated colonies of recombinant clones 1-13 and III-41 grown on LB containing carbenicillin (100 μg/ml) were grown overnight in the same liquid medium at 37°C with vigorous shaking. 100 μl of serial dilutions were titered on four LB plates which had carbenicillin (100 μg/ml), IPTG (1 mM), both, or neither added to the medium. The colonies were grown at 37°C overnight and the ratios of the colony counts after growth on the plates were calculated.

Both clones had the same plating efficiencies on LB and LB+carbenicillin, thus indicating that the plasmids conferring resistance to the antibiotic are retained by the majority of the cells in the populations. Clone III-41 showed the same plating efficiencies on media supplemented with IPTG, however, the plating efficiencies of clone 1-13 were drastically decreased when protein expression was induced with IPTG. Only about 0.004% of the cells could grow on the plates supplemented with IPTG and no or very few colonies (<0.00015%) grew on plates supplemented with both IPTG and carbenicillin.

In absolute agreement with the above data are growth curves determined by measuring absorbance at 600 nm of liquid cultures of the truncated clone, III-41, and two full-length clones, 1-1 and 1-13, growing with or without IPTG (Fig. 2). In the two latter clones, induction of protein expression results in complete cessation of growth between 2 and 3 hours after adding IPTG. On the contrary, IPTG has little effect on the culture of clone III-41, which continues to grow at about the same rate as all three non-induced cultures. This confirms that deletion of a part of the AgNOD-GHRP signal peptide results in non-toxicity or low toxicity of the protein obtained.

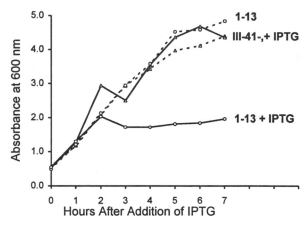

Figure 2. Growth curves for clones III-41 and 1-13 under inducing (+IPTG, —) and non-inducing (---) conditions. All clones were grown in LB containing carbenicillin (100 μg/ml) at 37°C, 300 rpm, to $OD_{600}$ ≈ 0.5 and then divided each into two aliquots (zero time cultures). To one aliquot was added IPTG to 0.3 mM, the other aliquot was grown as a control without the addition of IPTG. 1-ml aliquots of each cultures were taken every hour following the addition of IPTG, and the growth of each culture was determined by measuring the absorbance at 600 nm. The growth curves for clone 1-1 (data not shown) are nearly identical to those for 1-13.

*Discussion.* To a large degree, studying membrane-active proteins is hampered by difficulties in overexpression in *E. coli*, which may be due to toxicity resulting from insertion of large quantities of protein into the membrane. pMAL vectors have been used commonly for expression of soluble proteins or soluble domains of membrane proteins or hydrophobic membrane proteins. As shown with the very hydrophobic, toxic, and unstable bacterio-opsin, it becomes very stable when fused to MBP (Chen and Gouaux, 1996). The authors believe that fusion of the hydrophilic MBP to a membrane protein can reduce its toxicity and improve stability. However, while expression of the bacterio-opsin-MBP fusion resulted in formation of insoluble inclusion bodies in DH5α cells, which may be an additional factor of decreasing the protein toxicity, the behavior of the MBP-AgNOD-GHRP fusion protein is totally different in the same *E. coli* strain. The MBP-AgNOD-GHRP fusion is expressed as a soluble protein in the cell cytoplasm and even if the toxicity of full-length AgNOD-GHRP is decreased, to some extent, by MBP, it is not completely abolished.

*Acknowledgments.* BCM acknowledges support from the Department of Botany and USDA NRICGP grant #95-37305-3086. The Chancellor of the University of Tennessee and the University Computing Center are acknowledged for providing central and/or distributed computing facilities and services. We would like to thank Charles Murphy of the University of Tennessee Memorial Research Center for protein sequence analysis.

**LITERATURE CITED**

Chen, G.Q., Gouaux, J.E. 1996. Overexpression of bacterio-opsin in *Escherichia coli* as a water-soluble fusion to maltose binding protein: Efficient regeneration of the fusion protein and selective cleavage with trypsin. Protein Sci. 5: 456-467.

Conklin, D.S., McMaster, J.A., Culbertson, M.R. and Kung, C. 1992. COT1, a gene involved in cobalt accumulation in *Saccharomyces cerevisiae*. Molecular and Cellular Biology 12(9), 3678-3688.

Goetting-Minesky, M.P. and Mullin, B.C. 1994. Differential gene expression in an actinorhizal symbiosis: Evidence for a nodule-specific cysteine proteinase. Proc. Natl. Acad. Sci. USA 91, 9891-9895.

Mullin, B.C., Goetting-Minesky, M.P. and Twigg, P.G. 1993. Differential gene expression in the development of actinorhizal nodules. New Horizons in Nitrogen Fixation. pp 309-314.

Ribeiro, A., Akkermans, A., van Kammen, A., Bisseling, T. and Pawlowski, K. 1995. A nodule-specific gene encoding a subtilisin-like protease is expressed in early stages of actinorhizal nodule development. Plant Cell 7, 785-794.

Twigg, P.G. 1993. Isolation of a nodule-specific cDNA encoding a putative glycine-rich protein from *Alnus glutinosa*. PhD Dissertation (University of Tennessee, Knoxville, TN).

Wülfing, C., Lombardero, J. and Plückthun, A. 1994. An *Escherichia coli* protein consisting of a domain homologous to FK506-binding proteins (FKBP) and a new metal binding motif. J. Biol. Chem. 269(4), 2895-2901.

# Emerging areas and future prospects in the field of plant-microbe interactions

Andrzej B. Legocki

Institute of Bioorganic Chemistry, Poznan, Poland

Plant-microbe interactions represent a large field covering diverse research priorities on the genetics, physiology and molecular biology of bacteria, fungi and plants. Below are arbitrarily chosen examples of those research problems and questions which have drawn the attention of many laboratories for the recent several years. Although most of them represent well defined topics we are still missing important details and sometimes we cannot answer relatively simple questions.

## Microorganism competition and persistence

In natural soil conditions the occurence of effective interactions is usually attributed to the competition between microorganisms. This phenomenon depends on several physical factors such as uniform distribution of interacting bacteria and numerical superiority; it also reflects biological properties of microbial partners, their specific preferences or restrictions (Triplett, Sadowsky 1992). In the studies on symbiotic systems it has been observed that the competition for effective interaction with plants is most severe among genetically related populations of microorganisms. Contributions from related fields will be required to recognize habitat preferences and persistence factors for microorganisms as well as microbial restriction specifities.

## Signalling in plant-microbe interactions

The major breakthrough in the field of plant-microbe interactions in recent years has been connected with the identification of signal molecules, their generation, perception and transduction as well as with characterization of genes involved in various responses to signal molecules. The sessile nature of plants provides means for short-term responses to environmental alteration which subsequently mould long term acclimation to given conditions. The fact that several different types of microbial signals have already been characterized enables systematic studies on how these signals are perceived by plants and how subsequent signal transduction occurs within plant cells.

*Rhizobium* - legume symbiosis provides a unique model for studying plant morphogenesis induced by microbial signals. After preinfection in which flavonoids serve a role of chemoattractants inducing *Rhizobium* genes, at least two different signals are elicited in the system: one from *Rhizobium* (Nod factor) and the other generated by plant tissue in response to Nod factor. In the symbiotic system the host plant facilitates microbial invasion instead of fighting off the invader. Despite the progress in this area, one should be aware that there must be many still unrecognized microbe-generated signals to which plants respond. This especially applies to those types of interactions which are not very well characterized.

Although the structural arrangement of this bacterium-derived host-specific morphogen has already been recognized we still do not know which secondary signals occur during nodule development and nodule colonization. Since intracellular signalling between *Rhizobium* and host legume is likely not to stop after Nod factors it is important to identify the nature of such signalling and to recognize the role of phytohormones at further stages of nodule formation and development (Hirsch 1992). The role of the secondary signals for morphogenesis after the recognition of bacterial molecule may be played by phytohormones which function as endogenous morphogenetic regulators. An important question which draws the attention of several laboratories is the intra-and inter-species specificity of signalling. Compounds exuded from plant roots have multiple effects on soil microorganisms. These compounds may function directly or indirectly through metabolised products. Moreover, it has been recently evidenced that lipo-oligosaccharides may play a role as plant regulators also for nonleguminous plants.

Well defined but still unsufficiently recognized is the classical problem of signalling events that are crucial for distinguishing between beneficial and detrimental interactions (Vance 1983). As it was already shown, the initial infection leads to abortion with the symptoms of a protective hypersensitive response. One of the important questions which might distinguish between parasite and symbiotic/associative types of microbial interactions with the host plant is the extent of integration between plant and bacterial metabolism.

An iteresting finding about the mechanism of the plant-pathogen relationship has been recently revealed from the studies on interaction between *Arabidopsis thaliana* and various strains of *Pseudomonas syringe* (Reuber, Ausubel 1996; Ritter, Dangl 1996). It was shown that the products of two plant resistance genes may compete for a common element(s) in the signal transduction pathway leading to disease resistance. These studies might explain how the plant resistance genes are controlling pathogen recognition and activation of plant resistance genes. They also demonstrate the interference of a gene-for-gene interaction with another interaction showing the complexity of resistance gene signalling.

## PR proteins are differentially expressed during symbiosis

A defense system of legume plants is capable of distinguishing pathogenic bacteria from symbiotic ones and in the case of pathogens it is able to activate defense responses. These reactions may disturb microbial metabolism and limit the growth of the microorganism. They include the changes of cell degrading enzymes, the level of extracellular polysaccharides and other metabolites. Under extreme circumstances, they may even lead to host cell death. On the other other hand, specific recognition between leguminous plants and symbiotic bacteria evolved the mechanisms which can switch off the defense system in order to start symbiotic association in the proccesses of nodule morphogenesis and nitrogen fixation. Moreover, leguminous plants can utilize elements of their defense system to control symbiosis and developmentally regulated senesence of the symbiotic organ - root nodule.

Upon inoculation with rhizobia, legume plants express a group of symbiosis-specific proteins (nodulins). A comparison of the expression pattern of plant products that are characteristic of presymbiotic and symbiotic stages indicates that the initiation of symbiosis might be coupled with reduced synthesis of certain root proteins. Recently, following the course of nodule development in yellow lupin, Sikorski et al. observed that the expression of the pair of 18 kD acidic proteins was downregulated after the plant infection with *B.lupini* (Sikorski et al. 1996). The expression of these proteins was significantly reduced in root nodules while at the same time the expression of leghemoglobins was induced. Sequence analysis of these proteins and their cDNA clones revealed that they were highly homologous to pathogenesis-related proteins classified as proteins of PR10 class.

The function of PR10 proteins in the plant defense mechanism is not known. As it was shown by the Northern blot hybridization analysis, they are constitutively expressed in the roots of uninfected plants. Their downregulated expression during the symbiosis might indicate that they play a role in the effective symbiosis which is connected with the adjustement of the defense response in mature nitrogen-fixing nodules. This suggests that certain elements of the molecular recognition mechanism might be common to both pathogenic and symbiotic types of interactions.


## Coevolution of plant-microbe interactions

The symbiotic relationship between *Rhizobium* and legume host is perhaps the most effective strategy for plants to overcome biotic and abiotic stresses. It has been postulated that two separate nodulation events occurred during the evolution of legumes (Sprent 1994). One involved an ancestor of *Rhizobium* and concerned root infection. This interaction which was initially parasitic, turned out to be beneficial after the bacteria were released from infection threads and formation of new organ - root nodule was initiated. The other event involved a photosynthetic ancestor of *Bradyrhizobium* and has never evolved infection threads.

It is believed that genetic variation for plant resistance to pathogens is subject to selection by these pathogens. They could serve as selective agents to direct the evolution of plant morphological features or the composition of secondary metabolites in plant cells. There are two major strategies of coevolution between plants and pathogens: pairwise or diffuse coevolution (Janzen 1980). Although the available data are still very limited, coevolution between plants and pathogens seems to proceed as a diffused process rather than pairwise (Rausher 1996).

ACKNOWLEDGEMENTS

This paper was written within the grant No 6P 204 056 06 of Polish State Committee of Scientific Research.

REFERENCES:

Hirsch, A.M. 1992. Development biology of legume nodulation. New Phytol. 122:211-237.
Janzen, D.H. 1980. When is it coevolution? Evolution·34:611-612.
Mellor, R.B., Collinge, D.B. 1995. A simple model based on known plant defense reactions is sufficient to explain most aspects of nodulation. J. Experiment. Bot. 282:1-18.
Rausher, M.D. 1996 Genetic analysis of coevolution between plants and their natural enenemies. Trends in Genetics 12:212-217.
Reuber, T.L., Ausubel, F.M. 1996. Isolation of *Arabidopsis* genes that differentiate between resistance responses mediated by the RPS2 and RPM1 disease resistance genes. Plant Cell 8:241-249.
Ritter, C., Dangl, J.L. 1996. Interference between two specific pathogen recognition events mediated by distinct plant disease resistance genes. Plant Cell 8:251-257.
Sikorski, M.M., Szlagowska, A.E., Legocki, A.B. 1996. cDNA sequences encoding for two homologues of *Lupinus luteus* IPR-like proteins (accession Nos. X79974 and X79975 for LIR18A and LIR18B respectively) (PGR 95-114) Plant Physiol. 110:335.
Sprent, J. 1994. Evolution and diversity in the legume-rhizobium symbiosis: chaos theory? Plant and Soil. 161:1-10.
Triplett, E.W., Sadowsky, M.J. 1992. Genetics of competition for nodulation, Annu. Rev. Microbiol. 46:399-428
Vance, C.P. 1983. *Rhizobium* infection and nodulation: a beneficial plant disease. Ann. Rev. Microbiol. 37:399-424.

# Molecular Basis of Rhizosphere Colonization by *Pseudomonas* Bacteria

Ben Lugtenberg[1], Arjan van der Bij[1], Guido Bloemberg[1], Thomas Chin A Woeng[1], Linda Dekkers[1], Lev Kravchenko[2], Ine Mulders[1], Claartje Phoelich[1], Marco Simons[1], Herman Spaink[1], Igor Tikhonovich[2], Letty de Weger[1] and Carel Wijffelman[1]

[1]Leiden University, Institute of Plant Molecular Biology, Clusius Laboratory, 2333 AL  Leiden, The Netherlands
[2]Russian Institute of Agricultural Microbiology, St. Petersburg, Pushkin 6,  189620 Russia

## INTRODUCTION

Soil bacteria have an enormous impact on plant growth. In a study in which gnotobiotic wheat plants were inoculated with 150 single fluorescent *Pseudomonas* isolates, it appeared that 40% stimulated plant growth, 40% inhibited plant growth, and 20% had no effect. Bacteria can be used as inoculants for biofertilization and bioremediation, and as biopesticides and phytostimulants. In our study we use mostly *P.fluorescens* strain WCS365, an excellently root-colonizing strain which acts as a biopesticide in a cucumber-*Pythium aphanidermatum* system (J. Postma, pers. comm.). The mechanism behind its biocontrol property has not been identified. Other strains are *P.putida* strain WCS358 and *P.fluorescens* strain WCS374. The latter strains have biocontrol properties to which siderophore production contributes. Strain WCS374 in addition causes induced resistance (Schippers et al., 1995).

Bacterial control of plant diseases caused by fungi is based on (i) the production of one or more anti-fungal factors (AFFs), and (ii) colonization as the system which delivers the AFFs at the right time and sites on the root system (Tomashow and Weller, 1995). Colonization often is the limiting step in biocontrol (Schippers et al., 1987; Weller, 1988). Despite its importance, the molecular basis of colonization is not understood. Therefore we initiated a study on this topic. We followed two approaches. Firstly, we guessed which bacterial traits could be involved in colonization. We isolated mutants in these traits and tested their colonizing ability in comparison with that of the parental strain. Since this approach is limited by our imagination and our knowledge of bacterial physiology, we later followed a second approach in which individual random transposon mutants were screened on their ability to colonize the rhizosphere in competition with the parental strain. We used a gnotobiotic system for the latter approach. Rooted stem cuttings from potato or germinated seedlings from tomato, radish or wheat were inoculated and subsequently the root system was allowed to grow. Presence of the bacteria on the root tip was used as the criterium for colonization. The number of bacteria varied from $10^6$-$10^7$ cfu/cm root at the site just below the inoculation site to $10^2$-$10^4$ cfu/cm root near the root tip (van der Bij et al., *in preparation*). Scanning electron microscopical studies confirmed this picture and showed that the bacteria were not randomly distributed. Instead, a few days after inoculation micro-colonies were observed, especially in densely populated areas.

We propose that micro-colonies are important for the biocontrol properties of biocontrol strains since at these sites the concentrations of acyl-homoserine lactones, autoinducers which act as activators of transcription factors for the synthesis of many AFFs, can be expected to be sufficiently high for optimal AFF production (Chin-A-Woeng et al., *submitted*). Production of autoinducers in micro-colonies would also explain the unexpectedly high frequency of conjugation in the rhizosphere (van Elsas et al., 1988) since autoinducers were shown to play a role in conjugation (Zhang et al., 1993).

## COLONIZATION TRAITS IDENTIFIED BY PREDICTION

**Motility** mutants of strains WCS358 and WCS374 are impaired in colonization of the deeper root parts of potato when tested in a clay soil after inoculation of rooted stem cuttings. The mutants are normal in all other tested respects such as growth rate and LPS patterns (de Weger et al., 1987). Motility mutants also appeared to be defective in root colonization of tomato. Moreover, the random screening procedure for colonization mutants of strain WCS365 (see later on) yielded also motility mutants. Literature about the role of motility in colonization is not unequivocal. Therefore motility may not play a role under all conditions. We believe that chemotaxis towards one or more exudate components, rather than motility, is required for colonization.

The **O-antigen of lipopolysaccharide** (LPS) of strains WCS358 and WCS374 was also shown to be required for colonization. Colonization of potato roots was normal at the site of inoculation, but the mutants were apparently not able to colonize the deeper root parts (de Weger et al., 1989). Similar results were found on tomato. Moreover, among the *col* mutants of strain WCS365 identified by screening in the gnotobiotic system, several isolates appeared to lack the O-antigen. It is not clear why O-antigen mutants, which are motile, are defective in colonization. Although some of these mutants have a slower growth rate (see later on under **Mutant screening**), other O-antigen mutants were found to have a normal growth rate (de Weger et al., 1989). The possibility that motility- and/or O-antigen mutants are colonization-defective because they are supersensitive to toxic substances produced by the root can practically be excluded since normal levels of both types of mutants were found at the root base.

**Prototrophy for amino acids and vitamin B1** appeared to be essential for colonization of potato and tomato roots. Mutants in these traits were also defective in colonizing the lower root parts. Supplementation of the system with the appropriate amino acid restored colonization, usually to wild type levels.

As expected, a **high growth rate** is important for colonization. Putative colonization mutants resulting from the screening procedure were tested for growth in King's B medium in competition with the parental strain WCS365 for many generations. The growth rate of several of these mutants was lower than that of the wild type, suggesting that the defect was in household genes rather than in colonization. The nucleotide sequences of DNA regions surrounding the transposon insertions from one of the slower growing mutants, strain PCL1268, was analyzed. The insertion is located in a gene which is strongly homologous to *secB* of *E.coli*, a gene involved in protein folding and secretion. The growth rate of the mutant is 3 percent lower (50% less cells over 30 generations) than that of the parental strain WCS365. Assuming that the same growth defect occurs in the rhizosphere, it is uncertain whether this defect is sufficient to explain the 100- to 1000-fold decreased ability to reach the root tip in 7 days at 19°C. Also the fact that the mutation is located in *secB*, assuming that this is the reason for the colonization defect, does not answer the question whether this is a household gene or a real colonization gene.

**Growth on major seed and root exudate compounds** is assumed to play a major role in colonization. We have identified the major tomato exudate sugars (fructose, glucose, maltose) and organic acids (succinic, malic, citric, glycolic, fumaric and oxalic acid) and found four times more of the latter than of the former compounds (on weight basis). When succinate and glucose are offered as a mixture of C-sources, strain WCS365 appears to utilize the organic acid first. Mutants defective in the utilization of sugars appeared to colonize the tomato root tip to equal numbers as the parental strain WCS365. In contrast, mutants defective in the utilization of organic acids were not recovered from the root tip. These results indicate that the utilization of organic acids is a major colonization trait (Simons et al., *in preparation*). Moreover, this result is likely to provide the molecular basis of the "rhizosphere effect" of this strain, i.e. the ability of some strains to be present in much higher numbers in the rhizosphere than in the soil (Hiltner, 1904).

## COLONIZATION TRAITS IDENTIFIED AFTER MUTANT SCREENING

*Mutant screening.* Two mutant banks of *P. fluorescens* strain WCS365 were constructed, one using Tn5 and the other one using Tn5*lacZ* (constitutive). Assuming that most mutants show wild type behaviour with respect to colonization, two mutants, one out of each group, were tested against each other after inoculating four germinated tomato seedlings (cv. Carmello) in a 1:1 mixture ($10^7$ cfu/ml). The plants were allowed to grow for 7 days at 19°C in a gnotobiotic quartz sand system containing plant nutrient solution without added carbon source. The plants were removed from the sand. The roots, approximately 8 to 10 cm in length, contain a monolayer of sand grains. The number of bacteria on the root was 100-fold higher than on the sand grains. For mutant screening, one to two cm of the root tip was removed, vortexed, and the numbers of white (Tn5) and blue (Tn5*lacZ*) colonies were counted on KB medium supplemented with X-gal as described previously by Lam et al. (1990). Putative mutants were retested on ten plants. In a screening of 1300 mutants 141 mutants then remained which were subsequently tested on defects in the known colonization traits motility (15 mutants, 1.2%), prototrophy (13 mutants, 1.0%), O-antigen of LPS (6 mutants, 0.5%), growth rate in KB medium in competition with the wild type or with the Tn5*lacZ*-marked wild type (11 mutants, 0.9%, of which all 6 O-antigen mutants), cell envelope protein patterns (0 mutants) and Biolog patterns (utilization of 95 C-sources; 0 mutants). The results of the latter two tests also showed that the mutants were indeed WCS365 derivatives. The ten remaining mutants were tested statistically in triplicate for colonization on ten tomato plants. Together with three other mutants isolated during previous mutant hunts the procedure has yielded thirteen mutants impaired in novel colonization traits.

We plan to test these mutants, (i) for growth rate in exudate (in competition), (ii) for colonization behaviour in soil, and (iii) for host range of colonization (tomato, radish and wheat). Finally we want to identify the colonization genes, their regulation and the colonization traits they are involved in.

*Host range.* So far all wild type *col* genes seem to have a broad host range since the mutants behave as mutants on all three test plants, tomato, radish and wheat. In addition, strains PCL1210, PCL1233 and PCL1268 were isolated as colonization-defective mutants on potato.

*Genetic analysis.* The DNA-regions corresponding with three *col* genes have been analyzed in some detail so far.

In **mutant strain PCL1210** the insertion is present in a gene which is homologous with sensor kinases. Immediately upstream is a gene homologous with response regulators. Downstream in the same operon is an open reading frame whose significance is somewhat doubtful since its codon usage is not typical for *Pseudomonas*. The first two genes have been named *colS* and *colR* since they resemble genes of so called two-component systems which encode a sensor kinase and a response regulator. Since two-component systems have been implicated in the colonization traits chemotaxis and uptake of organic acids, we have carefully analyzed the ability of mutant PCL1210 to show chemotaxis towards the identified major tomato exudate components mentioned previously as well as for growth on these components as carbon sources as well as on exudate. In all these respects the mutant behaves like its parent. We conclude that the ColS/ColR system is involved in the perception and transduction of an unknown stimulus and that this results in the activation of a novel, so far also unknown, colonization trait. The nucleotide sequence of the promoter region indicates the presence of an IHF (integration host factor) binding site which overlaps with the putative -35 sequence (Dekkers et al,. *in preparation*).

In **mutant strain PCL1233** the insertion is located in *orf235* of the operon promoter-*llpA-lysA-dapF-orf235-xerC-orf238*, a gene with unknown function which shares homology with genes present in *E.coli* and *P.aeruginosa*. The gene immediately downstream is homologous to *xerC* from *E.coli* and to *sss* from *P.aeruginosa*. These genes encode site-specific recombinases belonging to the lambda integrase gene family which have been implicated in monomerization of plasmids, chromosome segregation and production of pyoverdin under the influence of the heavy metal ions $Zn^{2+}$ and $Cd^{2+}$ (Höfte et al., 1994). Moreover, they are homologous to *fimE* and *fimB*, whose gene products cause genetic inversion of DNA fragments involved in regulation and synthesis of fimbriae. DNA rearrangements have also shown to occur in the regulation of expression of flagella, in the *pheN* sensor kinase involved in the regulation of pathogenesis of *P.tolaasii*, in development of *Myxococcus xanthus*, in alginate production by cystic fibrosis-causing *P.aeruginosa*, and in antigenic variation related to LPS structure of the intracellular pathogen *Francisella tularensis*. Such rearrangements are presumed to enhance the ability of the organism to survive under different environmental conditions (Dybvig, 1993). If the *xerC*-like gene appears to be involved in colonization, it is conceivable that a certain genetic organization, which cannot be formed in the mutant, is required for optimal colonization.

A clone containing the promoter, *lppL, lysA, dapF, orf235* and *xerC* complements the mutation for colonization. Therefore *orf235* and/or *xerC* must be the gene(s) essential for colonization. This clone also improves the colonizing ability of the good-colonizing *P.fluorescens* biocontrol strain F113 ten-fold, suggesting the possibility of improving the biocontrol potential of this strain by the introduction of colonization genes. Also the putative promoter sequence of this second colonization operon contains a potential IHF binding site (Dekkers et al., *in preparation*). In these complementation studies we use plasmids that have shown to be stable in the rhizosphere (van der Bij et al., 1996).

The insertion of **mutant strain PCL1201** is present in a gene homologous with gene #4 of the 14-genes-containing *nuo* operon of *E.coli* NADH: ubiquinone oxidoreductase, an enormous enzyme complex involved in the generation of proton motive force (Weidner et al., 1993). *E.coli* contains two activities of this enzyme, which might explain why mutant PCL1201 grows perfectly well in KB medium. Apparently growth conditions in the rhizosphere under our test conditions are such that the activity of the oxidoreductase which is impaired in the mutant, is absolutely required for colonization. The regulation of this *col* gene will be subject of future research.

It appears that the analyses of these colonization mutants not only helps us to unravel the traits involved in colonization *perse* but also teaches us about the molecular basis of bacterial life in the rhizosphere.

ACKNOWLEDGMENTS

The work was partly supported by contracts BIO2-CT930053 (IMPACT) and BIO2-CT930196 (colonization) with the European Union and by grants form the Priority Programme of Leiden University and from the Dutch NWO-LNV Crop Protection Programme.

LITERATURE CITED

van der Bij, A.J., de Weger, L.A., Tucker, T.T. and Lugtenberg, B.J.J. 1995. Plasmid stability in *Pseudomonas fluorescens* in the rhizosphere. Appl. Environm. Microbiol. 62: 1076-1080.

Dybvig, K. 1993. DNA rearrangements and phenotypic switching in prokaryotes. Molecular Microbiology 10: 465-471.

van Elsas, J.D., Trevors, J.T. and Starodub, M.E. 1988. Bacterial conjugation between pseudomonads in the rhizosphere of wheat. FEMS Microbiol. Ecol. 53: 299-306.

Hiltner, L. 1904. Über neuere Erfahrungen und Probleme auf dem Gebiete der Bodenbakteriologie unter beßonderer Berücksichtigung der Gründüngung und Brache. Arbeiten der DLG, pages 59-78.

Höfte, M., Dong, Q., Kourambas, S., Krishnapillai, V., Sherratt, D., and Mergeay, M. 1993. The *sss* gene product, which affect pyovordin production in *Pseudomonas aeruginosa* 7NSK2, Is a site-specific recombinase. Molecular Microbiology 14: 1011-1020.

Lam, S.T., Ellis, D.M. and Lignon, J.M. 1990. Genetic approaches for studying rhizosphere colonization. Plant and Soil 129: 11-18.

Schippers, B., Lugtenberg, B.J.J., and Weisbeek, P.J. 1987. Plant growth control by fluorescent pseudomonads. Pages 19-39 in: Innovative approaches to plant disease control. I. Chet ed., Wiley, New York.

Schippers, B., Scheffer, R.J., Lugtenberg, B.J.J., and Weisbeek, P.J. 1995. Biocoating of seeds with plant growth-promoting rhizobacteria to improve plant establishment. Outlook on Agriculture 24: 179-185.

Thomashow, L.S. and Weller, D.M. 1995. Current concepts in the use of introduced bacteria for biological disease control: mechanisms and antifungal metabolites. Pages 187-235 in: Plant Microbe Interactions, Vol. 1, G. Stacey and N. Keen (eds), Chapman and Hall, New York.

de Weger, L.A., van der Vlugt, C.I.M., Wijfjes, A.H.M., Bakker, P.A.H.M., Schippers, B., and Lugtenberg, B.J.J. 1987. Flagella of a plant-growth-stimulating *Pseudomonas fluorescens* strain are required for colonization of potato roots. J. Bacteriol. 169: 2769-2773.

de Weger, L.A., Bakker, P.A.H.M., Schippers, B., van Loosdrecht, M.C.M., and Lugtenberg, B.J.J. 1989. *Pseudomonas* spp. with mutational changes in the O-antigenic side chain of their lipopolysaccharide are affected in their ability to colonize potato roots. Pages 197-202 in: Signal Molecules in plants and plant-microbe interactions. B.J.J. Lugtenberg, ed. NATO ASI series vol. H36, Springer Verlag, Heidelberg.

Weidner, U., Geier, S., Ptock, A., Friedrich, T., Leif, H. and Weiss, H. 1993. The gene locus of the proton-translocating NADH: ubiquinone oxidoreductase in *Escherichia coli*. Organisation of the 14 genes and relationship between the derived proteins and subunits of mitochondrial complex I. J. Mol. Biol. 233: 109-122.

Weller, D.M. 1988. Biological control of soilborne plant pathogens in the rhizosphere with bacteria. Annu. Rev. Phytopathol. 26: 379-407.

Zhang, L., Murphy, P.J., Kerr, A. and Tate, M.E. 1993. *Agrobacterium* conjugation and gene regulation by N-acyl-l-homoserine lactones. Nature 362: 446-448.

# The Biotechnology and Application of *Pseudomonas* Inoculants for the Biocontrol of Phytopathogens

Colum Dunne, Isabel Delaney, Anne Fenton, Scott Lohrke, Yvan Moënne-Loccoz and Fergal O'Gara*

Department of Microbiology, University College, Cork, Ireland

## Benefits of Biological Control

The negative interactions among microbial populations and between microbes and higher organisms form the basis for microbial based biological control of pests and pathogens (2,26). The use of biological inoculants for crop protection is compatible with the ideal of producing quality health promoting agricultural goods in an environmentally friendly manner. Fungicides traditionally used for the control of soil borne pathogens, such as metalaxyl or benomyl, are now perceived to be environmentally detrimental. These products have, in many cases, also lost their fungicidal abilities due to the selection of resistant pathogenic strains. Such fungicide treatments may also be toxic to key microorganisms that underpin soil fertility mechanisms (e.g. nitrogen fixing Rhizobium spp.).

An alternative chemical approach to control soil pathogens is fumigation of soil with agents such as methyl bromide (MeBr) or metam sodium. These fumigants are quite toxic, and may be lethal to microbes and many soil insects. Furthermore, metam sodium is dangerous to transport and use, and methyl bromide is harmful to the Earth's ozone layer and its use has been restricted by an international treaty (USEPA 1993). In addition, vegetables and small fruit crops grown on soil fumigated with MeBr often contain much higher inorganic bromide residues than the legal limit set by several countries. In an effort to solve these problems the EU commission has decided to reduce the production and use of MeBr to 1991 levels and in 1998 the levels of production and use must be 25% less than 1991. Biological control offers significant advantages over the use of chemical pesticides, but it is important to allay public and regulatory concerns about the commercial use of genetically modified biocontrol agents before these advantages can be fully achieved.

## Biotechnological Strategies for Biological Control

Disease suppressive soils are those in which the development of specific soil-borne plant diseases are impeded and they may provide a rich and valuable reserve of potential biocontrol agents. It is recognized that some pseudomonads have a positive role in pathogen inhibition in established suppressive soils and they have been demonstrated to be effective in protecting treated plants against fungal attack through production of a range of secondary metabolites (reviewed in 16). Antifungal compounds such as phenazines, pyrrol-type antibiotics, and 2,4- diacetylphloroglucinol (Phl) are produced by fluorescent *Pseudomonas* strains involved in plant protection and have been shown to mediate this protection (4,16,19,22,25). In addition, niche exclusion, competition for nutrients, production of iron scavenging siderophores, cyanide and lytic enzymes have all been implicated in plant protection by pseudomonad isolates (5,11,13,15,16). Extensive studies have resulted in the isolation and identification of a number of these secondary metabolites with the aim of cloning either biosynthetic or regulatory genes in order to improve or create novel biological control agents. Many, but not all, effective biocontrol strains produce Phl and hydrogen cyanide (HCN). Phl has been shown to be largely responsible for the prevention of "damping off" of sugar beet caused by the fungus *Phythium ultimum* (7, 20). However, both Phl and HCN are responsible for the control of black root rot of tobacco (caused by *Thelaviopsis basicola*) by *P. fluorescens* CHAO (10). In addition, the inhibition of phytopathogenic fungi in soil by lytic enzyme producing bacteria has proven to be a relatively successful biocontrol strategy. Chitinases and glucanases have a proven ability to degrade the chitin and glucan matrix integral to the structure of many fungal cell walls (13). Successful cloning of the biosynthetic genes responsible for the production of these enzymes has resulted in the development of transgenic microorganisms and plants with improved abilities to combat fungal disease due to the production and export of microbial enzymes.

### BIOCONTROL INOCULANTS MAY INDUCE SYSTEMIC ACQUIRED RESISTANCE (SAR)

Some *Pseudomonas* strains produce salicylic acid (SA) as a secondary metabolite. This is a precursor of pyochelin (1) and is a siderophore in its own right (24). SA is maximally produced under low iron conditions along with fluorescent siderophores which have a high affinity for iron. Indeed, such siderophores have been implicated in competition for iron between strains of *Pseudomonas* and as a possible mechanism of biological control. SA is also a plant hormone (18), and a metabolite implicated in the induction of systemic acquired resistance in plants (9). SA acts as a signal in the transduction chain during the induction of resistance by tobacco mosaic virus in hypersensitively reacting tobacco (23), and probably serves the same function in other plant species (6). Under those circumstances, the plant produces SA in response to an inducing pathogen. Concomitant with the appearance of local necrotic symptoms, resistance is induced in other plant parts not only to the inducing pathogen, but also to a wide variety of other pathogenic agents. This induced resistance, sometimes also referred to as "systemic acquired resistance" or "immunization", is long lived (12) and provides an enhanced level of protection against multiple diseases.

**SYNERGY OF BIOCONTROL MECHANISMS.**

Based on our current knowledge it would appear realistic that disease suppression can be enhanced when the activity of the pathogen is counteracted by more than one mechanism, e.g. production of anti-fungal metabolites and induced resistance. The rationale behind combining bacterial strains that, on the one hand, produce antifungal metabolites and, on the other hand, induce resistance in the plant, is that antifungal metabolites directly weaken the pathogen so that it will be less able to overcome the enhanced resistance in the plant. Evidence that combinations of strains are most effective stems from studies where the application of multiple bacterial strains improved control of take-all in wheat (17). However, the precise nature of the suppressing mechanisms remains to be elucidated. Finally, synergistic effects are critically dependent on compatibility between the strains as strains may antagonize each other through competition for nutrients or through antibiosis.

The widespread use of biocontrol agents requires a delivery system which will provide stability of the biocontrol agent during storage and effective delivery of the inoculant into the rhizosphere. Integrated biocontrol strategies must exploit developments in formulation and delivery systems in order to optimize the field efficacy of applied biocontrol agents.

## Enhanced biocontrol through genetic modification

**MODIFICATION OF ANTIBIOTIC PRODUCTION.**

The antifungal nature of secondary metabolites such as phenazines and phloroglucinols have made them prime targets for genetic manipulation. Alteration of production levels of such antifungal metabolites in pseudomonads has resulted in enhanced biocontrol ability. Genetically modified derivatives of *Pseudomonas fluorescens* strains with increased production of Phl and pyoluteorin (CHAO) and Phl (F113) have been obtained which exhibit enhanced antifungal activity *in vitro*. Treatment of cucumber with the improved CHAO strain resulted in increased protection against fungal attack (21).

In our laboratory, biosynthetic and regulatory genes for Phl have been identified and characterized from *Pseudomonas* strains. These characterized genes, however, have not been widely exploited in the construction of genetically modified strains. We have used a Phl biosynthetic locus and four regions, involved in the regulation of *Pseudomonas* secondary metabolism, in the construction of strains (i) with modified Phl production profiles and (ii) with the ability to produce secondary metabolites which have antimicrobial activity. A Phl biosynthetic locus, isolated from *Pseudomonas fluorescens* strain F113, was transferred into the wildtype background. The genetically modified strain was assessed for Phl production and significantly increased production levels observed (8). Further use of genetic modification for improved biocontrol has resulted in significantly enhanced antifungal abilities in a number of transgenic strains (3).

**LYTIC ENZYME PRODUCTION MEDIATES BIOCONTROL ABILITY IN SOME PSEUDOMONAD STRAINS**

A non-fluorescent pseudomonad strain, W81 (P), has been isolated from the sugarbeet rhizosphere on the basis of its biocontrol ability against "damping-off" caused by *Pythium ultimum* and due to the likelihood of it being preadapted to the sugarbeet rhizosphere (5). W81 (P) does not produce antifungal secondary metabolites typically found in pseudomonads with biocontrol ability. W81 (P) does, however, produce the lytic enzymes chitinase and protease. Mutagenesis with the transposable element Tn5-B50 allowed the isolation of a mutant deficient in extracellular enzyme production. Evaluation of wildtype W81 (P) and the enzyme deficient mutant strain under *in vitro* conditions on solid media and under *in vivo* microcosm conditions demonstrated that production of the extracellular lytic enzymes is required by W81 (P) for plant protection against "damping-off". Further genetic modification of W81 (P) has resulted in overproduction of the chitinase and protease enzymes. These mutants have been evaluated under *in vitro* conditions and increased fungal inhibition observed. The evaluation of the efficacy of the modified strain in the sugarbeet rhizosphere is in progress.

**IDENTIFICATION OF SIGNAL RESPONSIVE PROMOTERS IN BIOCONTROL STRAINS.**

The isolation of potential biocontrol strains from the environment in which they will be required to function may help to ensure better biocontrol efficacy. Similarly, the isolation and identification of bacterial promoters induced by signals present in the rhizosphere of crops may lead to the development of invaluable tools for the regulation of biocontrol traits in the soil environment. Recent studies in our laboratory using *Pseudomonas fluorescens* F113 have demonstrated that the strain responds to signals present in the exudate of sugarbeet seeds (unpublished data). The responsive regulatory region has been cloned from the bacterial genome and will be exploited to modify secondary metabolite production for greater control of metabolite production *in vivo*.

## Microbial inoculants in the Rhizosphere.

IMPACT OF *PSEUDOMONAS* INOCULANTS ON SOIL MICROFLORA.

The resident microbial population is an essential soil component and contributes to the fertility of soil and crop yield and health. Evaluation of the impact of biocontrol inoculants on key microflora is an essential aspect of biological control assessments. This may be evaluated through measurements carried out on a crop (sugarbeet) which serves as a "biosensor".

Field trials designed to evaluate the performance of microbial inoculants on sugarbeet were completed utilising the *P. fluorescens* biocontrol strain F113. Controls used in the study included a proprietary seed pelleting mix containing a commercial fungicide and another which was fungicide-free. There was no significant difference between F113 and the controls with regard to germination and plant parameters such as root yield, sugar content, and recoverable sugar. These results indicate that the inoculated *Pseudomonas* strain did not have any adverse effects on the indigenous microbial population as measured by plant performance.

A further evaluation of *Pseudomonas fluorescens* F113 was carried out at a field site to evaluate possible effects of F113 on the indigenous pseudomonad population. In addition to the evaluation of the biocontrol potential of the strain as measured by germination and plant yield, indigenous fluorescent pseudomonads were isolated from the rhizosphere and rhizoplane of inoculated and uninoculated plants. There were no significant differences detected in sugarbeet rhizosphere colonisation by total aerobic bacteria between the F113 treated and uninoculated plants.

In addition, rhizobial species were chosen as potentially useful indicator soil organisms to detect possible negative effects of the use of biocontrol inoculants. Red clover was used as a "biosensor" crop to detect perturbations in the indigenous population of *R. leguminosarum* bv. *trifolii*. Clover was sampled twice and several plant parameters examined including nodulation, crop yield and chemical analysis. No significant differences were detected with regard to yield parameters between plots inoculated with the *Pseudomonas fluorescens* strain F113 and those which received uninoculated controls. No obvious differences in nodulation were detected indicating that the use of the bacterial inoculant does not appear to adversely affect the native *Rhizobium* population as measured by plant yield, health and nodulation.

**TABLE I**

Bacteria currently used as agricultural inoculants in the field and new bacteria developed by genetic modification for improved performance

| Wild-type bacteria | Crop inoculated | Usage |
|---|---|---|
| *Anabaena-Azolla* | Rice | Biofertilizer |
| *Azospirillum brasilense* and *A. lipoferum* | Cereals | Biofertilizer |
| *Bradyrhizobium japonicum* | Forage and grain | Biofertilizer |
| *Rhizobium* spp. | legumes | Biofertilizer |
| *Frankia* spp. | Non-leguminous trees (*Alnus*) | Biofertilizer |
| *Bacillus subtilis* and *B. thuringensis* | Different Crops | Biopesticide |
| *Pseudomonas fluorescens* | Different Crops | Biopesticide |
| **Improved bacteria** | **Modified trait** | **Goal of modification** |
| *Agrobacterium radiobacter* | Deletion of transfer genes | Safe use for biological control of crowngall |
| *Clavibacter* sp. | Addition of endotoxin gene from *Bacillus thuringensis* | Biological control of insect damage to crops |
| *Bradyrhizobium japonicum* | Additional copies of *nif* gene | Increased nitrogen fixation |
| *Pseudomonas syringae* | Deletion of ice nucleation gene | Biological control of frost damage to crops |
| *Rhizobium meliloti* | Additional copies of *nif* and/or *dct* genes | Increased nitrogen fixation |

(From Dowling, D.N., O'Gara, F. and Nuti, M. P. 1995. Prokaryotes in agriculture. *In*: Biology of the Prokaryotes, vol. IX, Applied Microbiology, A. Pühler editor, in press).

## Acknowledgments

The authors would like to thank Pat Higgins for technical assistance and Jim Powell (Irish Sugar plc) for useful discussion. This work was supported in part by grants from the European Commission. Biotech BIO2-CT93-0196, BIO2-CT96-0053, BIO2-CT94-3001 and BIO-CT92-0084.

## References

1   Ankenbauer, R.J. and Cox, C.D. 1988. J. Bacteriol. 170:5364-5367.
2   Atlas, R.M. and Bartha, R. 1993. Microbial Ecology: Fundamentals and Applications. The Benjamin/Cummings Publishing Co., Inc.
3   Cook, R.J., Thomashow, L.S., Weller, D.M., Fujimoto, D., Mazzolo, M., Bangera, G. and Kim, D.S. 1995. PNAS. 92:4197-4201.
4   Dowling, D.N. and O'Gara, F. 1994. TIBTECH. 12 (4):133-141.
5   Dunne, C., Crowley, J., Moënne-Loccoz, Y., Dowling, D.N. and O'Gara, F. 1996 (Submitted).
6   Enyedi, A.J., Yalpani, N., Silverman, P. and Raskin, I. 1992. Cell 70:879-886.
7   Fenton, A.M., Stephens, P.M., Crowley, J., O'Callaghan, M. and O'Gara, F. 1992. Appl. Environ. Microbiol. 58:3873-3878.
8   Fenton, A.M., Delaney, I., Dowling, D.N. and O'Gara, F. 1996. (Submitted).
9   Gaffney, T., Friedrich, L., Vernooij, B., Negrotto, D., Nye, G., Uknes, S., Ward, E., Kessmann, H. and Ryals, J. 1993. Science 261:754-756.
10  Keel, C., Schnider, U., Maurhofer, M., Voisard, C., Laville, J., Burger, U., Wirthner, P., Haas, D. and Defago, G. 1992. MPMI 5:4-13.
11  Kobayashi, D.Y., Gugielmoni, M. and Clarke, B.B. 1995. Soil Biol. Biochem. 27 (11):1479-1487.
12  Kuc, J. 1982. Bioscience 32:854-860.
13  Lorito, M., Peterbauer, C., Hayes, C.K. and Harman, G.E. 1994. Microbiology 140:623-629.
14  O'Flaherty, S., Moënne-Loccoz, Y., Boesten, B., Higgins, P., Dowling, D.N., Condon, S. and O'Gara, F. 1995. Appl. Environ. Microbiol. 61:4051-4056.
15  O'Sullivan, D.J. and O'Gara, F. 1988. Appl. Environ. Microbiol. 54 (11):2877-2880.
16  O'Sullivan, D.J. and O'Gara, F. 1992. Microbiol. Rev. 56(4):662-676.
17  Pierson, E.A. and Weller, D.M. 1994. Phytopathology 84:940-947.
18  Raskin, I. 1992. Ann. Rev. Plant Mol. Biol. 43:439-463.
19  Scher, F.M. and Castagno, J.R. 1986. Can. J. Plt. Pathol. 8:222-224.
20  Shanahan, P., O'Sullivan, D.J., Simpson, P., Glennon, J.D,. and O'Gara, F. 1992. Appl. Environ. Microbiol. 58:353-358.

21 Schnider, U., Keel, C., Blumer, C., Troxler, J., Defago, G. and Haas, D. 1995. J. Bacteriol. 177:5387-5392.
22 Thomashow, L.S. and Weller, D.M. 1988. J. Bact. 170 (8):3499-3508.
23 Vernooij, B., Friedrich, L., Morse, A., Reist, R., Kolditz Jahwar, R. Ward, E., Uknes, S., Kessmann, H. and Ryals, J. 1994. Plant Cell 6:959-965.
24 Visca, P., Ciervo, A., Sanfilippo, V. and Orsi, N. 1993. J. Gen. Microbiol. 139:1995-2001.
25 Voisard, C., Bull, C.T., Keel, C., Laville, J., Maurhofer, M., Schnider, U., Defago, G. and Haas, D. 1994. In: O'Gara, F., Dowling, D.N and Boesten, B. (Eds.). Molecular ecology of rhizosphere microorganisms: Biotechnology and the release of GMOs. VCH.
26 Weller, D.M. 1988. Ann. Rev. Phytopathol. 26:379-407.

# Antifungal Metabolites Involved in Biological Control of Soilborne Plant Diseases by Rhizosphere Pseudomonads

Joyce E. Loper[1], Jennifer Kraus[2], Nathan Corbell[2], and Brian Nowak-Thompson[3]

[1]Horticultural Crops Research Laboratory, Agricultural Research Service, U.S.D.A.; [2]Department of Botany and Plant Pathology, and [3]Department of Biochemistry, Oregon State University, Corvallis, OR, USA.

Rhizosphere bacteria of diverse genera can suppress soilborne plant diseases, but the fluorescent pseudomonads have received the most attention from researchers evaluating potential biocontrol agents. This group of bacteria is abundant in the rhizosphere and many strains produce secondary metabolites that inhibit the growth of phytopathogenic fungi and bacteria (Thomashow and Weller 1995). A major contribution of molecular biology to the field of biological control has been in establishing the role of antifungal metabolite production by rhizosphere pseudomonads in disease suppression.

A common first step in evaluating the role of an antifungal metabolite in biocontrol is to compare the biological control activity of a mutant(s) deficient in antifungal metabolite production to that of a parental strain, which produces the antifungal metabolite. If a characterized mutation in a biosynthetic gene inactivates biocontrol activity of a bacterial strain, a role for antifungal metabolite production in disease suppression is inferred. For example, derivatives of *Pseudomonas fluorescens* strain Hv37a that are deficient in oomycin A production exhibit only 50% of the biocontrol activity of strain Hv37a against Pythium damping-off of cotton (Howie and Suslow 1991). Similarly, hydrogen cyanide (HCN), 2,4-diacetylphloroglucinol, phenazine-1-carboxylic acid, and pyrrolnitrin production contribute to the biocontrol activities of fluorescent pseudomonads (see Thomashow and Weller 1995 for a review). Antifungal metabolites generally contribute to, rather than account for all of, the biocontrol activity of parental strains. Residual levels of disease suppression exhibited by mutant strains are commonly attributed to induction of host resistance, production of other metabolites, or general nutrient competition. In contrast to the examples mentioned above, antifungal metabolite production does not always contribute detectably to the biocontrol activities of fluorescent pseudomonads. For example, pyoluteorin production contributes little to the suppression of pre-emergence damping-off of cucumber by *P. fluorescens* strain Pf-5 (Kraus and Loper 1992) although it contributes significantly to suppression of pre-emergence damping-off of cotton by the same bacterium (C. R. Howell, personal communication).

## Antifungal Metabolites Produced by *P. fluorescens* Pf-5

Strain Pf-5 produces pyoluteorin, pyrrolnitrin, 2,4-diacetylphloroglucinol, HCN, and a pyoverdine siderophore (Howell and Stipanovic 1980, Kraus and Loper 1992, Nowak-Thompson et al. 1994). The spectrum of antifungal metabolites produced by Pf-5 is remarkably similar to that of *P. fluorescens* CHA0, another rhizobacterium that suppresses soilborne plant diseases. Of the antifungal metabolites known to be produced by Pf-5 and CHA0, pyoluteorin is most toxic to *Pythium* spp. (Maurhofer et al. 1992). Pyoluteorin is active on seed surfaces in a soil environment (Howell and Stipanovic 1980) and accounts for the inhibition of *P. ultimum* by Pf-5 on certain culture media (Kraus and Loper 1992). To contribute to biocontrol, however, pyoluteorin must be produced by Pf-5 in the spermosphere in concentrations adequate to suppress *P. ultimum* before the fungus infects the seed.

### IN SITU TRANSCRIPTION OF PYOLUTEORIN BIOSYNTHESIS GENES BY PF-5 IN THE SPERMOSPHERE

Antifungal metabolite production by bacteria is affected profoundly by media composition and growth conditions in culture and presumably by the chemical composition and physical environment of microhabitats in the rhizosphere. The product of a reporter gene, if fused to a promoter of a biosynthetic operon, provides a reliable and convenient assessment of the expression of biosynthetic genes by bacteria inhabiting the rhizosphere. Transcriptional fusions, using $\beta$-galactosidase or ice-nucleation activities as reporter phenotypes, have enabled researchers to confirm the *in situ* expression of genes determining the biosynthesis of oomycin A (Howie and Suslow 1991) and phenazine-1 carboxylate (Georgakopoulos et al. 1994) in the spermosphere of cotton and wheat, respectively.

As described above, pyoluteorin production by *P. fluorescens* Pf-5 contributes differentially to suppression of pre-emergence damping-off on different plant hosts. Plt⁻ mutants of Pf-5 are as effective as the parental strain in suppression of pre-emergence damping-off of cucumber caused by *P. ultimum* (Kraus and Loper 1992), but the mutants are less effective than Pf-5 in protection of cotton seeds from Pythium damping-off (C. R. Howell, personal communication). Similarly, Plt⁻ mutants of *P. fluorescens* strain CHA0 are fully effective in protection of cucumber from Pythium damping-off but are less effective in protection of cress (Maurhofer et al. 1994b). Plant host could influence the population size or pyoluteorin production of *P. fluorescens* in the spermosphere and rhizosphere due to different composition and quantities of seed and root exudates. In support of this concept, CHA0 produces greater concentrations of pyoluteorin on cress than on cucumber, as determined by a bioassay based on the phytotoxicity of the metabolite (Maurhofer et al. 1994b). Both the population size of Pf-5 and its expression of pyoluteorin biosynthesis genes, evaluated with a transcriptional fusion to an ice nucleation reporter gene, are greater on cotton seed than on cucumber seed during the first 12 hours after seed are planted (Kraus and Loper 1995). Because *P. ultimum* infects cucumber seeds within 12 hr of seed imbibition (Nelson et al. 1986), transcription of pyoluteorin biosynthesis genes should be particularly important to suppression of pre-emergence damping off within that time period. Although the concentration of pyoluteorin in the spermosphere and consequent effects on Pythium damping-off are potentially affected by many factors, we speculate that temporal differences in expression of the pyoluteorin biosynthetic genes contribute to the differential role of pyoluteorin in disease suppression on the two host plants.

On surfaces of both cotton and cucumber seed, transcription of the *plt* gene(s) is low compared to transcription of the genes by Pf-5 grown in culture medium. To explore the possibility that biocontrol efficacy of Pf-5 could be improved by altering the *in situ* transcription of pyoluteorin biosynthesis genes, we initiated studies to evaluate regulation of pyoluteorin biosynthesis genes in Pf-5.

GLOBAL REGULATORS OF ANTIFUNGAL METABOLITE PRODUCTION

The production of antifungal metabolites by antagonistic strains of *Pseudomonas* spp. is controlled by complex regulatory cascades that respond to environmental and density-dependant signals (Thomashow and Weller, 1995) and are coupled to the physiological status of the bacterium (Sarniguet et al. 1995). We have identified three global regulators of antifungal metabolite production in Pf-5.

*The two-component regulatory system encoded by apdA and gacA.*
A mutation in the *apdA* (also called *lemA*) or *gacA* genes of *P. fluorescens* abolishes the production of pyoluteorin, pyrrolnitrin, 2,4-diacetylphloroglucinol, and hydrogen cyanide (Corbell and Loper 1995, Gaffney et al. 1994, Laville et al. 1992). Because ApdA⁻ and GacA⁻ mutants share common phenotypes and because the sequences of *apdA* and *gacA* are similar to the sequences of genes encoding known sensor kinases and response regulators, respectively, the two loci are likely to compose a classical two-component regulatory system controlling antifungal metabolite production by *P. fluorescens* (Corbell and Loper 1995, Rich et al. 1994).

The transmembrane sensor kinases of two-component regulatory systems are thought to autophosphorylate in response to a signal molecule(s), thereby mediating changes in gene expression in response to environmental signals. The existence or nature of environmental signals to which ApdA responds has not been established. Because phenotypes under the control of *apdA* are expressed by *P. fluorescens* in many culture media and in the rhizosphere (Kraus and Loper 1995), phosphorylation of ApdA is likely to be prompted by multiple signals, or by a single signal molecule that is commonly produced or encountered by *Pseudomonas* spp. occupying these diverse habitats. The identification of ApdA as a sensor kinase serving as a global regulator of antifungal metabolite biosynthesis genes provides an opportunity to identify the environmental cues to which *P. fluorescens* responds. Such information may have predictive value, allowing identification of those environmental conditions conducive to antifungal metabolite production, and may also lead to the genetic improvement of biological control agents.

*The sigma factor $\sigma^S$.* An RpoS⁻ mutant of Pf-5 does not produce $\sigma^S$, the sigma factor that controls the transcription of many genes expressed by bacterial cells in response to starvation or during the transition to stationary phase (Loewen and Hengge-Aronis 1994). An RpoS⁻ mutant of Pf-5 overproduces pyoluteorin and 2,4-diacetylphloroglucinol, but is deficient in pyrrolnitrin production (Sarniguet et al. 1995). *P. fluorescens* strain CHA0 containing multiple copies of *rpoD*, which encodes the principal "housekeeping" sigma factor $\sigma^{70}$, also overproduces pyoluteorin and 2,4-diacetylphloroglucinol (Schnider et al. 1995). From these data, we infer that promoters of pyoluteorin biosynthesis operons are not transcribed by the $\sigma^S$-RNA polymerase holoenzyme but are recognized by another sigma factor, probably $\sigma^{70}$. Genetic changes that reduce the ratio of $\sigma^S$ and $\sigma^{70}$ in the cell, such as a mutation in *rpoS* or multiple copies of *rpoD*, enhance pyoluteorin production. Therefore, antifungal metabolite production by *P. fluorescens* appears to be controlled by relative concentration of the $\sigma^{70}$ and $\sigma^S$ in the bacterial cell

(Sarniguet et al. 1995, Schnider et al. 1995). Derivatives of Pf-5 or CHA0 with a reduced $\sigma^S/\sigma^{70}$ ratio also are more effective than the parental strains in suppression of Pythium damping-off, presumably because they produce higher concentrations of pyoluteorin and 2,4-diacetylphloroglucinol (Maurhofer et al. 1994b, Sarniguet et al. 1995, Schnider et al. 1995). These data indicate that a mutation in single regulatory locus like *rpoS* can enhance the capacity of a bacterium to protect plants against infection.


## Pyoverdine Siderophores

Siderophores are low-molecular weight compounds that are produced by microorganisms under iron limiting conditions, chelate the ferric ion ($Fe^{3+}$) with a high specific activity, and serve as vehicles for the transport of $Fe^{3+}$ into a microbial cell. Pyoverdine siderophores (also called the fluorescent siderophores, pyoverdins, or pseudobactins) produced *in situ* by *Pseudomonas* spp. suppress Fusarium wilt (Lemanceau and Alabouvette 1993) and Pythium damping-off of some plant species (Buysens et al. 1996). Pyoverdines produced by *Pseudomonas* spp. in culture chelate available iron in a growth medium as ferric-pyoverdine complexes, which cannot be utilized as sources of iron by many phytopathogenic fungi and bacteria (Loper and Buyer 1991). Therefore, fluorescent pseudomonads can inhibit the growth of many phytopathogens in culture through pyoverdine-mediated iron competition.

Experiments comparing disease suppression by mutants deficient in pyoverdine production (Pvd⁻) and a near-isogenic parental strain (Pvd⁺) are useful in evaluating the contribution of pyoverdine production to biological control activity of *Pseudomonas* spp. (Loper and Buyer 1991). However, these experiments do not necessarily evaluate the importance of iron competition in biological control. Other possible contributions of pyoverdines to disease suppression exist, including a proposed role in inducing systemic resistance in the plant (Maurhofer et al. 1994b). Fluorescent pseudomonads commonly utilize heterologous ferric-pyoverdines (Raaijmakers et al. 1994) and ferric-siderophores that are produced by various genera of rhizosphere bacteria and fungi (Jurkevitch et al., 1992). Pvd⁻ and Pvd⁺ strains, which have the same capacity to utilize ferric-siderophores, are likely to place similar demands on the biologically-available pool of iron in the rhizosphere. Given this possibility, a conceptual model proposing that Pvd⁺ strains deplete such microhabitats of available iron whereas Pvd⁻ strains do not impoverish such environments of iron, may require reassessment. There is a clear need for new approaches in which the effect of a siderophore-producing rhizosphere bacterium on the iron status of a target pathogen can be evaluated. A microbial iron sensor, composed of a bacterial strain containing a fusion of an iron-regulated promoter to a reporter gene that can be monitored in natural habitats (Loper and Lindow 1994), provides a method for assessing iron availability to microorganisms in the rhizosphere. The iron sensor is proving useful in testing our current models of microbial iron competition.

## Conclusions

Fluorescent pseudomonads interact with plant pathogens in soil and on surfaces of seeds or roots, but we know little about the chemical nature of these habitats or the activities of biological control agents in these environments. On the surfaces of seeds and roots, bacteria do not achieve the cell densities found in culture media that are so much more conducive to our study. Instead, they exist as individual cells or in microcolonies on seed surfaces, at junctures of root cortical cells, or on the surfaces of root lesions where nutrients are relatively plentiful. They encounter fluctuating environmental conditions to which they must continually respond if they are to maintain viable populations. Molecular approaches such as reporter gene systems provide powerful ways to study the chemical composition of habitats that bacteria inhabit in nature and the activities of bacteria in these habitats. These approaches provide an unprecedented opportunity to evaluate the *in situ* expression of genes involved in biocontrol activity and the influence of edaphic factors on gene expression. Knowledge of expression of biological control genes and the limitations on this expression is likely to illuminate opportunities through which biological control activity can be improved.

## Literature Cited

Buysens, S., Heungens, K., Poppe, J., and Höfte, M. 1996. Involvement of pyochelin and pyoverdin in suppression of Pythium-induced damping-off of tomato by *Pseudomonas aeruginosa* 7NSK2. Appl. Environ. Microbiol. 62:865-871.

Corbell, N., and Loper, J. E. 1995. A global regulator of secondary metabolite production in *Pseudomonas fluorescens* Pf-5. J. Bacteriol. 177:6230-6236.

Gaffney, T. D., Lam, S. T., Ligon, J., Gates, K., Frazelle, A., Di Maio, J., Hill, S., Goodwin, S., Torkewitz, N., Allshouse, A. M.,

Kempf, H. J., and Becker, J. O. 1994. Global regulation of expression of anti-fungal factors by a *Pseudomonas fluorescens* biological control strain. Mol. Plant-Microbe Interact. 7:455-463.

Georgakopoulos, D. G., Hendson, M., Panopoulos, N. J., and Schroth, M. N. 1994. Analysis of expression of a phenazine biosynthesis locus of *Pseudomonas aureofaciens* PGS12 on seeds with a mutant carrying a phenazine biosynthesis locus-ice nucleation reporter gene fusion. Appl. Environ. Microbiol. 60:4573-4579.

Howell, C. R., and Stipanovic, R. D. 1980. Suppression of *Pythium ultimum*-induced damping-off of cotton seedlings by *Pseudomonas fluorescens* and its antibiotic, pyoluteorin. Phytopathology 70:712-715.

Howie, W. J., and Suslow, T. V. 1991. Role of antibiotic biosynthesis in the inhibition of *Pythium ultimum* in the cotton spermosphere and rhizosphere by *Pseudomonas fluorescens*. Mol. Plant-Microbe Interact. 4:393-399.

Jurkevitch, E., Hadar, Y., and Chen, Y. 1992. Differential siderophore utilization and iron uptake by soil and rhizosphere bacteria. Appl. Environ. Microbiol. 58:119-124.

Kraus, J., and Loper, J. E. 1992. Lack of evidence for a role of antifungal metabolite production by *Pseudomonas fluorescens* Pf-5 in the biological control of Pythium damping-off of cucumber. Phytopathology 82:264-271.

Kraus, J., and Loper, J. E. 1995. Characterization of a genomic region required for production of the antibiotic pyoluteorin by the biological control agent *Pseudomonas fluorescens* Pf-5. Appl. Environ. Microbiol. 61:849-854.

Laville, J., Voisard, C., Keel, C., Maurhofer, M., Défago, G., and Haas, D. 1992. Global control in *Pseudomonas fluorescens* mediating antibiotic synthesis and suppression of black root rot of tobacco. Proc. Natl. Acad. Sci. USA 89:1562-1566.

Lemanceau, P., and Alabouvette, C. 1993. Suppression of fusarium wilts by fluorescent pseudomonads: Mechanisms and applications. Biocontrol Sci. and Technol. 3:219-234.

Loewen, P. C., and Hengge-Aronis, R. 1994. The role of the sigma factor $\sigma^s$ (KatF) in bacterial global regulation. Annu. Rev. Microbiol. 48:53-80.

Loper, J. E., and Buyer, J. S. 1991. Siderophores in microbial interactions on plant surfaces. Mol. Plant-Microbe Interact. 4:5-13.

Loper, J. E., and Lindow, S. E. 1994. A biological sensor for iron available to bacteria in their habitats on plant surfaces. Appl. Environ. Microbiol. 60:1934-1941.

Maurhofer, M., Hase, C., Meuwly, P., Métraux, J.-P., and Défago, G. 1994a. Induction of systemic resistance of tobacco to tobacco necrosis virus by the root-colonizing *Pseudomonas fluorescens* CHA0: Influence of the *gacA* gene and of pyoverdine production. Phytopathology 84:139-146.

Maurhofer, M., Keel, C., Haas, D., and Défago, G. 1994b. Pyoluteorin production by *Pseudomonas fluorescens* strain CHA0 is involved in the suppression of Pythium damping-off of cress but not cucumber. Eur. J. Plant Pathol. 100:221-232.

Maurhofer, M., Keel, C., Schnider, U., Voisard, C., Haas, D., and Défago, G. 1992. Influence of enhanced antibiotic production in *Pseudomonas fluorescens* strain CHA0 on its disease suppressive capacity. Phytopathology 82:190-195.

Nelson, E. B., Chao, W. L., Norton, J. M., Nash, G. T., and Harman, G. E. 1986. Attachment of *Enterobacter cloacae* to hyphae of *Pythium ultimum*: possible role in the biological control of Pythium preemergence damping-off. Phytopathology 76:327-335.

Nowak-Thompson, B., Gould, S. J., Kraus, J., and Loper, J. E. 1994. Production of 2,4-diacetylphloroglucinol by the biocontrol agent *Pseudomonas fluorescens* Pf-5. Can J. Microbiol. 40:1064-1066.

Raaijmakers, J. M., van der Sluis, I., Koster, M., Bakker, P. A. H. M., Weisbeek, P. J., and Schippers, B. 1995. Utilization of heterologous siderophores and rhizosphere competence of fluorescent *Pseudomonas* spp. Can. J. Microbiol. 41:126-135.

Rich, J. J., Kinscherf, T. G., Kitten, T., and Willis, D. K. 1994. Genetic evidence that the *gacA* gene encodes the cognate response regulator for the *lemA* sensor in *Pseudomonas syringae*. J. Bacteriol. 176:7468-7475.

Sarniguet, A., Kraus, J., Henkels, M. D., Muehlchen, A. M., and Loper, J. E. 1995. The sigma factor $\sigma^S$ affects antibiotic production and biological control activity of *Pseudomonas fluorescens* Pf-5. Proc. Natl. Acad. Sci. USA 92:12255-12259.

Schnider, U., Keel, C., Blumer, C., Troxler, J., Défago, G., and Haas, D. 1995. Amplification of the housekeeping sigma factor in *Pseudomonas fluorescens* CHA0 enhances antibiotic production and improves biocontrol abilities. J. Bacteriol. 177:5387-5392.

Thomashow, L. S., and Weller, D. M. 1995. Current concepts in the use of introduced bacteria for biological control: Mechanisms and antifungal metabolites. pp. 187-235 *In*: G. Stacey and N. Keen (eds.), Plant Microbe Interactions, Vol. 1. Chapman and Hall, New York.

# Biocontrol: Genetic Modifications for Enhanced Antifungal Activity

James M. Ligon, Stephen T. Lam, Thomas D. Gaffney, D. Steven Hill, Phillip E. Hammer, and Nancy Torkewitz

Ciba-Geigy Corporation, Agricultural Biotechnology
Research Triangle Park, North Carolina USA

The use of antagonistic microbes to prevent infection of plants by plant pathogens, commonly known as biocontrol, has been well documented. Several studies that have investigated the biological mechanisms underlying this phenomenon indicate that biocontrol activity is due largely to the production of antifungal compounds by the biocontrol antagonist. These include the production of antibiotics such as phenazine-1-carboxylate (Thomashow and Weller 1988), 2,4-diacetylphloroglucinol (Keel et al. 1992), and pyrrolnitrin ([3-chloro-4-(2'-nitro-3'-chlorophenyl)-pyrrole], Prn) (Howell and Stipanovic 1979), hydrolytic enzymes such as chitinase (Shapira et al. 1989), hydrogen cyanide (Voisard et al. 1989), and siderophores (Becker and Cook 1988). We have recently characterized a *Pseudomonas fluorescens* strain, BL915, that is an effective biocontrol agent for the control of *Rhizoctonia solani* -induced seedling disease (Hill et al. 1994). This strain has been shown to produce several antifungal compounds, including Prn, 2-hexyl-5-propyl resorcinol (Res), chitinase, and cyanide. Furthermore, the production of these compounds is coordinately regLtlated by a bacterial two-component regulatory system consisting of a receptor-kinase and response regulator (Gaffney et al. 1994). Prn was first described by Arima et al. (1964) and is a highly antifungal metabolite produced primarily by pseudomonads. Prn has a safe toxicological profile and is used as a clinical antifungal agent for the treatment of skin mycoses (Tawara et al. 1989) and a phenylpyrrole derivative of Prn has been developed by Ciba-Geigy as an agricultural fungicide (Gehmann et al. 1990). Res, another metabolite known to be produced by pseudomonads, has antifungal and anti-Gram-positive bacterial activity (Kanda et al. 1975).

We have cloned and characterized the genes from strain BL915 that encode the receptor-kinase *(lemA)*, the response regulator *(gacA)*, chitinase, and enzymes involved in the synthesis of Prn (Hill et al. 1995). We have further attempted to use these genes and the current understanding of their role in biocontrol to genetically enhance the overall biocontrol activity of *P. fluorescens* strain BL915.

## Cloning and Characterization of the Prn Biosynthetic Genes

In an effort to isolate genes whose expression is regulated by the protein products of the *lemA* and *gacA* genes in a related *P. fluorescens* strain, strain BL914, Lam et al. (1995) in our group identified a Tn5/*lacZ* transposon mutant that was specifically affected in the synthesis of Prn. A cosmid clone from a genomic library containing DNA from strain BL915 was found that complemented the Prn⁻ phenotype of this mutant. Analysis of the cloned DNA in this cosmid by site-directed mutagenesis demonstrated that a 6.2 kb region centrally located in the cloned DNA of this cosmid was involved in Prn biosynthesis (Figure 1) (Hill et al. 1995). The nucleotide sequence of this region was determined and analysis of it revealed the presence of 4 colinear open reading frames (ORF) organized on a single transcriptional unit (Figure 1). Mutations in each of the ORFs were created in vitro by deletion of segments of DNA internal to the coding sequences and subsequent ligation of a kanamycin resistance gene (APH) into the site of the deletion. Each deletion mutation was introduced into the chromosome of strain BL915 by homologous replacement using the kanamycin resistance marker for selection. A mutation in any one of the

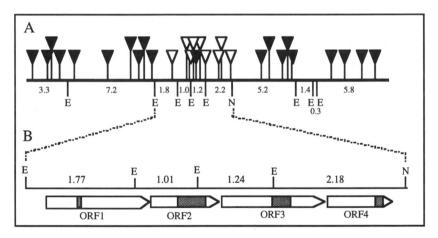

**Figure 1. (A)** Restriction map of the cloned DNA and the position of individual Tn5 insertions. Restrictions sites shown are *Eco*RI (E) and *Not*I (N), and the distance in kilobase pairs between restriction sites is indicated. Tn5 insertions are shown as filled or open triangles to indicate those with no affect on Prn synthesis and those resulting in no Prn synthesis, respectively. **(B)** The 4 identified ORFs and their direction of transcription, indicated by open arrows, and the position of the deletions constructed in each, indicated by shaded boxes, are shown.

identified ORFs abolished Prn synthesis in BL915. In addition, the coding sequences of each of the 4 ORFs, including putative ribosome binding sites, were amplified by PCR and fused to the *tac* promoter (from pKK223-3MCS, Pharmacia) in the proper juxtaposition to cause expression of the ORFs from this promoter. Each *tac* promoter/ORF fusion was introduced into the corresponding ORF deletion mutant and in every case, the mutant Prn⁻ phenotype was complemented.

The four ORF cluster was cloned from an *Xba*I site located about 100 bp 5' to the initial codon of ORF1 to the *Not*I site 3' to ORF4 and this fragment was fused to the *tac* promoter in the proper orientation to effect transcription of the four ORFs. This construction was introduced on a broad host range plasmid into several different Gram-negative bacteria that were incapable of Prn synthesis, including *E. coli, Enterobacter,* and other *Pseudomonas* strains. In each case, bacteria containing the *tac*/ORF1-4 fragment were shown to produce Prn (Hill et al. 1995). These results indicate that the four ORFs have a role in the synthesis of Prn and that together they represent the entire biosynthetic pathway for Prn synthesis. Therefore, we have assigned ORFs1-4 the genetic designations *prnA, prnB, prnC,* and *prnD,* respectively.

### Genetic Modifications for Increased Biocontrol

Plasmid pPRN-E11 that contains an 11 kb *Eco*RI fragment derived from the chromosome of strain BL915 (Hill et al. 1994) and known to contain the *gacA* gene (Gaffney et al. 1994) was transferred into the wild-type parent strain resulting in an increase of the copy number of the *gacA* gene. The production of Prn and Res in this modified strain, designated BL915(pPRN-E11),was quantified and the results are shown in Table 1 This strain produced 3.6-times more Prn than the parent strain. These results indicate that modifications of the *gacA* gene may result in increased metabolite biosynthesis. Subsequently, several other modifications of the

**Table 1.** Antifungal metabolite production by BL915 and derivatives

| Strain | Prn (mg/l) | Res (mg/l) |
|---|---|---|
| BL915 | 21.8 | 1691 |
| BL915(pPRN-E11) | 78.5 | 1675 |
| BL915-ATG/*gacA* | 47.0 | 1746 |
| BL915-*tac*/*gacA* | 43.8 | 1767 |
| BL915(*prnABCD*) | 83.3 | 1491 |

*gacA* gene were constructed and tested. The native *gacA* gene has an unusual TTG translation initiation codon (Gaffney et al. 1994). These are known to result in reduced translational efficiency relative to genes that have ATG as a translation initiation codon. Using PCR, we changed the first base in the *gacA* coding sequence to an adenine and this gene was used to replace the wild-type gene by homologous replacement. A second modification of the *gacA* gene in strain BL915 was constructed by replacing the native *gacA* promoter with the *tac* promoter which is expressed constitutively and at a high level in pseudomonads. The wild-type *gacA* gene of strain BL915 was replaced with the *tac/gacA* gene to create modified strain BL915-*tac/gacA*. Production of Prn and Res was assessed in BL915-ATG/*gacA* and BL915-*tac/gacA*. BL915-ATG/*gacA* and BL915-*tac/gacA* produced about twice as much Prn as BL915 (Table 1).

Another modified BL915 strain was constructed using the *prnABCD* gene cluster described above. The *prnABCD* gene fragment fused to the *tac* promoter and cloned in a broad host range plasmid was introduced into strain BL915 to create strain BL915(*prnABCD*). This derivative was shown to produce approximately four times the amount of Prn as BL915 and slightly less Res (Table 1). The explanation for the reduced production of Res in this derivative is not clear, but may be due to increased competition for common precursors in the biosynthesis of these two metabolites.

## Biocontrol Activity of BL915 Derivatives

The modified strains described above were tested for biocontrol activity in two pathosystems, including cucumber and impatiens, with *R. solani* as the pathogen. Since strain BL915 is very effective when used at the rate of $2 \times 10^8$ cells/g soil and we were interested in determining whether the BL915 derivative strains demonstrate increased biocontrol activity, we applied all modified strains at a rate one-tenth that of the normal rate ($2 \times 10^7$ cells/ml soil). The results shown in Table 2 demonstrate that in most of the pathosystems, the modified strains exhibited increased biocontrol activity relative to the parent strain. Strain BL915(*prnABCD*) at the low application rate demonstrated nearly as much biocontrol activity in the two pathosystems as strain BL915 at a 10-fold higher rate. Strain BL915(pPRN-E11) showed similar activity in the impatiens trials, but was not improved in the cucumber trials. Strains BL915-*tac/gacA* and BL915-ATG/*gacA* each had higher activity than the parent strain at the low rate in both pathosystems, but not as much as the parent strain applied at the higher rate. These results demonstrate that the genetic modifications of strain BL915 described herein have resulted in increased production of the antifungal metabolite Prn (Table 1) and in increased biocontrol activity (Table 2).

**Table 2.** Biocontrol activity of modified BL915 strains in two pathosystems against *R. solani.* Cucumber seeds or impatiens transplants were planted in infested soil except in the case of the healthy control plants which were planted in uninfested soil. All strains were applied as a drench at $2 \times 10^7$ cells/g soil, except for the high rate of BL915 which was applied at a 10-fold higher rate. Since the strains were tested in separate trials, results for the controls are shown with each test group. The number of plants planted for each treatment is shown in parentheses with the results for the healthy controls.

| Treatment | Stand (14 Days After Planting) | |
|---|---|---|
| | **Cucumber** | **Impatiens** |
| Healthy Control | 113 (120) | 120 (120) |
| Pathogen Control | 86 | 92 |
| BL915-Hi | 97 | 111 |
| BL915-Lo | 80 | 91 |
| BL915(pPRN-E11) | 75 | 103 |
| BL915(*prnABCD*) | 101 | 105 |
| | | |
| Healthy Control | 54 (60) | 119 (120) |
| Pathogen Control | 31 | 78 |
| BL915-Hi | 54 | 106 |
| BL915-Lo | 41 | 73 |
| BL915-*tac/gacA* | 46 | 80 |
| BL915-ATG/*gacA* | 44 | 98 |

## Literature Cited

Arima, K., H. Imanaka, M. Kousaka, A. Fukuda, and C. Tamura. 1964. Pyrrolnitrin, a new antibiotic substance, produced by *Pseudomonas*. Agr. & Biol. Chem. 28:575-576.

Becker, J.O., and Cook, R.J. 1988. Role of siderophores in suppression of *Pythium* species and production of increased growth response of wheat by fluorescent pseudomonads. Phytopathol. 78:778-782.

Gaffney, T., Lam, S., Ligon, J., Gates, K., Frazelle, A., Di Maio, J., Hill, S., Goodwin, S., Torkewitz, N., Allshouse, A., Kempf, H.-J., and Becker, J. 1994. Global regulation of expression of antifungal factors by a *Pseudomonas fluorescens* biological control strain. MPMI 7:455-463.

Gehmann, K., R. Neyfeler, A. Leadbeater, D. Nevill, and D. Sozzi. 1990. CGA173506: a new phenylpyrrole fungicide for broad-spectrum disease control. Proc. Brighton Crop Prot. Conf., Pest Dis. 2:399-406.

Hill, S., Lam, S., Hammer, P., and Ligon, J. 1995. Cloning, characterization, and heterologous expression of genes *from Pseudomonas fluorescens* involved in the synthesis of pyrrolnitrin. Phytopathol. 85:1187.

Hill, D.S., Stein, J., Torkewitz, N., Morse, A., Howell, C., Pachlatko, J., Becker, J., and Ligon, J. 1994. Cloning of genes involved in the synthesis of pyrrolnitrin from *Pseudomonas fluorescens* and role of pyrrolnitrin synthesis in biological control of plant disease. Appl. Environ. Microbiol. 60:78-85.

Howell, C., and Stipanovic, R. 1979. Control of *Rhizoctonia solani* on cotton seedlings with *Pseudomonas fluorescens* and with an antibiotic produced by the bacterium. Phytopathol. 77:480-482.

Kanda, N., Ishizaki, N., Inoue, N., Oshima, M., Handa, A., and Kitahara, T. 1975. DB-2073, a new alkylresorcinol antibiotic. I. Taxonomy, isolation, and characterization. J. Antibiot. 28:935-942.

Keel, C., Wirthner, P., Oberhansli, T., Voisard, C., Burger, U., Haas, D., and Défago, G. 1990. Pseudomonads as antagonists of plant pathogens in the rhizosphere: role of the antibiotic 2,4-diacetylphloroglucinol in the suppression of black rot of tobacco. Symbiosis 9:327-341.

Lam, S., Frazelle, R., Torkewitz, N., and Gaffney, T. 1995. A genetic approach to identify pyrrolnitrin biosynthesis and other globally regulated genes in *Pseudomonas fluorescens*. Phytopathol. 85:1163.

Shapira, R., Ordentlich, A., Chet, I., and A.B. Oppenheim, A.B. 1989. Control of plant diseases by chitinase expressed from cloned DNA in Escherichia coli. Phytopathol. 79:1246-1249.

Tawara, S., S. Matsumoto, T. Hirose, Y. Matsumoto, S. Nakamoto, M. Mitsuno, and T. Kamimura. 1989. In vitro antifungal synergism between pyrrolnitrin and clotrimazole. Japanese J. Med. Mycol. 30:202-210.

Thomashow, L., and Weller, D. 1988. Role of phenazine antibiotic from *Pseudomonas fluorescens* in biological control of *Gaeumannomyces graminis* var. *tritici*. J. Bacteriol. 170:3499-3508.

Voisard, C., Keel, C., Haas, D., and Défago, G. 1989. Cyanide production by *Pseudomonas fluorescens* helps suppress black root rot of tobacco under gnotobiotic conditions. EMBO J. 8:3 51-358.

# Phenazine Antibiotic Biosynthesis in the Biological Control Bacterium *Pseudomonas aureofaciens* 30-84 is Regulated at Multiple Levels

L. S. PIERSON III, D. W. WOOD, and S. T. CHANCEY
Department of Plant Pathology, University of Arizona, Tucson, AZ 85721.

Biological control in the rhizosphere is often the result of competition among microorganisms. Research on biological control has identified many useful competitive mechanisms, including antibiosis, scavenging compounds, toxic metabolites, etc. One approach to improving biological control is to understand the genetic and environmental factors that influence the expression of these competitive mechanisms.

We have been studying phenazine antibiotic production by *Pseudomonas aureofaciens* strain 30-84 as a model system to understand the molecular mechanisms that regulate the expression of genes involved in competition and hence, biological control. Strain 30-84 was isolated from wheat roots taken from a field where take-all disease caused by *Gaeumannomyces graminis* var. *tritici* had been suppressed. Strain 30-84 produces three phenazine antibiotics; phenazine-1-carboxylic acid (PCA), 2-hydroxy-phenazine-1-carboxylic acid (2-OH-PCA), and 2-hydroxy-phenazine (2-OH-PZ) (Pierson & Thomashow 1992). Phenazine production by strain 30-84 is responsible for pathogen inhibition (Pierson & Thomashow 1992) and for its rhizosphere competence (Mazzola et al. 1992). Understanding *phz* regulation on a molecular level will provide insights into the factors, both biotic and abiotic, that determine when a root-colonizing bacterium expresses the genes that result in pathogen suppression.

## Results and Discussion

### PHENAZINE PRODUCTION IS REGULATED BY A DIFFUSIBLE SIGNAL

Two genes (*phzR* and *phzI*) were identified that directly control *phz* regulation. PhzR is a member of the LuxR family of quorum-sensing transcriptional regulators (Pierson et al. 1994). PhzR activates *phz* expression in response to the accumulation of a diffusible *N*-acyl-homoserine lactone (*N*-acyl-HSL) signal produced by PhzI (Wood & Pierson, 1996). PhzI is a member of the LuxI family of *N*-acyl-homoserine lactone (*N*-acyl-HSL) synthases. Preliminary structural studies on the diffusible signal produced by 30-84 confirms that it is a member of the *N*-acyl-HSL class of signal molecules (Pierson et al, unpubl).

## *N*-ACYL-HSL SIGNALING OCCURS BETWEEN POPULATIONS IN THE RHIZOSPHERE

Although *N*-acyl-HSL signals are known to be produced by many bacteria (Fuqua et al, 1994), a true demonstration of the function of these signals *in situ* has been lacking. Recent work in our laboratory has shown that *N*-acyl-HSL-mediated crosstalk between isogenic bacterial populations occurs in the wheat rhizosphere (Wood and Pierson, unpubl.). Strain 30-84Ice (*phzB::inaZ, phzI⁺*) carries an ice nucleation (*inaZ*) fusion to *phzB*. This strain is PhzI⁺, produces InaZ protein in a manner analogous to phenazine production, and was used to measure the endogenous level of InaZ activity on roots. Strain 30-84I/Ice (*phzB::inaZ, phzI⁻*) is an isogenic *phzI* derivative of 30-84Ice which is unable to produce *N*-acyl-HSL. This strain served as a 'reporter' population and would detect the presence of *N*-acyl-HSL signal produced by a second bacterial population by expressing InaZ. Strain 30-84Z (*phzB::lacZ, phzI⁺*) served as the 'donor' population in these studies since it produces *N*-acyl-HSL but is InaZ⁻. These strains were applied to sterile pre-germinated 'Fielder' wheat seeds alone and in various ratios. The treated seeds were planted in sterile soil and grown 12-14 days. The bacteria were isolated from the roots and bacterial populations and InaZ activity were determined immediately.

The results demonstrate that InaZ activity in the 30-84I/Ice 'reporter' populations is restored to wild type levels by the 30-84Z *N*-acyl-HSL 'donor' populations (Fig. 1). These results demonstrate: 1) *N*-acyl-HSL production is required for *phz* expression on roots; 2) *N*-acyl-HSLs serve as regulatory signals in nature; and 3) this is the first demonstration of *N*-acyl-HSLs produced by one population influencing gene expression in a second population in the rhizosphere. This work, in addition to our studies which have identified a number of *N*-acyl-HSL producing bacteria from wheat roots in natural soils, provides support for our hypothesis that other rhizosphere bacteria are capable of influencing *phz* expression and may affect the biocontrol ability of strain 30-84 (Pierson & Pierson, 1996).

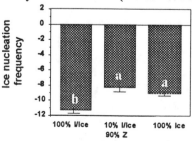

Fig 1. *N*-acyl-HSL-mediated cross-talk *in situ*. Y-axis: InaZ frequency of bacteria isolated from wheat roots (log nuclei/cell). X-axis: bacterial treatments. Values are reported as means with standard errors. Bars containing the same letter are not statistically different when compared using LSD and Tukey's analysis (P=0.0002). Each treatment was comprised of 6 replicates and the experiment was repeated three times.

DIFFUSIBLE SIGNAL PRODUCTION IS REGULATED BY A TWO COMPONENT SENSORY TRANSDUCTION PATHWAY.

Two pleiotrophic mutants (strains 30-84W & 30-84.A2) were isolated that were defective in the production of phenazines, protease, hydrogen cyanide and production of the *N*-acyl-HSL signal (Chancey & Pierson 1996). Neither mutation is complemented by *phzI*, *phzR* or the *phz* biosynthetic region. Therefore, these strains represent novel global regulatory mutations that affect phenazine biosynthesis in *P. aureofaciens*.

*Strain 30-84W*. Strain 30-84W was restored to phenazine, protease, HCN and *N*-acyl-HSL production by cosmid pLSP6-19 from a 30-84 genomic library. This activity was localized to a 4 kb region (pSTC121). DNA sequence analysis of this region revealed an open reading frame with extensive similarity to *gacA* genes isolated from *P. fluorescens* BL915 (83.4%), *P. fluorescens* CHA0 (79.9%), *P. syringae* pv. *syringae* (74.2%), and *P. aeruginosa* (70.7%). These results suggested that *P. aureofaciens* strain 30-84W was a *gacA* mutant.

To verify that strain 30-84W was a *gacA* mutant, cosmid pME3066 that contains *gacA* cloned from *P. fluorescens* strain CHA0 (Laville et al 1992) was introduced into strain 30-84W. The presence of the *gacA* gene from strain CHA0 restored strain 30-84W to Phz$^+$, Protease$^+$, HCN$^+$, and *N*-acyl-HSL$^+$, confirming that strain 30-84W was a *gacA* mutant.

*Strain 30-84.A2*. Strain 30-84.A2 was not complemented by pSTC121 or pME3066. However, the introduction of plasmid pEMH97 containing *lemA* from *P. syringae* pv. *syringae* (Willis et al. 1994) restored strain 30-84.A2 to Phz$^+$, Protease$^+$, HCN$^+$, and *N*-acyl-HSL$^+$. These results indicate that strain 30-84.A2 is a *lemA* mutant. Thus, mutations within both *gacA* and *lemA* have been identified in strain 30-84 and these genes are absolutely required for phenazine production.

*GacA/LemA is a Two Component Regulatory System*. In other bacterial systems, GacA and LemA are members of a two component sensory transduction system (Willis et al. 1994). LemA serves as a membrane-bound sensor that, in response to external environmental signals, activates the cytoplasmic GacA by phosphorylation. Activated GacA in turn stimulates target gene expression.

## GacA/LemA regulate the transcription of PHZI and N-acyl-HSL production

The reporter strain 30-84I/Z (*phzI*⁻, *phzB::lacZ*) can not synthesize *N*-acyl-HSL signal and is therefore Lac⁻. The presence of *N*-acyl-HSL signal can be detected in filter-sterilized culture supernatants by measuring the restoration of β-galactosidase activity to strain 30-84I/Z (Fig. 2). Neither strain 30-84W nor 30-84.A2 produce *N*-acyl-HSL signal as indicated by this assay. In contrast, wildtype 30-84, 30-84W (pSTC110) and 30-84.A2 (pEMH97) restore β-galactosidase activity to 30-84I/Z, indicating that *gacA/lemA* mediate the availability of *N*-acyl-HSL.

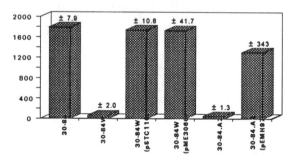

*Strains used to condition medium*

Fig. 2. GacA & LemA are required for *N*-acyl-HSL production. 'Conditioned' media from each test strain (x-axis) was assayed for the presence of *N*-acyl-HSL signal by its ability to restore β-galactosidase activity to the reporter strain 30-84I/Z (y-axis).

One hypothesis was that *gacA/lemA* regulate the transcription of *phzI*. Transcription of *phzI* was measured in strains 30-84Z, 30-84W and 30-84.A2 by introducing a plasmid-borne transcriptional *phzI::uidA* fusion and measuring β-glucuronidase activity (Table 1). These results indicate that *phz* expression is regulated by a two step signal cascade mechanism in which GacA/LemA regulate the PhzR/PhzI system by regulating expression of *phzI* and therefore the production of *N*-acyl-HSL signal.

| Strain | Relevant Phenotype | *phzI::uidA* expression | Ratio |
|---|---|---|---|
| 30-84Z | PhzB ⁻ | 4,666ᵃ | 1.0 |
| 30-84R | PhzR ⁻ | 1,590ᵇ | 3.0 ↓ |
| 30-84.A2 | LemA ⁻ | 117ᶜ | 40.0 ↓↓ |
| 30-84W | GacA ⁻ | cd | |

Table 1. Measurement of β-glucuronidase activity of a *phzI::uidA* reporter plasmid in strains 30-84Z, 30-84W, and 30-84.A2 after 24 h. Means with the same letter are not significantly different. cd = being verified.

MODEL FOR PHENAZINE GENE REGULATION IN *P. AUREOFACIENS*

Phenazine gene expression in *P. aureofaciens* 30-84 appears to be regulated at multiple levels (Fig. 4). At one level, the phenazine biosynthetic genes are regulated by PhzR which senses the intracellular concentration of the *N*-acyl-HSL signal produced by PhzI. At low *N*-acyl-HSL levels, *phzI* is basally expressed. However, when the intracellular signal reaches a threshold concentration, it interacts with PhzR to activate the expression of both the *phz* biosynthetic genes and *phzI*, presumably by PhzR binding to specific sequences known as 'Phz' boxes located within these promoter regions. Binding of PhzR to the *phzI* promoter increases *phzI* expression, resulting in further increases in *N*-acyl-HSL signal. The requirement for *phzI* can be bypassed by exogenous *N*-acyl-HSL signals provided by other rhizosphere bacteria (Pierson & Pierson 1996).

The second level of regulation involves a two component 'LemA/GacA' sensory transduction system. In *P. aureofaciens*, this system is required for the expression of *phzI* which synthesizes the *N*-acyl-HSL signal. We are currently addressing the nature of the environmental signal recognized by this two component sensory system.

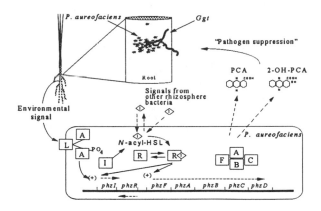

Fig. 4. Organization of the *P. aureofaciens* phenazine biosynthetic region.

Our use of *P. aureofaciens* as a model system is beginning to shed insights into the molecular basis for *phz* expression and the environmental and biotic factors that affect competitive gene expression. Our results should serve as a useful model with which to compare the regulation of competitive mechanisms in other biological control agents in the future.

ACKNOWLEDGMENTS

We would like to thank the following for their contributions to this work: Patricia Figuli, Fangcheng Gong, and Elizabeth Pierson.

REFERENCES

Chancey, S.T., and Pierson, L.S. III. 1996. Abstracts of the 96th Meeting of the American Society for Microbiology, New Orleans, LA.

Fuqua, W. C., Winans, S. C., and Greenberg, E. P. 1994. J. Bacteriol. 176:269-275.

Laville, J., Voisard, C., Keel, C., Maurhofer, M., Defago, G., and Haas, D. 1992. Proc. Natl. Acad. Sci USA 89:1562-1566.

Mazzola, M., Cook, R.J., Thomashow, L.S., Weller, D.M., and L.S. Pierson. 1992. Appl. Environ. Microbiol. 58:2616-2624.

Pierson, L.S. III, Gaffney, T., Lam. S., and Gong, F. 1995. FEMS Microbiol. Lett. 134:299-307.

Pierson III, L. S., Keppenne, V. D., and Wood, D. W. 1994. J. Bacteriol. 176:3966-3974.

Pierson, L. S. III, and Pierson, E.A. 1996. FEMS Microbiol. Letters 136:101-108

Pierson, L.S., and Thomashow, L.S.. 1992. Mol. Plant Microbe Interact. 5:330-339.

Thomashow, L.S., Weller, D.M., Bonsall, R.F., and L.S. Pierson, III. 1990. Appl. Environ. Microbiol. 56: 908-912.

Willis, D.K., Rich, J.J., Kinscherf, T.G., and Kitten, T. 1994. In Genetic Engineering. Vol. 5. Suslow, J.T., ed. Plenum Press, NY.

Wood, D.W., and Pierson, L. S. III. 1996. Gene 168:49-53.

# 2,4-Diacetylphloroglucinol, a Key Antibiotic in Soilborne Pathogen Suppression by Fluorescent *Pseudomonas* Spp.

L. S. Thomashow[*,1,2], M. G. Bangera[2], R. F. Bonsall[1], D. -S. Kim[1], J. Raaijmakers[1] and D. M. Weller[1]

[1]USDA-ARS and [2]Department of Microbiology, Washington State University, Pullman WA 99164-6430

The broad-spectrum polyketide antibiotic 2,4-diacetylphloroglucinol (Phl) is a major determinant in the biological control by fluorescent *Pseudomonas* spp. of many soilborne diseases (reviewed in Thomashow and Weller 1996). We previously showed that a 6.5-kb genomic DNA fragment from *P. fluorescens* Q2-87 was sufficient to transfer Phl biosynthetic capability to Phl-nonproducing recipient *Pseudomonas* strains. Genes required for Phl production were localized to at least two divergently oriented transcriptional units spanning approximately 5 kb (Bangera and Thomashow 1996). Sequences within the 5-kb region were conserved in all of 45 Phl-producing *Pseudomonas* strains of worldwide geographic origin, although the strains otherwise exhibited genotypic and phenotypic diversity (Keel et al. 1996). Most of the 45 strains originated from soils with natural suppressiveness to particular diseases including, for example, take-all of wheat. Biological control occurs spontaneously in such soils when "the pathogen does not establish or persist, establishes but causes little or no damage, or establishes and causes disease for a while but thereafter the disease is less important, although the pathogen may persist in the soil" (Baker and Cook 1974). The apparent prevalence of Phl producers in certain suppressive soils suggested that the bacteria might have a role in this natural biological control phenomenon. In this report we describe the Phl biosynthetic locus in greater detail and present evidence that indigenous Phl-producing populations can contribute significantly to the natural suppression of take-all (take-all decline).

## Structure of the Phl Biosynthetic Locus

DNA sequence analysis of the 6.5-kb fragment contained in pMON5122 (Vincent et al. 1991) revealed six open reading frames. Those designated *phlA*, *phlC*, *phlB*, and *phlD* are contained within a large transcriptional unit oriented from right to left and shown previously to be required for production of Phl (Bangera and Thomashow 1996). This cluster is flanked on the right by the divergently transcribed gene *phlF*, of which the 3' end is truncated in pMON5122. At the extreme left end is *phlE*, which is associated with the presence of a red pigment that usually is present in media when Phl is produced (Bangera and Thomashow 1996; Keel et al. 1996).

Several lines of evidence support our previous results indicating that genes within the core transcriptional unit encode Phl biosynthetic enzymes. Phl is thought to be synthesized via a polyketide pathway, with monoacetylphloroglucinol (MAPG) a probable intermediate (Shanahan et al. 1993). Polyketide assembly occurs via successive condensation reactions of short-chain carboxylic acids in a process resembling fatty acid biosynthesis (Hopwood and Sherman 1990; Shen et al. 1995). The predicted products of *phlA*, *phlC*, *phlD*, and perhaps *phlB*, have sequence similarity or homology with proteins that function in acyltransferase and acyl condensation reactions of the kind required for polyketide assembly. Thus, *phlA* encodes a 37.9 kDa product that resembles FabH (β-ketoacyl synthase III) in *Escherichia coli* (Tsay et al. 1992), and the *phlC* product has homology with the thiolase domain at the amino-terminal end of mammalian sterol carrier proteins (Seedorf et al. 1993). Colinear homologues of these genes, as well as *phlB*, have been reported in the archaeon *Pyrococcus furiosus*, where the designation *acaABC* suggests a role in acetoacetyl-CoA synthesis (Kletzin and Adams 1996). In the Phl locus, *phlACB* is cotranscribed with *phlD* (M. G. Bangera and L. S. Thomashow, manuscript in preparation), which has homology at the DNA and protein levels with chalcone synthases active in flavonoid phytoalexin production in higher plants. These Type III polyketide synthases catalyze all the condensation and cyclization reactions required for polyketide production (Martin 1993; Schröder and Schröder 1990), but are much smaller than the Types I and II PKS complexes present in *Streptomyces* spp. (Hopwood and Sherman 1990; Shen et al. 1995).

We hypothesize that PhlD is responsible for the condensation and cyclization steps leading to the formation of MAPG, and that the products of *phlACB*, probably acting in concert, catalyze the acetyltransferase reaction resulting in the conversion of MAPG to Phl (Shanahan et al.

1993). In support of this model, a *phlD* mutant of *P. fluorescens* Q2-87 produced neither MAPG nor Phl, but produced Phl when supplied with exogenous MAPG. Further, *E. coli* expressing *phlACBD* produced both MAPG and Phl, whereas the same strain expressing *phlACB* produced Phl when grown with MAPG, but not in its absence (M. G. Bangera and L. S. Thomashow, manuscript in preparation). It is presently unclear whether *phlD* alone is sufficient for production of MAPG, or if one or more of *phlA, phlC*, or *phlB* may be needed, perhaps to supply starter units for condensation reactions catalyzed by PhlD.

The gene *phlF* resides upstream of *phlACBD* and encodes a putative repressor of Phl biosynthesis. The predicted *phlF* product is a 23-kDa protein homologous with a repressor in the biosynthetic locus for the polyketide immunosuppressant rapamycin (Schwecke et al. 1996). Similarity to other, well-characterized repressors including TetR is concentrated within a conserved helix-turn helix domain typical of DNA-binding regulatory proteins. The *phlF* gene is truncated in pMON5122. This may render the putative repressor nonfunctional, and it may explain our earlier observation that pMON5122 can transfer Phl biosynthetic capability, whereas plasmids including the intact gene did not.

Distal to the 3'-end of *phlD*, a separate transcriptional unit containing *phlE* predicts a 45.2-kDa protein with homology to members of the transmembrane solute facilitator superfamily of membrane permeases including efflux pump proteins that confer antibiotic resistance (Saier 1994). PhlE retained conserved structural features of these membrane-spanning proteins including a hydrophilic central loop flanked on either side by six hydrophobic α-helices. The fact that antibiotic resistance determinants often are linked to their cognate biosynthetic genes, coupled with our earlier observation that certain mutations in *phlE* were correlated with reduced levels of inhibitory activity in vitro, indicate that while PhlE is not required for Phl synthesis it may function in resistance or export.

### 2,4-Diacetylphloroglucinol Has a Role in Take-All Decline

Take-all, caused by *Gaeumannomyces graminis* var. *tritici*, probably is the most important root disease of wheat worldwide. The phenomenon known as take-all decline (TAD) has been recognized for over 50 years as a natural biological control in which the incidence and severity of root lesions subside spontaneously after several consecutive, severe outbreaks during wheat monoculture in the presence of the pathogen. It has long been suggested that TAD occurs because of the buildup over years of antagonistic bacteria, particularly fluorescent *Pseudomonas* spp., on *G. g. tritici*-infected roots and infested plant residues. At least 20% of fluorescent *Pseudomonas* strains isolated from wheat roots grown in a TAD soil from Quincy, WA exhibited the red-pigmented phenotype characteristic of Phl producers, and up to 23% of 1,100 *Pseudomonas* isolates from wheat grown in a Swiss suppressive soil hybridized with a probe from the Phl biosynthetic locus (Keel et al. 1996). These results suggest that populations of Phl producers indigenous to TAD soils may have a key role in disease suppression.

As a test of this hypothesis, we have generated and used a 745-bp hybridization probe from *phlD* to identify and enumerate populations of *Pseudomonas* spp. capable of Phl production among isolates from a total of seven soils either suppressive or conducive to take-all (J. Raaijmakers et al. manuscript in preparation). Monoculturing of wheat in the greenhouse resulted in the enrichment of Phl-producing pseudomonads on the roots of seedlings grown only in the suppressive soils, where populations reached average maximum densities of $5 \times 10^5$ CFU/g (fresh weight) of root. Dose-response studies with the Phl-producing strain Q2-87 applied to seed indicated that take-all was suppressed at threshold populations of $10^5$ CFU/g of root. These results indicate that indigenous Phl-producing populations in TAD soils attained sizes large enough to suppress take-all, and are consistent with their involvement in other naturally suppressive soils as well.

Although biological control continues to be actively pursued as an alternative to the use of chemical pesticides for plant disease control, there is still concern about the nontarget effects of introduced agents on microflora indigenous to agroecosystems. Recent evidence that genes responsible for Phl production already are broadly distributed among root-associated pseudomonads from different continents, that such pseudomonads are comparatively abundant in naturally suppressive soils, and that they can be enriched further by monoculture cropping, should help to alleviate concern about the practical application of nonindigenous Phl producers or transgenic agents containing introduced Phl genes.

### Acknowledgements

This project was supported in part by the Storkan Hanes Research Foundation and grant number 94-37107-0439 from the USDA-CSREES-NRICGP. We thank Steve Thompson and the VADMS facility at

Washington State University for assistance in the use of the GCG sequence analysis package.

## Literature Cited

Baker, K. F., and Cook, R. J. 1974. Biological control of plant pathogens. W. H. Freeman & Co., San Francisco.

Bangera, M. G., and Thomashow, L. S. 1996. Characterization of a genomic locus required for synthesis of the antibiotic 2,4-diacetylphloroglucinol by the biological control agent *Pseudomonas fluorescens* Q2-87. Mol. Plant-Microbe Interact. 9:83-90.

Hopwood, D. A., and Sherman, D. H. 1990. Molecular genetics of polyketides and its comparison to fatty acid biosynthesis. Annu. Rev. Genet. 24:37-66.

Keel C., Weller, D.M., Natsch, A., Défago, G., Cook. R.J., and Thomashow, L.S. 1996. Conservation of the biosynthetic locus for 2,4-diacetylphloroglucinol among fluorescent *Pseudomonas* strains from diverse geographic locations. Appl. Environ. Microbiol. 62:552-563.

Kletzin, A., and Adams, M. W. W. 1996. Molecular and phylogenetic characterization of pyruvate and 2-ketoisovalerate ferredoxin oxidoreductases from *Pyrococcus furiosus* and pyruvate ferredoxin oxidoreductase from *Thermotoga maritima*. J. Bacteriol. 178:248-257.

Martin, C. R. 1993. Structure, function, and regulation of chalcone synthase. Int. Rev. Cytol. 147:233-283.

Saier Jr., M. H. 1994. Computer-aided analysis of transport protein sequences: gleaning evidence concerning function, structure, biogenesis, and function. Microbiol. Rev. 58:71-93.

Schröder, J., and Schröder, G. 1990. Stilbene and chalcone synthases: related enzymes with key functions in plant-specific pathways. Z. Naturforsch. 45c:1-8.

Schwecke, T., Aparicio, J. F., Molnar, I., König, A., Khaw, L. E., Haydock, S. F., Oliynyk, M., Caffrey, P., Cortés, J., Lester, J. B., Böhm, G., Staunton, J., and Leadlay, P. F. 1995. The biosynthetic gene cluster for the polyketide immunosuppressant rapamycin. Proc. Natl. Acad. Sci. USA 92:7839-7843.

Seedorf, U., P. Brysch, T. Engel, K. Schrag, and G. Assmann. 1993. Sterol carrier protein X is peroxisomal 3-oxoacyl coenzyme A thiolase with intrinsic sterol carrier and lipid transfer activity. J. Biol. Chem. 269:21277-21283.

Shanahan, P., Glennon, J. D., Crowley, J. J., Donnelly, D. F., and O'Gara, F. 1993. Liquid chromatographic assay of microbially derived phloroglucinol antibiotics for establishing the biosynthetic route to production, and the factors affecting their regulation. Anal. Chem. Acta 272:271-277.

Shen, B., Summers, R. G., Wendt-Pienkowski, E., and Hutchinson, C. R. 1995. The *Streptomyces glauescens tcmKL* polyketide synthase and *tcmN* polyketide cyclase genes govern the size and shape of aromatic polyketides. J. Am. Chem. Soc. 117:6811-6821.

Thomashow, L. S., and Weller, D. M. 1995. Current concepts in the use of introduced bacteria for biological disease control: mechanisms and antifungal metabolites. Pages 187-235 in: Plant-Microbe Interactions, vol. 1. G. Stacey and N. Keen, ed. Chapman & Hall, New York.

Tsay, J.-T., W. Ob, T. J. Larson, S. Jackowski, and C. O. Rock. 1992. Isolation and characterization of the β-ketoacyl-acyl carrier protein synthase gene (*fabH*) from *Escherichia coli* K-12. J. Biol. Chem. 267:6807-6814.

Vincent, M. N., Harrison, L. A., Brackin, J. M., Kovacevich, P. A., Mukerji, P., Weller, D. M., and Pierson, E. A. 1991. Genetic analysis of the antifungal activity of a soilborne *Pseudomonas aureofaciens* strain. Appl. Environ. Microbiol. 57:2928-2934.

# Zwittermicin A and Biological Control
# of Oomycete Pathogens

Elizabeth A. Stohl[1,2], Eric V. Stabb[1,3], and Jo Handelsman[1,2,3]

[1] Department of Plant Pathology, [2] Program in Cellular and Molecular Biology, [3] Department of Bacteriology, University of Wisconsin-Madison, Madison, WI 53706

## Introduction

*History of resistance in agriculture*

The recent history of agriculture is punctuated with the breakdown of pathogen and pest control strategies due to resistance of the target organisms to chemicals used for their control. Biocontrol has been hailed as a control measure that will be more robust than synthetic chemicals. But since antibiotics are critical to disease suppression by many biocontrol agents, breakdown of biocontrol could occur through development of antibiotic resistance in target pathogens. Antibiotic resistance is likely to be less of a problem in biocontrol than in pest and pathogen control strategies based on synthetic chemicals, since biocontrol agents often produce multiple antimicrobial compounds, making it less probable that the pathogen will develop resistance to the entire suite of inhibitory molecules involved in biocontrol. Furthermore, in biocontrol, antibiotic production is often localized on the surface of the plant, limiting the total exposure of pathogen populations to the antibiotic. However, these factors will merely slow the inevitable development of antibiotic resistance in pathogens; the past forty years of use of antimicrobials in agriculture and medicine have illustrated that if selection pressure is applied, resistant pathogens will evolve. Therefore, research is needed to understand mechanisms of antibiotic resistance before it develops in the field to contribute information to the development of models that predict appearance, spread, and evolution of resistance. Such models will provide the basis for management strategies that avoid or reduce the frequency and impact of resistance.

*Genetic bases of antibiotic resistance*

Antibiotic resistance develops in a sensitive population of microorganisms either by spontaneous mutation or horizontal gene transfer.

Mechanisms of antibiotic resistance mediated by mutation or gene transfer include antibiotic modification, target modification or replacement, and decreased net antibiotic uptake. Interestingly, the extrachromosomal resistance mechanisms found in pathogens often mirror the mechanisms of resistance found in antibiotic-producing organisms, which are known as "self-resistance" mechanisms. Increasing evidence suggests that self-resistance genes are a source of resistance in pathogens (Davies 1994). Antibiotic self-resistance genes from biocontrol agents could theoretically be transferred to the very target pathogens they are used to control. Therefore, to understand the scope of resistance mechanisms that may develop, it is necessary to study both mutations in the target organism and resistance genes in the producing organism.

*Biocontrol and zwittermicin A production by* Bacillus cereus *UW85*

Our work focuses on zwittermicin A (Fig. 1), a novel antibiotic, representing a new class of antibiotics known as aminopolyols (He et al. 1994). Zwittermicin A contributes to suppression of plant diseases by *B. cereus* UW85 (Silo-Suh et al. 1994). The novelty of this antibiotic coupled with its role in disease suppression makes resistance to zwittermicin A of particular interest. We have chosen to study spontaneous resistance to zwittermicin A in the model target organism *Escherichia coli* because of the wide array of genetic tools available for study of this bacterium. We have also isolated a gene for self-resistance to zwittermicin A from *B. cereus*, and we are studying its gene product in *E. coli*.

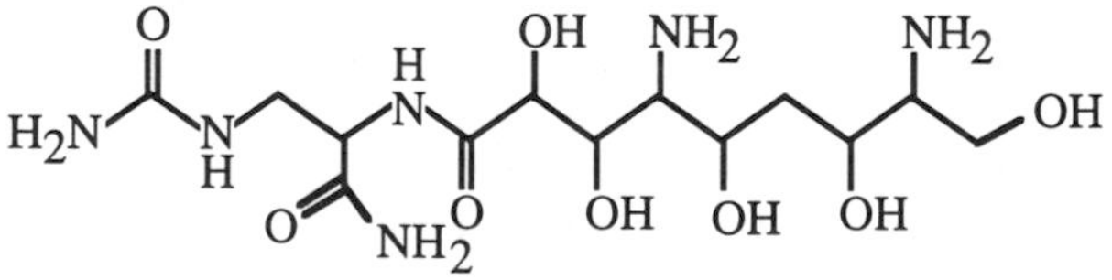

Fig.1. Structure of zwittermicin A.

## Results

*Spontaneous zwittermicin A resistance in* E. coli

The diverse microorganisms sensitive to zwittermicin A include important plant pathogens, such as the oomycetes, and well characterized model organisms such as *E. coli*. We have capitalized on the availability of powerful genetic tools (Singer et al. 1989) to study zwittermicin A resistance in *E. coli* with the intent to proceed rapidly to an understanding of zwittermicin A resistance in this organism that may be useful in understanding resistance in plant pathogens.  Therefore, as a first step toward understanding how zwittermicin A resistance arises in target organisms, we selected zwittermicin A-resistant mutants of *E. coli*. Mutants resistant to 150 µg/ml zwittermicin A arose spontaneously at a frequency of $2 \times 10^{-8}$.  Among twelve independent zwittermicin A-resistant mutants, the mutations conferring zwittermicin A resistance (ZmA$^R$) mapped to loci at 4, 9, 17, 27, 83, and 86 map units on the *E. coli* chromosome.  The mapping results, along with further genetic and biochemical studies, suggested that the mutations conferring ZmA$^R$ were in the *hemL, hemB, cyd, hemA, atp,* and *ubi* loci. (Fig. 2).

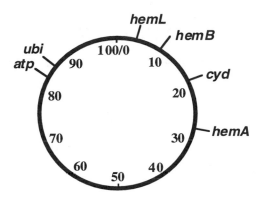

Fig. 2. Map positions of zwittermicin A-resistance loci in *E. coli*.

The *hemL, hemB, cyd, hemA,* and *ubi* genes contribute to functional electron transport chains, which generate a membrane potential ($\Delta\Psi$) by pumping protons out of the cell, while the membrane-bound ATPase product of the *atp* locus helps maintain $\Delta\Psi$ by regulating the flow of protons back into the cell.  Together, the products of these genes enable *E. coli* to generate and maintain $\Delta\Psi$ (Fig. 3).

$\Delta\Psi$ can drive the uptake of positively charged molecules.  Since zwittermicin A is positively charged, $\Delta\Psi$ may drive its uptake, and the mutants described above may be ZmA$^R$ due to a decreased $\Delta\Psi$ and decreased zwittermicin A uptake (Fig. 3).  Supporting this hypothesis, these ZmA$^R$ mutants showed increased resistance to aminoglycoside antibiotics, which are known to require $\Delta\Psi$ for maximum uptake by *E. coli* (Damper and Epstein 1981).  Furthermore, culture conditions that decrease $\Delta\Psi$, including lowering pH, raising osmotic potential (adding NaCl), or adding the uncoupler 2,4-dinitrophenol (Damper and Epstein 1981), decreased the zwittermicin A-sensitivity of wild type *E. coli* (Table 1).

Similarly, oomycetes, which are a key target for biocontrol by zwittermicin A-producing strains of *B. cereus*, maintain an electrical potential across their membranes. Although little is known about how culture conditions influence $\Delta\Psi$ in oomycetes, we found that conditions that decreased membrane potential and zwittermicin A sensitivity in *E. coli* (see above) also decreased zwittermicin A sensitivity in the oomycete *Pythium torulosum* (Table 1), suggesting a connection between membrane potential and zwittermicin A sensitivity in this plant pathogen.

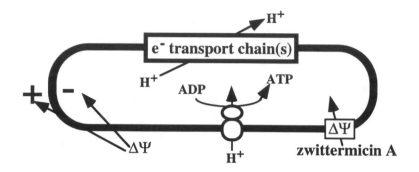

Fig. 3. Proposed model for zwittermicin A uptake by *E. coli*: Zwittermicin A requires a membrane potential for entrance into the cell.

Table 1. Effect of pH, osmolarity, and 2,4-dinitrophenol on minimum inhibitory concentration (MIC) of zwittermicin A.

| Medium | MIC |
|---|---|
| *E. coli*[a] | |
| MHMT8.1 (pH 8.1) | 30 |
| MHMT8.1 + 1 mM 2,4-dinitrophenol | 60 |
| MHMT7.1 (pH 7.1) | 100 |
| MHMT8.1 + 0.5 M NaCl | 125 |
| *P. torulosum*[b] | |
| MHMT7.5 (pH 7.5) | 10 |
| MHMT7.5 + 0.25 mM 2,4-dinitrophenol | 40 |
| MHMT6.5 (pH 6.5) | 20 |
| MHMT7.5 + 0.15 M NaCl | 60 |

[a] For *E. coli*, MIC ($\mu$g/ml) was determined by measuring the minimum concentration of zwittermicin A required to prevent visible growth in broth.

[b] For *P. torulosum*, MIC ($\mu$g/well) was determined by measuring the minimum quantity of zwittermicin A required to produce a zone of inhibition when added to a well cut in an agar plate.

Most antibiotic-producing organisms carry genes for resistance to the antibiotics they produce to prevent them from committing suicide. To identify a zwittermicin A resistance gene from *B. cereus* UW85 we constructed a genomic library from *B. cereus* UW85 and screened cosmid transformants of *E. coli* DH5α, which is sensitive to zwittermicin A, for resistance to zwittermicin A. Through subcloning and transposon mutagenesis we identified a genetic locus, designated *zmaR*, carried on a 1.2-kb fragment of DNA, that conferred zwittermicin A resistance on *E. coli,* as tested by a radial streak assay (Fig. 4). The DNA fragment carrying *zmaR* corresponded to a fragment of identical size from *B. cereus* UW85. Mutants of UW85 that were sensitive to zwittermicin A contained genomic deletions of *zmaR*, and resistance was restored by supplying *zmaR* on a plasmid. These results demonstrate that *zmaR* is functional in both *B. cereus* and *E. coli.*

Sequencing of the 1.2-kb DNA fragment identified an open reading frame, designated ZmaR, whose codon usage was typical of *B. cereus.* Cell extracts from an *E. coli* strain carrying *zmaR* contained a 43.5-kDa protein whose molecular mass and ten N-terminal amino acids matched those of the protein predicted by the *zmaR* sequence. Neither the nucleotide sequence nor the primary protein sequence had significant similarity to sequences in existing databases, suggesting that the mechanism of resistance mediated by ZmaR may be novel (Milner et al. 1996). Further work will focus on elucidating the mechanism by which ZmaR confers resistance on both *E. coli* and *B. cereus*, and determine whether ZmaR can also confer resistance on oomycetes.

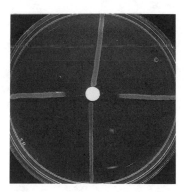

Fig. 4. Radial streak assay. 150 µg of zwittermicin A was applied to a sterile filter disk and *E. coli* cultures were tested for resistance to zwittermicin A. The strains (clockwise from top) are: Cosmid clone carrying *zmaR* in a 19-kb insert, cosmid with Tn*3* insertion in *zmaR*, 1.2-kb fragment containing *zmaR* carried on a plasmid, and the plasmid alone. A zone of inhibition around the disk indicates a sensitive culture.

## Conclusion

We studied resistance to zwittermicin A in a model organism, *E. coli*. Genetic and biochemical evidence suggests that zwittermicin A sensitivity requires a membrane potential and that resistance can be conferred by mutations in genes that contribute to electron transport or proton pumps. A gene, designated *zmaR*, from the zwittermicin A-producing organism, *B. cereus*, confers resistance to the antibiotic on sensitive strains of *E. coli* and *B. cereus*. Future work will focus on the mechanism by which *zmaR* confers resistance to zwittermicin A, the relationship between resistance and biosynthesis of the antibiotic, and on mutations in *E. coli* that confer resistance by altering the target for zwittermicin A. This work will contribute to our understanding of the biological effects of zwittermicin A and will aid in avoiding or managing antibiotic resistance to prolong the usefulness of *B. cereus* as a biocontrol agent for crop disease.

## References

Damper, D. P., and Epstein, W. 1981. Role of membrane potential in bacterial resistance to aminoglycoside antibiotics. Antimicrob. Agents Chemother. 20:803-808.

Davies, J. 1994. Inactivation of antibiotics and the dissemination of resistance genes. Science. 264:375-382.

He, H., Silo-Suh, L. A., Clardy, J., and Handelsman, J. 1994. Zwittermicin A, an antifungal and plant protection agent from *Bacillus cereus*. Tetrahedron Let. 35:2499-2502.

Milner, J. L., Stohl, E. A., and Handelsman, J. 1996. Zwittermicin A resistance gene from *Bacillus cereus*. J. Bacteriol. 178:4266-4272.

Silo-Suh, L. A, Lethbridge, B. J., Raffel, S. J., He, H., Clardy, J., and Handelsman, J. 1994. Biological activities of two fungistatic antibiotics produced by *Bacillus cereus* UW85. Appl. Environ. Microbiol. 60:2023-2030.

Singer, M., Baker, T. A., Schnitzler, G., Deischel, S. M., Goel, M., Dove, W., Jaacks, K. J., Grossman, A. D., Erickson, J. W., and Gross, C. A. 1989. A collection of strains containing genetically linked alternating antibiotic resistance elements for genetic mapping of *Escherichia coli*. Microbiol. Rev. 53:1-24.

# Plant Regulation of Bacterial Root Colonization

D.A. Phillips[1], W.R. Streit[1], H. Volpin[1], J.D. Palumbo[1,2]
C.M. Joseph[1], E.S. Sande[1], F.J. de Bruijn[3] and C.I. Kado[2]

[1]Department of Agronomy & Range Science and [2]Department of Plant Pathology,University of California, Davis, CA 95616; [3]Plant Research Laboratory and Department of Microbiology, Michigan State University, East Lansing, MI 48824

Root colonization by bacteria involves a number of processes including chemotaxis and attachment (Vande Broek and Vanderleyden, 1995), antibiotic production (Thomashow and Weller, 1988), and bacterial growth. We hypothesize that mechanisms directly increasing growth rate contribute to the specificity of root colonization and that plant signals affect such growth processes. Because over 400 naturally occurring potential signals exist in alfalfa (*Medicago sativa* L.) tissues and root exudates (Phillips and Streit, 1996), this plant serves as a useful model for studying competitive root colonization. Not all plant compounds have equally important effects on root-colonizing bacteria, but no data comparing the relative influence of different compounds have been reported from a single experimental system. Here we measure contributions by genes responsive to biotin, stachydrine, flavonoids and isoflavonoids in *Rhizobium meliloti* and *Agrobacterium tumefaciens* on alfalfa.

## Enzyme Cofactors Stimulate Root Colonization

Vitamin cofactors, such as biotin, thiamine, and riboflavin are known components of alfalfa root exudate (Rovira and Harris, 1961) that stimulate bacterial growth in pure culture (West and Wilson, 1939). Availability of these compounds in the rhizosphere can limit bacterial growth, and adding trace amounts produces a striking enhancement of root colonization by *R. meliloti* (Streit et al., 1996). Tests with auxotrophic mutants in that study showed clearly that *R. meliloti* benefits from plant-derived biotin because a biotin auxotroph doubled 3.2 times (10-fold increase in total cells) on the plant-derived biotin during the first 6 days of plant growth. Also, when wild-type *R. meliloti* 1021 was inoculated on germinating alfalfa seeds with the biotin-binding protein avidin, Rm1021 used its own biotin to double 7.9 times (Fig. 1). In the untreated control, bacterial cells doubled 10.4 times, and when 48 nanomoles of biotin were supplied to each root, bacteria doubled 13.2 times. Increasing biotin supplements to 240 nanomoles produced only marginally greater root colonization. We conclude from these results that both synthesis and uptake of biotin are important traits influencing bacterial colonization of alfalfa roots.

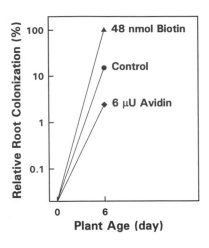

Fig. 1. Biotin effects on *R. meliloti* 1021 colonization of alfalfa roots. Biotin (nmol/plant) or avidin (μU/plant) was added on day 0, 2 and 4 to supply the total amount indicated. The 52 bacteria/plant inoculated on day 0 increased to $7.5 \times 10^5$ cells/root on day 6 in the control. Treatment effects were significantly different ($P \leq 0.01$) from the control on day 6 when root colonists reached the end of logarithmic growth in this system.

### Nutritional Substrate Effects on Root Colonization

Roots release a variety of sugars, organic acids and amino acids, but because many bacteria grow on these compounds, it is difficult to see how such molecules stimulate particular microbial species. One explanation may be that bacteria with superior uptake systems benefit more from those compounds than other bacteria. *Bradyrhizobium japonicum*, for example, grows faster than several *Enterobacter* species and *Pseudomonas fluorescens* at low concentrations of succinate (100 nM) (Humbeck et al., 1985). Novel plant compounds may enhance rhizosphere success of some bacteria. *R. meliloti,* for example, catabolizes the betaines trigonelline (Boivin et al., 1991) and stachydrine (Goldmann et al., 1994) which are released from many *Medicago* seeds (Phillips et al., 1995). In other cases like calystegins (Tepfer et al., 1988) and rhizopines (Murphy et al., 1987), normal host plants do not synthesize the compound so a microbial capacity to catabolize these compounds confers no clear benefit in the rhizosphere. Our group has measured the effect of one specialized carbon source from alfalfa seeds on root colonization by *R. meliloti*.

To test the effects of stachydrine, a mutant bank was generated in Rm1021 with *luxAB* transposable promoter probe pRL1063 (Lim et al., 1993, Wolk et al., 1991), and several stachydrine-inducible mutants were selected. Mutant Rm1021-S10, which grows poorly on stachydrine because of an impaired uptake system, grew normally on the root in the absence of any competing bacteria. However, in direct competitive colonization tests against parent strain Rm1021, the mutant was moderately impaired (Fig. 2A). When similar numbers of competing bacterial cells (1:1.1, wildtype: mutant) were inoculated on the seedling at day 0, the wildtype doubled

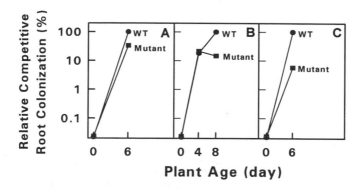

Fig. 2. Effects of mutations in plant-regulated bacterial genes on competitive colonization of alfalfa. On day 0 similar numbers of wild-type (WT) and mutant bacterial cells were inoculated on seedlings germinating in vermiculite. Mutants grew as well as the WT in single-strain colonization tests (Data not shown). Significantly different ($P \leq 0.05$) final recoveries of each pair of competing strains shown here indicates the mutated genes reduced competitive ability. A) *R. meliloti* 1021 (WT) and stachydrine-inducible, stachydrine uptake mutant Rm1021-S10 were inoculated at 500 and 560 cells/plant. B) *A. tumefaciens* 1D1609 (WT) and isoflavonoid-inducible mutant At1D1609-I1 were inoculated at 215 and 193 cells/plant. C) *R. meliloti* 1021 (WT) and flavonoid/betaine-inducible mutant TJ1A3 (Rm1021*nodC*::Tn5) (Jacobs et al., 1985) were supplied at 152 cells/plant.

12.1-fold by day 6, but the mutant doubled only 10.5 times. One can express this decrease in competitive ability either as a 13% decrease in doublings or as some larger factor that reflects the final 3:1 ratio of wild-type:mutant cells. In either case the results clearly indicate that this single mutation had a major, but not catastrophic, effect on the competitive ability of *R. meliloti* 1021 under these experimental conditions. Conversely, one might predict that adding stachydrine uptake and utilization genes to a potential rhizosphere bacterium would confer a measurable, but not overwhelming increase in the capacity to colonize alfalfa seedling roots.

## Isoflavonoid Regulation of Root Colonization

Plant isoflavonoids function both as positive inducers of bacterial genes and as negative growth factors in microbes, primarily fungi (Phillips and Kapulnik, 1995). Mutating the alfalfa pathogen *A. tumefaciens* 1D1609 with transposable promoter probe Tn5-B30 (Simon et al., 1989) tagged a chromosomal gene that is induced by two isoflavonoids, formononetin (10 µM) and coumestrol (1 µM). The mutant is still virulent on alfalfa, and it shows normal root colonization in single-strain tests. In competitive colonization tests against the isogenic wildtype, however, it is impaired (Fig. 2B). When similar numbers of cells (1.1:1, wildtype: mutant) were inoculated on the seedling at day 0, the two strains grew equally well until day 4, but then the wildtype doubled another 2.3 times while the mutant stopped growing. Our working hypothesis is that plant-derived isoflavonoids have their first effect on day 4 when they induce some physiological change favoring competitive colonization in the parent strain.

## Relationships Between Root Colonization and Nodule Formation

Studies of rhizobial competition often use nodule occupancy to measure competitive success (Triplett and Sadowsky, 1992). Nodule occupancy is a relevant test of rhizobial competitiveness, but using only that measure as a criterion for competitiveness may conceal useful information. For example, direct counts of *R. meliloti* show a positive correlation between bacterial numbers in root/rhizosphere populations and root nodule formation on several *Medicago* species (Young and Brockwell, 1992), but few data identify particular rhizobial genes that favor root colonization. Indeed, the absence of information on relationships between attachment and colonization in rhizobia contrasts markedly with evidence available for *A. tumefaciens* (Smit et al., 1992). This reasoning led us to test for direct relationships between competitive root colonization and nodule formation.

Initial experiments quantified root colonization by a mutant incapable of synthesizing Nod factor (TJIA3 = Rm1021*nodC*::Tn*5*, Jacobs, et al., 1985) in our standard alfalfa system. Single-strain trials showed no significant difference between TJ1A3 and parent Rm1021 in growth and colonization of the root. However, when equal numbers of mutant and wild-type cells were inoculated on day 0, the wildtype doubled 10 times while the mutant doubled only 5.9 times by day 6 (Fig. 2C). As a result, on day 6 there were 17.2 wild-type cells for every mutant cell on the root. It is possible that Nod factor(s) and other bacterial extracellular products facilitate attachment to the root and thus allow bacterial cells to benefit more from root exudates than comparable unattached bacteria. This is consistent with the hypothesis that firm attachment favors colonization (Smit, et al., 1992). However, our experiments detect impaired colonization in the *nodC* mutant only when related bacterial strains are competing for colonization sites. For this reason additional explanations are being sought.

## Interpretation and Practical Implications

Effects of biotin on alfalfa root colonization by *R. meliloti* were quite large (Fig. 1). Plant-derived biotin in the rhizosphere accounted for at least three bacterial cell doublings; bacterial synthesis of biotin produced another eight doublings; and three additional doublings were obtained with biotin supplements. Presumably these effects operate by increasing $CO_2$ uptake in a biotin-dependent enzyme like acetyl-CoA carboxylase. Rhizobia require $CO_2$ for growth (Lowe and Evans, 1962), and growth rate is closely linked to acetyl-CoA carboxylase activity in *Escherichia coli* (Li and Cronan, 1993). In the rhizosphere, increases in activity of this enzyme may enhance membrane formation and allow bacteria to compete more effectively for soluble catabolic substrates by increasing the surface/volume ratio.

Data reported here for *R. meliloti* and *A. tumefaciens* mutants permit a limited comparison of the extent to which different mechanisms affect alfalfa root colonization in these bacteria (Fig. 2). While none of the three mutations had a detectable effect on root colonization in single-strain tests, each impaired the capacity of the mutant to compete against the parent. Thus all three mutations truly showed competitive colonization phenotypes. In comparative terms based on final cell ratios, the *nodC* gene product made a six-fold greater contribution to competitive colonization than the capacity to take up stachydrine (17:1 vs. 3:1). Stachydrine uptake might provide a greater advantage if more of this betaine were available as an energy source, but we have no data on this point. The lesser contribution of stachydrine to competitive colonization by *Rhizobium* is consistent with data suggesting that carbon availability is not a major limitation to bacterial growth in the tobacco rhizosphere (Farrand et al., 1994).

A major problem preventing the introduction of beneficial bacteria to rhizosphere environments is competition from indigenous organisms (Thies et al., 1992). A genetic cassette that enhances root colonization by introduced bacteria may be one result of work described here.

**Acknowledgments:** This work was funded by NSF grant IBN-92-18567. W.R. Streit was supported in part by the Alexander von Humboldt Foundation; H. Volpin has a postdoctoral award from BARD, the US-Israel Binational Agricultural Research and Development Fund.

## References

Boivin, C., Barran, L. R., Malpica, C. A., and Rosenberg, C. 1991. Genetic analysis of a region of the *Rhizobium meliloti* pSym plasmid specifying catabolism of trigonelline, a secondary metabolite present in legumes. J. Bacteriol. 173:2809-2817.

Farrand, S. K., Wilson, M., Lindow, S. E., and Savaka, M. A. 1994. Modulating competition in the rhizosphere by resource utilization. Molec. Ecol. 3:619.

Goldmann, A., Lecocur, L., Message, B., Delarue, M., Schoonejans, E., and Tepfer, D. 1994. Symbiotic plasmid genes essential to the catabolism of proline betaine, or stachydrine, are also required for efficient nodulation by *Rhizobium meliloti*. FEMS Microbiol. Lett. 115:305-312.

Humbeck, C., Thierfelder, H., Gresshoff, P. M., and Werner, D. 1985. Competitive growth of slow growing *Rhizobum japonicum* against fast growing *Enterobacter* and *Pseudomonas* species at low concentrations of succinate and other substrates in dialysis culture. Arch. Microbiol. 142:223-228.

Jacobs, T. W., Egelhoff, T. T., and Long, S. R. 1985. Physical and genetic map of a *Rhizobium meliloti* nodulation gene region and nucleotide sequence of *nodC*. J. Bact. 162:469-476.

Li, S. J., and Cronan, J. E. 1993. Growth rate regulation of *Escherichia coli* acetyl coenzyme-A carboxylase, which catalyzes the first committed step of lipid biosynthesis. J. Bacteriol. 175:332-340.

Lim, P. O., Ragatz, D., Renner, M., and de Bruijn, F. J. 1993. Environmental control of gene expression: isolation of *Rhizobium meliloti* gene fusions induced by N- and C-limitation, Pages 97-100 in: Trends in Microbial Ecology. R. Guerrero and C. Pedros-Alio, ed. Spanish Society for Microbiology, Barcelona.

Lowe, R. H., and Evans, H. J. 1962. Carbon dioxide requirement for growth of legume nodule bacteria. Soil Sci. 94:351-356.

Murphy, P. J., Heycke, N., Banfalvi, Z., Tate, M. E., de Bruijn, F., Kondorosi, A., Tempe, J., and Schell, J. 1987. Genes for the catabolism and synthesis of an opine-like compound in *Rhizobium meliloti* are closely linked and on the Sym plasmid. Proc. Natl. Acad. Sci. USA 84:493-497.

Phillips, D. A., and Kapulnik, Y. 1995. Plant isoflavonoids, pathogens and symbionts. Trends in Microbiol. 3:58-64.

Phillips, D. A., and Streit, W. R. 1996. Applying plant-microbe signalling concepts to alfalfa: Roles for secondary metabolites, Pages In press in: Biotechnology and the Improvement of Forage legumes. B. D. McKersie and D. C. W. Brown, ed. CAB International, Wallingford.

Phillips, D. A., Wery, J., Joseph, C. M., Jones, A. D., and Teuber, L. R. 1995. Release of flavonoids and betaines from seeds of seven *Medicago* species. Crop Sci. 35:805-808.

Rovira, A. D., and Harris, J. R. 1961. Plant root excretions in relation to the rhizosphere effect V. The exudation of B-group vitamins. Plant and Soil 14:199-214.

Simon, R., Quandt, J., and Klipp, W. 1989. New derivatives of transposon Tn*5* suitable for mobilization of replicons, generation of operon fusions and induction of genes in gram-negative bacteria. Gene 80:161-169.

Smit, G., Swart, S., Lugtenberg, B. J. J., and Kijne, J. W. 1992. Molecular mechanisms of attachment of *Rhizobium* bacteria to plant roots. Molec. Microbiol. 6:2897-2903.

Streit, W. R., Joseph, C. M., and Phillips, D. A. 1996. Biotin and other water-soluble vitamins are key growth factors for alfalfa rhizosphere colonization by *Rhizobium meliloti* 1021. Molec. Plant-Microbe Interact. 9:330-338.

Tepfer, D., Goldmann, A., Pamboukdjaian, N., Maille, M., Lepingle, A., Chevalier, D., Dénarié, J., and Rosenberg, C. 1988. A plasmid of *Rhizobium meliloti* 41 encodes catabolism of two compounds from root exudate of *Calystegium sepium*. J. Bacteriol. 170:1153-1161.

Thies, J. E., Bohlool, B., and Singleton, P. W. 1992. Environmental effects on competition for nodule occupancy between introduced and indigenous rhizobia and among introduced strains. Canad. J. Microbiol. 38:493-500.

Thomashow, L. S., and Weller, D. M. 1988. Role of a phenazine antibiotic from *Pseudomonas fluorescens* in biological control of *Gaeumannomyces graminis* var. tritici. J. Bacteriol. 170:3499-3508.

Triplett, E. W., and Sadowsky, M. J. 1992. Genetics of competition for nodulation of legumes. Annu. Rev. Microbiol. 46:399-428.

Vande Broek, A., and Vanderleyden, J. 1995. The role of bacterial motility, chemotaxis, and attachment in bacteria-plant interactions. Molec. Plant-Microbe Interact. 8:800-810.

West, P. M., and Wilson, P. W. 1939. Growth factor requirements of the root nodule bacteria. J. Bacteriol. 37:161-185.

Wolk, C. P., Cai, Y., and Panoff, J. M. 1991. Use of a transposon with luciferase as a reporter to identify environmentally responsive genes in a cyanobacterium. Proc. Natl. Acad. Sci. USA 88:5355-5359.

Young, R. R., and Brockwell, J. 1992. Influence of soil pH on the development of symbiosis in field-grown acid-sensitive and acid-tolerant annual medics. Aust. J. Expt. Agric. 32:167-173.

# Molecular Genetic Approaches to Assessing Bacterial Habitat Composition, Modification, and Interactions on Leaves

Steven E. Lindow

Department of Plant and Microbial Biology, University of California, Berkeley, CA 94720-3110

A variety of bacteria that can be both beneficial and deleterious to plant health live on the surfaces of plants. Most plant pathogenic bacteria multiply as epiphytes on healthy plants for extended periods of time before disease is incited (Hirano and Upper 1983). Certain species such as *Pseudomonas syringae*, *Erwinia herbicola*, *Pseudomonas fluorescens*, and some pathovars of *Xanthomonas campestris* can catalyze ice formation (Gurian-Sherman and Lindow 1993; Lindow 1983; 1995). The incidence of frost damage to many frost-sensitive plant species is related directly to the numbers of epiphytic ice nucleation active bacteria (Lindow 1983; 1995). Many epiphytic bacteria can also produce the plant growth regulator, 3-indoleacetic acid (IAA). IAA production has been associated with alterations in epidermal cell growth and development which can lead to fruit russetting of pomaceous fruits such as pear and apple (Clark and Lindow 1989). Some epiphytic bacteria, by competing for limiting resources on plants and/or by producing antibiotic-like substances, can reduce the population size of deleterious bacteria, and hence contribute to the biological control of diseases and frost damage (Lindow 1995). Thus the surface of healthy leaves normally supports a relatively large ($> 10^6$ cells/cm$^2$) population of microorganisms which can influence plant health in various ways.

The nature of the leaf surface habitats occupied by epiphytic bacteria is largely unknown. It is generally believed that epiphytic bacteria exist in a commensalistic relationship with the plants on which they live, utilizing a variety of sugars, amino acids, and other organic compounds which leak from the plant as source of carbon and energy sources (Tukey 1970). The leaf surface however, is probably frequently stressful to bacterial cells, due to the high flux of ultraviolet light, frequent periods of desiccation, and other physical and chemical extremes that can occur on leaves (Dickinson 1986). The surface of most plants is highly torturous, providing a very irregular terrain at the size scale of bacterial cells. Individual cells may thus encounter physical and chemical factors which differ form cell to cell due to their high variability over small size scales. It is not possible to measure the chemical or physical characteristics of microbial habitats at the size scale of individual cells. We therefore need better tools to address the nature of such microhabitats.

Considerable evidence indicates that bacteria encounter physical or chemical signals on leaves that cannot be reproduced in culture. For example, cells of *P. syringae* exhibit considerable phenotypic plasticity since they exhibit different behaviors when on leaves than when in culture. For example, when cells were recovered from plants and re-applied to dry plants, they behaved differently than when they were grown in culture and applied to such plants (Wilson and Lindow 1993). This indicates that some component of the habitat occupied by bacteria on leaves induced expression of genes conferring adaptive traits while bacteria were on leaves but not in culture. Many genes of *P. syringae* are apparently expressed on plants but not in culture. Analysis of the bioluminescence of *P. syringae* mutants having random genomic insertions of the *lux* reporter transposon, Tn4431, revealed that up to 2% of the genes of *P. syringae* were expressed on plants but not in any of several different culture media (Cirvilleri and Lindow 1994).

The nature of microhabitats on leaves can be assessed at small scales using "biological sensors" consisting of bacterial cells harboring reporter genes fused to environmentally-responsive promoters. The product of such reporter genes must be easily measured while cells are in or on plants to be useful in such ecological studies. Thus such common reporter genes such as *lacZ* are not as useful in natural environments such as leaves as other reporter genes such as *lux*AB, *ina*Z, or *xyl*E.

Ice nucleation genes have several characteristics that have made them useful as reporter genes. Single genes confer ice nucleation in all bacterial strains examined (Warren 1995). Ice genes are all relatively large (> 3 kb) and confer the production of large outer-membrane-associated proteins which are comprised largely of tandemly-repeated domains (Gurian-Sherman and Lindow 1993; Lindow et al. 1989; Warren 1995; Wolber 1993). A 48-amino-acid tandem repeat, occurring with highest fidelity in ice nucleation proteins, apparently has a tertiary structure consisting of three paired b-hairpins (Kajava and Lindow 1993). The hydrogen bond donors and acceptor atoms on the surface of this protein domain apparently have close spatial complementarity with hydrogen bond donors and acceptors on the surface of an ice crystal (Kajava and Lindow 1993). While single ice nucleation proteins apparently are not active in ice nucleation, aggregates of this proteins catalyze ice at temperatures that increase with aggregate size (Govindarajan and Lindow 1988). Ice nucleating sites active at temperatures of about $-5$ C, a temperature commonly used in ecological studies, require the aggregation of approximately 20 ice nucleation proteins (Burke and Lindow 1990). As a consequence, the concentration of ice nuclei per cell increases exponentially with increasing concentration of ice protein in cell membranes (Lindgren et al. 1989). Therefore ice nucleation genes make very sensitive reporters of changes in the transcription of genes to which they are fused since ice nucleation activity will increase with approximately the second power of the rate of transcription of the target gene.

Biological sensors that use fusions to ice nucleation genes have proven useful for characterizing leaf surface habitats. The most extensive information on the use of such biological sensors has been obtained on the availability of ferric iron. $Fe^{+3}$ is essential for microbial growth but is expected to be in very low concentrations on leaves and on other habitats due to the insolubility of $Fe(OH)_3$ in aerobic environments. For this reason, nearly all bacteria have evolved one or more ferric ion acquisition systems involving siderophores for $Fe^+$ acquisition outside of the cell and cognate receptors for uptake of the ion-siderophore complex. The composition of microbial communities has been proposed to be determined in part by their competition for $Fe^{+3}$; those organisms most able to acquire $Fe^{+3}$ would presumably predominate in most $Fe^{+3}$-limited communities. $Fe^{+3}$-responsive promoters from a siderophore biosynthesis and uptake operon in *P. syringae* were fused to a promoterless *inaZ* gene. *P. syringae* and other *Pseudomonas* species which harbored this gene fusion exhibited iron-dependent ice nucleation activity (Loper and Lindow 1994). Application of cells that harbored this reporter gene fusion to plants revealed that $Fe^{+3}$, while not abundant, was apparently more available on plants than expected (Loper and Lindow 1994). It was possible to demonstrate however, that $Fe^{+3}$ availability could be altered by microbial processes on leaves. For example, the availability of $Fe^{+3}$ sensed by *P. syringae* cells on leaves co-inoculated with *P. fluorescens* strain A506 (which produces a siderophore with higher efficiency than *p. syringae*) was lower than on leaves in which only *P. syringae* predominated (Tsushima & Lindow, unpublished data). Thus while standard analytical methods cannot measure the availability of resources available to bacteria in their microhabitats, nor assess changes in such availability, biological sensors provide direct evidence for biologically-pertinent environmental constituents *in situ*.

A variety of other environmental constituents can now be assessed by using appropriate environmentally-responsive promoters fused to reporter genes. For example, we have produced reporter gene fusions capable of assessing the bioavailability of $Cu^{+2}$ (Rogers et al. 1994), tryptophan (Clark et al. 1992), $Zn^{+2}$ (Rogers et al. 1994), and sucrose and fructose (Miller & Lindow, unpublished data). These biological sensors promise to provide accurate estimates of the average abundance of specific environmental constituents available to cells in a sample. The use of new reporter genes such as green fluorescent protein (Chalfie et al., 1994), through which the fluorescence intensity of individual cells can be assessed as an indication of the transcriptional activity of the genes to which it is fused, promises to make it possible to estimate not only the average availability of environmental constituents but also the heterogeneity in availability of resources which cells in a population encounter. Obviously, as further environmentally-responsive promoters are identified, their fusion to appropriate reporter genes should provide the necessary tools to better understand the ecology of plant-associated bacteria.

Literature Cited

Burke, M. J., and Lindow, S. E. 1990. Surface properties and size of the ice nucleation site in ice nucleation active bacteria: theoretical considerations. Cryobiology 27:80-84.

Chalfie, M., Tu, Y., Euskirchen, G., Ward, W. W., and Prasher, D.C. 1994. Green fluorescent protein as a marker for gene expression. Science 263:802-805.

Cirvilleri, G., and Lindow, S. E. 1994. Differential expression of genes of *Pseudomonas syringae* on leaves and in culture evaluated with random genomic *lux* fusions. Molec. Ecol. 3:249-257.

Clark, E. and Lindow, S. E. 1989. Indoleacetic acid production by epiphytic bacteria associated with pear fruit russetting. Phytopathology 79:1191.

Clark, E., Brandl, M., and Lindow, S. E. 1992. Aromatic aminotransferase genes from an indoleacetic acid-producing *Erwinia herbicola* strain. Phytopathology 82:1100.

Dickinson, C. H. 1986. Adaptations of micro-organisms to climatic conditions affecting aerial plant surfaces, Pages 77-100 in: Microbiology of the phyllosphere. N. J. Fokkema and J. van den Heuvel eds. Cambridge University Press, New York.

Govindarajan, A. G., and Lindow, S. E. 1988. Size of bacterial ice nucleation sites measured *in situ* by gamma radiation inactivation analysis. Proc. Natl. Acad. Sci. (USA) 85:1334-1338.

Gurian-Sherman, D., and Lindow, S. E. 1993. Bacterial ice nucleation: significance and molecular basis. FASEB J. 9:1338-1343.

Hirano, S. S., and Upper, C. D. 1983. Ecology and epidemiology of foliar bacterial plant pathogens. Annu. Rev. Phytopathol. 21:243-269.

Kajava, A., and Lindow, S. E. 1993. A molecular model of the three-dimensional structure of bacterial ice nucleation proteins. J. Mol. Biol. 232:709-717.

Lindgren, P.B., Frederick, R., Govindarajan, A. G., Panopoulos, N. J., Staskawicz, B. J., and Lindow, S.E. 1989. An ice nucleation reporter system: Identification of inducible pathogenicity genes in *Pseudomonas syringae* pv. *phaseolicola*. EMBO J. 8:1291-1301.

Lindow, S. E., Lahue, E., Govindarajan, A. G., Panopoulos, N. J., and Gies, D. 1989. Localization of ice nucleation activity and the *iceC* gene product in *Pseudomonas syringae* and *Escherichia coli*. Mol. Plant-Microbe Interact. 2:262-272.

Lindow, S.E. 1995. Control of epiphytic ice nucleation-active bacteria for management of plant frost injury. Pages 239-256 in: Biological Ice Nucleation And Its Applications. R. E. Lee, G. J. Warren, and L. V. Gusta, eds. American Phytopathological Society Press, St. Paul, MN.

Lindow, S. E. 1983. The role of bacterial ice nucleation in frost injury to plants. Annu. Rev. Phytopathology 21:363-384.

Loper, J. E., and Lindow, S. E. 1994. A biological sensor for iron available to bacteria in their habitats on plant surfaces. Appl. Environ. Microbiol. 60:1934-1941.

Rogers, J.S., Clark, E., Cirvilleri, G., and Lindow, S. E. 1994. Cloning and characterization of genes conferring copper resistance in epiphytic ice nucleation active *Pseudomonas syringae* strains. Phytopathology 84:891-897.

Tukey, H. B. J. 1970. The leaching of substances from plants. Annu. Rev. Plant Physiol. 21:305-324.

Warren, G.J. 1995. Identification and analysis of *ina* genes and proteins. Pages 85-100 in: *Biological Ice Nucleation And Its Applications*. R. E. Lee, G. J. Warren, and L. V. Gusta, eds. American Phytopathological Society Press, St. Paul, MN.

Wilson, M., and Lindow, S. E. 1993. Effect of phenotypic plasticity on epiphytic survival and colonization by *Pseudomonas syringae*. Appl. Environ. Microbiol. 59:410-416.

Wolber, P. K. 1993. Bacterial ice nucleation. Adv. Microb. Physiol. 34:203-237.

# Exploring the Microbial Diversity and Soil Management Practices to Optimize the Contribution of Soil Microorganisms to Plant Nutrition

**Mariangela Hungria[1], and Milton A. T. Vargas[2]**
[1]EMBRAPA-CNPSo, Londrina, PR, Brazil; [2]EMBRAPA-CPAC, Planaltina, DF, Brazil.

The last decade has been characterized by great progresses towards the understanding of the molecular basis of several plant-microbe interactions. The exciting idea about the knowledge obtained in this area is that, in a near future, it will be probably possible to manipulate the genes of both host plants and microsymbionts, increasing the benefits obtained from processes such as biological $N_2$ fixation and association with mycorrhizal fungi. For example, the elucidation of the mechanisms associated with the molecular communication between legumes and rhizobia can lead to the enhancement of nodulation, either by the inoculation with a strain that would require less or no plant inducers for transcription of nodulation (*nod*) genes, or by extending the nodulation ability to other non-host plants. However, it is important to remember always that nature can provide, very often and with less effort, efficient and competitive microsymbionts. It is also important to remind that a proper soil management is essential to keep the biodiversity and to optimize the benefits which will result from microbiological processes.

*Looking for microbial diversity as a source of agriculturally important microorganisms*. The great diversity and capacity of bacteria to adapt to a wide range of environmental conditions make the soil an important reservoir of strains and, whenever necessary, the selection for specific characteristics can be proceeded. A successful selection occurred in the Brazilian savannas, called "Cerrado", where soybean (*Glycine max*) was introduced as a new crop in the 60's. The area was originally free of soybean bradyrhizobia and the first inoculants carried the strain SEMIA 566, which was recommended until 1978. In the last decade, the $N_2$ fixing capacity with the available strains started to decrease in the soybean fields, urging the selection of more efficient and competitive strains. Since the success of new strains was often restricted by the "Cerrado" stressed environment, which is characterized by long periods of water stress and high temperatures, a research program was started aiming the isolation of strains "adapted" to these inhospitable conditions. Soybean was then sowed in an area which had been inoculated with the

SEMIA 566 strain, about 15 years ago, with no further inoculation. Bacteria were isolated from the good size and color nodules and tested individually for $N_2$ fixation capacity under greenhouse conditions and, subsequently, for grain yield in field experiments. From several isolates, one was selected, the CPAC 15, which consistently increased yield, and this strain began to be commercially recommended from 1992 (Vargas et al. 1992). CPAC 15 belongs to the same serogroup as SEMIA 566, so it must be a natural variant obtained by several years of soil stressing conditions, which caused morphological, physiological and genetic changes (Nishi et al. 1996), as well as in the Hai (hair induction) root phenotype (Table 1).

*Obtaining more efficient strains with traditional methods.* Another approach to solve the decrease in soybean $N_2$ fixation rates in the "Cerrado" region consisted of searching for natural variability in a very efficient but low competitive strain, CB 1809, brought from Australia to Brazil in 1966. Dozens of individual colonies from a pure culture were tested in Leonard jars for competitiveness, and a promising isolate was identified, the CPAC 7. With more nodules occupied by this efficient strain, the rates of $N_2$ fixation increased and so did the yield in several field trials. Consequently, the strain also started to be commercially recommended since 1992 (Vargas et al. 1992). A positive and important remark in this traditional breeding method is that the stability of the strain is very high. When the natural variant CPAC 7 was compared with the parental strain CB 1809, differences were detected not only in Hai phenotype (Table 1), but also in the profile of Nod metabolites (lipo-chitin oligosaccharides, LCO's) (Hungria et al., this Congress). Consequently, these distinct diffusable molecules may be responsible for the differences detected in both Hai phenotype and competitiveness. Other morphological, physiological and genetic differences between the pair of strains CB 1809 and CPAC 7 were also reported (Nishi et al. 1996).

**Table 1**. Root Hai phenotype (number of root hairs/field of a Neubauer chamber) induced by the inoculation with filter sterilized (2 µm) bacteria exudates, produced in N free minimum medium supplied with 2 µM of genistein, and effect of the inoculation with these strains, in a soil of the State of Paraná, Brazil, with established population ($10^4$ cells of bradyrhizobia/g soil), on the yield of cultivar BR-37 of soybean.

| Treatment | Hai  phenotype | Yield (kg/ha) |
|---|---|---|
| Control | 22.1 c[a] | 2,305 c[a] |
| Control + N[b] | 20.2 c | 3,450 a |
| SEMIA 566 | 55.7 b | 2,708 b |
| CPAC 15 | > 100 a | 3,402 a |
| CB 1809 | 44.6 b | 2,355 c |
| CPAC 7 | > 100 a | 3,444 a |
| CV (%) | 23 | 18 |

[a] Means of six replicates and values followed by the same letter are not statistically different (Tukey, P<0.05).
[b] 20 mM of N for Hai phenotype and 200 kg of N/ha at the field.

*Looking for soil management practices which will maintain microbial diversity and improve the benefits of microbial processes.* Agricultural sustainability relies on practices that must include an adequate habitat for soil microorganisms, maximizing the contribution of microbiological processes to plant nutrition. In this context, some agricultural practices, as no-tillage and crop rotation with legumes, have proved to be microbiologically and economically attractive. Indeed, the comparison between plow-till and no-till systems, in several experiments performed in the State of Paraná, in the South Region of Brazil, have shown the benefits of sowing directly through the mulch on soil microbiological communities. In one of these experiments, carried out in soils cropped with bean (*Phaseolus vulgaris*), grown under plow-till and no-till systems, for three years, quantitative differences were detected in several populations of microorganisms (Table 2). Qualitatively, in the no-till system there were more species of vesicular arbuscular mycorrhizal fungi (VAMF), higher diversity of nitrifying bacteria and a higher percentage of *Rhizobium tropici* and *R. etli*, which are more efficient in the bean $N_2$ fixation process (Table 2).

**Table 2.** Effects of different tillage systems on some quantitative (microbial soil biomass, soil rhizobia population, ß-galactosidase activity of soil extracts, nodulation and yield of bean, cultivar IAPAR-54) and qualitative (distribution of rhizobial species, number of species of vesicular arbuscular mycorrhizal fungi, VAMF, and number of DNA profiles detected in 50 nitrifying bacteria isolates, N-bact.) parameters. Experiment performed in the State of Paraná, Brazil.

| Tillage system | Soil biomass ($\mu g\ C/g^a$) | Rhizobia (cells/$g^a$)[b] | ß-gal. activity ($U/g^a$) | Nodulation (mg/pl) | Yield (kg/ha) |
|---|---|---|---|---|---|
| | | Quantitative differences | | | |
| Plow-till | 34 b[c] | $2.5 \times 10^3$ b | 40 b | 80 b | 1,882 b |
| No-till | 55 a | $3.9 \times 10^4$ a | 65 a | 113 a | 2,324 a |
| CV (%) | 19 | 31 | 28 | 11 | 14 |
| | | Qualitative differences | | | |
| | *R. tropici* | *R. etli* | *R.* spp.[d] | VAMF (number) | N-bact. (number) |
| | ( ---------------- % ---------------- ) | | | | |
| Plow-till | 64.8 b[c] | 3.5 b | 31.7 a | 3 | 11 |
| No-till | 80.3 a | 9.8 a | 9.9 b | 5 | 7 |
| CV (%) | 17 | 9 | 21 | - | - |

[a] Per g of dry soil.

[b] Rhizobia able to nodulate bean or leucaena (*Leucaena leucocephala*).

[c] Means of six replicates and values for each parameter, in the same column, followed by the same letter, are not statistically different (Tukey, P<0.05).

[d] Rhizobia able to nodulate exclusively leucaena.

*Application of the information obtained for agronomic benefits*. In the last few years, reports from several laboratories helped to elucidate the mechanisms related with the mutual exchange of diffusable signal molecules between rhizobia and the host plants. In a first step, each legume host exudes signals, mostly flavonoids, which induce the transcription of bacterial *nod* genes, whose protein products are essential for the infection process. However, only limited efforts were made to bring this basic information to practical use, although a few studies have indicated that nodulation could be limited by the availability of *nod* gene inducers. In Brazil, increases in nodulation, $N_2$ fixation activity, and grain yield, due to the addition of 40 µM of genistein to the inoculant, were observed under field conditions for both soybean (Table 3), and bean. Also, under stress, such as high temperatures and low pH, the decrease in nodulation can be alleviated by an extra supply of *nod* gene inducers (Hungria and Stacey 1996). Consequently, it may be possible to add inducers to the inoculants or to select host varieties which would release more *nod* gene inducers. Furthermore, certain soil management practices, such as no-tillage (Table 2) and crop rotation or intercrop with legumes (Hungria and Stacey 1996), allow higher accumulation of *nod* gene inducers in soils, what could explain the superior nodulation and $N_2$ fixation rates reported in several experiments performed with these systems. Consequently, the information obtained in basic understanding should be more frequently used for obtaining agronomic benefit.

**Table 3**. Effects of inoculation with *Bradyrhizobium* strains CPAC 15 and CPAC 7, with or without a supply of 40 µM of genistein, on nodule dry weight (NDW) at 25 days after sowing and yield of cultivar BR-37 of soybean. Experiment performed in the State of Paraná, Brazil, in a soil with $10^4$ cells of bradyrhizobia/g of soil.

| Treatment | NDW (mg/pl) | Yield (kg/ha) |
|---|---|---|
| Control | 15.3 b[a] | 3,281 c |
| Control + 200 kg N/ha | 7.2 c | 3,935 a |
| Inoculation | 17.1 ab | 3,582 b |
| Inoculation + Inducer | 19.4 a | 3,913 a |
| CV (%) | 18 | 14 |

[a] Means of six replicates and values followed by the same letter are not statistically different (Tukey, P<0.05).

## LITERATURE CITED

Hungria, M., and Stacey, G. 1996. Molecular signals exchanged between host plants and rhizobia: Basic aspects and potential application in agriculture. Soil Biol. Biochem. (in press).

Nishi, C. Y. M., Boddey, L. H., Vargas, M. A. T., and Hungria, M. 1996. Morphological, physiological and genetic characterization of two new *Bradyrhizobium* strains recently recommended as Brazilian commercial inoculants for soybean. Symbiosis 20: 147-162.

Vargas, M. A. T., Mendes, I. C., Suhet, A. R., and Peres, J. R. R. 1992. Duas novas estirpes de rizóbio para a inoculação da soja. EMBRAPA-CPAC, Planaltina, DF. (Comunicado Técnico 62).

# Rep-PCR Genomic Fingerprinting of Plant-Associated Bacteria and Computer-Assisted Phylogenetic Analyses

F.J. de Bruijn[1,2,3]*, J. Rademaker[2], M. Schneider[1,2]
U. Rossbach[1] and F.J. Louws[1,2,4]

[1]MSU-DOE Plant Research Laboratory, [2]NSF Center for Microbial Ecology, [3]Dept. of Microbiology, Michigan State University, E. Lansing, MI 48864, USA; [4]Dept. of Plant Pathology, N. Carolina State University, Raleigh, NC 27695

The utility of a recently developed method to classify bacteria on the basis of their genomic fingerprint patterns was investigated, using collections of both symbiotic and pathogenic plant-associated bacteria. The genomic fingerprinting method employed is based on the use of DNA primers corresponding to naturally occurring interspersed repetitive elements in bacteria, such as the REP, ERIC and BOX elements, and the PCR reaction (rep-PCR). We have been able to show that rep-PCR fingerprinting is a highly reproducible and simple method to distinguish closely related strains, to deduce phylogenetic relationships between strains and to study their diversity in a variety of ecosystems. We have also used computer assisted pattern analysis programs (e.g. GelCompar) for microbial identification and phylogenetic analysis of complex data sets, and describe the application of these methods for the creation of databases for bacterial diagnosis.

## Introduction

The identification and classification of symbiotic and pathogenic plant-associated bacteria are important both in terms of their agricultural applications, as well as for basic studies on plant-microbe interactions. A variety of phenotypic methods has been traditionally used to type these bacteria, including serotyping, phage typing, microscopic identification, substrate utilization screening (e.g. BIOLOG), multilocus enzyme electrophoresis (MLEE), fatty acid methyl ester analysis (FAME), 2-D PAGE of total proteins, and intrinsic antibiotic resistance profiling.

In addition, in the case of symbiotic bacteria, such as rhizobia, plant nodulation tests are performed, and in the case of pathogenic bacteria, pathogenicity tests on specific hosts are carried out. The majority of these techniques require purification and cultivation of the bacteria and/or can be quite laborious and time consuming.

More recently, DNA-based (genotypic) approaches have increasingly been applied to microbial identification and classification. In fact, these "molecular" approaches have resulted in the birth of a new ecology subspecialty, Molecular Microbial Ecology (see Akkermans et al. 1995). Generally, these methods tend to be less dependent on bacterial growth variables, more 'stable', less time-consuming and are very useful for determining phylogenetic relationships between microbial isolates and for assigning strains into specific groups. Such methods include DNA-DNA hybridization studies (which constitute the basis for bacterial genus/species designations), characterization of rRNA sequences, and other methodologies, including digestions of total genomic DNA with infrequently (rare) or frequently cutting restriction enzymes, followed by pulse field gel electrophoresis (PFGE) or hybridization with specific probes (RFLP), respectively, and plasmid profiling (see Louws et al. 1996 for a discussion of the relative utilities of these techniques and references).

Clearly, the utility of DNA-based approaches has been enhanced tremendously by the application of PCR. A most useful, application of PCR to bacterial identification and classification has been in the area of genomic fingerprinting using, for example, primers corresponding to endogenous interspersed repetitive sequences (Lupski and Weinstock, 1992; de Bruijn, 1992; Versalovic et al. 1991). The latter approach has been named rep-PCR genomic fingerprinting and will be discussed here.

## Interspersed repetitive DNA elements and rep-PCR fingerprinting.

As mentioned above, naturally occurring interspersed repetitive DNA elements, found in many (if not all) bacteria, can serve as primer sites for genomic DNA amplification (Versalovic et al. 1991;1994; de Bruijn 1992). Several families of repetitive sequences are interspersed throughout the genome of diverse bacterial species (see Lupski and Weinstock 1992). Three families of repetitive sequences have been studied in most detail, including the 35-40 bp repetitive extragenic palindromic (REP) sequence, the 124-127 bp enterobacterial repetitive intergenic consensus (ERIC) sequence, and the 154 bp BOX element (see Versalovic et al. 1994). These sequences appear to be located in distinct, intergenic positions all around the chromosome. The repetitive elements may be present in both orientations on the chromosome, and PCR primers have been designed to "read outward" from the inverted repeats in REP and ERIC, and from the boxA subunit of BOX (Versalovic et al. 1994). The use of the above primer(s) and PCR leads to the selective amplification of distinct genomic regions located between REP, ERIC or BOX sequences. The corresponding protocols are referred to as REP-PCR, ERIC-PCR and BOX-PCR, respectively, and **rep-PCR** collectively (Versalovic et al. 1991;1994). Amplified bands are size fractionated through a gel matrix to yield fingerprint patterns resembling "bar codes" (Lupski, 1993), analogous to UPC codes used in grocery stores, and function as a signature for specific bacterial strains (see Figure 1).

# rep-PCR genomic fingerprinting protocol

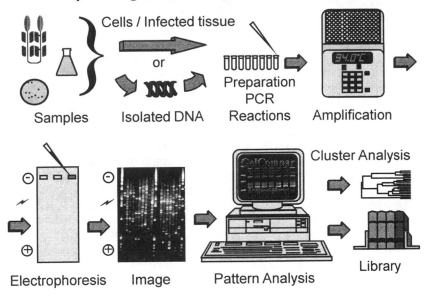

Figure 1. General rep-PCR genomic fingerprinting protocol using different templates.

## Rep-PCR genomic fingerprinting of plant-associated bacteria using different templates

For the identification and classification of rhizobial strains from soils or from nodules induced on plants, or pathogenic bacteria from lesions on plant leaves and fruits, it is particularly useful to have a rapid, simple and highly reproducible method that is not absolutely dependent on purified DNA. We, and others, have been able to show that the rep-PCR fingerprinting patterns of purified rhizobial DNA, single colonies from a plate, liquid cultures or direct extracts of nodule tissues are identical in most cases (de Bruijn et al. 1992; Nick and lindstrom 1994; Schneider and de Bruijn 1996). The same applies to purified DNA versus whole cells and extracts from plant lesions caused, for example, by xanthomonads (Louws et al. 1994; 1995; 1996). Thus, rep-PCR can be carried out with different templates (see Figure 1).

## Computer assisted phylogenetic pattern analysis programs based on rep-PCR generated genomic fingerprints.

The banding patterns generated by rep-PCR genomic fingerprinting analysis of large collections of strains are too complex to analyse by eye. Therefore, we have tested the utility of several computer assisted banding pattern determination and phylogenetic analysis programs.

We previously reported the use of the AMBIS System (Scanalytics, Waltham, Mass. USA) for phylogenetic analysis of rep-PCR generated genomic fingerprints of different rhizobia (Rossbach et al. 1995; Schneider and de Bruijn 1996).

More recently, we have been using the GelCompar system (Applied Math, Kortrijk, Belgium) for our analyses, since it appears to be more user friendly, more versatile and allows the construction/screening of rep-PCR generated "bar codes" for diagnostic purposes (see below), as well as for phylogenetic analyses.

We have applied these analyses to the identification and classification of plant pathogenic xanthomonads (e.g. Louws et al. 1994;1995). In order to test our hypothesis that rep-PCR generated fingerprints directly reflect genomic structure, we carried out a cluster analysis of 19 *Xanthomonas* strains belonging to 6 homology groups (Vauterin et al. 1995). Using the GelCompar program, we analysed the combined REP, ERIC and BOX fingerprints of the 19 strains (kindly provided by Drs. Swings and Vauterin, University of Ghent, Belgium) in duplicate. The results of this analysis is shown in Figure 2 and reveals that: 1. Virtually all duplicates ended up alligned next to eachother; 2. The rep-PCR generated groupings correspond directly to the DNA homology groups (numbered on top); 3. Even very closely related strains can be distinguished by this method.

We are also interested in being able to use the rep-PCR fingerprinting method as a tool for strain diagnosis. For this purpose, we are constructing a database of rep-PCR patterns of a large collection (600-800) of *Xanthomonas* isolates using the GelCompar (library search) program, in collaboration with the laboratory of Prof. Swings (Ghent, Belgium). A limited database has been constructed and pilot searches with unknown input strains conducted. The fingerprint pattern of an "unknown" (*X. cassavae*) isolate was run through the library of fingerprints, and the four most closely related patterns were found to be indeed from *X. cassavae* strains present in the database.

## Conclusions

In conclusion, we believe that rep-PCR genomic fingerprinting coupled to computer assisted phylogenetic analysis and library search programs will constitute a useful method for the identification or diagnosis of plant pathogenic, as well as symbiotic bacteria (see also Versalovic et al. 1994; Schneider and de Bruijn 1996; Louws et al. 1996 and references contained therein). This technique has already been successfully used to study the population structure of plant pathogens, to follow green house infections, to examine nodule occupancy, to identify (brady)-rhizobial strains that could not be distinguished by any other method, and to carry out phylogenetic analyses of world-wide collections of rhizobia and plant pathogens (for references see Versalovic et al., 1994; Louws et al. 1996). It therefore is another useful molecular approach in molecular microbial ecology.

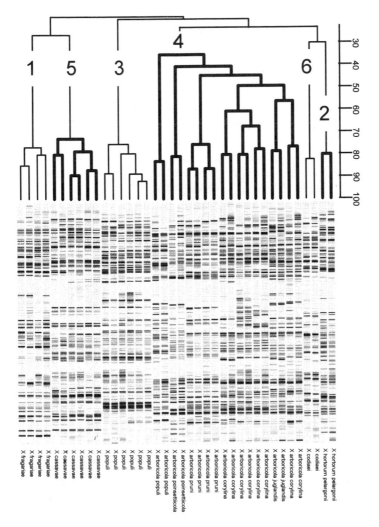

Figure 2. Cluster analysis of 19 *Xanthomonas* strains belonging to 6 DNA homology groups using combined REP, BOX and ERIC fingerprints (Pearson correlation; UPMGA).

**Acknowledgements**

This work has been supported by the DOE (DE FG 0290ER20021, The NSF Center for Microbial Ecology (DIR 8809640) and the Consortium for Plant Biotechology Research.

# Literature Cited.

Akkermans, A.D.L., van Elsas, J.D., and de Bruijn, F.J. 1995. Molecular Microbial Ecology Manual. Kluwer Academic Publishers, Dordrecht, The Netherlands, pp 1–488

de Bruijn, F.J 1992. Use of repetitive (repetitive extragenic palindromic and enterobacterial repetitive intergenic consensus) sequences and the polymerase chain reaction to fingerprint the genomes of *Rhizobium meliloti* isolates and other soil bacteria. Appl. Environ. Microbiol. 58: 2180-2187

Louws, F.J., Fulbright, D.W., Stephens, C.T. and de Bruijn, F.J. 1994. Specific genomic fingerprints of phytopathogenic *Xanthomonas* and *Pseudomonas* pathovars and strains generated with repetitive sequences and PCR. Appl. Environ. Microbiol. 60:2286-2295.

Louws, F.J., Fulbright, D.W., Stephens, C.T., and de Bruijn, F.J. 1995. Differentiation of genomic structure by rep-PCR fingerprinting to rapidly classify *Xanthomonas campestris* pv. *vesicatoria*. Phytopath. 85:528-836.

Louws, F.J., Schneider, M., and de Bruijn, F.J. 1995. Assessing genetic diversity of microbes using repetitive-sequence-based PCR (rep-PCR). In: Toranzos, G., Ed., Nucleic Acid Amplification Methods for the Analysis of Environmental Samples. Technomic Publishing Co., In Press.

Lupski, J.R. 1993. Molecular epidemiology and its clinical application. JAMA 270:1363-1364.

Lupski, J.R. and Weinstock, G.M.. 1992. Short, interspersed repetitive DNA sequences in prokaryotic genomes. J. Bacteriol. 174:4525-4529.

Nick, G. and Lindstrom, K. 1994. Use of repetitive sequences and the polymerase chain reaction to fingerprint the genomic DNA of *Rhizobium galegae* strains and to identify the DNA obtained by sonicating the liquid cultures and root nodules. Syst. Appl. Microbiol. 17:265-273.

Rossbach, S.R., Rasul, G., Schneider, M., Eardley, B., and de Bruijn, F.J. (1995) Structural and functional conservation of the rhizopine catabolism (*moc*) locus is limited to selected *Rhizobium meliloti* strains and unrelated to their geographical origin. Mol. Plant–Microbe Interact. 8: 549–559

Schneider,M., and de Bruijn, F.J. 1996. Rep–PCR–mediated genomic fingerprinting of rhizobia and computer–assisted phylogenetic pattern analysis. World J. Microbiol. and Biotechnol., 12: 163-174.

Vauterin, L., Hoste, B., Kersters, K., and Swings, J. 1995. Reclassification of *Xanthomonas*. Intl. J. Systematic Bacteriol. 45: 472–489

Versalovic, J., Koeuth, T., and Lupski, J.R. 1991. Distribution of repetitive DNA sequences in eubacteria and application to fingerprinting of bacterial genomes. Nucl. Acids Res. 19: 6823-6831

Versalovic, J., Schneider, M., de Bruijn, F.J., and Lupski, J.R. 1994). Genomic fingerprinting of bacteria using repetitive sequence based PCR (rep-PCR). Meth.Cell. Mol. Biol. 5: 25-40

# COMMENTS ON *RHIZOBIUM* SYSTEMATICS. LESSONS FROM *R. TROPICI* AND *R. ETLI.*

E. Martínez-Romero. Centro de Investigación sobre Fijación de Nitrógeno, UNAM. Ap. P. 565-A. Cuernavaca, Morelos, México.

*Rhizobium* species form nitrogen fixing nodules in the Leguminosae. There are around 17000-19000 legume species, yet only a few *Rhizobium* species have been reported. To propose new species a comprehensive description is requested, but what does that description mean in terms of the biology of the bacteria? *R. etli* and *R. tropici* will be taken as examples to illustrate the complexities of *Rhizobium* systematics. These species form nitrogen fixing nodules on *Phaseolus vulgaris* bean.

## *R. tropici*

*R. tropici* strains were first designated as *R. phaseoli* type II strains since they differed from the other *R. phaseoli* strains by their *nif* gene organization and their host range (Martínez *et al.* 1988). Originally, *R. tropici* strains were isolated from nodules of bean plants in acid soils in South America. They are clearly separated from *R. etli* and from other *Rhizobium* sp. by a variety of criteria, including multilocus enzyme electrophoresis (MLEE), total DNA-DNA homology, the sequence of ribosomal genes, patterns of hybridization to different probes, the type of Nod factor produced and by other characteristics (reviewed in Martínez-Romero and Caballero-Mellado, 1996). The thorough characterization of *R. tropici* was considered as a model to follow when describing new *Rhizobium* species (Graham *et al.* 1991) since its proposal was the result of sequential analyses reported in many papers.

Among *R. tropici* strains, two groups designated type A and type B, could be distinguished (Martínez-Romero *et al.* 1991). Their status as two subspecies or even as two separate species has been discussed (Geniaux *et al.* 1995). In addition to type B strains, there are other *R. tropici* strains that could not be clearly classified as either type A or type B strains.

By MLEE, each type constitutes an independent cluster (Martínez-Romero *et al.* 1991), type B strains being more diverse (Mean genetic diversity, H=0.379), than type A isolates (H=0.108) (J. Caballero-Mellado, personal communication). Type A and type B *R. tropici* strains share low

DNA-DNA homology (aprox. 36%) (Martínez-Romero, 1994) and they have differences in the sequence of their 16S rRNA genes. *R. tropici* type A strains have an additional insertion sequence in the 16S rRNA gene that is uncommon or not present in type B strains (Amarger *et al.* 1994; Martínez-Romero, unpublished). Type A and Type B strains have distinct glutamine synthetase GSII isoforms and different hybridization patterns with *glnA*. Polymorphisms in GSII are useful as markers of different *Rhizobium* species (Taboada *et al.* 1996).

By numerical taxonomy, type A and type B groups are as separated from one another as they are from *R. etli* (Martínez-Romero *et al.* 1991). Our new results on carbon usage with 98 carbon sources showed that type A and type B strains were so separated that *R. meliloti*, *R. fredii*, *R. loti* and *R. leguminosarum* strains were intermediate among them. A derivative of CFN299 (type A) with a large deletion of the symbiotic plasmid (Poupot *et al.* 1993), had a different carbon usage pattern than CFN299. Catabolic genes borne on plasmids seem to be a common feature in *Rhizobium* (M.F. Hynes, personal communication) and may partially explain discrepancies of phenotypic and genotypic analysis.

All of the *R. tropici* strains tested harbor megaplasmids (>1500 Kb), and those megaplasmids carrying *exo* genes are type specific (Geniaux *et al.* 1995) A plasmid smaller that the symbiotic plasmid (called pb) with a size of around 200 Kb has been found in *R. tropici* type A strains but not in the type B strains analyzed. In bean plants, pb promotes an increase in nodule number of the *R. etli* transconjugants that acquire it (Martínez-Romero and Rosenblueth, 1990) and it improves the symbiotic performance of the *Agrobacterium tumefaciens* strains that also carry the *R. tropici* symbiotic plasmid (Martínez *et al.* 1987). Genes whose expression is increased with bean exudates (but not with flavonoids) have been located in pb.

In contrast, the symbiotic plasmids of *R. tropici* type A and type B strains are highly similar. The origin of replication of the sym plasmids of both *R. tropici* type A and type B strains seems to be the same (N. Toro, personal communication). Both types possess an extra copy of a plasmid borne citrate synthase gene. The *nifH* and *nodABC* gene hybridization patterns are identical. The organization of the *nodHPQ* genes is the same and the sequence of these genes is practically identical with only two nucleotide substitutions when CFN299 (type A) and CIAT899 (type B) were compared (Laeremans *et al.* 1996). In addition, Nod factors from both strains (Poupot *et al.* 1993; Folch-Mallol *et al.* 1996) have many common features.

The host range of CIAT899 and CFN299 is almost identical (Hernández-Lucas *et al.* 1995). The fact that *R. tropici* A and B share a highly similar sym plasmid brings them together again constituting a genospecies. It seems that closely related lineages may be exchanging genetic material more easily.

By the sequence of 16S ribosomal genes, it was determined that the closest relatives to *R. tropici* are *Agrobacterium rhizogenes* (bv. 2) and *Agrobacterium* spp. strains. The latter were isolated in Japan from kiwi and cherry tumors. Two of these strains have been analyzed by us. Chag3 and Kag4 have megaplasmids smaller in size than those of *R. tropici*. *R. tropici* type A strains, but not type B strains, produce a new type of auxinic compound in the presence of tryptophan. CFN299 strains cured of the symbiotic plasmid or of the pb plasmid are not affected in the synthesis of this compound. Chag4, but not KAg3, is capable of producing this compound as well, Chag4 shares characteristics both of type A and type B strains and thus may be an example of recombination of the two types. The *R. tropici-R. leguminosarum-R. etli* cluster is mainly symbiotic, the existence of related agrobacteria could be explained if a plasmid from *Agrobacterium* colonized some of these bacterial backgrounds (those more related to *R. tropici* in this case).

*R. tropici* has not been encountered in Mexican soils in areas where bean diversified but it has been isolated in France (Amarger *et al.* 1994) and in Kenyan acid soils from bean nodules (Giller *et al.* 1994). It is not known if it was introduced to these areas upon introduction of bean plants. The true host for *R. tropici* is not known. Bacteria identical to *R. tropici* by protein profiles have been isolated from *Bolusanthus* and *Spartium* in South Africa (Dagutat and Steyn, 1995).

## *R. etli*

The majority of isolates from nitrogen fixing nodules of *Phaseolus vulgaris* bean in Mesoamerica are *R. etli* strains, etl means bean in the Aztecs' language (Segovia *et al.* 1993). *R. etli* was proposed to designate the formerly identified *R. leguminosarum* bv. phaseoli type 1 strains. They have multiple copies of *nif* genes and a peculiar organization of *nod* genes. Melanin and gum production and large numbers of plasmids (4-7) are quite common features in these strains. Loss of symbiotic properties due to genetic instability is also characteristic of the species (Romero *et al.* 1991). Some *R. etli* isolates from the bean rhizosphere lack the symbiotic plasmid and these populations may be more abundant than the symbiotic ones (Segovia *et al.* 1991).

There is a large genetic diversity among *R. etli* strains, but all the strains tested have a common isoform of GS, distinct from other rhizobia species (Taboada *et al.* 1996). By MLEE, *R. etli* is distinguished from *R.*

*leguminosarum*. However, *R. etli* and *R. leguminosarum* share LPS patterns and by fatty acid analysis *R. etli* and *R. leguminosarum* are not so clearly different (BDW Jarvis, personal communication). A plasmid carrying LPS determinants is conserved in *R. etli* and *R. leguminosarum* (A. García de los Santos and S. Brom, personal communication).

By DNA-DNA homology *R. etli* strains are more similar among themselves than with the *R. leguminosarum* strains tested. The sequence of a fragment of 16S rRNA genes served to separate *R. etli* from *R. leguminosarum* (Segovia *et al.* 1993). The complete sequence of the gene places *R. etli* more closely related to *R. leguminosarum* (van Berkum *et al.* 1996). Some *R. etli* strains (as defined by MLEE) have the allele corresponding to that of *R. leguminosarum* 16S rRNA genes (Eardly *et al.* 1995). This raises the possibility of recombination of ribosomal genes and reinforces the need to analyze a broader part of the genome.

Different *Rhizobium* spp. were isolated from native legumes (*Clitoria*, *Dalea* and others) grown inside bean fields. By the partial sequence of their 16S rRNA genes, they are closely related to *R. etli* (Hernández-Lucas *et al.* 1995). These bacteria constitute, together with genomic species 1 (Laguerre *et al.* 1993), OR191 and FL27 (Eardly *et al.* 1992), a cloud of rhizobia surrounding *R. etli*. OR191 was isolated from ineffective alfalfa nodules in the USA (Eardly *et al.* 1992), genomic species 1 is from bean rhizobia in France (Laguerre *et al.* 1993) and FL27 was isolated from bean nodules grown in *Leucaena* fields. FL27 falls inside *R. etli* strains by carbon source utilization. However, FL27 and the other rhizobia share very low DNA-DNA homology with *R. etli* (Martínez-Romero, 1994). FL27, and related bacteria may have a different plasmid content than *R. etli* and that may account for the large differences in DNA-DNA homology, since the latter takes into consideration the total bacterial genome. It would be very difficult to ascribe FL27 or OR191 and related bacteria to a single different species. A continuum of strains may fill the gap that separates different "bona fide" *Rhizobium* species.

## Conclusions

The phenotype is not a linear function of the genotype, and for this reason there is not clear agreement in many cases. In addition, different parts of the genome may have different rates of change. Plasmids seem to correspond to hypervariable parts that may bring drastic genetic changes in single steps by loss or acquisition. Even deletions or rearrangements of large parts of chromosomes could be a driving force to generate large divergence of otherwise closely related lineages. The *Rhizobium* genome may be modular and different parts may have a different evolutionary origin.

Acknowledgements to D. Romero and M. Dunn for reading the manuscript, to J. Martínez and J. Espiritu for computing assistance. Partial financial support was from VLIR-ABOS grant from Belgium.

## Literature cited

Amarger, N., Bours, M., Revoy, F., Allard, M.R., and Laguerre, G. 1994. *Rhizobium tropici* nodulates field-grown *Phaseolus vulgaris* in France. Plant Soil. 161:147-156.

Dagutat, H., and Steyn, P.L. 1995. Taxonomy and distribution of rhizobia indigenous to South African soils. p. 683 in: Nitrogen Fixation: Fundamentals and Applications. I.A. Tikhonovich et al. Eds. Kluwer Academic Publishers. Netherlands.

Eardly, B.D., Wang, F.S., Whittam, T.S., and Selander, R.K. 1995. Species limits in *Rhizobium* populations that nodulate the common bean (*Phaseolus vulgaris*). Appl. Environ. Microbiol. 61:507-512

Eardly, B.D., Young, J.P.W., and Selander, R.K. 1992. Phylogenetic position of *Rhizobium* sp. strain Or 191, a symbiont of both *Medicago sativa* and *Phaseolus vulgaris*, based on partial sequences of the 16S rRNA and *nifH* genes. Appl. Environ. Microbiol. 58:1809-1815.

Folch-Mallol, J.L., Marroquí, S., Sousa, C., Manyani, H., López-Lara, I.M., van der Drift, K.M.G.M., Haverkamp, J., Quinto, C., Gil-Serrano, A., Thomas-Oates, J., Spaink, H.P., and Megías, M. 1996. Characterization of *Rhizobium tropici* CIAT899 nodulation factors: the role of *nodH* and *nodPQ* genes in their sulfation. Mol. Plant Microbe Interact. 9:151-163.

Geniaux, E., Flores, M., Palacios, R., and Martínez, E. 1995. Presence of megaplasmids in *Rhizobium tropici* and further evidence of differences between the two *R. tropici* subtypes. Int. J. Syst. Bacteriol. 45:392-394.

Giller, K.E., Anyango, B., Beynon, J.L., and Wilson, K.J. 1994. The origin and diversity of rhizobia nodulating *Phaseolus vulgaris* L. in African soils. Pages 57-62 in: Advances in Legume Systematics 5: The Nitrogen Factor, Sprent, J.I. and McKey, D., ed. Royal Botanic Gardens, Kew.

Graham, P.H., Sadowsky, M.J., Keyser, H.H., Barnet, Y.M., Bradley, R. S., Cooper, J.E., De Ley, D.J., Jarvis, B.D.W., Roslycky, E.B. Strijdom, B.W., and Young, J.P.W. 1991. Proposed minimal standards for the description of new genera and species of root- and stem-nodulating bacteria. Int. J. Syst. Bacteriol. 41:582-587.

Hernández-Lucas, I., Segovia, L., Martínez-Romero, E., and Pueppke, S.G. 1995. Phylogenetic relationships and host range of *Rhizobium* spp. that nodulate *Phaseolus vulgaris* L. Appl. Environ. Microbiol. 61:2775-2779.

Laeremans, T., Caluwaerts, I., Verreth, C., Rogel, M.A., Vanderleyden, J., and Martínez-Romero, E. 1996. Isolation and characterization of the *Rhizobium tropici* Nod factor sulfation genes. Mol. Plant Microbe Interact. 9: (in press).

Laguerre, G., Fernandez, M.P., Edel, V., Normand, P., and Amarger, N. 1993. Genomic heterogeneity among French *Rhizobium* strains isolated from *Phaseolus vulgaris* L. Int. J. Syst. Bacteriol.43:761-767.

Martínez, E., Flores, M., Brom, S., Romero, D., Dávila, G., and Palacios, R. 1988. *Rhizobium phaseoli*: a molecular genetics view. Plant Soil. 108:179-184.

Martínez, E., Palacios, R., and Sánchez, F. 1987. Nitrogen-fixing nodules induced by *Agrobacterium tumefaciens* harboring *Rhizobium phaseoli* plasmids. J. Bacteriol. 169:2828-2834.

Martínez-Romero, E. 1994. Recent developments in *Rhizobium* taxonomy. Plant Soil. 161:11-20.

Martínez-Romero, E., and Caballero-Mellado, J. 1996. *Rhizobium* phylogenies and bacterial genetic diversity. Crit. Rev. Plant Science. 15:113-140.

Martínez-Romero, E., and Rosenblueth, M. 1990. Increased bean (*Phaseolus vulgaris* L.) nodulation competitiveness of genetically modified *Rhizobium* strains. Appl. Environ. Microbiol. 56:2384-2388.

Martínez-Romero, E., Segovia, L., Martins Mercante, F., Franco, A.A., Graham, P., and Pardo, M.A. 1991. *Rhizobium tropici*: a novel species nodulating *Phaseolus vulgaris* L. beans and *Leucaena* sp. trees. Int. J. Syst. Bacteriol. 41:417-426.

Poupot, R., Martínez-Romero, E., and Promé, J.C. 1993. Nodulation factors from *Rhizobium tropici* are sulfated or non sulfated chitopentasaccharides containing a *N*-methyl-*N*-acyl glucosaminyl terminus. Biochemistry. 32:10430-10435.

Romero, D., Brom, S., Martínez-Salazar, J., Girard, M.L., Palacios, R., and Dávila, G. 1991. Amplification and deletion of *nod-nif* region in the symbiotic plasmid of *Rhizobium phaseoli*. J. Bacteriol. 173:2435-2441.

Segovia, L., Piñero, D., Palacios, R., and Martínez-Romero, E. 1991. Genetic structure of a soil population of nonsymbiotic *Rhizobium leguminosarum*. Appl. Environ. Microbiol. 57:426-433.

Segovia, L., Young, J.P.W., and Martínez-Romero, E. 1993. Reclassification of American *Rhizobium leguminosarum* biovar phaseoli type I strains as *Rhizobium etli sp. nov*. Int. J. Syst. Bacteriol. 43:374-377.

Taboada, H., Encarnación, S., Vargas, M.C., Mora, Y., Martínez-Romero, E., and Mora, J. 1996. Glutamine synthetase II constitutes a novel taxonomic marker in *Rhizobium etli* and other *Rhizobium* species. Int. J. Syst. Bacteriol. 46:485-491.

van Berkum, P., Beyene, D., and Eardly, B.D. 1996. Phylogenetic relationships among *Rhizobium* species nodulating the common bean (*Phaseolus vulgaris* L.). Int. J. Syst. Bacteriol. 46:240-244.

# Root Border Cells

Martha C. Hawes, L. A. Brigham, H.-H. Woo, Y. Zhu, F. Wen.
Department of Plant Pathology, University of Arizona

Central to the science of plant-microbe interactions is the question of how plants sense and respond to stimuli to produce appropriate adaptive responses. To our knowledge, only one adaptive mechanism allows plants to not merely respond to signals from the environment, but to change it, by releasing thousands of healthy somatic cells from the tip of the root (Fig. 1). We refer to these unique cells as root "border" cells, formerly called "sloughed peripheral root cap cells," for two reasons: 1. To emphasize that these cells by definition are not part of the root cap, and 2. To emphasize that under natural conditions populations of the cells constitute a physical and biological interface, or border, between the root surface and the soil environment. Border cells are those cells which are physically separated from each other and the root so that they disperse into suspension upon immersion of roots into liquid. When border cells are thus removed, border cell separation is immediately reinitiated; new cells can be collected within one hour, and a complete new set of cells separates from the cap into the soil within 24 hours (Hawes & Lin 1990).

The function of border cells is unknown, but based on their properties we have proposed that they protect plant health by the controlled delivery of biologically active chemicals that regulate microbial populations in the rhizosphere (Fig. 2, Hawes & Brigham 1992 and references cited therein). Our approach to test predictions of this model is to compare normal roots with roots that are identical except that (1) their border cells are incapable of producing specific chemicals or (2) their border cells are not produced at the usual rate, time, or level. Transgenic roots with directed changes in border cell viability and number are being developed using two categories of target genes: border cell specific genes and genes controlling border cell production.

## Border Cell Specific Gene Expression

In general, border cells can carry out the same functions as other cell types (i.e. they synthesize proteins, they produce defense structures in response to attack, they divide and grow into organized tissue in culture), but in some cases border cells do things that other tissues do not (reviewed in Hawes & Brigham 1992). The most obvious example is that border cells by definition violate the fundamental expectation that a tissue of a given plant will exist in nature as an integrated part of that plant (Fig. 1). Border cells also can act as cell-specific reservoirs of products such as chemoattractants

(Fig. 2). The simplest hypothesis to account for border cell specific phenotypes is that, as in any tissue differentiated for specialized functions, gene expression in border cells differs from that in progenitor cells. A prediction of this hypothesis is that proteins synthesized by border cells are distinct from those made by the root cap. Recent results reveal that border cells, which incorporate labelled amino acids 260% more efficiently than cells of the root cap, synthesize a set of proteins that is holistically distinct from those made by progenitor cells of the cap (Brigham et al 1995b). Most

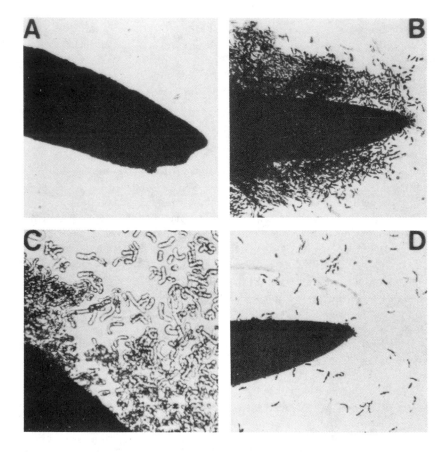

Fig. 1. **Root border cells.** A. Root tip of a seedling germinated for 2 days on water agar overlaid with filter paper to prevent direct exposure to water, or penetration of the agar; B. The same root tip 60 seconds after adding a droplet of distilled water and C. At higher magnification. D. After gentle agitation, nearly all border cells are released into suspension. *Reproduced with permission from Hawes & Pueppke (1986) American Journal of Botany 73, 1466-1473.*

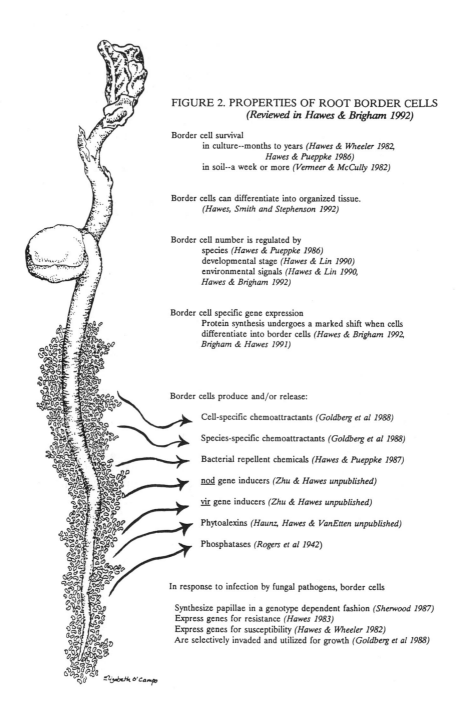

## FIGURE 2. PROPERTIES OF ROOT BORDER CELLS
### *(Reviewed in Hawes & Brigham 1992)*

Border cell survival
> in culture--months to years *(Hawes & Wheeler 1982,*
> > *Hawes & Pueppke 1986)*
> in soil--a week or more *(Vermeer & McCully 1982)*

Border cells can differentiate into organized tissue.
> *(Hawes, Smith and Stephenson 1992)*

Border cell number is regulated by
> species *(Hawes & Pueppke 1986)*
> developmental stage *(Hawes & Lin 1990)*
> environmental signals *(Hawes & Lin 1990,*
> *Hawes & Brigham 1992)*

Border cell specific gene expression
> Protein synthesis undergoes a marked shift when cells
> differentiate into border cells *(Hawes & Brigham 1992,*
> *Brigham & Hawes 1991)*

Border cells produce and/or release:

Cell-specific chemoattractants *(Goldberg et al 1988)*

Species-specific chemoattractants *(Goldberg et al 1988)*

Bacterial repellent chemicals *(Hawes & Pueppke 1987)*

<u>nod</u> gene inducers *(Zhu & Hawes unpublished)*

<u>vir</u> gene inducers *(Zhu & Hawes unpublished)*

Phytoalexins *(Haunz, Hawes & VanEtten unpublished)*

Phosphatases *(Rogers et al 1942)*

In response to infection by fungal pathogens, border cells

Synthesize papillae in a genotype dependent fashion *(Sherwood 1987)*
Express genes for resistance *(Hawes 1983)*
Express genes for susceptibility *(Hawes & Wheeler 1982)*
Are selectively invaded and utilized for growth *(Goldberg et al 1988)*

of the new proteins are rapidly exported into the external environment. Surprisingly, in fact, considering that the root cap has long been known to be a secretory organ (Rougier 1981), 25% of the proteins synthesized during a one hour period by border cells are released extracellularly, compared with only 2% of the proteins from cells of the root cap. The dramatic change in protein synthesis and export that occurs upon differentiation of root cap cells into border cells is controlled at least in part at the level of transcription: the observed change in protein profiles is correlated with a similarly dramatic change in profiles of expressed genes (Brigham et al 1995b). Border cell specific genes isolated to date share no sequence homology to known eucaryotic or procaryotic genes, and their functions are unknown, but promoters from such genes provide a tool to alter the ability of border cells to synthesize specific chemicals to be released into the rhizosphere. For example, transgenic roots whose border cells are programmed to commit suicide upon separation from the root can be created by the border cell specific expression of genes encoding cytotoxic products.

## Gene Expression During Border Cell Development

Border cell development begins with cell division in the root cap meristem, and culminates when peripheral cells of the cap differentiate into border cells by separating from each other (Feldman 1984). Border cell separation is fundamentally distinct from other known processes involving cell wall degradation in that it involves the complete dissociation of individual cells from each other and from root tissue, without causing functional injury to the cell wall. Unlike abscission and fruit ripening, which are terminally differentiated processes that occur over days, border cell separation is dynamic and very rapid, and the product is a population of single cells with intact walls that are osmotically stable in distilled water. Border cell separation must be very tightly regulated, to allow the complete separation of a defined layer of cells from each other within one hour, without resulting in the death of any cells.

The mechanism of border cell separation is unknown, but obvious candidates are hydrolytic enzymes that degrade pectin, the primary component of middle lamellae. Border cell production can be induced and synchronized, independently of root development *per se*, by the simple expedient of removing existing cells (as in Fig. 1). This phenomenon has been exploited to define a molecular framework for border cell development (Fig. 3), and to test the hypothesis that the action of pectinmethylesterase (PME) in peripheral cells of the root cap triggers activity of polygalacturonase (PG) and other pectolytic enzymes that cause border cell separation. PME activity is correlated with border cell separation (Stephenson & Hawes 1994), and is regulated at least in part at the level of transcription. Thus, transcription of PsPME, a root cap localized gene

## FIGURE **3.** WORKING MODEL FOR BORDER CELL SEPARATION

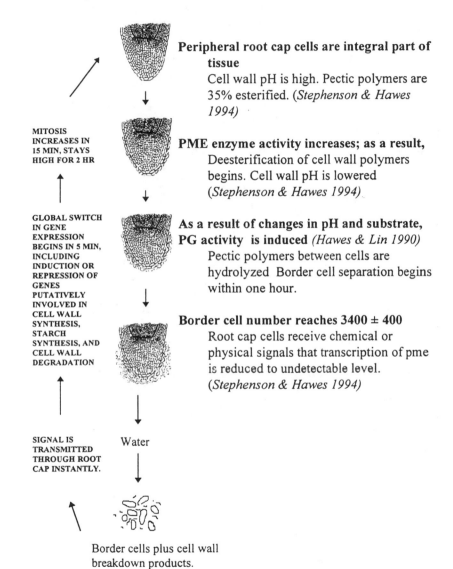

**Peripheral root cap cells are integral part of tissue**
Cell wall pH is high. Pectic polymers are 35% esterified. (*Stephenson & Hawes 1994*)

MITOSIS INCREASES IN 15 MIN, STAYS HIGH FOR 2 HR

**PME enzyme activity increases; as a result,**
Deesterification of cell wall polymers begins. Cell wall pH is lowered (*Stephenson & Hawes 1994*)

GLOBAL SWITCH IN GENE EXPRESSION BEGINS IN 5 MIN, INCLUDING INDUCTION OR REPRESSION OF GENES PUTATIVELY INVOLVED IN CELL WALL SYNTHESIS, STARCH SYNTHESIS, AND CELL WALL DEGRADATION

**As a result of changes in pH and substrate, PG activity is induced** (*Hawes & Lin 1990*)
Pectic polymers between cells are hydrolyzed Border cell separation begins within one hour.

**Border cell number reaches 3400 ± 400**
Root cap cells receive chemical or physical signals that transcription of pme is reduced to undetectable level. (*Stephenson & Hawes 1994*)

SIGNAL IS TRANSMITTED THROUGH ROOT CAP INSTANTLY.

Water

Border cells plus cell wall breakdown products.

encoding PME in pea, is induced within 5 minutes of removing border cells, and remains high until border cell separation is complete (Zhu et al 1996). As predicted for a gene involved in border cell production, expression of PsPME in root caps is localized to peripheral cells undergoing border cell separation (Brigham et al 1995c). Transgenic roots expressing antisense mRNA are being exploited to test the role of PME and other induced cell wall degrading enzymes in border cell separation. Plants with defined changes in the ability to deliver border cells to the rhizosphere can be used to test the hypothesis that the presence of border cells controls the ability of microorganisms to associate with plant roots.

## Literature Cited

Brigham, L. A., Woo H. H., Hawes, M. C. 1995a. Root border cells as tools in plant cell studies. Methods Cell Biol. 49: 377-386.

Brigham, L. A., Woo, H. H., Hawes, M. C. 1995b. Differential expression of proteins and mRNAs from border cells and root tips of pea. Plant Physiol.109: 457-463.

Brigham, L. A., Woo, H. H., Hawes, M. C. 1995c. Developmentally regulated gene expression in root border cells. ISPMB Vol. 4.

Feldman, L. J. 1984. Regulation of root development. Ann. Rev. Plant Physiol. 35: 223-242.

Hawes, M. C., Brigham, L. A. 1992. Impact of root border cells on microbial populations in the rhizosphere. Adv. Plant Pathol. 8: 119-148.

Hawes, M. C., Brigham, L. A., Nicoll, S. M., Stephenson, M. B. 1993. Plant genes controlling the release of root exudates. Biotechnol. Plant Protection 4: 1-10

Hawes, M. C., Lin, H. J. 1991. Correlation of pectolytic enzyme activity with programmed release of cells from the root cap of pea. Plant Physiol. 94: 1855-1859.

Rougier, M. 1981. Secretory activity of the root cap. 542-574 in Encyclopedia of Plant Physiol 13B.

Stephenson, M. B., Hawes, M. C. 1994. Correlation of pectinmethylesterase activity in root caps of pea with root border cell separation. Plant Physiol.106: 739-745.

Woo, H. H., Brigham, L. A., Hawes, M. C. 1995a. Molecular cloning and expression of mRNAs encoding H1 histone in root tips of pea. Plant Mol. Biol. 28: 1143-1147.

Woo, H. H., Brigham, L. A., Hawes, M. C. 1994. Primary structure of the mRNA encoding a 16.5 kDa ubiquitin conjugating enzyme of pea. Gene 148: 369-370.

Zhu, Y, Wen, F., Hawes, M. C. 1996. Expression of a unique PME gene whose activity is synchronized with production of root border cells. MPMI Annual Meeting, Knoxville TN.

# Vesicular-Arbuscular Mycorrhizae: Molecular Approaches to Investigate Phosphate Nutrition in the Symbiosis.

Maria J. Harrison, Stephen H. Burleigh, Henry Liu and Marianne L. van Buuren.
Plant Biology Division, Samuel Roberts Noble Foundation, Ardmore, Oklahoma, USA.

The symbiotic associations formed between VA mycorrhizal fungi and plant roots have been identified in natural ecosystems throughout the world and such interactions are estimated to occur in over 80% of all terrestrial plant species (reviewed in Smith and Gianinazzi-Pearson 1988). The soil-borne fungi that form these associations belong to the order Glomales (Zygomycetes) and the molecular data suggests that they originated around 400 million years ago (Simon et al. 1993). VA fungal structures have been identified in fossils from Early Devonian plants and it has been hypothesised that the association may have been instrumental in the origin of land flora (Remy 1981).

The VA mycorrhizal association is a mutualistic one. The fungi are obligate symbionts and colonise the cortex of the root in order to obtain carbon from their plant host, while assisting the plant with the uptake of mineral nutrients, predominately phosphate, from the soil. Phosphate is essential for plant growth and development and in many soils the available phosphate is present at low concentrations which may be limiting for growth. Thus, the additional supply of phosphate obtained as a consequence of the mycorrhizal association, can be extremely significant for plant growth and nutrition. Increases in plant growth and health due to the mycorrhizal association have been clearly documented for a wide range of plant species (reviewed in Jeffries 1987; Bethlenfalvay 1992).

Numerous microscopy studies of mycorrhizal roots have resulted in an in depth knowledge of the morphological events that occur during the development of the association. In general, fungal hyphae originating from chlamydospores in the soil contact the root surface, differentiate to form an appressorium and then penetrate the root, frequently entering between two epidermal cells. Once inside the root the hyphae ramify inter- and intracellularly throughout the root cortex and differentiate to form two distinct structures; arbuscules and vesicles. Arbuscules are dichotomously branched, terminal hyphae that penetrate the cortical cells but remain separated from the plant cell cytoplasm by the plant plasma

515

membrane which invaginates around the hyphal branches (reviewed in
Bonfante and Perotto 1995). Although not directly demonstrated, it is
generally believed that nutrient transfer occurs at this arbuscular/ cortical
cell interface. In addition to the internal structures the fungus also
maintains a network of hyphae external to the root via which phosphate
and other mineral nutrients are initially absorbed (Fig.1). These hyphae
are considerably smaller than the plant root and can penetrate soil
micropores not acessible by the plant root system. In addition, they can
extend beyond the depletion zones of the plant root and are able to access
phosphate that might otherwise be unavailable to the plant host. Recent
measurements of phosphate uptake have  demonstrated that a significant
part of the total phosphate uptake of the mycorrhizal root may occur via
the fungal hyphae. These, and other studies have indicated that there is
considerable variation between mycorrhizal fungi in their ability to take
up phosphate and to translocate it to the host plant, as well as variation in
the functional compatibility between a single species of fungus and
different plant hosts (Jakobsen 1995).

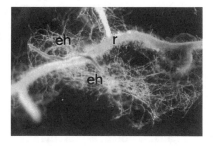

Figure 1.  Light micrograph of a *M. truncatula* root (r) colonised with *G. versiforme*. External
hyphae (e h) extend out of the root.

*Experimental System.*   In order to analyse the molecular and genetic
mechanisms underlying the formation and functioning of the VA
mycorrhizal symbiosis we have initiated analyses of the association
formed between *M. truncatula* and *G. versiforme. M. truncatula* is a
diploid, autogamous relative of the agriculturally important forage legume
*Medicago sativa*, and is currently used by a number of laboratories for the
investigation of the *Rhizobium* symbiosis. *Glomus versiforme* is known to
colonise *Medicago* species  (Lackie et al. 1988) and is one of the few VA
mycorrhizal fungi that produces spores in above ground sporocarps, a
useful feature for the aquisition of pure fungal material.  The association
formed between *M. truncatula* and *G. versiforme* is a typical VA

mycorrizal association and appressoria, arbuscules and vesicles are formed in the roots.

## STRATEGIES FOR ISOLATING GENES INVOLVED IN PHOSPHATE NUTRITION IN THE MYCORRHIZAL SYMBIOSIS.

We have taken both random and targetted approaches to identify genes that are involved in phosphate nutrition in the mycorrhizal symbiosis. Differential screening of cDNA libraries prepared from RNA isolated from mycorrizal roots has resulted in the identification of a number of cDNA's representing genes whose expression is altered following colonisation. One such cDNA isolated by this approach represents a *M. truncatula* gene whose expression is decreased following colonisation by *Glomus versiforme*. Similar expression patterns are observed in *M. sativa*, and in both species the transcript levels in non mycorrhizal roots also decrease following the addition of phosphate (Fig. 2). This cDNA clone may be informative as to molecular events occuring during changes in the phosphate status in plant roots. Future studies will determine whether the decrease in transcript levels in the mycorrhizal association is a phosphate mediated effect, or whether the initial interaction with the mycorrhizal fungus can also result in down regulation of gene expression.

Figure 2.
Northern blot of RNA from non-colonised
*M. truncatula* (Mt) and *M. sativa* (Ms) roots
and *M. truncatula* and *M. sativa* roots colonised
with *G. versiforme* (Gv) probed with
cDNA 42 (upper panel, 0.5kb transcript) and
18S rRNA (lower panel). The rRNA blot was
used as a control for loading and transfer.

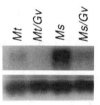

## ISOLATION OF A PHOSPHATE TRANSPORTER FROM MYCORRHIZAL ROOTS OF *M. TRUNCATULA*

A second approach to isolate genes involved in phosphate nutrition during the symbiosis has been to use information from other systems to target the identification of specific cDNA's.

In *Saccharomyces cerevisiae* the *PHO84* gene encodes a high affinity phosphate transporter which is responsible for phosphate uptake under low phosphate conditions (Bunya et al. 1991). A second low affinity system mediates phosphate uptake under high P conditions. A similar system in present in *N. crassa*, and physiological measurements of phosphate uptake in plants and VA fungi suggest that both of these organisms also have low and high affinity uptake systems which function to mediate phosphate uptake in different P environments (Beevers and Burns 1980).

The *PHO84* gene from *S. cerevisiae* was used as a probe to screen a cDNA library prepared from *M. truncatula* roots colonised by *G. versiforme*. Eight cDNA clones showing hybridisation to the *PHO84* probe were selected and on the basis of their sequences it was determined that they all represented the same gene. One full length clone (GvPT) was selected for further analyses. GvPT is 1932 bp in length and is predicted to encode a protein of 521 amino acids (Harrison and van Buuren 1995). The protein shares 47.9% amino acid sequence identity with *PHO84* and 45% identity with *PHO-5* from *Neurospora* (Versaw 1995*)*. All three proteins are predicted to form the same secondary structure, consisting of 12 membrane spanning $\alpha$-helices with a hydrophilic loop between domains 6 and 7. This structural motif has been observed in many membrane transporters from both eukaryotes and prokaryotes and members of this superfamily termed the 'major facilitator superfamily' are hypothesised to have arisen from a common ancestor (Marger and Saier 1993).

*Complementation of S. cerevisiae PHO84 mutant.* In order to determine whether GvPT was able to function as a phosphate transporter we transferred the sequence to yeast expression vector and introduced the construct into *S. cerevisiae PHO84* mutant cells. Initial experiments in which the complete GvPT cDNA sequence was used were not successful, and *PHO84* cells transformed with this construct retained the mutant phenotype. As it has been previously noted that long 5' untranslated leader sequences may negatively affect the expression of heterologous proteins in yeast, a second attempt was made in which the 5' untranslated sequence (112bp) was deleted prior to ligation to the expression vector (Fig. 3).

Figure 3.
Untranslated sequence (112bp) at 5' end of GvPT cDNA (bold text), omitted for successful complementation of the yeast *PHO84* mutant.

[1]**AATCGACCTTTCTATTAACTATAAATATAATATATAATCA
GATAATTATTTTCAAAAAAAAAAAAAGAAGAAAATAAAAGA
CCTTTAAACCTTTAATTCCTTTATCATACATA**ATGTCTAC
[1]M   S   T

This construct was successful and it was noticeable that *PHO84* cells carrying the GvPT clone grew more rapidly than *PHO84* cells carrying the vector without an insert. Phosphate uptake measurements indicated that *PHO84* mutant cells carrying GvPT were able to accumulate phosphate from a low phosphate media confirming that the GvPT clone is capable of transporting phosphate and therefore complementing the mutant (Harrison and van Buuren 1995).

*Genomic origin of GvPT.* As the cDNA clone was obtained from a library containing sequences of both plant and fungal origin it was necessary to determine the origin of the gene. Southern blot analysis suggested that the cDNA did not represent a plant gene. Due to our inability to culture the VA mycorrhizal fungi in the absence of a plant host it is difficult to obtain sufficient quantities of pure DNA for a *G. versiforme* genomic Southern blot, and instead PCR analysis was used to demonstrate the origin of the clone. Primers were designed to the GvPT sequence and used in a polymerase chain reaction to amplify a region of the GvPT gene from DNA isolated from *G.versiforme* spores. The primers did not amplify DNA from the *M. truncatula* genome. The sequence of the amplified fragment from the spore sample matched the phosphate transporter sequence, indicating that GvPT represents a *G. versiforme* gene (Harrison and van Buuren 1995).

*Expression of GvPT is localised to the external hyphae.* Northern blot analyses indicated that GvPT transcripts were in present in mycorrhizal roots. Reverse transcription coupled with PCR amplification was used to further define the site of expression and demonstrated the presence of GvPT transcripts specifically in the external hyphae and not in the fungal structures within the roots (Harrison and van Buuren, 1995). The expression studies in yeast indicate that GvPT encodes an active transporter and the site of expression of the gene, in the external hyphae is consistent with a function in the uptake of phosphate from the environment.

To assist in further analyses of the phosphate transporter, we have initiated the preparation of antibodies to a central cytoplasmic domain of the phosphate transporter. The DNA sequence encoding the 61 amino acid region occurring between the predicted membrane spanning domains 6 and 7, was ligated to the β-galactosidase gene under the control of the *LacZ* promoter. Recombinant fusion protein was produced in *E. coli*, purified by affinity chromatography and used to immunise mice. Initial tests suggest that the immune antisera contains antibodies which recognise the 61 amino acid segment of the phosphate transporter (Fig.4).

Future studies using these molecular and immunological tools, will be aimed at elucidating the mechanisms of regulation of the transporter in the mycorrhizal symbiosis.

Acknowledgements
This work was supported by the Samuel Roberts Noble Foundation.

Figure 4.

Western blot of β-galactosidase/phosphate transporter fusion protein (lane 1) and β-galactosidase (lane 2) probed with the immune serum from a mouse injected with β-galactosidase/phosphate transporter fusion protein. The serum has been purified to remove the antibodies that cross react to β-galactosidase.

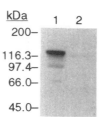

References

Beevers, R.E. and Burns, D.J.W. (1980) Phosphate Uptake, Storage and Utilisation by Fungi. Advances in Botanical Research 8, 127-219.

Bethlenfalvay, G.W. (1992) Mycorrhizae and Crop Productivity. In Mycorrhizae in Sustainable Agriculture, (Bethenfalvay, G.J. and Linderman, R.G., eds), ASA, Special Publication 54, 1-27.

Bonfante, P and Perotto, S. (1995) Strategies of arbuscular mycorrhizal fungi when infecting host plants. New Phytologist, 130, 3-21.

Bun-ya, M., Nishimura, M., Harashima, S. & Oshima, Y. (1991) The PHO84 gene of Saccharomyces cerevisiae encodes an inorganic phosphate transporter. Mol. Cell. Biol. 11, 3229-3238.

Harrison, M.J. and van Buuren, M.L. (1995) A phosphate transporter from the mycorrhizal fungus Glomus versiforme. Nature 378, 626-629.

Jakobsen, I. (1995) Transport of phosphorous and carbon in VA mycorrhizas in Mycorrhiza, Structure, Function, Molecular Biology and Biotechnology. p297-324. Springer-Verlag Berlin. Ed, A Varma, B Hock.

Jeffries, P. (1987) Uses of mycorrhizae in agriculture. Crit. Rev. in Biotech. 5, 319-357.

Lackie S.M., Bowley, S.R. and Peterson, R.L. (1988) Comparison of colonisation among half-sib families of Medicago sativa L. by Glomus versiforme (Daniels and Trappe) Berch. New Phytologist 108, 477-482.

Marger, M.D. and Saier, Jr., M.H. (1993) A major superfamily of transmembrane facilitators that catalyze uniport, symport and antiport. Trends Biochem. Sci. 18, 13-20.

Remy, W., Taylor, T.N., Hass, H. and Kerp, H. (1994) Four hundred-million-year-old vesicular arbuscular mycorrhizae. PNAS 91, 11841-11843.

Simon, L., Bousquet, J., Levesque, R.C and Lalonde, M. (1993) Origin and diversification of endomycorrhizal fungi and coincidence with vascular land plants. Nature 363, 67-69.

Smith, S. E. and Gianinazzi-Pearson, V. (1988) Physiological interactions between symbionts in vesicular-arbuscular mycorrhizal plants. Ann. Rev. Plant Physiol. Plant Mol. Biol. 39, 221-244.

Versaw, W.K. (1995) A phosphate-repressible, high affinity phosphate permease is encoded by the pho-5+ gene of Neurospora crassa. Gene 153, 135-139.

# Root-Knot Nematode Induced *TobRB7* Expression And Antisense Transgenic Resistance Strategies

Charles H. Opperman and Mark A. Conkling
Associate Professors of Plant Pathology and Genetics, respectively
North Carolina State University
Raleigh, North Carolina, USA

Although plant parasitic nematodes collectively cause over $77 billion in crop losses worldwide, the vast majority of the damage is caused by sedentary endoparasitic forms, particularly *Meloidogyne* spp. (Sasser and Freckman, 1987). Sedentary endoparasitic nematodes establish a variety of feeding sites within the host root. Feeding sites induced by the root-knot (*Meloidogyne* spp.) nematodes are the most elaborate. The specialized nature of the sedentary endoparasitic nematode-host relationship makes it susceptible to exploit for resistance.

Root-knot nematodes have a very broad host range, encompassing over 2,000 plant species (Sasser, 1980). Most cultivated crops are attacked by at least one species of *Meloidogyne* (Sasser, 1980). The feeding site, or giant cells, of the root-knot nematode is initiated after the infective second-stage juvenile has penetrated the host root, generally near the root tip, and has migrated to the developing vascular cylinder. The 5-7 giant cells formed become the permanent feeding site of the nematode. The giant cells undergo repeated nuclear divisions without cytokinesis, and nuclei become enlarged and lobate (Jones, 1981). A characteristic feature of giant cells is the highly invaginated and thickened cell wall, similar to cell walls observed in transfer cells (Jones, 1981). The nematode is totally dependent upon the giant cells for its survival and reproduction.

## Alterations in plant gene expression induced by root-knot nematodes

Plant gene expression patterns are quantitatively and qualitatively altered during nematode feeding site initiation. Bird and co-workers (Wilson et al., 1994) isolated over 200 tomato cDNA's representing genes up-regulated in induced giant cells. Many of these genes are not normally expressed in root tissue (Bird and Wilson, 1994a; Bird and Wilson, 1994b), suggesting that the nematode is able to induce alterations in gene expression patterns. There have also been preliminary reports of genes encoding enzymes and structural proteins, such as extensins (van der Eycken et al., 1992; Niebel et al., 1993; Bird and Wilson, 1994a), that are up-regulated during either root-knot or cyst nematode feeding site establishment.

Control of gene expression at the nematode feeding site is of great significance in designing genetically-engineered resistance. There have been few studies on promoters involved in feeding site gene expression. One study suggests that many genes are down-regulated during feeding site formation (Goddjin et al., 1993). Characterization of these genes, however, has not yet been reported. There have been several reports of promoters that are up-regulated during feeding site formation by root-knot nematode. In one study, the promoter of the hydroxymethylglutaryl CoA reductase gene (*hmg*2) fused to GUS was observed to be strongly expressed in the developing giant cells and gall tissue shortly after infection by *M. incognita* or *M. hapla* (Cramer et al., 1993). The tomato *hmg*2 is a defense related gene induced by both fungal and bacterial pathogens. Computer comparisons of available regulatory region nucleotide sequences involved in nematode-induction or suppression do not reveal any obvious relationships (Goddjin et al., 1993).

## Potential Strategies for Designed Nematode Resistance

There are several different approaches to design nematode-resistant transgenic crop cultivars. The simplest to implement is the expression of a nematode-specific toxin gene under the control of a constitutive promoter. The advantage of this strategy is that it is a preformed defense and does not require nematode-induction to be activated. Theoretically, any susceptible nematode which feeds for a suitable period upon cells containing the toxin moieties will not survive. A second approach would be directed expression of the toxin by a tissue-specific or nematode inducible promoter. However, very few peptide toxins that act specifically upon plant parasitic nematodes have been identified. Even so, nematicidal plants could be produced if a suitable toxin gene and promoter could be identified. The search for nematode-specific toxins has only recently begun, but several invertebrate-specific toxins do exist and can be tested against nematodes.

There are several disadvantages to these approaches, however. Peptide toxins useful in this approach typically are narrow in their toxic spectra, as is the case with the BT toxins. The use of transgenic plants expressing toxins with activity against only certain nematode species may end up selecting for non-sensitive species, resulting in narrow resistance. The constitutive expression of any "toxin" gene may place upon the sensitive nematode population very strong selective pressure for resistance, placing the durability of this type of defense in doubt. In addition, the global, constitutive expression of toxin genes guarantees that non-target species, including humans, will be exposed to the protein products.

The combinatorial nature of plant gene promoters so far characterized (Benfey, et. al., 1989) suggested to us that induction during nematode infection could be uncoupled from normal gene expression control. If this is the case, novel approaches for control become available. We believe the most durable method to genetically engineer nematode resistance is based upon nematode-induced expression of a molecule interfering with proper feeding site formation. This strategy is the most desirable of the three, in that selective pressure for toxin resistance is alleviated. One approach might be to interfere with the plant gene expression patterns in the developing giant cells, disrupting the proper formation and function of the feeding site. Ribozymes or antisense constructs to nematode-responsive plant genes could be used in this approach. A nematode attempting to feed on cells carrying a nematode-inducible promoter fused to a gene encoding one of these molecules would initiate gene expression resulting in the degradation of the feeding site. This type of engineered resistance is not dissimilar to the natural hypersensitive response to nematode parasites in resistant plants.

## *TobRB7* promoter analysis.

Work in our laboratories has centered on the *TobRB7* gene, first identified from tobacco (Conkling, *et al.*, 1990; Yamamoto et al., 1991). *TobRB7* is a root-specific gene expressed in the developing vascular cylinder. This is also the region of the root that root-knot nematode establishes its feeding site.

Fusions between a deletion series of the *TobRB7* 5'-flanking region and the bacterial reporter gene, ß-glucuronidase (GUS) (Jefferson, 1987) revealed that *cis*-acting elements necessary for root-specific expression of *TobRB7* reside between 636 and 299 nucleotides ($\Delta$0.6 - $\Delta$0.3) 5' of the site of transcription initiation (Yamamoto et al., 1991).

GUS expression patterns in root-knot nematode-infected transgenic tobacco plants carrying the promoter deletion-reporter constructs demonstrated that *TobRB7* is one of the plant genes up-regulated by nematode parasitism. Within 4 days after infection, significant levels of GUS activity in and around the developing feeding site were detected (Opperman *et al.*, 1994). The spatial and temporal shift in gene expression indicates that root-knot nematode infection has resulted in significant alterations in the control of *TobRB7* expression.

This finding is further confirmed by results obtained with the deletion series of *cis*-acting sequences. Although no transgene expression is observed in uninfected plants with deletions containing only the $\Delta$0.3 region, following root-knot nematode infection GUS accumulated in the developing giant cells and appeared to be regulated by the nematode infection (Opperman et al., 1994). The most significant aspect of these findings is that the nematode-responsive element (NRE) of the *TobRB7* promoter is not the same as the root-specific element, and can be uncoupled (Opperman et al., 1994). We have extended these findings to other crops, demonstrating nematode-inducible transgene expression in giant cells of transgenic tomato roots and in giant cells in both roots and tubers of transgenic potato (Song, Saravitz, Opperman, and Conkling, unpublished).

All races and species of *Meloidogyne* examined thus far induce expression of the Δ0.3 reporter, but cyst nematodes do not (Opperman et al., 1994). Recently, we have demonstrated that the reniform nematode, *Rotylenchulus reniformis*, is capable of inducing the *TobRB7* promoter in the region of its feeding site (Saravitz, Opperman, and Conkling, unpublished).

### TobRB7-based transgenic resistance strategies in tobacco.

We have taken several independent approaches to designing transgenic nematode-resistant plants using the *TobRB7* NRE. Our most successful approach has been to inhibit proper feeding site development or function using an antisense *TobRB7* construct. pRB7 functions as a water channel (Li, Saravitz, Song, and Conkling, unpublished) and we believe its role in giant cells is to maintain osmotic balance. Full-length *TobRB7* cDNA antisense constructs driven by either the Δ0.3 promoter or the CaMV 35S promoter were transformed into tobacco and tested for nematode resistance (Opperman, Acedo, and Conkling, unpublished). Some plants containing constructs driven by the 35S promoter exhibited stress-like phenotypes,

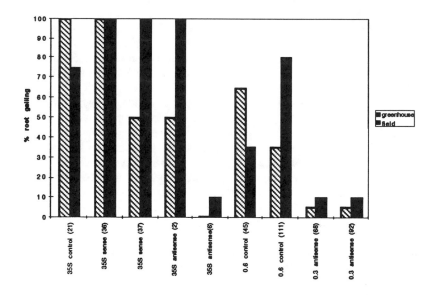

Figure 1. *TobRB7* antisense plants showing resistance to root-knot nematodes. Data is expressed as the percentage of the root system with galls (damaged) after 60 days (greenhouse) or 125 days (field). Data are compiled from 2 independent field trials and four independent greenhouse trials.

while others exhibited normal phenotypes. Some normal 35S transgenic plants exhibited substantial reductions in galling and egg production. Histochemical examination of roots revealed that numerous male nematodes had formed, and the feeding sites were small compared to the control plants. Transgenic plants containing the antisense construct driven by the Δ0.3 promoter showed no observable phenotype. Greenhouse and field trials with the Δ0.3 promoter antisense plants provided strong evidence that nematode infection was reduced substantially (Fig. 1). In both trials, root galling was reduced by approximately 70% compared to control treatments. The field trial indicated that protection lasted for the entire growing season (approximately four months). Galls that were observed on the roots of antisense plants were small and tended to appear as discrete entities compared to the large and clustered galls on the controls. Although these results provide only circumstantial evidence that *TobRB7* is essential to giant cell formation, they are one of the first indications that interference with feeding site development may be a viable approach to engineering transgenic nematode resistance. Importantly, nematode-resistance has maintained through three years of field trials and five generations of transgenic tobacco, and transgene is inherited in a normal Mendelian fashion.

The availability of promoters (such as *TobRB7*) capable of directing expression of transgenes precisely in the tissues targeted in root-knot nematode infection allow the development of novel strategies towards nematode control that overcomes most, if not all of the disadvantages of other transgenic nematode control strategies described previously. We have demonstrated that the Δ0.3 *TobRB7* promoter can be used to produce transgenic plants resistant to root-knot nematodes. We have shown further that the strategy of disruption of gene expression is viable and functions as predicted.

Nematode resistance strategies for food crops must meet three criteria: 1) They must be effective and persist through multiple generations; 2) Resistant plants must be available to the end-user in a timely fashion (within the next five years); and 3) The transgene conferring resistance must have a high probability of public and regulatory acceptance. Gene disruption strategies using antisense technology are based on transformation of a plant gene as opposed to a foreign sequence. Regulatory agencies have approved antisense technologies for food crops such as *FlavorSaver*™ tomatoes, and public acceptance is increasing. Therefore, we have chosen to concentrate the majority of our efforts upon gene disruption strategies.

### References

Benfey, P.N., Ren., L., and Chua, N.-H., 1989, The CaMV 35S enhancer contains at least two domains which can confer different developmental and tissue-specific expression patterns, EMBO J. 8:2195-2202.

Bird, D. McK. and M. A. Wilson. 1994a. DNA sequence and expression analysis of root-knot nematode elicited giant cell transcripts. Molec. Plant-Microbe Interact. 7: 419-424.

Bird, D. McK. and M. A. Wilson. 1994b. Plant molecular and cellular responses to nematode infection. In: NATO ARW: Advances in Molecular Plant Nematology, F. Lamberti, C. De Georgi and D. McK. Bird (Eds), pp 181-195. Plenum Press, New York, NY.

Conkling, M. A., Cheng, C.-L., Yamamoto, Y.T., and Goodman H.M., 1990, Isolation of transcriptionally regulated root-specific genes from tobacco, Plant Physiol. 93:1203-1211.

Cramer, C.L., Weissenborn, D.L., Cottinghan, C.K., Denbow, C.J., Eisenback, J.D., Radin, D.N., and Yu, X., 1993, Regulation of defense-related gene expression during plant-pathogen interactions, J. Nematol. 25:507-518.

Goddjin, O.J.M., Lindsey, K., van der Lee, F.M., Klap, J.C., and Sijmons, P.C., 1993, Differential gene expression in nematode-induced feeding structures of transgenic plants harbouring promoter-*gus*A fusion constructs, Plant J. 4:863.

Jefferson, R. A., 1987, Assaying chimeric genes in plants: the GUS gene fusion system. Plant Mol. Biol. Rep. 5:387-405.

Jones, M.G.K., 1981, Host cell responses to endoparasitic nematode attack: Structure and function of giant cells and syncytia, Annals Appl. Biol. 97:353-372.

Niebel, A., de Almeida Engler, J., Tiré, C., Engler, G., Van Montagu, M., and Gheysen, G., 1993, Induction patterns of an extensin gene in tobacco upon nematode infection, Plant Cell 5:1697-1710.

Opperman, C.H., Taylor, C.G., and Conkling, M.A., 1994, Root-knot nematode directed expression of a plant root-specific gene, Science 263:221-223.

Sasser, J.N., 1980, Root-knot nematodes: A global menace to crop production, Plant Dis. 64: 36-41.

Sasser, J.N., and Freckman, D.W., 1987, A world perspective on nematology. *in:* "Vistas In Nematology," D.W. Dickson and Veech, J.A., eds., Society of Nematologists, Hyattsville.

van der Eycken, W., Niebel, A., Inze, D., van Montagu, M., and Gheysen, G., 1992, Molecular analysis of the interaction between *Meloidogyne incognita* and tomato, Nematologica 38:441 [Abstr.].

Wilson, M. A., Bird, D. McK. and E. van der Knaap. 1994. A comprehensive subtractive cDNA cloning approach to identify nematode-induced transcript in tomato. Phytopathol. 84: 299-303.

Yamamoto, Y.T., Taylor, C., Acedo, G.N., Cheng, C.-L., and Conkling, M.A., 1991, Characterization of *cis*-acting sequences regulating root-specific gene expression in tobacco, Plant Cell 3:371-382.

# Use of Phytoremediation Strategies to Bioremediate Contaminated Soils and Water

Michael J. Sadowsky and Daniel R. Smith

Department of Soil, Water, and Climate, and Department of Microbiology, University of Minnesota, St. Paul, Minnesota (USA)

Plant-based bioremediation technologies have received recent attention as strategies to "clean-up" contaminated soils and water. These strategies have collectively been termed phytoremediation and refer to the use of plants for the *in situ* treatment of soil, sediment, and ground water. This definition, however, is rather limited in scope and has recently been re-defined to mean "the use of green plants and their associated microbiota, soil amendments, and agronomic techniques to remove, contain, or render harmless environmental contaminants" (Cunningham et al. 1996). The proposition of using plants to remove pollutants is not a novel idea and phytoremediation strategies were first used in Germany over 300 years ago to treat municipal sewage waste (Cunningham et al. 1996). Since this time, however, many new and potentially better strategies have been developed to use plants and their attendant microorganisms to extract or render harmless environmental pollutants.

Remediation of soils, water, and sediments contaminated with environmental pollutants is now, and continues to be, of major concern. It has been estimated that it will require over $750 billion over a 30 year period to remediate contaminated sites to today's legal standards (Stomp et al. 1994. The dollar amount and time required, however, is an estimate based on today's figures and technologies, and these could be miscalculated by a factor of 2 to 3 times.

Historically, non biological-based methods have been used to remediate contaminated waste sites. The remediation technologies which have been used can be divided into two major types depending on whether contaminant remediation is done *in situ* or *ex situ*. The technique used is largely determined by the extent and location of contaminants within soils. While *ex situ* techniques can be used when contaminants are close to the soil surface horizon and affect limited areas, *in situ* techniques are more useful for more deeply located contaminants that occupy a greater land area. Due to higher costs associated with the *ex situ* treatment of hazardous waste-contaminated soils, and the fact that many contaminants are making their way deeper into soil profiles, more research effort and technology advancement is being made in the area of *in situ* treatment processes. The ultimate type of remediation technology used, however, is largely dependent on the contaminants physical, chemical, and biological properties. It can be argued that the availability of pollutants in natural systems is one of the major factors affecting all bioremediation processes. Soil organic matter and clays can have a major impact on the bioavailability of contaminants in all natural systems (McBride 1994; Pignatello 1989).

Biologically-based remediation strategies (bioremediation) have received much recent attention as means to clean-up contaminated soils and water. Phytoremediation is just one type of bioremediation technique. Phytoremediation strategies have been examined as a means to "clean-up" a number of hazardous and recalcitrant organic and inorganic pollutants. Biologically based remediation strategies, including phytoremediation, have been estimated to be four to 1000 times cheaper, on a per volume basis, than current non-biological technologies (Cunningham et al. 1996). Compounds targeted for phytoremediation strategies include: heavy metals (Salt et al. 1995; Kumar et al. 1995; Brown et al. 1994), chlorinated solvents (Haby and Crowley 1996; Walton and Anderson 1990), polycyclic aromatic hydrocarbons (April and Sims 1990; Reilly et al. 1996), polychlorinated biphenyls (Brazil et al. 1995; Donnelly and Fletcher 1995), pesticides (Anderson and Coats 1995; Anderson et al. 1994), munitions (Schnoor et al. 1995) and radionuclides (Entry et al. 1995). While some of these contaminants are more readily degraded or detoxified than others, plants or their attendant rhizosphere microbes have been shown in several instances to transform these compounds to some degree. Generally speaking, soluble organic contaminants, which can move into plant roots or the rhizosphere by mass flow or diffusion, are the most amenable to phytoremediation processes (Cunningham et al. 1996; Schnoor et al. 1995). Plants showing great promise as phytoremediation agents include: grasses, legumes, trees and several other monocots and dicots (Cunningham et al. 1996; Dushenkov et al. 1995; Salt et al. 1995; Schnoor et al. 1995).

The ultimate goal of all phytoremediation technologies is to either remove the contaminant from the affected area (phytodecontamination) or to stabilize the compound to prevent off-site movement. The latter strategy, referred to as phytosequestration, is useful for contaminants having low biodegradation potential. Phytodecontamination processes can occur above or below ground. Above ground processes include: phytovolatilization, phytodegradation (via plant metabolism), and phytoextraction (the use of accumulator plants to transport materials into plant tissue for eventual destruction). Below ground phytodecontamination processes, on the other hand, rely on rhizosphere degradation activity (either plant enzyme- or microbially-driven) to transform hazardous waste materials.
Phytostabilization processes occur chiefly below ground and involve the sequestration of the contaminant into soil particles, cell wall lignins, or into the soil humus fraction (Cunningham et al. 1996). These processes reduce the bioavailability of contaminants (Salt et al. 1995).

Phytoremediation of inorganic or organic compounds from soils or water can occur *ex planta* or *in planta*. *Ex planta* phytoremediation processes can occur by the action of plant- or microbially-derived soil enzymes (Schnoor et al. 1995) or by plant associated microorganisms (Anderson and Coats 1995; Anderson et al. 1993; Haby and Crowley 1996). Halogenated hydrocarbons, nitroaromatic compounds, and anilines have been shown to be degraded by plant-derived soil enzymes (Schnoor et al. 1995) and the enzymes responsible for *ex planta* biodegradation (oxido-reductases, dehalogenases, nitroreductases, nitrilases, and laccases) have been investigated in some detail (Schnoor et al. 1995). Of particular interest have been peroxidases, which have been shown to biodegrade phenolic compounds in aquatic systems (Adler et al. 1994).

*Ex planta* phytoremediation can also occur via the degradative activity of rhizosphere microorganisms. The rhizosphere is operationally defined as the "soil-root interfacial area" and relatively large numbers of diverse species of microorganisms live in association with plant roots (Curl and Truelove 1986). The intimacy of the association between soil microorganisms and plant roots is determined, in part, by the types and concentrations of compounds exuded by roots. Root exudations are thought to have a stimulatory affect on rhizosphere microbes, which in turn, are purported to accelerate biodegradation in the rhizosphere (Anderson and Coates 1995; Anderson et al. 1993; Anderson et al. 1994; April and Sims 1990; Haby and Crowley 1996; Reily et al. 1996). Recently, Brazil and coworkers (1995) described the genetic construction of rhizosphere-competent pseudomonads which were engineered to contain the bacterial *bph* genes for biodegradation of PCBs. These strains have the potential to degrade PCBs in the rhizosphere and could be useful for bioremediation purposes.

In contrast to degradation external to the plant proper, *in planta* phytoremediation processes require that the pollutant is taken up into the plant. Generally speaking, organic pollutants are taken up by roots, and either sequestered or translocated to shoots and leaves (Devine and Vanden Borden 1991). Plants usually uptake organic compounds in the aqueous phase, by diffusion or mass flow processes, although in some instances vapor phase transport can occur (Cunningham et al. 1996). Once inside plant roots, organic and inorganic compounds can be transported to other portions of the plant apoplastically or symplastically (Salt et al. 1995). Ultimately, the compound is either metabolized within the plant, in some instances by conjugation to glutathione (Field and Thurman 1996) or sequestered.

Sequestration of pollutants within plants is the basis for phytoremediation of soils and water contaminated with heavy metals. Once inside the plant, the metals are removed from the environment by a process termed phytoextraction (Kumar et al. 1995). Metals targeted for this type of phytoremediation process include Cd, Pb, Zn, Cu, Cr, Ni, Se, and Hg. Phytoextraction, using "hyperaccumlating" plants is proving to become an effective method to remediate metal contaminated soils and water (Baker et al. 1991) and plant species such as *Thlaspi caerulescens* have been shown to accumulate very high levels of Zn and Cd from soils (Baker and Brooks 1989; Brown et al. 1994). *Brassica juncea* has also been found to be an excellent accumulator plant for metals in soils, such as Cd, Cr, Ni, Zn, and Cu (Salt et al. 1995; Kumar et al., 1994) and several plant species have been shown to accumulate Pb (Dushenkov et al. 1995). Plants, such as *Eichhornia crassipes, Hydrocotyle umbellata, Lemna minor*, and *Azolla pinnata* and are also effective at removing metals from aquatic systems (see Salt et al. 1995). Plant shoots and roots containing metals are subsequently harvested and treated as hazardous waste or the metals are recovered as ore.

There are several biotechnological strategies for enhancing phytoremediation in the future. These include enlarging root mass, using *Agrobacterium rhizogenes*, to increase root adsorption area (Tepfer et al. 1989), direct genetic engineering of plants for altered biodegradation potential (see Stomp et al. 1994 and Cunningham et al. 1996 for reviews), and the genetic engineering of rhizosphere microorganisms (Brazil et al. 1995).

In summary, phytoremediation processes hold great promise as means to remediate contaminated soils and water. As with any technology, however, there are advantages and disadvantages associated with phytoremediation strategies. Cost effectiveness, non-invasiveness, and a visually-pleasing technology are often cited as the major advantages of phytoremediation strategies over other currently available techniques. However, phytoremediation processes are relatively time consuming and ineffective at remediating sites containing pollutants located deep into the soil profile. In addition, costs associated with the disposal or recycling of toxic vegetation and the migration of contaminants off site, due to either plant-mediated solubilization of contaminants or the blowing of contaminated leaves, are problems which must be considered when deciding to employ phytoremediation strategies to bioremediate contaminated soils and water.

## Literature Cited

Adler, P. R., Arora, R., El Ghaouth, A., Glenn, D. M., and Solar, J. M. 1994. Bioremediation of phenolic compounds from water with plant surface peroxidases. J. Environ. Qual. 23:1113-1117.

Anderson, T. A., and Coats, J. R. 1995. Screening rhizosphere soil samples for the ability to mineralize elevated concentrations of atrazine and metolachlor. J. Environ. Sci. Health B30:473-484

Anderson, T. A., Guthrie, E. A., and Walton, B. T. 1993. Bioremediation in the rhizosphere. Environ. Sci. Technol. 27:2630-2636.

Anderson, T. A., Kruger, E. L., and Coats, J. R. 1994. Enhanced degradation of a mixture of three herbicides in the rhizosphere of a herbicide-tolerant plant. Chemosphere 28:1551-1557.

April, W., and Sims, R. C. 1990. Evaluation and use of prairie grasses for stimulating polycyclic aromatic hydrocarbon treatment in soil. Chemosphere 20:253-265.

Baker, A. J. M., and Brooks, R. R. 1989. Terrestrial higher plants which hyperaccumulate metallic elements - a review of their distribution, ecology, and phytochemistry. Biorecovery 1:81-126.

Baker, A. J. M., Reeves, R. D., and McGrath, S. P. 1991. *In situ* decontamination of heavy metal polluted soils using crops of metal accumulating plants - a feasibility study. Pages 600-605 in: *In situ* Bioreclamation. R. L. Hinchee and R. F. Olfenbuttel, eds. Butterworth-Heinemann, Boston, MA.

Brown, S. L., Chaney, R. L., Angle, J. S., and Baker, A. J. M. 1994. Phytoremediation potential of *Thalspi caerulescens* and bladder campion for zinc and cadmium-contaminated soil. J. Environ. Qual. 23:1151-1157.

Brazil, G. M., Kenefick, L., Callanan, M., Haro, A., de Lorenzo, V., Dowling, D. N., and O'Gara, F. 1995. Construction of a rhizosphere pseudomonad with potential to degrade polychlorinated biphenyls and detection of bph gene expression in the rhizosphere. Appl. Environ. Microbiol. 61:1946-1952.

Cunningham, S. D., Anderson, T. A., Schwab, A. P., and Hsu, F. C. 1996. Phytoremediation of soils contaminated with organic pollutants. Adv. Agron. 56:55-114.

Curl, E. A. and Truelove. B. 1986. The Rhizosphere. Springer-Verlag, Berlin.

Devine, M. D., and Vanden Borden, W. H. 1991. Absorption and transport in plants. Pages 119-140 in: Environmental Chemistry of Herbicides. R. Grover and A. J. Cessna, eds. CRC Press, Boca Raton, FL.

Donnely, P. K., and Fletcher, J. A. 1995. PCB metabolism by ectomycorrhizal fungi. Bull. Environ. Toxicol. 54:507-513

Dushenkov, V., Kumar, P. B., Motto, Ah., and Raskin, I. 1995. Rhizofiltration: the use of plants to remove heavy metals from aqueous streams. Environ. Sci. & Technol. 29:1239-1245.

Entry, J. A., Vance, N. C., Hamilton, M. A., Zabowski, D., Watrud, L. S., and Adriano, D. C. 1995. Phytoremediation of soil contaminated with low concentrations of radionuclides. Water, Air, and Soil Poll. 88:167-176.

Field, J. A., and Thurman, E. M. 1996. Glutathione conjugation and contaminant transformation. Environ. Sci & Technol. 30:1413-1418.

Haby, P. A., and Crowley, D. E. 1996. Biodegradation of 3-chlorobenzoate as affected by rhizodeposition and selected carbon substrates. J. Environ. Qual. 25:304-310.

Kumar, P. B., Dushenkov,, V., Motto, H., and Raskin, I. 1995. Phytoextraction: the use of plants to remove heavy metals from soils. Environ. Sci. & Technol. 29:1232-1238.

McBride, M. B. 1994. Environmental Chemistry of Soils. Oxford University Press, New York.

Pignatello, J. J. 1989. Sorption dynamics of organic compounds in soil and sediments. in: Reactions and Movement of Organic Chemicals in Soils. B. L. Sawhney and K. Brown, eds. Soil Science Society of America Special Pub. No. 22, Madison, WI.

Reilley, K. A., Banks, M. K., and Schwab, A. P. 1996. Organic chemicals in the environment: dissipation of polycyclic aromatic hydrocarbons in the rhizosphere. J. Environ. Qual. 25:212-219.

Salt, D. E., Blaylock, M., Kumar, N. P., Dushenkov, V., Ensley, B. D., Chet, I., and Raskin, I. 1995. Phytoremediation: a novel strategy for the removal of toxic metals from the environment using plants. Bio/Technol. 13:468-474.

Schnoor, J. L., Licht, L. A., McCutcheon, S. C., Wolfe, N. L., and Carreira, L. H. 1995. Phytoremediation of organic and nutrient contaminants. Environ. Sci. & Technol. 29:318-323.

Stomp, A.-M., Han, K.-H., Wilbert, S., Gordon, M. P., and Cunningham, S. D. 1994. Genetic strategies for enhancing phytoremediation. Ann. N.Y. Acad. Sci. 721:481-491.

Tepfer, D., Metzger, L., and Prost, R. 1989. Use of roots transformed by *Agrobacterium rhizogenes* in rhizosphere research: applications in studies of cadmium for sewage sludges. Pl. Mol. Biol. 13:295-302.

Walton, B. T., and Anderson, T. A. 1990. Microbial degradation of trichlorethylene in the rhizosphere: potential application of biological remediation of waste sites. Appl. Environ. Microbiol. 56:1012-1016.

# The Possible Links between RNA-Directed DNA Methylation (RdDM), Sense and Antisense RNA, Gene Silencing, Symptom-Induction upon Microbial Infections and RNA-Directed RNA Polymerase (RDRP)

H. L. Sänger, W. Schiebel, L. Riedel, T. Pelissier and M. Wassenegger

Max-Planck-Institut für Biochemie, Abteilung Viroidforschung, D-82152 Martinsried, Germany

The following contribution to the session IXB „Emerging Areas" is devoted to RNA-mediated gene regulation. This is, in fact, an area that develops rapidly at present. Therefore one can now start to speculate on the possible involvement of various biomolecules and mechanisms in gene regulation because this will stimulate new experimental approaches, new models and concepts.

## RDDM AND GENE SILENCING UPON MICROBIAL INFECTIONS

Previous work in our laboratory has shown that potato spindle tuber viroid (PSTVd)-specific cDNA integrated into the genome of *Nicotiana tabacum* SR1 becomes specifically methylated as soon as an autonomous viroid RNA-RNA replication has taken place in these plants (Wassenegger *et al.*, 1994). From these observations we inferred that RNA in general is capable of inducing and directing sequence-specific *de novo* methylation of genomic DNA. Although certain fully methylated plant genes are known to be well expressed, DNA methylation can nevertheless lead to gene silencing. Thus it is conceivable that a pathogen-induced RNA-directed DNA methylation (RdDM) results in a subsequent plant gene silencing. This interference with the normal regulation and expression of host genes could be responsible for the initiation and phenotypic expression of disease symptoms such as for example localized lesions, general growth retardations and discolorations, specific mal-formations of plant organs and overall decreases in yield.

Provided that the above mechanism actually exists, one can assume that not only pathogen-specific PSTVd RNA (Wassenegger *et al.*, 1994) but also RNAs in general can trigger RdDM. Infact, it was recently shown that transgenically expressed GUS-specific sequences are able to induce a sequence-specific RdDM (English *et al.*, 1996). Thus certain plant-endogenous RNA species should be capable of down-regulating various genes in healthy plants. The PSTVd RNA case suggests that possibly those RNAs are particularily efficient in inducing RdDM which are untranslatable. Considering gene regulation in uninfected healthy plants, this would mean that the corresponding cellular mRNAs are devoid of a cap-structure and/or a poly(A)-tail. Such aberrant forms of mRNA could be generated by premature termination of transcription, by irregular splicing, and by partial degradation. If these aberrant RNAs are able to induce RdDM they could be involved in transcriptional gene regulation (Fig. 1).

EFFECTS OF VIRAL SENSE AND ANTISENSE RNAS

Recent studies on transgenically mediated virus resistance in plants have shown that virus-specific sense and antisense RNA constructs of the corresponding virus-specific RNA can cause gene silencing at the post-transcriptional level. The observation that this is also possible with non-translatable truncated sense RNA demonstrated that the RNA itself and not its gene product is responsible for the induction of plant resistence against viruses (Smith *et al.*, 1994; English *et al.*, 1996). Both authors postulated that virus-specific antisense RNA is responsible for this effect and that whenever the transgene codes for a sense RNA, the corresponding antisense RNA is most probably synthesized by a host-encoded RdRP.

THE PROPERTIES OF RdRP AND ITS POSSIBLE ROLE IN GENE REGULATION

In figure 1 the presumed keyrole of the plant-encoded RdRP in this system is indicated. It is noteworthy in this context that RdRPs are so far considered to be unique for higher plants, and that their biological functions are still enigmatic because neither their templates nor their products are known. The only polymerase of this type that has been well characterized with respect to its physicochemical (Schiebel *et al.*, 1993a) and catalytic properties *in vitro* (Schiebel *et al.*, 1993b) is that of tomato.

To further characterize and to elucidate the still enigmatic *in vivo*

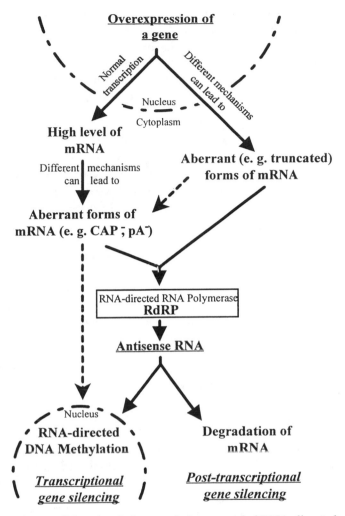

Fig. 1.   The possible role of plant- and virus-encoded RNA-directed
RNA polymerase (RdRP) in gene silencing

functions of this RdRP from tomato leaf tissue, we have subjected the
purified enzyme to microsequencing. From a series of the resulting
peptide sequences two were chosen, and a primer-pair deduced from
them was utilized for PCR amplification. With the aid of the resulting
PCR product cDNA libraries were screened but the maximum size of
hybridizing cDNA fragments was about 2200 bp, and it did not represent
the expected full-length clone with about 3300 bp. Therefore the 5'-part

of the putative RdRP-specific cDNA was isolated with the help of the RACE-technique. In this way we obtained a clone consisting of an open reading frame of 3345 bp which corresponds to a protein with a calculated molecular weight of 127 kD. This is in good agreement with the values experimentally determined in our laboratory by SDS-PAGE (128 kD) and sucrose gradient centrifugation (119 kD) (Schiebel *et al.*, 1993a).

Because neither the presumed RdRP-specific DNA sequence nor the deduced AS-sequence could be assigned to a published sequence, the final verification of the RdRP nature requiers further biochemical and biological characterizations. In case that we have the cloned tomato leaf RdRP in our hands it is planned to demonstrate by transgene experiments that this enzyme plays an important role in RNA-mediated gene regulation.

RNA-MEDIATED RESISTANCE (RMR) OF PLANTS AGAINST VIRUSES

Regarding the RMR of plants against viruses, the following points should be considered: RMR can most probably be induced only by antisense RNA. In principle, viral antisense RNA can be directed against the viral RNA genome and against the virus-specific mRNAs transcribed from it. From these two alternatives the genome-directed antisense RNA is functionally the most effective one because it has to cope only with the low copy numbers of the RNA of the infecting virus. Therefore we will discuss only this latter case. It could be observed in transgenic plants in which the corresponding transgenes consists of sequences that were also present in the viral genome. The transgene RNA or the RNA synthesized by the RdRP are assumed to function as antisense RNAs in that they are hybridising to the viral RNA sequences. These hybridization products are known to be immediately degraded by plant RNases (Fig. 2). In this way the viral RNA genomes are inactivated which can be regarded as a kind of post-transcriptional gene silencing. Naturally, this can function only as long as the corresponding antisense RNA is supplied. Therefore continous antisense RNA production would guarantee lifelong resistance of host plants against viruses.

In contrast, the RdDM mechanism (Fig. 2) is less suitable for conferring RMR because it is acting at the level of plant-specific mRNA transcription. Nevertheless the RdDM mechanism might be important for our investigations on the molecular interaction in the PSTVd-host plant system. The reason for this is clear, because, so far, the viroid-specific cellular target and the mechanism of viroid pathogenesis are still unknown. Moreover, viroid-RNA is not translated into proteins or

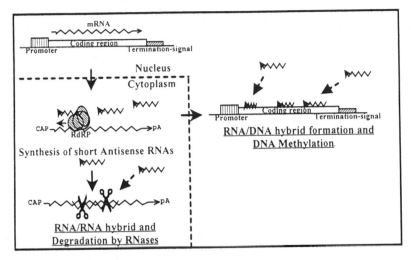

Fig. 2.    Possible interactions of antisense RNAs that can cause gene
silencing

peptides, so that this RNA itself must exert the pathogenic function. The
possible cellular targets for viroids can be proteins and nucleic acids and
our research interest is directed towards the latter.

IS RDDM RESPONSIBLE FOR VIROID PATHOGENICITY ?

From the data of our earlier experiments (Wassenegger *et al.*, 1994) we
assume, that in infected plants viroid RNA could lead to the RdDM of
important plant genes. Such a mechanism would demand plant-specific
sequences that are complementary to the PSTVd RNA. With standard
techniques it has not been possible, as yet, to detect in the tomato or
tobacco genome PSTVd-complementary DNA sequences of sizes larger
than 100 bp. Therefore, it is one of our present projects to determine the
minimal viroid sequence that is still capable of inducing RdDM. For this
porpose viroid-specific subfragments down to 30 bp in lenght are
integrated into the plant genome. Then the resulting transgenic plants will
be viroid-inoculated and the subsequently occuring pattern of DNA
methylation will be determined.

Parallel to this approach the search for host plant-specific sequences
which are complementary to the viroid RNA sequence has been
intensified. For this purpose PCR with viroid-specific DNA primers is

performed, and the methylation pattern of genomic sequences corresponding to the resulting PCR products are analysed in infected and healthy plants. In additional experiments we are looking for genes that are only silenced in viroid-infected symptom-bearing tomato plants. The methylation pattern of their genes is subsequently compared with that in healthy plants and the correlation between gene silencing and the appearance of symptoms will be analysed.

SEARCHING FOR SILENCED GENES

To find the genes that have been switched off, the RNA display technique will be used. There are two experimental systems which will be studied. In the first one, the comparison between RT-PCR products of viroid-free and viroid-infected tomato plants will show genes which are either silenced or induced as a result of infection (Fig. 3).

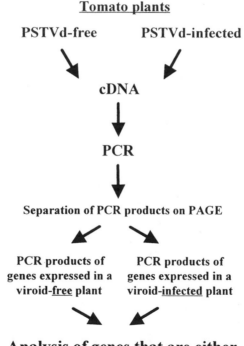

**Tomato plants**

**PSTVd-free**          **PSTVd-infected**

**cDNA**

**PCR**

**Separation of PCR products on PAGE**

**PCR products of genes expressed in a viroid-free plant**          **PCR products of genes expressed in a viroid-infected plant**

## Analysis of genes that are either silenced or induced

Fig. 3.    Investigation of the molecular plant-viroid interaction in PSTVd-infected and healthy tomato plants by RNA display

In the second system two different PSTVd-infected tomato cultivars (cv) are used, one of which reacts with severe symptoms of disease (cv "St. Pierre") whereas the other one (cv "Basket Pak") is virtually free of symptoms. The RNA display will again reveal the specific expression of genes in each cultivar (Fig. 4).

With these approaches we have isolated so far 24 different PCR products corresponding to cDNA fragments with open reading frames. Out of these, 13 exhibit homologies to known structural genes and six of those show clearly differential expression. For example, in symptom-bearing "St. Pierre" plants the porphobilinogen (PBG)-deaminase is downregulated, whereas in the symptom-free "Basket Pak", and in the non-infected "St. Pierre" plant normal expression is observed. In addition to the analysis of the cDNAs, we are also trying to clarify whether viroid-complementary sequences are found in regulative parts of the host plant

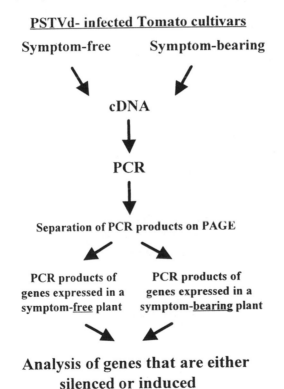

Fig. 4.  Investigation of the molecular plant-viroid-interaction in symptom-bearing and symptom-free PSTVd-infected tomato cultivars by RNA display

genome and whether subtle differences occur in the methylation pattern of these regions.

The further analysis of the 24 plant genes is in progress, including the isolation of their full-lenght cDNA and their genomic clones. We hope that these investigations will shed light on the elusive mechanisms of viroid pathogenesis and on the biological role of RdRP and RdDM in higher plants.

ACKNOWLEDGEMENT

We thank the Deutsche Forschungsgemeinschaft for partial support of these studies by grant Wa 1019/1-1 given to M. W.

REFERENCES

English JJ, Mueller E und Baulcombe DC. 1996. Suppression of Virus Accumulation in Transgenic Plants Exhibiting Silencing of Nuclear Genes. Plant Cell. 8:179-188.

Schiebel W, Haas B, Marinkovic S, Klanner A und Sänger HL. 1993a. RNA-directed RNA Polymerase from Tomato Leaves: I. Purification and physical properties. J Biol Chem. 263:11851-11857.

Schiebel W, Haas B, Marinkovic S, Klanner A und Sänger HL. 1993b. RNA-directed RNA Polymerase from Tomato Leaves: II. Catalytic *in vitro* properties. J Biol Chem. 263:11858-11867.

Smith HA, Swaney SL, Parks TD, Wernsman EA and Dougherty WG. 1994. Transgenic plant virus resistance mediated by untranslatable sense RNAs: expression, regulation, and fate of nonessential RNAs. Plant Cell 6:1441-1453.

Wassenegger M, Heimes S, Riedel L und Sänger HL. 1994. RNA-directed *de novo* Methylation of Genomic Sequences in Plants. Cell, 76, 567-576.

# Evolution of *Epichloë* Species Symbioses with Grasses

Christopher L. Schardl,[1] Huei-Fung Tsai,[1] Kuang-Ren Chung,[1] Adrian Leuchtmann,[2] and Malcolm R. Siegel [1]

[1] Department of Plant Pathology, University of Kentucky, Lexington, Kentucky, U.S.A.; and [2] Geobotanisches Institut, ETH, Zürich, Switzerland

Grass symbionts of genus *Epichloë*, including related asexual fungi (formally, form genus *Neotyphodium*), can be important for biological protection of their hosts under stresses imposed biotically (e.g. herbivores and parasites) or abiotically (e.g. drought) (Clay 1990). Symbioses with these fungal "endophytes" can be categorized into three types (Schardl 1996). The most benign (nonpathogenic) types have asexual endophytes which only vertically transmit by systemic infections of seeds, and tend to be highly beneficial to their hosts. In contrast, the most antagonistic, sexual *Epichloë* species only transmit horizontally (i.e. contagiously) and nearly completely suppress host seed production (choke disease). Of particular interest are symbioses of the pleiotropic type, whereby the symbionts abort only some host inflorescences and are transmitted both horizontally and vertically (Tsai et al. 1994). In many pleiotropic symbioses the benefits to the plant host probably far outweigh the cost of occasional infertile inflorescences, so these associations are almost certainly mutualistic. Although the asexual endophytes are also commonly regarded as mutualists, it is conceivable (but not generally testable) that some are asexual because their sexual cycles are suppressed by the host (i.e. the host antagonizes the endophyte), just as the antagonistic *Epichloë* species suppress host sex (Schardl 1996; H.H. Wilkinson, pers. com.). We used molecular phylogenetic analysis to compare *Epichloë* and grass phylogenies for evidence of cospeciation, and found that the phylogeny of pleiotropic and some antagonistic *Epichloë* species mirrored that of the hosts. In contrast, most asexual species were interspecific hybrids with genetically diverse ancestral *Epichloë* species.

## Materials and Methods

Mating tests were conducted as described by Leuchtmann et al. (1994) to determine the sex (mating type) and biological species (mating population) of each isolate. Some isolates could not be used for mating tests and were tentatively assigned to mating populations based on isozyme analysis. For molecular phylogenetic analysis, noncoding portions of the β-tubulin genes (*tub2*) and nuclear rDNA were amplified and sequenced as described by Schardl et al. (1994). When isolates had multiple *tub2* copies each was separately sequenced (Tsai et al. 1994). Phylogenetic analysis was by parsimony, using branch-and-bound searches (exact solutions) (Swofford 1993).

## Results

Mating tests placed all known sexual strains into nine biological species of *Epichloë* (designated as mating populations MP-I through MP-IX), although assignments of some isolates to MP-IV, MP-VIII and MP-IX are tentative pending more complete tests. In keeping with observations by Leuchtmann et al. (1994), all sexual strains of *Epichloë* isolated from any one host species belonged to the same biological species. Furthermore, each host genus except *Agrostis* and *Brachypodium* was host to only one sexual *Epichloë* species. Conversely, host ranges of most *Epichloë* mating populations were narrow, confined to individual grass species, genera or tribes. MP-II was identified only in *Festuca* species (tribe Poeae), MP-III in *Elymus* species (Triticeae), MP-IV and MP-V in several genera of tribe Aveneae, MP-VI in *Bromus erectus* (Bromeae), MP-VII in *Brachypodium sylvaticum* (Brachypodieae), MP-VIII in *Glyceria striata* (Meliceae) and MP-IX in *Brachyelytrum erectum* (Brachyelytreae). Only MP-I (*Epichloë typhina*) had a broad host range. Strains that tested as MP-I were isolated from grasses belonging to tribes Poeae and Aveneae.

Asexual, vertically transmitted endophytes are known from many grass genera and species (Leuchtmann and Clay 1990), including several that are also hosts of different *Epichloë* species. With one exception, asexual isolates exhibited mating-type activity: spores of each isolate, when used as spermatia, elicited development of barren fruiting structures on *Epichloë* stromata of either sex (mating type), but not both. This observation was also typical of interspecific matings of sexual *Epichloë* strains of opposite sex. The difference was that most of the asexual species also failed to elicit even rare stromata on their hosts. An exception was a seed-transmissible endophyte (isolate E187) which occasionally developed stromata on its host, *Poa ampla*, but which failed to mate or even elicit barren perithecia when used either as a male (spermatial) or female (stromal) parent with any *Epichloë* strain tested. Thus, the *Poa ampla* endophyte appears also to be asexual because it lacks mating-type activity.

A striking difference in genotypes of sexual and asexual endophytes was noted in studies of the β-tubulin genes (*tub2*) (Schardl et al. 1994; Tsai et al. 1994). Each of the sexual isolates had only a single sequence, probably indicating only a single *tub2* copy. In contrast, multiple copies were identified in almost all asexual endophytes of hexaploid tall fescue (*Festuca arundinacea* var. *genuina*) (Tsai et al. 1994). This is a particularly interesting grass because of the very diverse endophyte genotypes for which it is a host, and because the grass is well known to gain numerous fitness enhancements from the endophyte *Neotyphodium coenophialum* (Schardl 1996). Two *tub2* copies were also identified in the *P. ampla* symbiont E187, and in a rare endophyte genotype from perennial ryegrass (*Lolium perenne*). Only single *tub2* copies were identified in isolates of asexual species *Neotyphodium lolii* (from perennial ryegrass) and *N. uncinatum* (from *Festuca pratensis*).

Relationships among sexual *Epichloë* strains and species were assessed by molecular phylogenetic analysis of *tub2* 5'-regions, including three of the four *tub2* introns. A single most parsimonious tree was resolved (Fig. 1). In all cases except MP-I, isolates of each mating population grouped with each other. The sequence of isolate E430 was particularly divergent from those of other MP-I isolates. With this sequence included, MP-I appeared paraphyletic to MP-VII.

Inclusion of sequences from *N. coenophialum* and other tall fescue symbionts confirmed the hybrid nature of these asexual species (Tsai et al. 1994). Most isolates had two or three *tub2* gene sequences, and in all cases the multiple copies failed to group together in the phylogenetic tree. Nevertheless, clear relationships were indicated between the *tub2* copies of the tall fescue endophytes and those of sexual *Epichloë* species. For example, *N. coenophialum* had three *tub2* copies: one (designated *tub2-4*) related to the single copy in *Epichloë baconii* (MP-V), another (*tub2-2*) to that in *E. festucae* (MP-II), and the third (*tub2-3*) identical to that in *N. uncinatum* (Fig. 1).

The stroma-forming symbiont of *P. ampla* had two distinct *tub2* sequences (designated *tub2-3* and *tub2-6*). The 5'-region of *tub2-6* was amplified with a copy-specific primer (Tsai et al. 1994) and sequenced; the other copy, *tub2-3*, was amplified, cloned and sequenced. Inclusion of the two sequences with those of the *Epichloë* mating populations in parsimony analysis indicated that isolate E187 was a probable hybrid with ancestors related to MP-I and MP-IV (Fig. 1). One *tub2* copy was particularly similar to that of MP-I isolate E430. Likewise, the rDNA internal transcribed spacer sequences of E430 and E187 (both of which had only a single detectable sequence for this region) were more closely related to each other than to those of the other MP-I isolates (data not shown).

The *tub2* gene phylogenies of *Epichloë* species mirrored that of the hosts (Davis and Soreng 1993) in most aspects. In particular, the phylogeny of the five pleiotropic species (MP-II, MP-III, MP-IV, MP-VII, and MP-IX) and two antagonistic species (MP-V and MP-VI) matched that of the host

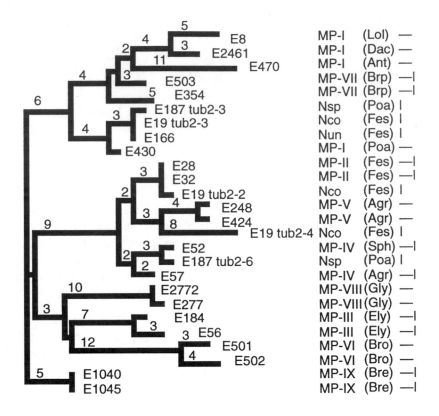

Fig. 1. Most parsimonious phylogram on *tub2* intron sequences of *Epichloë* and *Neotyphodium* species. The left edge of the phylogram is the root inferred by midpoint rooting. Isolates are indicated at the termini, with *tub2* copy designations when multiple copies are present. Right columns list mating populations (MP-I to MP-IX) or asexual species (where Nco=*N. coenophialum*, Nun = *N. uncinatum*, and Nsp=*Neotyphodium* sp.), host genera (abbreviated in parentheses as: *Lolium, Dactylis, Poa, Anthoxanthum, Brachypodium, Festuca, Agrostis, Sphenopholis, Glyceria, Elymus, Bromus*, and *Brachyelytrum*), and whether the symbiont is transmitted horizontally (—), vertically (l), or both (—l). Horizontal branch lengths indicate the numbers of nucleotide substitutions (given above each branch of length >1). The vertical lines are merely illustrative, and have no quantitative value.

tribes. The two exceptions to the mirror host and fungus phylogenies involved two antagonistic species: the broad host range species MP-I, and MP-VIII from tribe Meliceae. No appropriate outgroup has been identified with sufficient sequence similarity to root the *Epichloë* gene tree, so the position of the root was estimated by midpoint rooting. The inferred root of the *Epichloë* *tub2* tree (placed at the left edge of the phylogeny in Fig. 1) is near the position expected if it mirrors the phylogeny of the host tribes (Davis and Soreng 1993). Brachyelytreae (host of MP-VII) and Brachypodieae (host of MP-IX), together with Meliceae (host of MP-VIII), are very basal among C3 grasses relative to the tribes Aveneae (hosts of MP-IV and MP-V), Poeae, Triticeae, and Bromeae (hosts of MP-II, MP-III and MP-VI, respectively). Thus, the *tub2* tree, at the midpoint, reflects phylogenies of host tribes for seven of the nine *Epichloë* species.

## Discussion

Pleiotropic symbioses of *Epichloë* species and grass hosts can be regarded as totipotent, in that they exhibit both the life cycle with vertical transmission of the asexual (*Neotyphodium*) species, and the sexual life cycle and horizontal transmission of the more antagonistic *Epichloë* species. For this reason, we propose that the pleiotropic symbioses represent the most developmentally balanced interactions of the symbiotic partners, reflective of a high degree of coadaptation. If so, it is reasonable to hypothesize that the pleiotropic species have a long term history of cospeciation with their hosts. Because grasses of the same tribe (though often different genera) hybridize in nature, it is appropriate to consider "cospeciation" of symbiont species with host tribes. The phylogenetic analysis of *tub2* introns presented here support cospeciation in evolution of the pleiotropic symbioses.

Except for the pleiotropic symbioses, there is not an overall pattern indicative of grass-*Epichloë* or grass-*Neotyphodium* cospeciation. Where it might have been expected is in those symbioses in which the endophyte only transmits vertically. However, such asexual endophyte may have arisen recently, most by a complicated path involving interspecific hybridization. We proposed that the clonal nature of these endophytes makes them susceptible to the effects of accumulated deleterious mutations (Muller's ratchet), and potentially evolutionary dead-ends (Schardl et al. 1994). Their way out of this quandary—perhaps the reason they persist at all—may be that they can obtain fresh genetic inputs by hybridizing with meiotic progeny of their sexual relatives. Cospeciation cannot be discerned for the hybrid endophytes, many of which have *Epichloë* ancestors not known to infect their hosts or related hosts.

The other extreme of the symbiotic continuum, involving the

antagonistic *Epichloë* species, also did not exhibit overall cospeciation. On the present evidence we can propose three evolutionary origins of the antagonists. First, they may have evolved from pleiotropic symbionts in ecological settings where maximal horizontal transmission is most beneficial to them. Possible examples are MP-VI symbiotic with *B. erectum* and MP-V with tribe Aveneae. Second, they may have arisen by infection of hosts to which they are not highly adapted by cospeciation, and the new associations lack the developmental balance of pleiotropism. A possible example is MP-VIII in *G. striata*. The third possibility seems to be exemplified by *E. typhina* (MP-I), which may never have specialized to a group of related host genera. The broad host range of such a species may come at the cost of poor coadaptation to any one host. Thus, we suggest that some or all of the antagonistic *Epichloë*-grass symbioses arose by various evolutionary pathways that disrupted or prevented the coadaptation necessary for pleiotropic symbiosis.

## Literature Cited

Clay, K. 1990. Fungal endophytes of grasses. Ann. Rev. Ecol. Syst. 21: 275-295.

Davis, J. I., and Soreng, R. J. 1993. Phylogenetic structure in the grass family (Poaceae) as inferred from chloroplast DNA restriction site variation. Amer. J. Bot. 80: 1444-1454.

Leuchtmann, A., and Clay, K. 1990. Isozyme variation in the *Acremonium/Epichloë* fungal endophyte complex. Phytopathology 80: 1133-1139.

Leuchtmann, A., Schardl, C. L., and Siegel, M. R. 1994. Sexual compatibility and taxonomy of a new species of *Epichloë* symbiotic with fine fescue grasses. Mycologia 86: 802-812.

Schardl, C. L. 1996. *Epichloë* species: fungal symbionts of grasses. Ann. Rev. Phytopathol. 34: 109-130.

Schardl, C. L., Leuchtmann, A., Tsai, H.-F., Collett, M. A., Watt, D. M., and Scott, D. B. 1994. Origin of a fungal symbiont of perennial ryegrass by interspecific hybridization of a mutualist with the ryegrass choke pathogen, *Epichloë typhina*. Genetics 136: 1307-1317.

Swofford, D. L. 1993. PAUP: Phylogenetic Analysis Using Parsimony, Version 3.1.1. Illinois Natural History Survey, Champaign, Illinois.

Tsai, H.-F., Liu, J.-S., Staben, C., Christensen, M. J., Latch, G. C. M., Siegel, M. R., and Schardl, C. L. 1994. Evolutionary diversification of fungal endophytes of tall fescue grass by hybridization with *Epichloë* species. Proc. Natl. Acad. Sci. USA 91: 2542-2546.

# Meeting Summary

## R. James Cook

**Research Leader, Root Disease and Biological Control Research Unit, U.S. Department of Agriculture, Agricultural Research Service, Pullman, WA 99164**

This meeting can be summarized in one word: *outstanding*! The science is outstanding, the local arrangements and meeting organization are outstanding, and the attendance at roughly 950 people from 40 countries is outstanding. I am especially impressed with the international representation and the obvious international cooperation represented in the work presented here. I am also impressed with the number of young people present, which bodes well for the future of research on molecular plant-microbe interactions as an area of science.

I should say a few words about my own interests in science and technology development, because it has some bearing on the approach I will take to summarize this meeting. I am a plant pathologist still in my first permanent job after 31 years. I am part of a team that includes Linda Thomashow, David Weller, and me as lead scientists and Jos Raaijmakers from the Netherlands as an ARS post doctoral fellow. Together, we are responsible for the development of the knowledge base and technology needed to manage root diseases of wheat and barley. My role is to work out the etiology, ecology, and epidemiology of the root diseases within the various farming systems, including "no-till" systems, characteristic of modern cereal-based agriculture. It is also my role to test and deliver any new technology for root disease control into application. In addition, I recently completed 2½ years (October 1, 1993 through March 31, 1996) as Chief Scientist of the USDA's National Research Initiative Competitive Grants Program (NRI). While my personal research is very applied, I am a passionate defender of fundamental research needed to open new directions for applied research.

## "It's a Jungle in There"

A recent issue of *BioScience* (Stevens 1996) has an article entitled: "It's a Jungle in there." The article is about the hundreds of species of microorganisms that live in our mouths. As new-borns, our mouths are sterile but then quickly become colonized by a succession of microorganisms from many sources. In adulthood, our mouths harbor a "climax" microbiota, of sorts, made up of an incredible diversity of microorganisms present in different habitats created by the chemical and physical environment of our mouths. The great majority of species exist in a benign relationship with the tissues and structures available to them. Some even provide us with protection against infections. Unfortunately for us, a few are detrimental to us.

Every form of life is a support system for microorganisms, some beneficial, some along for ride, but always a few that can cause harm to their supporting life-form. Moreover, while we may be critical of ourselves for having produced so little solid scientific information about molecular plant-microbe interactions, I doubt very much if any other supporting life-form as a habitat for microorganisms is understood at the molecular level any better than we understand molecular plant-microbe interactions.

We can visualize the distribution patterns of microorganisms suppported by plants the next time we gaze down on the distribution of homes, villages, and cities over a landscape as viewed from an airplane. Looking down on an Arizona desert would be like looking down on an area of leaf or root where the numbers of colonies and the kinds of microbes are few and far between, because conditions are too harsh or the carrying capacity is too limited to support more than very sparse colonization. Then there is Phoenix--the human equivalent of a thriving and expanding network of colonies of microorganisms--perhaps still in log-phase growth. Flying over Washington DC is like looking down on the surface of a mature leaf supporting complex interrelated and interdependent communities of microorganisms responding to perturbations but otherwise in a climax state ecologically--like virtual gridlock. And while the majority of plant-associated microorganisms carry out part or all of their life cycles on plant surfaces, as epiphytes, a few are specialized to carry out their life cycles "underground," that is, within plant tissues as endophytes.

Like the mouth of the new-born baby, microbial colonization of plants must start sometime and somewhere. Some species are natural inhabitants on or within the seeds. They go wherever the seed goes. As the seed begins to swell and germinate in the soil, the seedborne microorganisms multiply and some then spread or are physically transported as the *pioneer colonists* upward on the emerging shoot, downward on the elongating root, or both. Most information on this process has been obtained from research with seedborne plant pathogens, but the colonization process is the same whether the seedborne microorganisms are pathogenic or nonpathogenic.

Most inhabitants on leaves (or the phyllosphere) get there by emigration from other plants or plant remains. They arrive as windborne and water-splashed spores, cells, and aggregates of cells; in aerosols; or carried by birds and insects. These emigrants multiply in response to substrates provided by the plant, and gradually they displace, dilute, or complement the pioneer colonists as part of the natural succession of epiphytes and endophytes over time and as the plant matures.

In contrast to colonists of above ground plant parts, most inhabitants of roots (or the rhizosphere) lie dormant in the soil waiting for the root to come to them, and then multiply in response to substrates provided by the root. Gradually these microorganisms, like their above-ground counterparts, diplace, dilute, or complement the pioneer colonists that arrived with the seed--again, as part of that natural succession over time and as the root matures.

Working within this framework of principles, whether on the seed, leaf, flower, stem, or root-- It's a jungle! Little wonder that it's so difficult to establish a given strain of microbial biocontrol agent or nitrogen-fixing bacterium in the rhizosphere or phyllosphere, even when introduced on the seed to assure their status as pioneer colonists. The positive side is that pathogens must also compete in this jungle before they can infect, and some must continue to compete with nonpathogens--the secondary colonists of lesions--after infection. This gives some explanation for the pathogen-suppressive soils, wherein disease declines with crop monoculture or is mild in spite of a fully susceptible host, virulent inoculum of the pathogen, and favorable physical environment. I will have more to say about this subject later when I discuss work presented at this meeting on strategies for defense against root pathogens.

## Communication and Conservation

The information presented at this meeting is beginning to give us a glimpse of the kinds of molecular communications that take place, not just between corresponding genotypes of microbes and plants, but also between and among the communities, colonies, and individual cells of the microorganisms themselves. Quorum sensing, the homoserine lactone autoinduction system, and cross-feeding are the more prominent examples of the kinds of interactions reported here to take place between and among plant-associated microorganisms.

It is also clear that microorganisms perform quite differently *in planta* than in culture--something very familiar to those of us that work with plant pathogens. When it comes to unravelling the secrets of how plant-associated microorganisms actually perform and interact in nature, I always marvel at the genius of Steve Lindow, who continues to come up with both the methods and the models to reveal the kinds of interactions between plants and microorganisms under natural conditions.

It is especially fascinating to see the growing list of examples of gene conservation by microorganisms, including the apparently highly-conserved *gacA/lemA* two-component signal transduction system, the *rpoS* global regulator system, and genes and associated mechanisms used by so many different kinds of microorganisms. We are duly impressed with the diversity of molecular mechanisms, but we must be just as impressed by the consistency in the molecular mechanisms.

## Coevolution and Gene-for-Gene Relationships

The interdependence of plants and their microbial associates makes sense only if we think of the interactions as an outcome of *coevolution*. As Sharon Long put it in her presentation--"the stories that can be told as the result of plants and microorganisms having lived together for tens of millions of years."

There may be no better example of the product of coevolution than the gene-for-gene relationship between host and parasite. The unravelling of this story in among the most exciting accomplishments in all of plant biology this century, starting with the work of H.H. Flor in the 1950s. The first breakthroughs in our understanding of this relationship at the molecular level came with the cloning of the *avr* genes from pathogens. This has been rapidly followed by the amazing advances during just the past 2 years in cloning R genes from the host. Again, the consistency of the mechanisms and conservation of genetic information is remarkable. I'm not sure that any of us were quite prepared for these results.

The cloning and characterization of disease-resistance genes from plants are just as significant scientifically as are the recent breakthoughs in cloning and characterization of genes that condition susceptibility of humans to certain genetic diseases. I would also complement the leaders of this remarkable work, especially Fred Ausubel, Barbarba Baker, Pierre de Wit, Jeff Ellis, Greg Martin, and Brian Staskawicz, for the way they and their groups have worked together and supported one another. This field of research is highly competitive, yet these investigators have found ways to work and publish together.

Some of the more exciting new information presented at this meeting suggests that *avr* gene products interact directly with their corresponding R gene products, possibly by entering plant cells by way of the harpin secretion apparatus (e.g., see poster abstract of J.R. Alfano, D.W. Bauer, and A. Collmer). I find this work especially interesting, since it brings full circle, at least for bacterial pathogens, what was proposed without direct evidence years ago: that the *avr* gene product interacts directly with the corresponding R-gene product for recognition of the invader and initiation of the plant defense response in the gene-for-gene plant-pathogen interactions.

We must still account for why the vast majority of plant-microbe interactions are neither detrimental nor beneficial to the plant--why most plant-associated microorganisms are limited to a saprophytic existence on of superficially within the plant but apparently at no significant cost to the plant. I was particularly interested in the paper by Thomas Boller on perception by the plant of specific molecules or conserved domains of molecules common to many different kinds of microorganisms. For fungi, he identified some common extracellular proteins along with chitin fragments and ergosterol, each of which was recognized by the plant at very low threshold concentrations. For bacteria, he identified common surface proteins present in genera as different as *Pseudomonas, Bacillus, and Escherichia*. It makes sense biologically that, through coevolution with microorganisms, plants have come to recognize a few traits, molecules, or portions of molecules common to these microorganisms as part of an efficient defense mechanism.

### Enter the "Stealth" Fungi

Among the diversity of apparently benign plant-microbe interactions, I would call your attention to a group of fungi that I call the "stealth" fungi, because of their apparent ability to colonize plant tissues to a considerable extent "under the radar" system of microbe detection used by plant cells. I am referring to the saprophytic or very weakly parasitic species of *Fusarium*, *Cladosporium*, and *Alternaria*, as prime examples, but there are many others. Once established, they live endophytically in the apoplast, waiting for the plant--or tissues--to die. They are then poised to be the first to take possession of the dead or dying tissues. These fungi, and there may be bacteria such as *Pseudomonas syringae* that do this as well, are key to the early stages of decomposition of plant debris.

Pierre de Wit described an aspect of this kind of plant-microbe interaction in his paper on the life cycle of *Cladosporium fulvum*, except that this fungus has abilities as a pathogen. In fact, each of the genera or species representive of these fungi include species or subspecies that are pathogenic. If they are pathogens, it is usually because they produce a toxin or colonize the xylem and cause vascular wilt. It would not be surprising if the saprophytic fungi that I have in mind carry *avr* genes as historical records of past interactions with plants, but I am not aware of evidence that they trigger a defense response typical of *avr*-R gene-for-gene interactions. Barbara Valent reported here that 58% of the apressoria formed by the avirulent *Magniporthe grisea* did not even attempt penetration, but I am not aware that the fungi I call the stealth fungi even form appressoria. While among the most successful of all plant colonists, they do little if any damage, other than possibly accelerating senescence or death of already senescent tissues.


## Strategies for Defense Against Root Pathogens

Virtually all genes for resistance to plant pathogens known today are for pathogens that attack above-ground plant parts. This could be because less attention has been paid to finding genes for resistance to root diseases, but I doubt that this is the case. Rather, it could be that plants use a different strategy to defend themselves below ground--the strategy that takes advantage of antagonistic microbe-microbe interactions beneficial to and possibly even supported by the plant.

Just as the addition of a specific substrate selectively enriches for microorganisms with the enzymes to use that substrate, so the repeated exposure of the soil microbiota to roots of the same plant species can be expected to enrich for microorganisms with enzymes and other properties that favor their ability to compete in the rhizosphere of that plant species. The more intense this competition, the more hazardous it will be for pathogens that depend on these same roots or root exudates for their prepenetration growth or secondary spread from root to root or plant to plant. For this and other reasons, the opportunities for protection by antagonists are greater for roots in soil environments than for above-ground plant parts in aerial environments. Plants benefiting from this kind of protection might, therefore, have less selection pressure to evolve the kinds of mechanisms needed for defense against foliar pathogens.

Research reported by Linda Thomashow and in the poster by Jos Raaijmakers et al. provides clear evidence that wheat benefits from this mechanism of defense against the root disease known as take-all. The take-all pathogen, *Gaeumannomyces graminis* var. *tritici*, as a member of the pyrenomycetes, is taxonomically related to *Magniporthe grisea*. However, *G. graminis* var. *tritici* has a much wider host range within the grass family and is highly suited to survival in infested host debris in soil. Resistance exists within some distant relatives of wheat, e.g., the diploid *Dasapyrum villosum* (R.J. Cook and Steve Jones, unpublished), but not within the hexaploid wheats or tetraploid durums. In direct constrast to the dearth of genetic resistance, it is commonly observed that soils initially conducive to the development of take-all become suppressive and take-all declines after one or more outbreaks of severe disease and continued monoculture of the host.

Our work has revealed that the increased suppressiveness associated with take-all decline is accompanied by qualitative and quantitative shifts in rhizosphere bacteria (rhizobacteria) with ability to produce antibiotics inhibitory to the pathogen. The antibiotic 2,4-diacetylphloroglucinol produced by fluorescent *Psuedomonas* species is a key determinant in the natural suppression of the wheat take-all pathogen in soils following take-all decline. An analysis of bacterial genes required for biosynthesis of 2,4-diacetylphloroglucinol has revealed at least six open reading frames, including one with homology to the chalcone/stilbene synthase family of plant genes. Again, we see evidence of the dependence of organisms on the same or similar trait to do a job. If the plant made this antibiotic within the lesions caused by the take-all fungus, it would be called a *phytoalexin*. Raaijmakers' poster reports further that the frequency of bacteria with ability to produce this antibiotic increases dramatically and coordinately with increased suppressiveness to take-all in response to planting wheat repeatedly in the same soil, which is how wheat grows in the wild.

I would also call your attention to the paper by Charles Opperman on the role of *TobRB7*, a root-specific gene involved in directing formation of the feeding apparatus (starting with formation of giant cells) for the root knot nematode, *Meloidogyne incognita*. Expression of this plant gene in the antisense direction prevented formation of giant cells and resulted in more than 90% control of the nematode. His work with Mark Conkling at NC State not only reveals a new direction for development of varieties of crop plants with resistance to root knot nematode, it could open an entirely new genetic approach to control of plant parasites more generally.

Finally, I would call your attention to the report by Martha Hawes on root border cells. These cells, once thought to be dead or dying cells sloughed from roots by the abrasive effects of roots extending through soil, are now shown to be metabolically very active cells released from the root cap region by a highly controlled process. These cells, in turn, are recognized specifically by fungal and bacterial pathogens as well as microbial symbionts. This work potentially is the most significant contribution to our understanding of the role of roots in directing their own colonization by microorganism since A.D. Rovira's work on root exudates published starting more than 40 years ago.

### Enhanced Host-Plant Susceptibility to Disease

A great many host pathogen interactions become destructive to the plant only when the plant is predisposed by some environmental stress. These diseases represent apparently compatible host-pathogen interactions but are still limited in the vigorous host, presumably by the defense mechanisms characterized by production of PR proteins, phenolics, lignin, and other substances or barriers limiting to growth of the pathogen. In contrast to the situation in the healthy vigorous host, these disease develop at a much faster rate in plants under some form of environmental stress--what has been referred to as "super susceptibility." Plant diseases favored by environmental stresses on the host are very important to crop production, because environmental stresses are the norm and not the exception in agriculture. They currently are managed primarily by changing the cultural practices to provide a less stressful environment for the crop, or by breeding varieties that can tolerate or escape the environment stress. It would also be useful to know something about the genetic control of this enhanced susceptibility as still another means to reduce the effects of these diseases. In his Hertel Foundation lecture at the opening of this meeting Fred Ausubel described what he calls enhanced disease susceptibility or *eds* genes. It would be interesting to know whether expression of *eds* genes plays a role in the super-susceptible phenotype characteristic of certain host-pathogen interactions conditioned by an environmental stress.

## So What?

This brings me to the issue of usefulness of the results of research on molecular plant-microbe interactions. Who cares about all this fundamental information? And, with so much of this work supported by public funds, why should we ask the citizens of our respective countries to pay for the assemblage of all this information?

First, as scientists, we should never cease attempts to make the case for the importance of new knowledge to the health and well-being of society. This knowledge is key to better decisions by consumers and better policies by our governments. We should also take note--and heart--that intellectual understanding is a fundamental human need; witness the success of National Geographic magazine and the many television shows on science that focus on how nature works. People are curious! Unfortunately, I must report from my experience with the USDA's NRI that, in spite of the obvious benefits of knowledge to society, and the strong societal support for science, selling our research to society strictly on this basis of knowledge for knowledge sake is extremely difficult if not virtually impossible!

Second, the information on molecular plant-microbe interactions specifically is helping to reveal fundamentally new information on plant processes and plant-microbe communications. For example, the work presented here by William Lucas on virus movement through plasmodesmata in plants is revealing previously unrecognized mechanisms of symplastic trafficking of large molecules by plants. Similarly, information on molecular signals and events leading to successful infection of root hairs in legume-*Rhizobium* (or *Bradyrhizobium*) interactions has become the model and and a source of new tools for research on molecular root-microbe interactions more generally.

Third, while rarely mentioned, we should consider the importance of work on molecular plant-microbe interactions towards understanding the success of native plant species, including threatened and endangered species, and the means to better or wiser ecosystem management. This suggestion may sound strange coming from such a traditional agriculturalist, but I would submit that understanding the basis for survival of plants in their native habitats is key to not only their preservation but also to improving the productivity of their descendants used in agriculture.

Finally, and obviously of greatest practical importance to humankind, are the implications of this work for production of food and other plant products important to people. The advances are forthcoming on three broad fronts: biological control of pests and diseases with plant-associated microorganisms; more sustainable plant nutrition through nutrient uptake facilitated by mycorrhizae and biological nitrogen fixation; and new sources of more sustainable resistance to diseases, plant parasitic nematodes, and insect pests. We learned at this meeting, for example, of the accomplishments and plans of the Monsanto Agricultural Company and Ciba Plant Protection for the deployment of new genes for pest and disease resistance and a new product to promote systemic acquired resistance. These developments are the direct outcome of fundamental research of the kind presented here this week.

The United States along with many other countries has gained enormous experience over most of this century through a network of public and private plant breeding programs in a genetics approach to solving problems for agriculture. Today, I would submit that developed and developing countries alike are poised to take advantage of any new technology that will broaden the base of useful germplasm or the speed with which new genes can be put to use in plants for the benefit of society. Consider the use of the disarmed Ti plasmid of *Agrobacterium tumefaciens* for transformation of crop plants. Following the demonstration in 1979-1980 by Gene Nester, Mary Dell Chilton, and others at the University of Washington, that *A. tumefaciens* produces crown gall by inserting a segment of it own DNA into the host plant genome, it quickly became apparent that the transfer DNA (T-DNA) from this pathogen could serve to introduce virtually any gene into a suitable host. Within just five years, Roger Beachy using T-DNA as a vector had demonstrated the potential for coat-protein-mediated resistance of plant to their viruses, and very soon thereafter, Monsanto had field tests in progress with tomato transformed to express the coat protein gene from tomato mosaic virus. Today, Asgrow is marketing a variety of squash transformed to express coat-protein mediated virus resistance.

In closing, I would like to quote the most famous microbiologist of all time.

"... there does not exist a category of science to which one can give the name 'applied science'. There are science and the applications of science, bound together as the fruit to the tree that bears it."

Louis Pasteur

## Literature Cited

Stevens, J.E. 1996. It's a jungle in there. BioScience 46(5) 314:317.

# Satellite Meeting Report:
# Emerging Model Legume Systems: Tools
# and Recent Advances
# July 12 - 14, 1996

Kathryn VandenBosch[1], Douglas Cook[2], Frans de Bruijn[3], and Thierry Huguet[4]

[1]Department of Biology, Texas A&M University, College Station, TX 77843-3258, USA; [2]Department of Plant Pathology and Microbiology, Texas A&M University, College Station, TX, 77843-2132, USA; [3]DOE Plant Research Laboratory, Michigan State University, East Lansing, Michigan, 48824, USA; [4]Laboratoire de Biologie Moléculaire, CNRS-INRA, BP27, 31326 Castanet-Tolosan Cédex, France

Within the last decade, some investigators in the field of nitrogen fixation research have selected two primary legume species that meet criteria for tractable model systems for genetic and molecular analyses (Barker et al., 1990; Handberg and Stougaard, 1992). Since their introduction, both *Medicago truncatula* and *Lotus japonicus* have attracted additional investigators studying rhizobial and mycorrhizal symbioses and from other areas where legumes offer unique research opportunities. Due to the recent efforts of several research groups, tools are now available for routine genetic and molecular analysis in these legume species.

No meeting has taken place previously with a focus on the current status of these model systems and the prospects for their application. With this in mind, this meeting on "Emerging Model Legume Systems" was organized as a satellite to the 8th International Congress on Molecular Plant-Microbe Interactions. In addition to the general goal of fostering communication and interactions among researchers working with these legume species, primary objectives of the meeting were (1) to present the current status of tools for classical and molecular genetic analysis in the selected species, (2) to review recent applications of these tools to biological questions for which these legume systems are well suited, (3) to explore the potential for the expanded application of these model legumes to issues of basic plant biology, and (4) to establish goals for continued development of a research infrastructure. This meeting report summarizes the progress on many of these topics presented in lectures, posters, and related discussions.

Past work on nitrogen fixing symbioses has emphasized major crop species. Although these species have economic importance and some have distinguished histories as study organisms for genetics, many crop species have drawbacks for molecular analysis, such as large genome sizes with abundant repetitive DNA, tetraploidy, or difficulties in transformation and regeneration. Both *Medicago truncatula* and *Lotus japonicus* possess many of the attributes that have made *Arabidopsis thaliana* a successful model for genome analysis in flowering plants: they are diploid plants with genomes among the smallest of the legume family; they are self-fertile annuals, with short generation times; and they are transformable and regenerable. To further *M. truncatula* and *L. japonicus* as genetic systems, meeting participants deemed it necessary to develop efficient mutagenesis protocols and make available mutagenized seed bulks; to establish tools for genome analysis and map-based cloning of genes; and to have efficient transformation and regeneration systems to study gene function *in vivo*. Advances towards these goals are reported.

Genetic maps have been initiated in both *Lotus japonicus* and *Medicago truncatula*. The *M. truncatula* map, in the laboratory of T. Huguet, is based on crosses between cultivar Jemalong and three natural populations. Over 100 molecular markers have been mapped, comprising 10 linkage groups spanning >90% of the genome. The *L. japonicus* map being executed in the laboratory of P. Gresshoff is based on a mapping population derived from a cross between cultivars Gifu and Funakura and the linkage of morphological and molecular (DNA amplification polymorphisms) markers is being determined. The mapping of additional markers remains a primary goal for both model legume species, with a view towards gene isolation by positional cloning. Synteny between legume genomes was recognized as an opportunity to exploit the existing maps in crop legumes such as pea, alfalfa and soybean.

Several mutagenesis approaches have been taken with model legume species. Chemical mutagenesis, using ethylmethane sulfonate, has successfully generated a wide variety of symbiotic and other mutants in *M. truncatula* (research groups of D. Cook and T. Huguet) and *L. japonicus* (research groups of F. de Bruijn and K.J. Webb), and also in a third species proposed as a model legume, *Melilotus alba* (A. Hirsch, T. LaRue and collaborators). Irradiation with γ-rays has generated genetically stable symbiotic mutants in *M. truncatula* (T. Huguet). Several reports (K.J. Webb, J. Stiller, M. Chiurazzi) were made of progress towards T-DNA mutagenesis using promoterless reporter gene fusions in *L. japonicus*. Finally, J. Stougaard reported preliminary results from gene targeting and random insertion mutagenesis approaches in *L. japonicus*.

The availability of a growing number of symbiotic genes known only by phenotype emphasizes the necessity of developing map-based cloning methods in model legume species. Towards this goal, Y-W. Nam (laboratory of D. Cook) reported on progress with the construction of a bacterial artificial chromosome (BAC) library in *M. truncatula*. The current library represents roughly three haploid genome equivalents, with an average insert size of $\geq$ 100 kb. A BAC library in *L. japonicus* is also under construction in the laboratory of P. Gresshoff.

A second approach to identifying host genes important to symbiotic development is the analysis of differential gene expression. F. de Bruijn and colleagues reported on the use of RNA differential display to construct a library of expressed sequence tags (ESTs) of late nodulins in *L. japonicus*. A similar approach is being employed by R. Voegeli-Lange to identify unique ESTs in mycorrhizal interactions of *Lotus* and *Medicago*. P. Gamas also described the identification, using subtractive hybridization, of new nodulin families in *M. truncatula*.

Analysis of gene expression and function relies upon simple and efficient transformation and regeneration protocols, and advances were reported for both *M. truncatula* and *L. japonicus*. M. Harrison described techniques for rapid direct organogenesis of *M. truncatula* cotyledon explants transformed with *Agrobacterium tumefaciens*, and success with a similar approach for *L. japonicus* was described by P. Oger. T-H. Trinh (lab of E. Kondorosi) reported on *A. tumefaciens* transformation of *M. truncatula* leaf explants with high regeneration capacity via somatic embryogenesis. Lastly, transformation of *L. japonicus* with *A. rhizogenes* was reported by J. Stiller as a short-cut method for producing transformed, nodulated hairy roots for gene trapping of nodulins based on promoterless reporter gene fusions.

APPLICATIONS OF MODEL SYSTEMS TO LEGUME BIOLOGY

Platform and poster presentations emphasized the host plant's role in development of nitrogen-fixing symbioses. Reports included genetic and molecular assessment of the plant's contributions to infection and nodule organogenesis (T. Huguet, R.V. Penmetsa, K. Szczyglowski, A. Hirsch, R. Prabhu), carbon and nitrogen metabolism (G. Hernandez, K.J. Webb), cell wall modifications (M. Djordjevic, R. Prabhu, M. Gonzales) and characterization of nodule-specific genes (P. Gamas, R. Dickstein, H. Peng, and M. Djordjevic). In addition to these avenues of inquiry, presentations and discussion sections stressed the potential contributions of model legumes to elucidating such basic aspects of plant biology as root development (K. Szczyglowski, L. Subramanian), hormonal signal transduction (R. V. Penmetsa, K. Szczyglowski, L. Subramanian), production of microbial signals and their transduction in plants (F. Côté, H. Stotz, D.J. Sherrier, L. Smith), and cellular organization and growth.

An emerging area of emphasis for model legume research is fungal interactions. Three research groups reported recent studies of associations of model legume species with vesicular/arbuscular (VA) mycorrhizal fungi (M. Harrison, M. Parniske, and R. Voegeli-Lange). Previously, little genetic and molecular analysis on plant contributions to VA mycorrhizae has been carried out, in part due to the lack of a good study system. The progress from these groups indicates that model legumes show promise for defining the plant's role in the development and function of this important rhizosphere symbiosis. *M. truncatula*, has also been adopted as a genetic system for analysis of host perception of the elicitor of the pathogenic fungus *Phytophthora sojae* (F. Côté), and for assessing the role of plant chitinases in fungal disease resistance and in nodulation (D. Kim). In addition to the development of molecular and genetic tools, a factor in the choice of this system is the knowledge base from the close relative alfalfa (*M. sativa*) on downstream defense responses, including synthesis of antimicrobial phytoalexins (M. Hahn). This pathway gives rise to isoflavonoids uniquely in legumes. Therefore, model legumes provide an excellent genetic platform for studying biosynthesis of this class of molecules that include phytoestrogens, compounds touted for anti-cancer and anti-nematode activities, as wells as antimicrobial agents (N. Paiva).

In summary, the development of these diploid legume species as model genetic organisms has the potential to define plant genetic control of nodule development and function, and to make an impact in multiple additional areas of plant growth and development. In particular, the focus on root-based plant-microbe interactions makes these legume systems a logical choice for the detailed analysis of rhizosphere biology and root biology generally. Because the development and implementation of the technology to support these goals will require the efforts of many research groups, coordination and communication among groups is considered paramount to the success of model legumes. Continued interactions among the participants and other members of the research community, via future meetings and electronic communications, will help to provide the necessary continuity in setting and reaching goals for these model systems.

## Program

FRIDAY, JULY 12, 1996

*Opening Session*
> Opening Remarks: K. VandenBosch (Texas A&M University)
> Keynote Lecture: "What's the use of models?" P. Gresshoff
> (University of Tennessee)

SATURDAY, JULY 13

*Lectures I: Genetic Maps.* Chair: T. LaRue (Cornell University)
> T. Huguet (Laboratoire de Biologie Moléculaire, INRA-CNRS):
> Mutants and natural variation of the model plant *Medicago truncatula* for studying the *Rhizobium*-legume symbiosis.

P. Gresshoff (University of Tennessee): Linkage of DNA amplification polymorphisms to a morphological character in a segregating mapping population of the model legume *Lotus japonicus*.

*Lectures II: Mutagenesis.* Chair: Thierry Huguet (INRA-CNRS)
  J. Stougaard (University of Aarhus): Gene targeting in *Lotus japonicus*: Approaches and preliminary results.
  R. V. Penmetsa (Texas A&M University): Developing a genetic system in *Medicago truncatula:* characterization of non-nodulating and hyper-nodulation mutants.
  J. Webb (Institute of Grassland and Environmental Research, Aberystwyth): T-DNA mutagenesis of *Lotus japonicus*.
  K. Szczyglowski (Michigan State University): Genetic analysis of *Lotus japonicus* symbiotic mutants.

*Discussion I: Genetic Analysis.* Chair: F. de Bruijn (Michigan State Univ.)

*Lectures III: Molecular Genetic Analysis.* Chair: D. Cook (Texas A&M)
  Frans de Bruijn (Michigan State University): Towards the construction of a *Lotus japonicus* nodule specific EST library of late nodulin genes.
  Young-Woo Nam (Texas A&M University): Construction of a bacterial artificial chromosome library in *Medicago truncatula.*
  Sam Cartinhour (Texas A&M University): Resources for plant genome informatics.

*Discussion II: Infrastructure and Communication.* Chair: P. Gresshoff

*Poster session*

SUNDAY, JULY 14

*Discussion III: Objectives for Applications of Model Legumes.* Chair: K. VandenBosch. Panel: N. Pavia (Noble Foundation); M. Hahn (University of Georgia); M. Parniske (Sainsbury Laboratory)

*Lectures IV: Transformation and Regeneration.* Chair: J. Stougaard
  M. Harrison (Noble Foundation): Recent advances in transformation and regeneration of *Medicago truncatula.*
  J. Stiller (University of Tennessee): Trapping of symbiotic genes in the model legume *Lotus japonicus.*
  T.H. Trinh (Institut des Sciences Végétales, CNRS): Rapid and efficient somatic embryogenesis and *Agrobacterium tumefaciens*-mediated transformation of *Medicago truncatula* cv. 108-1.
  P. Oger (University Illinois-Urbana). A simple technique for the direct transformation and regeneration of the diploid legume species *Lotus japonicus*.

*Discussion IV: Culture and Cultivation.* Chair: J. Stougaard

*Lectures V: Research Advances.* Chair: M. Harrison (Noble Foundation)
  F. Côté (University of Georgia): Identification of a hepta-β-glucoside binding site in model legume *Medicago truncatula* cv. Jemalong.
  M. Parniske (Sainsbury Laboratory): Mycorrhizal mutants of *Lotus japonicus.*
  A. Hirsch. (University of California-Los Angeles): Studies on the diploid indeterminate nodule-forming plant *Melilotus alba* Desr., white sweet clover.
  B. Rolfe (Australian National University): Separation and characterization of *Trifolium subterraneum* proteins by two-dimensional gel electrophoresis.
  P. Gamas (Laboratoire de Biologie Moléculaire, INRA-CNRS) Identification and characterization of new *Medicago truncatula* nodulin genes.
  R. Dickstein (Drexel University): Characterization and sequence analysis of *Medicago truncatula* ENOD8 genes.

*Closing Session.* Chair: D. Cook (Texas A&M University)

## Acknowledgements

We wish to thank Dr. Gary Stacey for helping to coordinate the satellite meeting with the main meeting, and all the conference participants for their contributions in lecture and discussion sections. Special thanks are due to Ms. Susan Davis and other members of the UT Conference Center Staff for their help in on site organization. Lastly, we gratefully acknowledge support from the Integrative Plant Biology Program of the National Science Foundation, the Nitrogen Fixation and Nitrogen Metabolism Panel of the USDA National Research Initiative and Competitive Grants Program, and the Energy Biosciences Program at the United States Department of Energy.

## Literature Cited

Barker, D.G., Bianchi, S., London, F., Dattée, Y., Duc, G., Essad, S., Flament, P., Gallusci, P, Génier, G., Guy, P., Muel, X., Tourneur, J., Dénarié, J., and Huguet, T. 1990. *Medicago truncatula*, a model plant for studying the molecular genetics of the *Rhizobium*-legume symbiosis. Plant Mol. Biol. 8:40-49.
Handberg, K., and Stougaard, J. 1992. *Lotus japonicus*, an autogamous, diploid legume species for classical and molecular genetics. Plant Journal 2:487-496.

**International Symposium on Molecular Plant-Microbe Interactions**

**Original Music**

# One more control

Late at work, lookin' round for my pen,
the lab is dark, and the phone rings again.
Oh no, no I can't come to you,
I've got one more control to do.

Sunday night, and you wait at my desk,
I close my eyes, cos' I know what you'll ask.
Oh no, no I can't come with you,
I've got one more control to do.

> It happened one time, it happened twice,
> but I learned the hard way not to trust my eyes.
> Is it a wrong track, or am I right?
> Oh Lord, help me to decide.

'n Empty glass, and I call for a cab,
don't push me now, gotta go to the lab.
Oh no, no I can't stay with you,
I've got one more control to do.

> It happened one time, it happened twice,
> but I learned the hard way not to trust my eyes.
> Is it for certain, or premature?
> Oh Lord, help me to be sure.

Late at night, and you knock on my door,
please go, I've heard it all before.
Oh no, no I can't love you,
I've got one more control to do.
Oh no, no I can't love you,
I've got one more control to do.

(✍ and ♫: Kijne)

# Lonesome ligand

They say that I'm important,
got some credits to my name.
I'm a mitogenic factor,
nodulation brought me fame.
I can curl young root hairs,
induce gene expression too.
But if I'm so damned important,
how come I ain't got you?

chorus:
I ain't got you, I ain't got you.
I ain't got you, I ain't got you.
I'm a lonesome ligand but I don't know what I do
without you. How come I ain't got you?

All these binding proteins,
they don't mean a thing.
Non-specific one-night stands,
without signalling.
You're made for me, my darling,
and I am made for you.
Some strange domains reach out for me,
but you know, I will be true.

chorus.

May be you're a lectin,
and may be you are not.
My sweet and shy receptor,
what is it I ain't got?
I've got a fatty acid,
and some pretty side groups too.
But if they're so damn specific,
how come I ain't got you?

(✍: VandenBosch, ♪: Kijne)

# Tears on a labcoat

Blue
stains on my hands
and my labcoat is blue.
Everything drops, now I'm thinkin' of you.
You broke my dreams, and left me the slivers.
And I, I'm feeling my tears comin' through.

> But I don't cry, don't cry,
> for tears on a labcoat are too hard too hide.
> No, I don't cry, don't cry,
> cos' tears on a labcoat never dry.

Red
stains on the floor
and my fingers are red.
I'm loosing control, thinkin' back on what you've said.
I clean the bench, and cover your traces.
But how can I wipe you out of my head?

> But I don't cry, don't cry,
> for tears on a labcoat are too hard too hide.
> No, I don't cry, don't cry,
> cos' tears on a labcoat never dry.

(✍ and ♫: Kijne)

## Coomassie Blues

He:
All day Friday at the bench,
can't get rid of this mercapto stench.
Full of Tris and full of booze,
see I got the Coomassie Blues.

I ain't doing very well,
can't get rid of this awful smell.
Stink from by head down to my shoes,
can't get rid of my Coomassie Blues.

Got no woman, got no belle,
got no proteins on my gel,
no Sue-Anne's, no Betty-Lou's,
just the smell of the Coomassie Blues.

She:
I can help you bear that load,
hook a lead up to your electrode.
Send a current from your head to your shoes,
help you loosin' the Coomassie Blues.

He:
I found a woman, I got a pal,
resolved the proteins on my gel.
No more trouble, I've paid my dues,
and got rid of the Coomassie blues.


(✍: Rosenberg, VandenBosch, ♫: Kijne)

# SUBJECT INDEX

pv. *tomato,*
9,52,71,
149,174,213
pseudonodules, 351
pterocarpans, 81
*pth* genes, 197
Pto kinase, 40,73
pulsed-field electrophoresis, 247,
249
pyochelin, 442

QTL, 372
quorum sensing, 116,174,188,550

*Ralstonia (Pseudomonas,*
*Burholderia) solanacearum,*
165,205
rapamycin, 471
RAPD, 373
reactive oxygen, 15,33,39,84
receptor, 149,239,255
reoviridae, 293
REP-PCR, 497
resistance genes, 39, 48,71,93,
148,197
　　　Bs3, 192,203,205
　　　Bs3-2, 203,207
　　　Cf2, 40,57
　　　Cf4, 57,72,253
　　　Cf9, 40,57,72,253
　　　Hm, 71,223,245
　　　L, 40,65,71
　　　N, 40,65,71,271
　　　Prf, 9,40,71,241
　　　Pto, 9,70
　　　Rpg1, 162
　　　Rpg4, 148
　　　Rpm1, 40,71,162
　　　Rpp5, 40
　　　RPS2, 40,50,71
　　　Tm, 65
　　　Xa-7, 192
　　　Xa-21, 40,72
restriction-enzyme-mediated

integration (REMI), 220,242
retinoblastoma, 289
RFLP, 372
*Rhizobium,* 5,7,99,173,307
　　　*R. leguminosarum* bv.
　　　　*viciae,* 300,311,
　　　　343,400,505
　　　*R. leguminosarum* bv.
　　　　*trifolii,* 353,445
　　　*R. etli,* 337,503
　　　*R. fredii,* 332,413
　　　*R. loti,* 301
　　　*R. meliloti,*174,300,325,
　　　　332,405,411,446,
　　　　481,504
　　　*R. tropici,* 503
　　　strain NGR234, 320,332
rhizosphere, 319,4449,463,485,
514,560
*Rhynchosporium secalis,* 89
ribgrass, 274
riboflavin, 481
rice, 117,191,293
rice dwarf virus, 293
rj1, 372
Rj2, 372
RNA helicase, 282
root cap, 512
rooted leaves, 354
*rpf* genes, 209
*rsm*A gene, 187
RT-PCR, 390
rye, 117
rye grass, 213,274

*Saccharomyces cerevisiae,* 139,
284,517
*sai*1gene, 37
salicylic acid, 27,65,72,223
salicylate hydroxylase, 34
*Salmonella,* 160,165
　　　*S. enterica,* 327
saponins, 233

580

# AUTHOR INDEX